Ultrafast Dynamics at the Nanoscale

Ultrafast Dynamics at the Nanoscale

Biomolecules and Supramolecular Assemblies

edited by
Irene Burghardt
Stefan Haacke

Pan Stanford Publishing

Published by

Pan Stanford Publishing Pte. Ltd.
Penthouse Level, Suntec Tower 3
8 Temasek Boulevard
Singapore 038988

Email: editorial@panstanford.com
Web: www.panstanford.com

British Library Cataloguing-in-Publication Data
A catalogue record for this book is available from the British Library.

Ultrafast Dynamics at the Nanoscale: Biomolecules and Supramolecular Assemblies

ISBN 978-981-4745-33-8 (Hardcover)
ISBN 978-981-4745-34-5 (eBook)

Printed in the USA

Contents

Preface

Over the past decade, research on ultrafast (bio)molecular dynamics has evolved significantly, from spotlighting isolated molecular species toward focusing on molecular assemblies, including chromophore–protein complexes, biopolymers like DNA and RNA, and functionalized systems like combinations of biomolecules and inorganic nanoparticles, plasmonic nanostructures, and semiconducting polymer materials. Potential applications range from light harvesting to biosensing, artificial signal transduction, and organic photovoltaics. The functionality and control of these systems are currently under intense investigation in view of developing a detailed understanding of ultrafast nanoscale energy and charge transfer, as well as fostering novel technologies based on sustainable energy resources. At the same time, fundamental issues like the role of quantum coherence versus decoherence have moved into focus. The book aims to illustrate this evolution—which is far from complete at the time of publication—with contributions from top researchers in the field.

The book groups together 13 chapters, comprising 6 contributions from spectroscopy and 7 from theory and computation. Within a broad range of topics, from DNA photostability and repair to light-harvesting complexes and novel hybrid materials, a common denominator is the key importance of ultrafast quantum effects at the border between the molecular scale and the nanoscale. Contributions include cutting-edge developments in ultrafast nonlinear optical spectroscopies and quantum dynamical simulations of the observed dynamics, including direct simulations of two-dimensional optical experiments. Taken together, these techniques attempt to elucidate whether the quantum coherent nature of the ultrafast events persists in the presence of strong vibronic interactions and electrostatic couplings to the environment. Moreover, it is still an open issue to date whether the coherent nature of the elementary events possibly enhances the efficiency of the relevant processes and where the quantum-classical boundary sets in in these highly complex biological and material systems.

The first experimental contribution, by B. Kohler (Chapter 1), offers a didactic introduction to the fundamental concepts of ultrafast transient absorption spectroscopy in the context of investigating isolated DNA bases or single-stranded DNA. This review emphasizes the recent progress in understanding how base stacking influences the excited-state electronic structure and modulates nonradiative processes such as exciton and charge transfer involving neighboring nucleic acid bases. These studies are also driven by the general interest in ultrafast nonradiative internal conversion processes, which are believed to protect DNA from photodamage by UV light.

Active photoinduced repair of DNA is carried out by photolyases, that is, photoreactive enzymes which were discovered more than 50 years ago. Two contributions in this book highlight the recent advances made in dissecting the molecular details and reaction steps of enzyme photoactivation, involving ultrafast intraprotein charge transfer processes, and repair of different types of lesions. K. Brettel, M. Byrdin, and M. Vos (Chapter 2) devised a special transient absorption experiment with high sensitivity dedicated to the investigation of how DNA photolyases repair so-called CPD lesions, involving forward and backward electron transfer between the photolyase enzyme and a defective covalently linked thymine pair site. D. Zhong and his collaborators (Chapter 3) report results for the same type of system and complete this topic by reviewing their present understanding of the electron and proton transfer processes between photolyase cofactors and the nucleobases involved in repair of mutagenic so-called 6-4PP lesions. Both studies demonstrate the power of ultrafast multiwavelength transient absorption spectroscopy in revealing reaction kinetics and quantum yields, but also the particular effort involving the investigation of many mutants and molecular variants—a tour de force that is required to identify the often overlapping spectral signatures of the multiple species involved in these complex photoreactions.

The photoactive yellow protein (PYP) is a prototypical photosensor protein used by some bacteria to avoid harmful UV light (negative phototactic response). M. Groot and K. Hellingwerf (Chapter 4) have worked out the intimate details of the primary photoreactions involving a subpicosecond photoisomerization of the *p*-coumaric acid chromophore, the covalently bound cofactor of PYP. Besides providing a review of the different complementary time-resolved spectroscopy and X-ray diffraction techniques, the authors' report on the investigation of point-mutated PYP provides critical information on how the protein environment tunes and controls the different kinetic steps in the PYP photocycle.

Inspired by the concept of light harvesting through distributed multichromophore absorption, I. Nabiev and his team (Chapter 5) have implemented new nanobio hybrid materials, where semiconductor quantum dots with highest extinction coefficients are covalently bound to bacteriorhodopsin or photosynthetic reaction centers. The process of Förster resonant energy transfer (FRET) was studied in detail for these novel donor–acceptor systems, along with the key question of how the functional efficiency of these hybrids is enhanced as compared to the bare proteins. This work exemplifies an important emerging research direction aimed at exploiting the synergy of biomolecules and synthetic nanostructures in hybrid materials for improved photon harvesting and energy conversion.

In the final experimental chapter, A. Cannizzo et al. (Chapter 6) present different approaches to directly probe the spectroscopic response of the protein environment interacting and modulating the ultrafast photochemistry of the reactive cofactor. The authors summarize their work on using Trp residues as local reporters through intraprotein Stark effects or energy transfer. Two-dimensional femtosecond spectroscopy techniques are in principle the method of choice for the investigation of coupled multichromophoric systems. The present progress of their challenging implementation in the near-UV spectral region, targeting amino acids or nucleobases, is reviewed.

On the theory side, the Fenna–Matthews–Olson (FMO) complex, a pigment–protein complex appearing in green sulfur bacteria, has served as a paradigm system over recent years for studying excitation energy transfer in biological light-harvesting systems. The chapters by P. Nalbach (Chapter 7) and P. Huo and D. Coker (Chapter 8) both address this system, using path integrals, master equation techniques, and novel mixed quantum-classical approaches to solve the highly non-Markovian, correlated quantum dynamics of this multichromophoric system. Key issues include the role of spatial and temporal correlations in inducing long-lived excitonic coherences and the relevance of these coherences for the efficiency of the light-harvesting process. While earlier contributions in this field were strongly suggestive of an enhanced quantum efficiency of biological light harvesting, the viewpoint by Nalbach, Huo, and Coker tends to be critical in this regard.

In the following contribution, S. Jang (Chapter 9) develops a quantum master equation in the polaron picture, which is able to interpolate between the coherent and incoherent regimes of excitation energy transfer in multichromophoric systems. Together with the approaches mentioned above, this work illustrates the importance of adapting existing quantum propagation techniques to the treatment of high-dimensional systems that do not allow for standard approximations like weak-coupling limits or separation of timescales. Instead, alternative zeroth-order pictures need to be found—as exemplified by the polaron transformation—that are able to capture the strong vibronic (electron–phonon) coupling effects that are ubiquitous in biological systems and organic materials.

The theoretical treatment of hybrid assemblies involving metal nanoparticles and single molecules or molecular assemblies is an important and challenging aspect accompanying recent experimental developments in this area (cf. Chapter 5 by I. Nabiev). Indeed, plasmonics—relating to the specific spectroscopic and transport properties of collective electronic excitations in noble metal nanoparticles—has become a rapidly emerging area within the field of nanotechnology, sensing, and biophotonics. Against this background, V. May and collaborators (Chapter 10) develop a consistent time-dependent quantum description, at a density matrix level, of resonant energy transfer involving multipole collective electronic excitations of the metal nanoparticle and molecular transitions of the acceptor system. This elegant approach naturally connects to the class of master equation approaches mentioned above.

The contribution by H. Tamura et al. (Chapter 11) addresses photoinduced energy and charge transfer in organic photovoltaic materials, using a combination of electron–hole model Hamiltonians parametrized by electronic structure calculations, and high-dimensional quantum dynamics using efficient multiconfigurational methods. This approach offers an alternative to master equation and path integral strategies, attempts to build a maximum of molecular information into a first-principles parametrized model Hamiltonian, and systematically exploits the lattice structure of the relevant materials. As a result, a precise picture of highly nonexponential, non-Markovian transfer processes in composite donor–acceptor systems can be obtained, as exemplified by a study of the fullerene-based P3HT–PCBM paradigm system.

The final theory chapters by J. Yuen-Zhou, A. Aspuru-Gúzik and collaborators (Chapter 12) as well as A. Dijkstra and Y. Tanimura (Chapter 13) are concerned with the simulation of nonlinear optical spectroscopic experiments with a focus on multichromophoric assemblies. The interpretation of such experiments, which encode information on transport processes in spectral signatures (e.g., cross-peaks in two-dimensional spectroscopies) is often complex and critically relies on theoretical analysis. In this context, Yuen-Zhou et al. give a didactic introduction to pump–probe spectroscopy, connecting to a quantum information theory setting, while Dijkstra and Tanimura focus on two-dimensional spectra and explore the quantum-classical transition from the viewpoint of spectroscopic observables.

While being far from comprehensive, we believe that the present collection of contributions provides a combined experimental and theoretical spotlight on the molecular-level investigation of light-induced quantum processes in biological systems and nanostructured assemblies. We hope that this collection will inspire future experiment–theory connections at this frontier between molecular and nanoscale quantum phenomena.

We would like to express our sincere gratitude to the authors, colleagues, and friends who joined this book project with the aspiration to present their research in a creative and pedagogical style. Finally, we are most grateful to the editorial team at Pan Stanford Publishing for its invaluable help with publishing this volume.

Irene Burghardt
Stefan Haacke
Frankfurt and Strasbourg, July 2016

Section I

Experiment

Chapter 1

Excited States of Single-Stranded DNA Revealed by Femtosecond Transient Absorption Spectroscopy

Bern Kohler
Department of Chemistry and Biochemistry, The Ohio State University, Columbus, OH 43210, USA
kohler.40@osu.edu

Damage to the genome by UV light has fueled interest in excited electronic states of DNA for over 50 years. DNA is also a fascinating model system for understanding how two widespread supramolecular self-assembly motifs, π-π stacking and hydrogen bonding, influence exciton dynamics in multichromophoric assemblies. The importance of minimizing photochemical damage has endowed DNA with remarkable photophysical properties that could inspire new approaches to controlling exciton and charge transport in nanomaterials. Although most excitations in single DNA bases decay nonradiatively in hundreds of femtoseconds, much longer-lived excited states are observed in femtosecond transient absorption experiments on single- and double-stranded DNAs. Experiments on single-stranded nucleic acids isolate effects due to base stacking. Base stacking is shown to be of paramount importance for the

Ultrafast Dynamics at the Nanoscale: Biomolecules and Supramolecular Assemblies
Edited by Irene Burghardt and Stefan Haacke

ISBN 978-981-4745-33-8 (Hardcover), 978-981-4745-34-5 (eBook)
www.panstanford.com

slow, nonradiative decay channels. This chapter provides a detailed discussion of the femtosecond transient absorption technique with emphasis on considerations relevant to studies of DNA excited-state dynamics. A rate equation analysis of transient absorption signal strengths is presented and used to reach conclusions about excited states from experiments on single-stranded adenine homo-oligonucleotides.

1.1 Introduction

Many of the excited electronic states produced in DNA model compounds by the absorption of UV light decay nonradiatively to the ground state in less than one picosecond [1]. The ability to directly observe the dynamics of short-lived excited states in DNA by femtosecond spectroscopy has been instrumental in advancing the field of DNA photophysics since 2000 [2–7]. Excited states are observed on femtosecond timescales either by monitoring their absorption or their emission. Absorption detection is done with the femtosecond transient absorption (TA) technique, while the fluorescence upconversion technique [8] has been used to monitor emission on subpicosecond timescales beginning in the 1980s [9–12]. The first accurate measurements of the excited-state lifetimes of single nucleosides were made using the TA technique and published in 2000 [1]. Shortly afterward, the fluorescence lifetimes were measured in femtosecond fluorescence up-conversion experiments and shown to be in excellent agreement with ones obtained from TA experiments [13–15].

In this contribution, the TA technique will be discussed in depth with a focus on the insights recent experiments have provided into excited states of single-stranded DNA oligonucleotides. The chapter is organized in several sections: The TA technique is first described along with some relevant experimental details. Next, expressions are given for the strength of TA signals from solutions that contain multiple absorbers—a description that is essential for the multichromophoric DNA molecule. These expressions, which are based on a rigorous rate equation analysis of light–matter interactions in the incoherent limit, are derived in Appendix A.

The results are used to make some general observations about TA signals and to illustrate a few specific considerations that apply to experiments on DNA model compounds. Finally, results from some recent TA experiments on single DNA strands are presented and interpreted with help from the theory presented in the first section.

1.1.1 Origins

Norrish and Porter pioneered the recording of absorption spectra using short light pulses soon after World War II [16, 17]. The flash photolysis technique, which they invented and which was recognized by the 1967 Nobel Prize in Chemistry, introduced the "double flash" or pump–probe method that is the basis of most ultrafast spectroscopic techniques today. In the 1950s, the time resolution was adequate for studying absorption by triplet states of organic molecules in fluid solution on the microsecond timescale [18].

The invention of the laser around 1960, followed by the generation of progressively shorter laser pulses [19], allowed flash photolysis or TA experiments to be performed with ever-higher time resolution. In the 1970s, researchers at Bell Laboratories first produced femtosecond laser pulses using passively mode-locked dye lasers [20, 21]. The first TA experiments with subpicosecond time resolution were performed by Ippen and Shank and coworkers in the 1970s [22, 23], but it was the invention of the colliding pulse mode–locked dye laser [24] that enabled the routine generation of pulses below 100 fs and allowed femtosecond spectroscopy to flourish [25].

The techniques behind TA experiments were developed in the 1980s, and the schematic of a broadband TA spectrometer published by Shank in 1983 (Fig. 9 in Ref. [25]) could describe with only a few minor adaptations the instrumentation in use 30 years later. Improvements since that time have been made to the laser sources, which provide more stable and more widely tunable pulses than ever before. The titanium sapphire laser systems that are used nearly exclusively can be purchased commercially from many vendors and provide pulses that are readily tunable from the deep UV to the mid- and even far-infrared (IR) spectrum. Improved multichannel detectors for broadband spectroscopy, particularly in the mid-IR, are also available now.

1.2 The TA Experiment

In a TA experiment, the output from an amplified femtosecond laser system is divided by a beam splitter into two beams for sample excitation (pump) and measurement of the induced absorbance change (probe) (Fig. 1.1). Because pump and probe pulses are derived from the same initial optical pulse, jitter is minimized. Transient species created by the pump pulse change in time according to a kinetic mechanism that describes their evolution and the evolution of species formed from them. These species are detected through changes in the transmission of the probe pulse as a function of the delay time between pump and probe pulses.

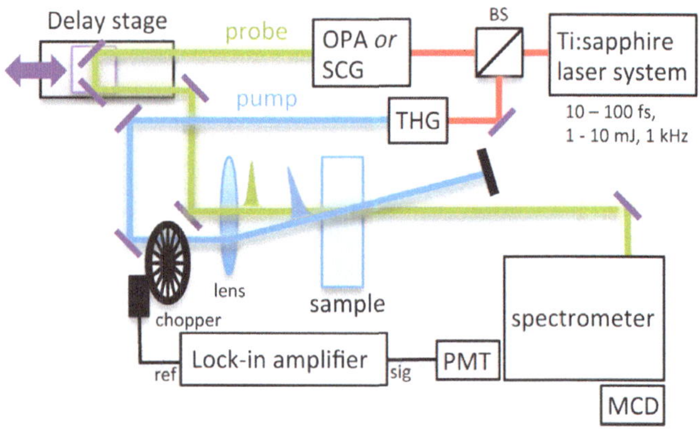

Figure 1.1 Schematic of a femtosecond TA spectrometer. The output of a titanium:sapphire laser system is divided into pump and probe beams using a beam splitter (BS). The third harmonic generation (THG) produces deep UV pump pulses, while visible probe pulses are produced with an optical parametric amplifier (OPA) or supercontinuum generation (SCG). The transmitted probe beam is detected after the sample either in broadband mode using a multichannel detector (MCD) or in single-wavelength mode using a photomultiplier tube (PMT). The PMT signal is read out with a lock-in amplifier that detects weak transmission changes at the modulation frequency set by an optical chopper positioned in the pump beam.

Delay time is defined as the arrival time of the probe pulse minus the arrival time of the pump pulse at a common plane such as the front face of the sample.

Delay time is thus negative when the probe precedes the pump pulse, positive when the probe lags the pump, and zero when pump and probe have maximum temporal overlap. A retroreflector (corner cube) mounted on a computer-controlled delay stage is used to precisely vary the time delay between pump and probe pulses. In double-pass geometry, each micrometer of stage motion corresponds to 6.67 fs of delay time.

Most TA setups begin with a commercial titanium sapphire laser system that uses chirped pulse amplification [26–28] to produce multi-mJ pulses that can be as short as 10 fs at kHz repetition rates. Even when the laser amplifier is designed to allow some degree of wavelength tuning within the range of near-IR wavelengths that fall under the titanium sapphire gain curve, tuning can be tedious, and the laser system is often used as a fixed wavelength source of pulses near 800 nm.

Pump and probe pulses at other wavelengths are derived from the fixed fundamental laser output by nonlinear optical frequency conversion. Optical parametric amplifiers (OPAs) can produce tunable femtosecond pulses with wavelengths that span the deep UV to the mid-IR. Frequency doubling and tripling of the fundamental output is a simpler and less expensive method for obtaining a few discrete wavelengths. The third harmonic of a titanium sapphire laser (e.g., 265 nm for a fundamental wavelength of 795 nm) falls near the absorption maximum of DNA and is particularly convenient for studies of DNA model compounds.

Pump and probe pulses are typically linearly polarized and the mutual orientation of their polarization directions is an experimental variable. Time-dependent absorption anisotropy is determined by signals recorded with parallel and perpendicular pump–probe polarizations. These signals provide information about molecular reorientation and energy transfer [29, 30]. To follow population kinetics without interference from rotational diffusion in fluid solution, the angle between linearly polarized pump and probe pulses is set to the magic angle ($\cos^{-1}\left(\frac{1}{\sqrt{3}}\right) = 54.7^\circ$) [31, 32]. The focus here is on population dynamics, so only signals recorded under magic angle conditions will be considered.

The pump beam is focused in the sample to increase the signal strength, which depends on the fluence of the pump pulse (see below). The probe beam is spatially overlapped with the pump beam in the sample and focused to a spot size several times smaller than the pump to ensure that it samples a relatively uniform distribution of excited states and to prevent the signal from changing rapidly due to motion of either beam, which could reduce their overlap.

1.2.1 The TA Signal

In TA spectroscopy, the absorbance change (ΔA) induced by the pump pulse is measured as a function of the delay time. The signal, $\Delta A(d, \lambda_{pr})$, at delay time d and probe wavelength λ_{pr}, is equal to the (decadic) absorbance of the sample at the probe wavelength with the pump pulse on, $A(d, \lambda_{pr})$, minus the sample absorbance at the same wavelength with the pump pulse blocked, $A_{off}(\lambda_{pr})$,

$$\Delta A(d, \lambda_{pr}) = A\left(d, \lambda_{pr}\right) - A_{off}(\lambda_{pr}). \tag{1.1}$$

Hereafter, a tilde will be used to indicate a wavelength-dependent quantity that is to be evaluated at the probe wavelength. This notation allows Eq. 1.1 to be written simply as,

$$\Delta\tilde{A}(d) = \tilde{A}\,(d) - \tilde{A}_{off}. \tag{1.1$'$}$$

The absorbance is related to the transmission, T,

$$A = -\log T = -\ln T / \ln 10. \tag{1.2}$$

Equation 1.1′ then becomes,

$$\Delta\tilde{A}\,(d) = -\log\left(\frac{\tilde{T}\,(d)}{\tilde{T}_{off}}\right). \tag{1.3}$$

The transmission of a laser pulse is given by the ratio of its energy after the sample to its energy before the sample. However, because the pulse energy before the sample is the same with the pump pulse blocked or unblocked the transmission ratio in Eq. 1.3 is equal to the energy of the probe pulse measured after the sample divided by the probe pulse energy measured after the sample with the pump blocked.

An optical chopper placed in the pump beam alternately passes or blocks the pump pulse and a lock-in amplifier is used to detect the modulated signal from any square-law photodetector that provides a signal that varies linearly with pulse energy. Transmission changes due to optical components placed in the probe beam path or due to Fresnel losses at interfaces do not contribute to the signal as long as they are the same with the pump on and off. Any light scattering by the probe pulse is usually independent of the pump pulse and will also cancel out.

As shown in Appendix A, TA signals are proportional to the pump pulse fluence. The fluence is the intensity (energy flux) of the pulse, I, integrated over all time and is equal to,

$$F = \int_{-\infty}^{\infty} I(t)dt. \tag{1.4}$$

Fluence is measured in photons per unit area and is proportional to the pulse energy, allowing Eq. 1.3 to rewritten as,

$$\Delta\tilde{A}(d) = -\log\left(\frac{\tilde{F}(d)}{\tilde{F}_{\text{off}}}\right) \approx \frac{1}{\ln 10}\left(1 - \frac{\tilde{F}(d)}{\tilde{F}_{\text{off}}}\right), \tag{1.5}$$

where the approximation $-\ln(y) \approx 1-y$ has been used because the fluence ratio is very close to one for small ΔA signals. Equation 1.5 is used in Appendix A to derive analytical expressions for pump–probe signals.

$\tilde{F}(d)$ and $\tilde{F}_{\text{off}}$ in Eq. 1.5 both fluctuate due to laser noise. By measuring these quantities after every laser pulse with fast detectors the fluctuations in their ratio can be reduced, improving the signal sensitivity. With an optical chopper and a lock-in amplifier, the fluence ratio in Eq. 1.5 is effectively measured a few pulses at a time, again reducing noise from drift in the laser intensity that occurs on slower timescales. Regardless of how the quantities in Eq. 1.5 are measured, the TA signal at each delay time is typically averaged over thousands of pump and probe pulse pairs often by averaging together separate scans recorded one after the other over the entire delay range.

The photodetectors used in TA experiments are time-integrating detectors that produce signals proportional to pulse fluence

(Eq. 1.4). The time resolution of the TA technique is not determined by their response time, but by the temporal widths of pump and probe pulses. The photodetector should be fast enough to respond to each probe pulse for the pulse-by-pulse detection just mentioned, but this requires only submillisecond time resolution for a kilohertz pulse train. The actual time resolution of the pump–probe measurement can easily be higher by a factor of 10^9 or more.

The TA signal in Eq. 1.5 can be measured either in broadband or single-wavelength mode. For broadband measurements, the sample response at a range of wavelengths is determined simultaneously at each delay time by spectrally resolving a broadband probe pulse using a spectrometer and a multichannel detector. In-depth discussion of many of the technical issues related to broadband TA spectroscopy together with detailed descriptions of state-of-the-art spectrometers are available elsewhere [33, 34].

For broadband transient electronic spectroscopy (200 nm $\leq \lambda_{pr} \leq$ 1500 nm), the probe pulse is usually a supercontinuum pulse [35] generated by tightly focusing a femtosecond pulse into a dielectric solid (e.g., CaF_2 glass), liquid (e.g., D_2O) or, more recently, a photonic crystal fiber [36]. The bandwidth of a supercontinuum pulse may span several hundred nm. Mid-IR continuum probe pulses can be generated using semiconductors [37, 38] or other approaches [39, 40].

Many mid-IR TA spectrometers forgo use of a continuum probe pulse and simply disperse the mid-IR probe pulse obtained from a difference-frequency-mixing OPA. A Gaussian, 100 fs mid-IR pulse has a bandwidth of ~150 cm^{-1} (Eq. 1.6 below) and consequently has sufficient bandwidth to probe multiple vibrational resonances at one time. In single-wavelength TA measurements, the probe pulse is not spectrally dispersed and a single-channel detector such as an amplified photodiode or PMT replaces the spectrometer and multichannel detector (Fig. 1.1).

The solution under study is usually recirculated through a wire-guided jet [41] or a flow cell with optical windows. The jet can eliminate unwanted signal contributions from glass windows or artifacts due to adhesion of permanent photoproducts on the surface of a flow cell. Recirculating the solution under study is essential for eliminating thermal artifacts produced by laser heating (see below) and for ensuring that successive pump and probe pulse pairs interact with a solution of constant composition.

Cumulative effects due to re-excitation of long-lived transient species or photoproducts formed by an earlier pump pulse are eliminated by flowing the sample at a high enough rate to completely replace the pumped volume with fresh solution before the arrival of the next pump pulse. The sample volume need not be large if photoproduct formation is minimal. For example, the use of microtubing and a peristaltic pump make it possible to work with as little as 2 mL of solution, facilitating work with expensive, defined-sequence DNA oligonucleotides.

Several considerations dictate the choice of sample thickness for a fluid solution. If the path length is too great, then the absorbance at the probe wavelength can be too high (i.e., >2), making it difficult to detect the weak probe pulse, much less measure the change in absorbance it experiences. This consideration is particularly important for time-resolved vibrational spectroscopy due to strong mid-IR absorption by the solvent (D_2O) in these experiments.

A further issue is that a long sample path length can lead to dispersive pulse broadening or temporal walk-off of pump and probe pulses. These effects can compromise time resolution. If the path length is too short, then limited solubility may make it impossible to achieve a high enough absorbance to maximize the signal strength (see below). Even if the sample is sufficiently soluble, the higher concentration needed to produce the target absorbance could lead to aggregate formation.

1.2.2 Dispersive Pulse Broadening and Temporal Walk-off

Femtosecond pulses, unlike longer laser pulses, can be dramatically reshaped by propagation even through transparent dielectric materials. A femtosecond laser pulse is narrow in time but spectrally broad as required by the time uncertainty principle. Let $\Delta\tilde{\nu}$ be the full-width at half-maximum (FWHM) of the pulse spectrum in wavenumbers. Let τ be the pulse width, defined as the FWHM of the intensity envelope $I(t)$. For a transform-limited Gaussian pulse, $\Delta\tilde{\nu}$

and τ are related by,

$$\Delta\tilde{\nu} = \frac{\ln 4}{\pi c \tau} = \frac{1.47 \times 10^4\ \mathrm{cm}^{-1}}{(\tau/\mathrm{fs})}. \tag{1.6}$$

Accordingly, a 75 fs Gaussian pulse has $\sim$200 cm^{-1} of bandwidth. This value is less than 1% of the average frequency of a UV pulse and 10% as large as the frequency of a mid-IR pulse at 5.0 μm. Because the ratio of the pulse bandwidth to the average frequency is less than one for mid-IR and shorter-wavelength pulses, femtosecond pulses are quasi-monochromatic and can be accurately represented by a carrier frequency and a slowly varying envelope function. The carrier frequency is often chosen to be the center wavelength, that is, the wavelength with maximum amplitude when the frequency spectrum of the pulse is symmetric.

It is customary to refer to an ultrashort pulse by its center wavelength (i.e., a 265 nm pump pulse) even though it has spectral components that extend to shorter and longer wavelengths. Absorption spectra of molecules in solution almost always exhibit strong inhomogeneous broadening and the dynamics detected in pump–probe experiments are often insensitive to shifts in the center wavelength of either pulse by a few nm.

Although the pulse bandwidth is usually small compared to the carrier frequency, the spread of frequencies in an ultrashort pulse leads to effects not encountered with longer pulses. Due to dispersion, the refractive index of a dielectric material such as that found in a lens or polarizer or the sample solution changes with wavelength. Dispersion alters the relative arrival times of different spectral components of the pulse, leading to a broadened pulse with a chirped frequency spectrum.

Dispersive pulse broadening can be significant even for short material path lengths, and the effects become more severe as the transform-limited pulse width decreases (Fig. 1.2). For example, the pulse duration of a transform-limited 40 fs Gaussian pulse with a center wavelength of 265 nm doubles after propagating through 4.8 mm of water. In contrast, 22 mm of water are needed to double a 40 fs Gaussian pulse at 600 nm, showing that dispersion is much more important at UV wavelengths.

Although dispersion is unavoidable, a grating [42] or prism [43, 44] compressor can be used to remove frequency chirp and restore

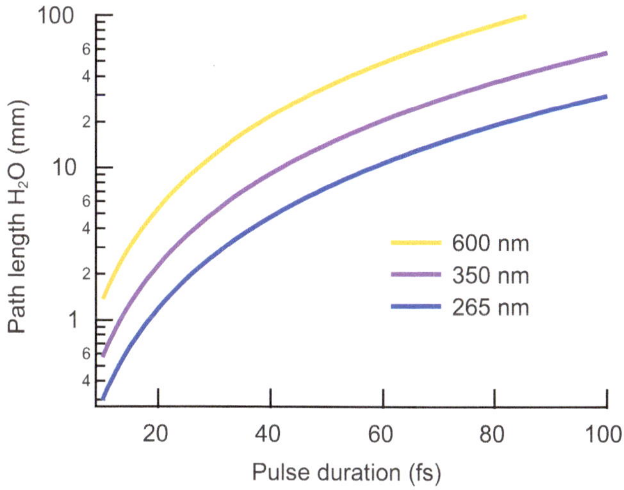

Figure 1.2 Propagation distance in neat water needed to double the pulse duration of an initially transform-limited Gaussian pulse with the indicated center wavelength. Refractive index dispersion data used in the calculations is from Ref. [46].

a pulse to its transform-limited pulse width after passage through material. Alternatively, a compressor can be used to prechirp a pulse such that it will reach its minimum duration at a particular location such as the sample position. Dispersion control is also critical for the generation and amplification of ultrashort pulses.

A pulse of temporal width τ traveling with group velocity c has a spatial extent $\Delta z \approx c\tau$, which is usually small compared to the sample thickness in the TA experiment. The thin pump pulse loses energy by absorption as it travels through the optically thick sample, creating a traveling wave of excited states, which initiate the photochemical dynamics that are subsequently read out by the delayed probe pulse. Because the sample is optically thick, it is necessary to simultaneously solve photon transport equations that describe the propagation of pump and probe pulses taking into account their group velocities in addition to rate equations that describe the population kinetics (see Appendix A). In contrast, nanosecond pump pulses are much longer in space than the sample path length and all molecules in the sample experience the same pumping rate at the same time because the intensity of the light pulses changes minimally during propagation [45].

It takes 5 ps for a pump pulse with a center wavelength of 265 nm to traverse 1 mm of water. Considering that the excited-state lifetime of adenosine is approximately 200 fs in aqueous solution [1], adenosine excited states created near the front of a sample return to the electronic ground state long before the pump pulse reaches the back of an optically thick sample.

Fortunately, the experimental time resolution can be orders of magnitude higher than the propagation time, if pump and probe pulses travel through the sample in unison. The latter condition is satisfied when pump and probe pulses have identical group velocities as is true when pump and probe wavelengths are degenerate. When pump and probe pulses have different group velocities, the group velocity mismatch, GVM (Eq. 1.7), determines the separation in time between pulses that accrues with distance through the sample.

$$\mathrm{GVM} = \frac{1}{v_{g1}} - \frac{1}{v_{g2}}. \tag{1.7}$$

GVM can be significant when pump and probe wavelengths are very different. For pump and probe wavelengths of 265 nm and 570 nm, respectively, GVM is calculated to be 420 fs mm^{-1} in water using refractive index data in Ref. [46]. This means that two 100 fs pulses cannot maintain overlap after propagating through 1 mm of water. GVM affects both the signal strength and the time resolution. In practice, the effects are mitigated somewhat by the fact that much of the signal is generated near the front of the cell before there is significant walk-off.

1.3 Transient Absorption Signal Strength

A formalism for modeling TA signals in the incoherent limit, which holds for most experiments and is of greatest interest for studies of photochemical kinetics, is presented in Appendix A. Cho et al. recently published a somewhat more detailed model for predicting absolute TA signal strengths [47]. In the incoherent limit, absorption is determined by the intensity (energy flux) and not by the amplitude or phase of the electric field of the pump pulse. In this limit, quantum optical interactions [48–50] can be neglected. Absorption and stimulated emission can then be modeled using rate equations and a kinetic model for the time-dependent concentrations of the various transient species.

A rate equation analysis is appropriate in the rapid dephasing limit when the pulse duration τ is longer than the dephasing (T_2) time of the sample molecules. Equivalently, the pulse bandwidth should be narrow compared to the width of a homogeneously broadened absorption band. This limit is easily satisfied by the strongly allowed UV transitions of DNA nucleobases, the absorption bands of which are on the order of several thousand wavenumbers (FWHM) [51] or more than an order of magnitude greater than the $\sim$150 cm^{-1} bandwidth of a 100 fs pump pulse.

The solution under study contains multiple one-photon absorbers. Each absorber can undergo a transition to a single excited state at a rate that depends on its absorption cross section. The model assumes that the intensity envelopes of pump and probe pulses are unchanged by propagation through the sample. This is the case if broadening due to dispersion and absorption can be neglected. Dispersive broadening can be minimized by appropriate choice of the sample path length.

If the pump fluence is low enough, only a small fraction of the absorbing molecules is excited and absorption saturation can be neglected. When the pump pulse is strong enough to saturate the transition, the front edge can create so many excited states that the back edge of the pulse experiences reduced absorption, and the pulse duration is altered by propagation. Saturation is also undesirable because it causes the TA signal to vary nonlinearly with the pump pulse energy and can distort the temporal evolution of TA signals [29].

Absorption saturation occurs when the fluence of the pump pulse is too high. The product of the cross section times the fluence, σF, gives the fraction of absorbers transferred to the excited state in the limit of small population transfer ($\sigma F \ << \ 1$). As a general rule, saturation is unimportant if no more than 5% of molecules are excited. At 260 nm, the molar absorption coefficient of adenosine is

15,400 M^{-1} cm^{-1}, corresponding to $\sigma = 5.9 \times 10^{-17}$ cm^2. To keep σF below 0.05, the fluence must be kept below 8.5×10^{14} photons cm^{-2}. For a pump pulse that is Gaussian in space and time, this value corresponds (see Eq. B.6 in Appendix B) to the peak fluence of a pulse with an on-axis peak intensity of 6.1 GW cm^{-2}. A 1 μJ, 100 fs pump pulse exceeds this intensity when focused to a spot that is less than or equal to 300 μm (see Eq. B.3 in Appendix B).

This is a conservative criterion for avoiding saturation because the saturating fluence will only be reached very briefly during the most intense part of the pulse and at the center of the focused spot ($r = 0$). In addition, the pump pulse is attenuated exponentially with distance when the Beer–Lambert law is obeyed, making it difficult to saturate the absorption away from the front entrance window.

The initial excited states created by the pump pulse may pass through one or more states before repopulating the electronic ground state or forming permanent photoproducts. A kinetic scheme, which differs in general for each type of excited state i, describes how initial excited states created by the pump pulse evolve at later times.

The traveling wave of excitation initiates dynamics that unfold over timescales that are usually much longer than the time during which the pump pulse interacts with each absorber. The dynamically evolving species of interest depend on spatial and temporal coordinates. Fortunately, motion by the species themselves can be neglected because even a rapidly diffusing small molecule travels no more than a few nm during the maximum time delay of several nanoseconds in a typical TA experiment. This is completely negligible compared to the spatial extent of the pump pulse.

The TA signal at delay time d from a sample containing multiple absorbers is given by Eq. A.32 in Appendix A,

$$\Delta\tilde{A}(d) = \frac{F_{\text{ave}}L}{\ln 10}\sum_i N_i\sigma_{\text{A},i} \times \left(\left[R_{\text{B},i}\Delta\tilde{\sigma}_{\text{B},i} + R_{\text{C},i}\Delta\tilde{\sigma}_{\text{C},i} + \cdots\right] * G'\right)(d). \quad (1.8)$$

Here, F_{ave} is the pump pulse fluence averaged over the path length through the sample, L, and N_i is the number density (molecules per unit volume) of absorber species i, which has ground-state absorption cross section $\sigma_{A,i}$ at the pump wavelength. The response functions $R_{X,i}$ ($X = $ B, C, ...) describe the kinetics of the initial excited state B and further transient states (C, D, ...) for each species i in response to an infinitely short excitation pulse. The difference cross section $\Delta\tilde{\sigma}_{X,i}$ is the absorption cross section of state X minus the absorption cross section of ground state A, both measured at the probe wavelength. The weighted response functions are convoluted with the normalized Gaussian function G' to account for the finite time resolution. All of these quantities are described in more detail in Appendix A.

When the Beer–Lambert law is obeyed (i.e., when only one-photon absorbers are present and there is no absorption saturation) the average fluence can be written as,

$$F_{\text{ave}} = \frac{1 - 10^{-A}}{A \ln 10} F(0), \tag{1.9}$$

where A is the absorbance at the pump wavelength,

$$A = L \sum_i N_i \sigma_{A,i} / \ln 10 = \sum_i A_i. \tag{1.10}$$

In Eq. 1.10, A_i is the absorbance that would be measured if species i were present by itself in solution,

$$A_i = L N_i \sigma_{A,i} / \ln 10. \tag{1.11}$$

Using Eqs. 1.9 and 1.10, Eq. 1.8 can be written,

$$\Delta\tilde{A}(d) = \frac{F(0)\left(1 - 10^{-A}\right)}{\ln 10} \times \sum_i \frac{A_i}{A} \left(\left[R_{B,i} \Delta\tilde{\sigma}_{B,i} + R_{C,i} \Delta\tilde{\sigma}_{C,i} + \cdots\right] * G'\right)(d). \tag{1.12}$$

Equation 1.12 indicates that the contribution of each species to the TA signal is weighted by its contribution to the total absorbance. When the pump pulse can create only a single kind of excited state, then the subscript i can be dropped,

$$\Delta\tilde{A}(d) = \frac{F(0)\left(1 - 10^{-A}\right)}{\ln 10} \times \left(\left[R_B \Delta\tilde{\sigma}_B + R_C \Delta\tilde{\sigma}_C + \cdots\right] * G'\right)(d). \tag{1.13}$$

According to Eqs. 1.12 and 1.13, the only experimentally adjustable parameters that affect the signal strength are the sample absorbance at the pump wavelength and the incident pump fluence. The signal does not depend explicitly on the path length, L, so there is no advantage to using a longer path length once an appropriately high value of A has been achieved. The TA signal reaches 90% of the maximum possible value when the absorbance of the pump pulse is unity. A longer path length may allow a targeted value of A(e.g., $A = 1$) to be achieved that might not be possible in a shorter cell because a high solute concentration could either be unattainable because of solubility limits, or lead to aggregate formation.

On the other hand, a longer path length results in reduced time resolution due to temporal walk-off of the pump and probe pulses (see Section 1.2.2). In practice, a path length of 1 mm is a reasonable choice for pulse durations near 100 fs. Because the signal increases linearly with the pump pulse fluence, the signal will grow quadratically as the pump spot size is decreased (Eq. B.6), but the fluence must be low enough to avoid saturation or deleterious nonlinear optical effects such as self-focusing or continuum generation.

1.3.1 Bleach Recovery Signals

Bleach signals are ones for which ΔA is negative. To observe a bleach signal, the probe wavelength must be absorbed by the sample. The only quantities in Eq. 1.12 that can be less than zero are the difference cross sections, $\Delta\tilde{\sigma}_{X,i}$. Thus, bleach signals are observed when ground-state species absorb the probe wavelength more strongly than excited states.

Any state that affects the kinetics of ground-state repopulation will contribute to the dynamics seen in a bleach recovery signal. Bleach recovery experiments thus provide a powerful and general method for observing the dynamics of "invisible" excited states that may absorb only weakly. For example, bleach recovery measurements first identified a long-lived excited state in single pyrimidine nucleobases [52]. As will be discussed in Section 1.4, bleach recovery experiments have also provided important insights into the dynamics of long-lived excited states in DNA strands.

The largest possible bleach signal at time zero for a system of one absorber is obtained when only state A absorbs at the probe wavelength. In that case, $\Delta\tilde{\sigma}_{\mathrm{B}} = \Delta\tilde{\sigma}_{\mathrm{C}} = \cdots = -\tilde{\sigma}_{\mathrm{A}}$ and inserting these relations in Eq. 1.8 yields,

$$\Delta\tilde{A}\,(0) = -\frac{F_{\mathrm{ave}}L}{\ln 10}N\sigma_{\mathrm{A}}\tilde{\sigma}_{\mathrm{A}}\left([1 - R_{\mathrm{A}}] * G'\right)(0), \tag{1.14}$$

where R_{A} is defined in Eq. A.25 in Appendix A. When the IRF is faster than any of the other timescales, this maximum signal is equal to,

$$\Delta\tilde{A}(0) = -\sigma_{\mathrm{A}}F_{\mathrm{ave}}\frac{NL\tilde{\sigma}_{\mathrm{A}}}{\ln 10} = -\sigma_{\mathrm{A}}F_{\mathrm{ave}}\tilde{A}, \tag{1.15}$$

where $\tilde{A}$ is the absorbance at the probe wavelength. If the observed bleach signal near time zero is larger than about $-0.05\tilde{A}$, then this may indicate that the signal is approaching saturation.

1.3.2 DNA TA Experiments

Previous sections have focused on general aspects of TA spectroscopy. In this section some specific considerations relevant to experiments on DNA samples are presented. Specifically, thermal effects due to laser heating and consequences of the generally weak ESA observed for both DNA base monomers and multimers are explored. Thermal effects must be considered because the quantum yield of nonradiative decay is close to one for compounds made of natural nucleobases. Consequently, the energy of nearly every absorbed photon is dissipated in the solution. Weak ESA means that nucleobase TA signals at visible probe wavelengths can be easily overwhelmed by signals due to solvated electrons produced by the two-photon ionization of water.

Some simple calculations illustrate the importance of avoiding laser heating of the sample. Consider a 1 μJ pump pulse that is focused to a spot size of 200 μm in a 1 mm path length cell. Only thermal effects due to the pump pulse need to be considered because the probe pulse is typically one hundred times weaker in average power. If the solution absorbance is 1.0 at the pump wavelength, then ~90% of the pump pulse energy is deposited as heat in a volume of 0.3 μL. If the pumped solution is not exchanged with fresh solution, and if the rate of heat transport out of the pumped volume can be neglected, then the temperature will rise at a rate of about 0.7°C per second in aqueous solution at 25°C, assuming a 1.0 kHz pulse repetition rate.

To reduce the rate of laser heating, the pump energy must be dissipated in a larger volume of solution. A spinning cell was used for a time in our laboratory in which a 1 mm thick slab of the solution under study is held between two windows and the entire cell is spun about an axis perpendicular to the laser beams by an electrical motor [53]. By positioning the overlapping pump and probe beams approximately 1 cm off the rotation axis it is straightforward to spin the cell at a high enough angular frequency to ensure that the next pump pulse does not overlap the solution volume excited by the previous pulse.

Unfortunately, inadequate circulation of the solution inside the cell and slow heat diffusion out of the annular region excited by the pump beam during many revolutions of the spinning cell leads to unacceptably high steady-state temperatures. As discussed elsewhere [53, 54], temperatures of 40°C, or even higher, can be produced by laser heating in the spinning cell. These temperatures are high enough to denature low-melting double-stranded DNA samples such as AT-rich sequences.

The problem of laser heating is completely eliminated by flowing the solution, effectively dispersing the heat throughout a larger sample volume. For example, the rate of temperature increase is predicted to be just 0.9°C per hour when the volume is increased to 1 mL of water, but with all other parameters mentioned above unchanged. Consequently, even a total solution volume of just 1 mL is large enough to make laser heating negligible when the solution that is recirculating through the flow cell is well mixed. The combination of a recirculating flow cell system and a temperature bath for maintaining constant temperature has allowed accurate measurements to be carried out on some substrates that were previously affected by temperature artifacts [54].

Excited states of both DNA base monomers and multimers exhibit weak ESA at visible probe wavelengths. The molar absorption coefficient is estimated to be on the order of ~100 M^{-1} cm^{-1} for single nucleosides near 600 nm [55]. Weak ESA makes it necessary to increase the pump fluence to increase the signal strength, but this has the undesirable effect of increasing two-photon absorption by the aqueous solvent. Water is a reasonably strong two-photon absorber at deep UV wavelengths [56–58] with a two-photon absorption coefficient of approximately 1 GM (1 GM, or Goeppert–Mayer, equals 1.0×10^{-50} cm^4 s) at 264 nm [58]. The bases themselves have two-photon absorption cross sections of similar magnitude near 500 nm [59, 60], but the much higher number density of solvent molecules gives rise to a significant two-photon excitation rate of water molecules.

At a wavelength of 267 nm, two-photon excitation produces a water molecule with an energy of 9.3 eV, which ionizes to yield a solvated electron or dissociates into a hydrogen atom plus a hydroxyl radical. Each of these channels has approximately equal quantum efficiency of $\sim$0.4 [61]. The solvated electron produced by ionization is of greatest concern due to its strong absorption that spans UV to near-IR wavelengths. Using the measured extinction coefficient for the solvated electron near the band maximum of ε (720 nm) = 22,700 M^{-1} cm^{-1} [62] together with the parameterization of the solvated electron band shape of Jou and Freeman [63], it can be shown that the extinction coefficient of the solvated electron exceeds 5,000 M^{-1} cm^{-1} between 470 and 1030 nm.

Interfering absorption from solvated electrons can completely obscure the dynamics of the DNA bases, if the pump intensity is too high [55]. As discussed in Ref. [2], the use of low fluences in the TA experiments of Pecourt et al. [1] was a critical factor behind the first successful measurements of subpicosecond excited-state lifetimes for single DNA bases in aqueous solution.

When the pump pulse experiences significant two-photon absorption, then the Beer–Lambert law no longer describes the decrease in its intensity with distance through the sample. In this case, the pump pulse energy will partition between one and two-photon absorbers, and the model in Appendix A for absolute TA signal strengths must be modified. Nevertheless, the results given there still correctly predict the relative magnitudes of signals arising from one-photon absorbers. Pecourt et al. [55] found that a pump peak intensity of 3.5 GW cm^{-2} for purine bases and 8.3 GW cm^{-2} for pyrimidines struck a reasonable compromise between maximizing the ESA signal at 570 nm, while minimizing two-photon ionization of water.

The problems of weak signals and interfering absorption by solvated electrons are greatly reduced by recording bleach recovery signals at UV probe wavelengths. These experiments offer several advantages over ESA measurements at visible wavelengths. First, the signals are strong because ground-state cross sections are much larger than the weak ESA cross sections in the visible. Second, solvated electrons have minimal absorption below 300 nm, alleviating the need for correction [64]. Third, it may be possible to determine the quantum yields with which some excited states are formed from bleach recovery signals (see Section 1.4.3). Of course, one drawback of bleach recovery signals is that the signals at times less than about 10 ps are dominated by vibrational cooling dynamics after ultrafast internal conversion [65]. This can obscure the fastest events, but does not pose a problem for observing long-lived excited states in oligomers.

1.4 Excited-State Dynamics of Single DNA Strands

Previous sections in this chapter introduced the TA technique, presented general expressions for the strength of TA signals from samples containing multiple one-photon absorbers, and discussed some practical issues that arise in TA experiments on DNA samples. In this section, a few results from TA experiments on single-stranded adenine homo-oligonucleotides are presented and analyzed using the rate equation model for TA signal strengths introduced in Section 1.3 and described in more detail in Appendix A.

The kinetic models chosen for this analysis allow conclusions to be drawn about both the nature and the quantum yields of the various decay channels. Excited states in double strands will not be covered, but are discussed elsewhere [3, 4, 66]. This choice reflects the central role that the π-π-stacking interactions found in single strands play in DNA excited-state dynamics [67]. It also acknowledges that base stacking effects on excited states of DNA are currently better understood than base pairing effects.

A DNA oligonucleotide is a multichromophoric molecule with a multitude of overlapping transitions at wavelengths of photobiological importance (260–320 nm). These transitions populate different kinds of excited states such as delocalized excitons, excimers, exciplexes, charge transfer (CT) states, and locally excited states of single nucleobases (Fig. 1.3). The latter states, which might be considered to be the simplest and most tractable excitations in a DNA strand are nonetheless complex on account of multiple relaxation channels [52]. Ultimately, the diverse excitations of DNA, which depend on base sequence [68] and on subtle conformational characteristics, create a combinatorially complex electronic structure problem that makes the correct interpretation of TA signals from base multimers highly challenging.

Structure influences the ensemble of possible excitations and decay pathways because the spatial separation between nucleobases determines the coupling strength between pairs of chromophores (Fig. 1.3). Base stacking and base pairing—the dual architectural motifs of DNA—bring pairs of bases into intimate, van der Waals contact. These noncovalent interactions are responsible for the interchromophoric interactions that cause the electronic structure of DNA to differ from that of the separated nucleobases. Excitations that involve two or more bases such as excitons and excimers arise when the pairwise distances between bases are sufficiently small [51, 69–71].

The vertical distance between stacked bases is on the order of 3.4 Å, the same distance found in coplanar stacks of other aromatic compounds. While base stacking defines a minimum separation between bases in single-stranded DNA, the covalent backbone limits the maximum distance between neighboring bases that are unstacked. Unstacked bases may still be close enough to experience significant Coulombic interactions due to the long-range nature of dipole–dipole coupling, which varies as $1/R^3$, where R is the separation between bases, but CT interactions arising from orbital–orbital overlap decay exponentially with R, and may be completely eliminated when two nucleobases are separated by just a single solvent molecule [72]. Understanding the precise interactions and

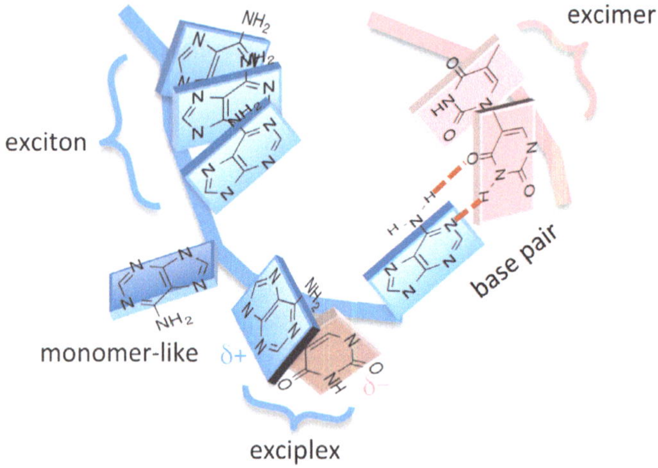

Figure 1.3 Structural relationships among nucleobases govern the types of excitations that can form in DNA, illustrated here by A (blue) and T (red) bases. Here, exciton, excimer, and exciplex refer to classes of excited electronic states that extend over more than one nucleobase. "Exciton" refers to Frenkel exciton states that result from dipolar coupling among the π–π^* states of single bases. Excitons have a relatively low degree of charge transfer (CT) character. Excited states of two bases that have strong CT character and that often cause the two bases to approach each other more closely are termed excimers if the two bases are the same, or exciplexes, if they differ. These collective excited states are limiting cases, and general excited states can have a variable amount of excitonic and CT character. Note also that CT states can be delocalized over more than just the two bases implied by the terms "excimer" and "exciplex."

couplings responsible for the observed decay channels is the goal of ongoing experimental and theoretical work.

Base stacking effects on excited-state dynamics are isolated through studies of single strands with base sequences that do not permit base pairing, that is, non-self-complementary sequences that cannot form hairpins or other base-paired structures. To date, TA experiments on single strands with as few as two and as many as several hundred bases have revealed long-lived excited states with lifetimes that can be orders of magnitude longer than the ones encountered in single bases [6, 67, 73]. Many of these experiments have examined single-stranded adenine homo-oligonucleotides [64,

67, 73–75]. Oligonucleotides made of adenine are of interest because of adenine's high propensity to stack with itself [76]. Although the excited-state lifetime of adenosine ($\sim$290 fs [1]) is the shortest among the DNA nucleobase monomers, $(dA)_n$ sequences with $n \geq 2$ exhibit a long-lived decay component of roughly 200 ps [75].

The remaining sections of this chapter are organized as follows: The structures adopted by single strands in aqueous solution are discussed first because of the profound effect that substrate conformation, namely π-π stacking, has on excited-state dynamics. Next, TA signals recorded for $(dA)_n$ sequences with a variable number n of residues are described. Guided by the formalism for modeling TA signals introduced in Section 1.3, the signals are then interpreted using kinetic models appropriate for the heterogeneous excited-state populations expected in disordered single strands in solution. The dynamics of DNA excitons are discussed in a final section, followed by a few concluding remarks.

1.4.1 Structure of $(dA)_n$ Single Strands

Many different techniques have demonstrated base stacking by successive adenines in single-stranded oligonucleotides. Nuclear magnetic resonance (NMR) and circular dichroism (CD) experiments on adenine dinucleotides demonstrated many years ago that the two bases adopt a right-handed helical arrangement similar to that found in double-stranded DNA [77]. It is also well known that homoadenine sequences have a tendency to form stacked helices [78]. Other physical techniques provide evidence that runs of consecutive adenines form helical base stacks with distinctive physical properties. For example, sedimentation velocity analysis indicates that dA homo-oligonucleotides have more extended shapes than their dT and dC counterparts [79]. Mills et al. [80] studied the electrophoretic mobility of so-called gapped duplexes in which a single-stranded region is located between two double-stranded sections. These substrates exhibit reduced mobility compared to full duplex structures, but higher mobility is seen for a $(dA)_n$ gap than for one made of consecutive thymines, consistent with greater rigidity in the former sequence. Single-molecule atomic force microscopy experiments also demonstrate that poly(dA) single strands have mechanical properties that differ dramatically from those of more flexible poly(dT) single strands [81].

The above experiments provide evidence of increased rigidity and other unusual mechanical properties in $(dA)_n$ sequences due to base stacking, but it is important to keep in mind that the bases do not form a single, monolithic stack at room temperature. Goddard et al. [82] studied single-stranded DNA hairpin sequences that contained a fluorophore and a quencher at either end. They measured the time needed for these "molecular beacons" [83] to open and close using fluorescence correlation methods. They found that an initially open hairpin folds more slowly when the loop region of the hairpin is made of 21 A residues compared to a loop containing 21 T residues [82]. Base stacking by runs of adenines retards, but does not eliminate the folding that results in fluorescence quenching, indicating that the A bases in a $(dA)_n$ oligonucleotide do not form a perfectly rigid stack.

The stacked, yet flexible nature of $(dA)_n$ sequences is consistent with models that predict that base stacking in oligo-adenylates is weakly stabilized and largely noncooperative [84, 85]. In the noncooperative limit, the probability of forming a stack between any two bases is independent of whether either base is already part of an existing stacked domain.

There are $n - 1$ positions in an oligomer of n residues at which stacking can occur between adjacent bases. Applequist and Damle used a one-dimensional Ising model to calculate the distribution of stacks, assuming that stacking at any of these $n - 1$ positions is described by a two-state equilibrium (unstacked $\rightarrow$ stacked) with equilibrium constant K (Fig. 1.4a,b). Their model treats both cooperative and noncooperative stacking, but comparison with experiment indicates that base-stacking cooperativity is low and only the noncooperative solutions will be discussed here.

A reasonable question is whether the equilibrium constant K really has the same value at all $n - 1$ stacking positions in an oligomer. A terminal base could be less likely to stack with a neighbor than an interior base given the ubiquitous end-fraying phenomenon in double-stranded helices [86–89]. Breslauer and Sturtevant questioned whether terminal bases in a single strand

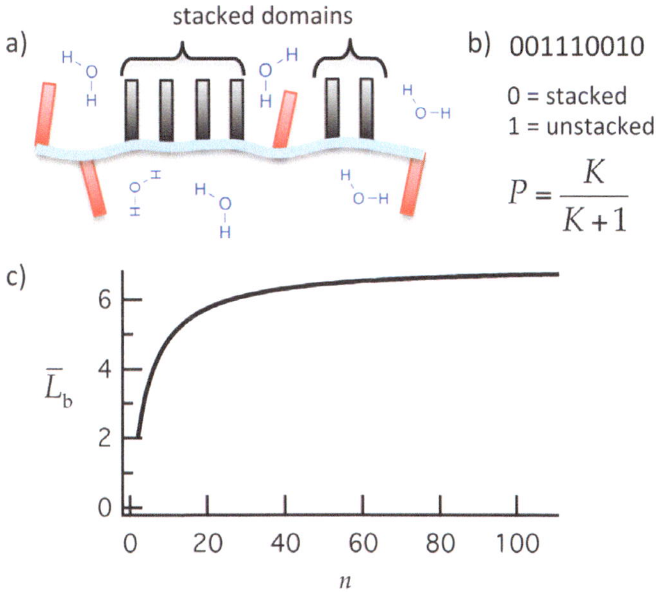

Figure 1.4 (a) Schematic of base stacking in a DNA decamer. Stacked bases are shown in black and unstacked bases in red. (b) Each configuration can be represented by a binary number with $n-1$ digits, where digit i indicates stacking between bases i and $i+1$. (c) $\bar{L}_b$ (Eq. 1.16) versus n for $K = 5$.

are characterized by the same thermodynamic stacking parameters as interior residues [90]. Earlier, Brahms et al. [91] suggested that base stacks are slightly more stable in a longer polymer than in a dimer due to decreased entropy gain when stacks are broken in the interior of longer chains. Buhot and Halperin also assumed that terminal bases, which can only stack with a single neighbor, have a smaller reduction in their configurational entropy compared to interior bases, which can stack with two neighbors [92].

Although it is not unreasonable to expect that K could be smaller for stacking between a terminal base and its nearest neighbor, experimental evidence is lacking and it will be assumed hereafter that K is the same at all positions where stacking may occur [84]. In fact, it is difficult to answer this question experimentally because terminal bases are already less likely to be found in stacks even when K is the same everywhere (see below).

Table 1.1 Equilibrium constant (K) and thermodynamic parameters for the unstacked-to-stacked transition of $(dA)_2$ in aqueous solution at 25°C

Method[a]	$\Delta H^{\circ,b}$	$\Delta S^{\circ,c}$	$\Delta G^{\circ,b}$	K	Ref.
TA	−5.4	−15.	−0.93	4.8	[95]
CD	−6.7	−20.7	−0.53	2.4	[93]
CD	−7.4	−20	−1.4	11	[94]
UV	−6.7	−19	−1.0	5.7	[94]

[a] TA, transient absorption; CD, circular dichroism; UV, ultraviolet hypochromism
[b] Units of kcal mol^{-1}
[c] Units of cal mol^{-1} K^{-1}

The magnitude and temperature dependence of K have been determined for the $(dA)_2$ dinucleotide using CD spectroscopy [93, 94], UV melting experiments [94], and very recently, TA measurements [95]. Values of K for the base stacking transition in the dinucleotide $(dA)_2$ together with thermodynamic parameters determined from van't Hoff analyses are summarized in Table 1.1. The small value of ΔG° (ca. -1 kcal mol^{-1} at 25°C) indicates that base stacking is marginally stable as expected for biomolecules that must be capable of easily changing their structure at room temperature.

The probability of stacking, P, is given by $K/(K+1)$, and varies from roughly 70 to 90% according to the K values in Table 1.1. If the same probability characterizes stacking at each of the $n-1$ positions in an n-mer, then the probability that all bases in an oligomer of n residues are stacked is equal to P^{n-1} [84]. The average K value of 6 from Table 1.1 then implies that only 7% of all $(dA)_{18}$ oligomers are fully stacked at 25°C. In an oligomer of n residues, there will be a distribution of base-stacked domains containing between 2 and n bases. The average stack contains $\bar{L}_b$ bases, where

$$\bar{L}_b = \frac{n(K+2)-2}{n-1+K}. \tag{1.16}$$

For $K = 6$, the average stacked domain in an ensemble of $(dA)_{18}$ oligomers is predicted to span ~6 bases. The comparatively small value of K means that unstacked bases frequently interrupt runs of stacked bases [84].

Equation 1.16 also indicates that the average length of a stacked domain decreases with n from a maximum value of $K + 2$ for the infinite homopolymer to a minimum stack size of two bases when $n = 2$ (Fig. 1.4c). This comes about because the average domain length observed in a very long oligomer must decrease as the oligomer's length becomes commensurate with the domain size. This confinement effect on the size of stacked domains could limit the size of delocalized excitons in DNA [75].

An important consequence of the model is that even though the same value of K describes stacking everywhere in the oligomer, terminal bases are still less likely to be found in stacks. This occurs because interior bases can stack with either of two neighbors, while terminal bases can stack with only one. This causes the fraction of all bases that are found in stacked domains with at least two bases, f_b, to vary with n as [75],

$$f_b = P\,(2 - P) + \frac{2P\,(P - 1)}{n}, \qquad (1.17)$$

where P is the stacking probability defined in Fig. 1.4b.

Fluorescence excitation experiments provide evidence for decreased stacking by terminal bases in DNA single strands. Ramreddy et al. [96] recorded fluorescence excitation spectra from $(dA)_{30}$ single strands in which a single adenine residue at a defined position was replaced by 2-aminopurine (2AP), a highly fluorescent isomer of adenine. 2AP fluorescence can be initiated by energy transfer from a proximal adenine [97] and this energy transfer is strongly enhanced by base stacking [98]. Ramreddy et al. [96] found that energy transfer from adenine to 2AP was much less efficient when 2AP was the terminal 5′ base compared to when 2AP was the second residue from the 5′ terminus. The most efficient energy transfer occurred when 2AP was placed in the middle of the 30-mer, particularly at low temperature. These observations suggest that interior residues are more likely to be stacked than terminal residues in agreement with the model discussed above.

In summary, base-stacking disorder in single DNA strands at room temperature causes stacked domains to be relatively modest in length, even when those strands are made solely of adenine, the base with the highest stacking propensity. The noncooperative stacking model predicts that the fraction of bases in stacked domains increases with oligomer length (Eq. 1.17). Also, as the number of residues in an oligomer is reduced, confinement causes the average length of a stacked domain to decrease. This structural heterogeneity decisively affects the TA signals as shown in the next section.

1.4.2 TA Signals from (dA)$_n$ Single Strands

Long-lived excited states with lifetimes that are orders of magnitude longer than the short-lived $^1\pi\pi^*$ states of single bases are ubiquitous in single and double-stranded DNA helices [6]. These states can be seen in minimum stacks consisting of just two bases [99], but are only observed when the bases are intimately π-stacked [72, 75].

The bleach recovery signal of the adenine mononucleotide dAMP recovers with a time constant of approximately 2 ps (Fig. 1.5a) due to vibrational cooling following subpicosecond relaxation from S_1 to the vibrationally excited ground state. Long-lived excited states are responsible for the 180 ps decay component seen in the dinucleotide $(dA)_2$ at pH 7 (blue circles, Fig. 1.5b).

Coulombic repulsion between the protonated residues of $(dA)_2$ at low pH disrupts base stacking and eliminates the long-lived excited states (red circles, Fig. 1.5b) [75]. Addition of an organic cosolvent such as methanol (80% by volume) also disrupts stacking and eliminates the slow decay channel (Fig. 1.5c) [72]. The elimination of long-lived excited states at low pH and by an added denaturant (methanol) implies that these states only form when there is orbital–orbital overlap between the bases.

The importance of π-π stacking interactions is also evident from time-resolved photoionization experiments performed on small clusters of adenine molecules. Smith et al. [100] observed only ultrashort decay times ($\tau_1 \leq 100$ fs, $\tau_2 = 1.3$ ps) for adenine and for the hydrogen-bonded adenine dimer, but a lifetime of >1 ns was observed for the microhydrated adenine dimer clustered with between one and five H_2O molecules. Calculations indicate that adenine dimers clustered with a small number of water molecules adopt π-stacked structures [101].

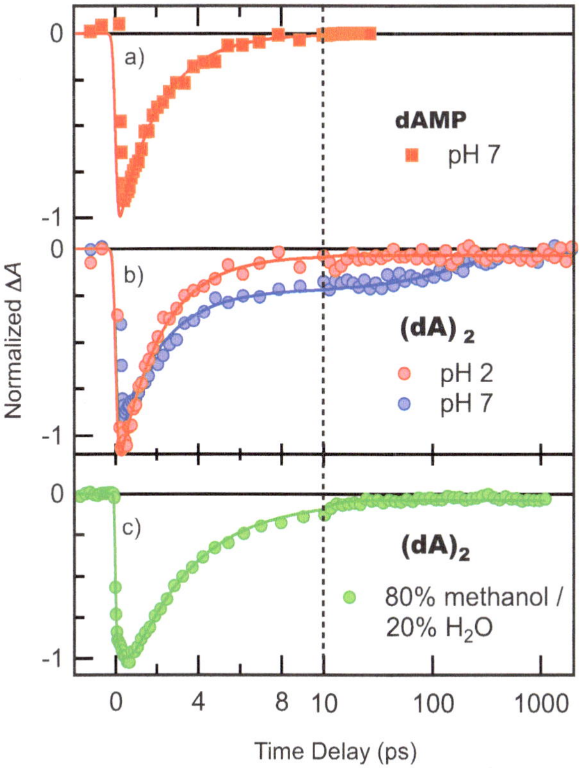

Figure 1.5 Bleach recovery signals (265 nm pump/250 nm probe) for (a) dAMP and for the dinucleotide $(dA)_2$ in (b) acidic and neutral pH solution, and (c) in the presence of a denaturing cosolvent.

Bleach recovery signals from adenine homo-oligonucleotides in aqueous buffer solution are biphasic (Fig. 1.6). Virtually all relaxation occurs on two timescales: a shorter timescale of less than a few picoseconds, and a slower decay is observed with a time constant of between 100 and 200 ps. A small amplitude offset is also present that may be due to photodegradation or kinetics beyond the timescale of the measurements [4]. The fast decay component is assigned to vibrational cooling following subpicosecond internal conversion to the electronic ground state. The slow component reveals the existence of long-lived excited states with lifetimes that are orders of magnitude longer than excited states of monomeric bases.

Assignment of the slow component has been controversial and one of the most vigorously discussed issues in DNA photophysics during the past decade. Crespo-Hernández et al. assigned the slow component seen in single-stranded $(dA)_{18}$ to singlet excimers [67]. Buchvarov et al. assigned the long-lived TA signals to absorption by delocalized excitons [74]. Conti et al. proposed instead that the long-lived decay is due to excited states localized on single adenine residues, which decay slowly due to steric hindrance that impedes motion to the deactivating conical intersection [102].

Although this question has also been addressed by other researchers, the three preceding proposals illustrate the contradictory interpretations discussed in the literature. Each proposal assigns the long-lived signal to a different type of excitation, which is known to exist in DNA. Thus, low-temperature luminescence experiments provided evidence already many years ago for excimers in DNA strands [103], while circular dichroism spectroscopy [104] and fluorescence anisotropy experiments [105] capture the spectroscopic signatures of delocalized excitons. Finally, it is certainly plausible that an excitation in a DNA strand could localize on a single base, depending on the strength of the base's interactions with its neighbors, and subsequently follow a monomer-like decay pathway. It is not the existence of these different excitations that is in doubt, but rather the dynamical signatures of these states in TA experiments.

The notion that nearest-neighbor bases could impede out-of-plane motions by a photoexcited base is intuitively appealing, but long-lived excited states in dinucleotides decay with identical lifetimes as ones in considerably longer strands (Fig. 1.6). This observation suggests—somewhat illogically—that a base in a dinucleotide, which can only stack with one of its two faces against a neighboring base, would experience the same steric hindrance as a base in a longer strand that is sandwiched between two neighbors. Additionally, small lifetime effects due to changes in solvent viscosity imply that only very small atomic displacements are needed to reach the relevant conical intersection(s) [106]. Nonadiabatic dynamical simulations have been used to explore this hypothesis, but the results to date predict excited-state lifetimes for

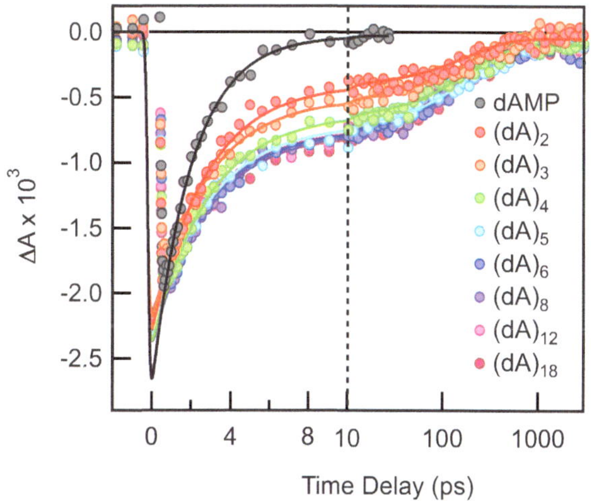

Figure 1.6 Bleach recovery signals (266 nm pump/250 nm probe) for dAMP and $(dA)_n$ oligonucleotides. Fits were made to data points at delay times greater than 0.8 ps to avoid the strong coherent spike present near $t = 0$. This is why the fit curves, which do not attempt to model the short-time coherent signal, deviate from the data points immediately after $t = 0$. Adapted with permission from Ref. [75]. Copyright (2012) American Chemical Society.

adenine [107] (or an adenine mimic [108]) in a solvated single-strand environment that are considerably shorter than the 200 ps lifetime seen experimentally [75]. The long lifetimes are better explained by a fundamentally new deactivation pathway arising from a change in electronic structure.

Buchvarov et al. [74] explored the possibility that TA spectroscopy can be used to determine the size of excitons in DNA. These authors recorded TA signals for $(dA)_n$ oligonucleotides as a function of n and interpreted the length-dependent changes in TA signal amplitudes in terms of excitons that increase in the absorption cross section as the strand length increases. They concluded that excitons in stacked helices have a "1/e delocalization length" of 3.3 ± 0.5 bases.

Su et al. re-investigated excited-state dynamics in $(dA)_n$ sequences by TA spectroscopy [75]. These authors showed that the amplitudes of bleach recovery signals vary with oligomer length in the same manner as the signals derived from pump–probe measurements made at longer wavelengths by Buchvarov et al. According to Eq. 1.8, pump–probe signals depend on both populations and cross sections. Whereas Buchvarov et al. argued that absorption cross sections change with n, while populations remain roughly constant, Su et al. presented evidence that the opposite is true. They proposed that the long-lived excited states are excimer or CT states that are localized on just two stacked bases. These states are formed from initial excitons, which could be delocalized over several bases, but that decay to excimer or CT states localized on two bases faster than the experimental time resolution of several hundred femtoseconds. In the Su et al. model, which will be presented in more detail in the following sections, the absorption cross sections of these states are the same for oligomers of any length, but their populations vary because the fraction of stacked bases increases with n (Eq. 1.17).

1.4.3 Estimating Quantum Yields from Bleach Signals

The kinetic mechanism connecting the various transient states and other species is the key to interpreting TA signals (Eq. 1.12). In an unbranched, sequential model [109] such as B $\rightarrow$ C $\rightarrow$ D $\rightarrow$ ... $\rightarrow$ A, every initial excited state passes through every later state in the kinetic mechanism. However, if population can branch to two or more states, then the fraction of all excited states that pass through alternative decay channels is of interest. Such quantum yields are difficult to estimate from fluorescence experiments, but can be obtained from TA measurements in favorable cases as shown in this section.

The two kinetic models illustrated in Fig. 1.7 are used to analyze the signals in Fig. 1.6. Both models postulate that two distinct populations of excited states are responsible for the dual decays seen in the $(\mathrm{dA})_n$ transients. In the model in Fig. 1.7a, a single initial excited-state population branches to short- and long-lived excited states faster than the time resolution of the pump–probe spectrometer. Although a homogeneous excited-state population is inconsistent with the heterogeneity expected from the structural model presented in Section 1.4.1, it is nevertheless useful to consider

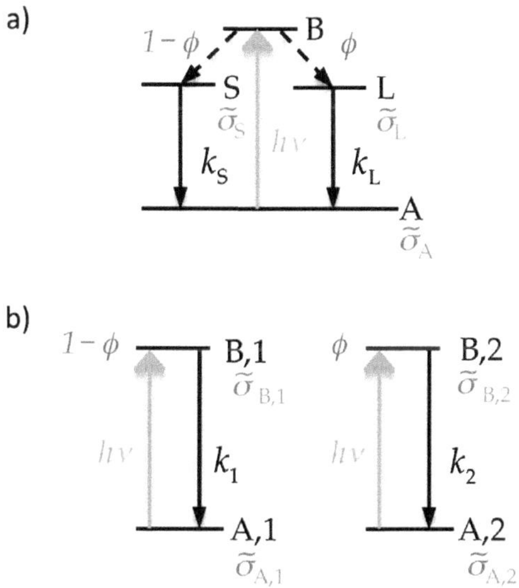

Figure 1.7 The two kinetic schemes used to analyze bleach recovery signals from $(dA)_n$ oligomers. Pump photons are shown by gray arrows. Black arrows representing nonradiative transitions are either dashed, indicating a step that occurs faster than the instrument response function, or solid, indicating that decay occurs with the resolvable rate constant next to each arrow. Cross sections for each state at the probe wavelength are shown in green. The scheme in (a) assumes a homogeneous excited-state population that branches to short- and long-lived excited states, while the scheme in (b) postulates distinct ground- and excited-state species.

the predictions of this simple model, which can describe parallel decay pathways of $^1\pi\pi^*$ excited states in pyrimidine base monomers [52].

The fraction of all initial excited states that branch to the short-lived (subscript S) or the long-lived (subscript L) excited states is given by $1 - \phi$ and ϕ, respectively, in the kinetic scheme in Fig. 1.7a. The short-lived excited states decay to the ground state with rate k_S, and their population in response to an infinitely short excitation pulse decays in time as,

$$R_S(t) = (1 - \phi)\exp(-k_S t), \tag{1.18}$$

while the long-lived excited states decay at the slower rate k_L,

$$R_L(t) = \phi\exp(-k_L t). \tag{1.19}$$

Substituting these functions into Eq. 1.13, the signal at delay time d can be written

$$\Delta\tilde{A}(d) = \frac{F(0)\left(1-10^{-A}\right)}{\ln(10)} \times \left(\left[(1-\phi)\,\Delta\tilde{\sigma}_{\mathrm{S}}e^{-k_{\mathrm{S}}d} + \phi\Delta\tilde{\sigma}_{\mathrm{L}}e^{-k_{\mathrm{L}}d}\right] * G'\right)(d). \quad (1.20)$$

The signal can only be negative at all delay times *and* decay monotonically toward zero, in agreement with experiment, if $\Delta\tilde{\sigma}_{\mathrm{S}} = \tilde{\sigma}_{\mathrm{S}} - \tilde{\sigma}_{\mathrm{A}}$ and $\Delta\tilde{\sigma}_{\mathrm{L}} = \tilde{\sigma}_{\mathrm{L}} - \tilde{\sigma}_{\mathrm{A}}$ are both less than zero. This is the case when the ground-state cross section is greater than those of both the short- and the long-lived excited state (all cross sections are evaluated at the probe wavelength), that is, $\tilde{\sigma}_{\mathrm{S}}, \tilde{\sigma}_{\mathrm{L}} < \tilde{\sigma}_{\mathrm{A}}$. The experimental signal is fit to the convolution of the fitting function $\alpha_{\mathrm{S}}\exp(-k_{\mathrm{S}}t) + \alpha_{\mathrm{L}}\exp(-k_{\mathrm{L}}t)$ with the instrument response function, G'. Then, the *relative* amplitude of the slow decay component, α'_{L}, is equal to,

$$\alpha'_{\mathrm{L}} \equiv \frac{\alpha_{\mathrm{L}}}{\alpha_{\mathrm{S}} + \alpha_{\mathrm{L}}} = \frac{\phi}{(1-\phi)\frac{\Delta\tilde{\sigma}_{\mathrm{S}}}{\Delta\tilde{\sigma}_{\mathrm{L}}} + \phi}. \quad (1.21)$$

If short- and long-lived excited states have identical absorption cross sections at the probe wavelength, then $\Delta\tilde{\sigma}_{\mathrm{S}} = \Delta\tilde{\sigma}_{\mathrm{L}}$, and α'_{L} is equal to the quantum yield of forming the long-lived excited state, ϕ. If there is no ESA at the probe wavelength for either excited state ($\tilde{\sigma}_{\mathrm{S}} = \tilde{\sigma}_{\mathrm{L}} = 0$), then the signal is determined purely by bleaching of the ground state. In this case, the fractional amplitudes of the exponentials are equal to the quantum yields of each decay channel. If normalized bleach signals are identical at multiple probe wavelengths, then this could indicate that ESA is negligible for both short- and long-lived excited states. Conversely, if normalized bleach signals recorded at different probe wavelengths differ in their relative amplitudes then there must be an ESA contribution to the signal from at least one of the two excited states.

In general, $\phi > \alpha'_{\mathrm{L}}$, if $\tilde{\sigma}_{\mathrm{A}} > \tilde{\sigma}_{\mathrm{L}} > \tilde{\sigma}_{\mathrm{S}}$, but $\phi < \alpha'_{\mathrm{L}}$, if $\tilde{\sigma}_{\mathrm{A}} > \tilde{\sigma}_{\mathrm{S}} > \tilde{\sigma}_{\mathrm{L}}$. Thus, α'_{L} can be a lower or upper bound on the quantum yield of the long-lived excited state in this kinetic model, depending on whether $\tilde{\sigma}_{\mathrm{L}}$ is greater or smaller than $\tilde{\sigma}_{\mathrm{S}}$, respectively. A bleach signal at a single probe wavelength is therefore an unreliable predictor of quantum yields whenever there is uncertainty about ESA by the states in the mechanism.

The short decay component seen for $(dA)_n$ oligomers has a similar lifetime and kinetic isotope effect as the mononucleotide of adenine dAMP [75], suggesting that the fast component should be assigned to relaxation by monomer-like excited states that decay via ultrafast internal conversion to the electronic ground state. In this case, the signal strength of the short-lived signal component seen for an oligonucleotide can be compared with the signal from monomeric adenine. When the TA signals are recorded on equal absorbance solutions with identical laser parameters, then the same number of excited states is formed in each and the factor $F(0)\left(1-10^{-A}\right)$ in Eq. 1.13 is equal for both solutions. If excitation of the monomer yields short-lived excited states, which are identical to those formed in the oligomer, then the ratio of the oligomer signal at time zero to the monomer signal at time zero is equal to,

$$\frac{\Delta A\,(0)}{\Delta A^{\mathrm{m}}\,(0)}=1-\phi+\phi\frac{\Delta\tilde{\sigma}_{\mathrm{L}}}{\Delta\tilde{\sigma}_{\mathrm{S}}}=1-\phi\left(1-\frac{\Delta\tilde{\sigma}_{\mathrm{L}}}{\Delta\tilde{\sigma}_{\mathrm{S}}}\right), \qquad (1.22)$$

assuming that the cross section for dAMP is equal to the per nucleotide absorption cross section of the oligonucleotide. Back-to-back TA experiments on equal absorbance solutions have shown that the absolute value of the monomer signal is greater than the absolute value of the oligomer signal [67, 75]. This can be seen in the solid fit curves in Fig. 1.6. In this case, the left-hand side of Eq. 1.22 is less than one and $\tilde{\sigma}_{\mathrm{L}} > \tilde{\sigma}_{\mathrm{S}}$ when both are less than $\tilde{\sigma}_A$. This implies that α_{L}' is a lower bound for the long-lived state quantum yield as stated in Ref. [67].

A weakness in the above model is the assumption that monomer and oligomer have the same per nucleotide absorption cross section. In fact, the absorption cross section of the A monomer is higher than the per nucleotide cross section of the oligomer due to hypochromism, the reduction in absorbance at UV wavelengths due to nucleobase stacking [110]. Hypochromism is a short-range interaction that decreases rapidly with distance between the bases and is therefore largest for stacked bases. Thus, the unstacked form of the RNA dinucleotide $(rA)_2$ at low pH shows little residual hypochromism [111].

A further weakness is the assumption that the pump pulse creates a single excited-state population in the oligomer. An improved model recognizes that structural heterogeneity results in different initial excitations. The biphasic character of the signals suggests that just two states are needed, which are formed from two distinct ground-state populations as shown in the kinetic scheme in Fig. 1.7b.

If the heterogeneous excited-state population decays via parallel channels, then the fraction of excited states that decay via any particular channel is given by the ratio of that channel's absorbance to the total absorbance. The signal arising from two decay channels, corresponding to short- and long-lived excited states, can then be written,

$$\Delta\tilde{A}\,(d) = \frac{F\,(0)\,\left(1-10^{-A}\right)}{A\cdot\ln\,(10)} \times \left(A_1 R_{\mathrm{B},1}\Delta\tilde{\sigma}_{\mathrm{B},1} + A_2 R_{\mathrm{B},2}\Delta\tilde{\sigma}_{\mathrm{B},2}\right) * G'\,(d)\,, \quad (1.23)$$

where subscript 1 denotes short-lived excited states, 2 denotes long-lived excited states, $R_{\mathrm{B},1}$ and $R_{\mathrm{B},2}$ are single-exponential response functions, and A_1 and A_2 are the absorbances due to each population. If the short-lived excited state of the oligonucleotide is identical to the excited state of the adenine monomer, then the signal from an equal absorbance solution of dAMP can be written,

$$\Delta\tilde{A}\,(d) = \frac{F\,(0)\,\left(1-10^{-A}\right)}{\ln\,(10)}\left(R_{\mathrm{B},1}\Delta\tilde{\sigma}_{\mathrm{B},1}\right) * G'\,(d)\,. \quad (1.24)$$

If the oligonucleotide signal is fit to $\alpha_1 \exp(-k_1 t) + \alpha_2 \exp(-k_2 t)$ and the signal from an equal absorbance solution of the adenine monomer is fit to $\alpha^{\mathrm{m}} \exp(-k_1^{\mathrm{m}} t)$, then the amplitude ratio of the fast components of both signals can be used to determine the fractional absorbance by the short-lived excited state in the oligonucleotide,

$$\frac{\alpha_1}{\alpha_1^{\mathrm{m}}} = \frac{A_1}{A} = 1 - \phi. \quad (1.25)$$

The quantum yield of long-lived excited states, ϕ, is therefore equal to [75],

$$\phi = 1 - \frac{\alpha_1}{\alpha_1^{\mathrm{m}}}. \quad (1.26)$$

The ratio of A_1, the oligomer absorbance due to absorbers that populate short-lived excited states, to A, the total absorbance of the oligomer can be written,

$$\frac{A_1}{A} = \frac{N_1 \sigma_{A,1}}{N \sigma_A}, \tag{1.27}$$

where σ_A is the per nucleotide absorption cross section of the oligomer and N is the number density of oligonucleotides in the oligomer solution. The number densities of short-lived and long-lived absorbers are given by N_1 and N_2, respectively, and $N_1 + N_2 = N$. If the fraction of all nucleotides in the oligomer that produce long-lived states when excited is defined to be ϕ', then,

$$\frac{N_1}{N} = 1 - \phi'. \tag{1.28}$$

Substituting Eq. 1.27 into Eq. 1.28 yields,

$$\phi' = 1 - \frac{\sigma_A A_1}{\sigma_{A,1} A}. \tag{1.29}$$

The definition of the hypochromicity, h, at the pump wavelength,

$$h = 1 - \frac{\sigma_A}{\sigma_{A,1}}, \tag{1.30}$$

and Eq. 1.25 can be used to obtain,

$$\phi' = 1 - \frac{\alpha_1}{\alpha_1^{\mathrm{m}}} (1 - h). \tag{1.31}$$

Values of ϕ and ϕ' from the TA amplitudes reported in Ref. [75] are listed versus n in Table 1.2.

To test the hypothesis that long-lived excited states are formed in bases that are stacked, the ϕ' versus n values were fit to the

Table 1.2 Values of ϕ (Eq. 1.26) and ϕ' (Eq. 1.31) from TA data for the $(dA)_n$ oligomer [75]

n	ϕ	ϕ'
2	0.31	0.46
3	0.39	0.56
4	0.37	0.58
5	0.41	0.62
6	0.43	0.64
8	0.42	0.65
12	0.46	0.68
18	0.51	0.71

fraction of bases in stacked domains of two or more bases calculated from Eq. 1.17. The fit, which depends on the single parameter P, the probability of forming a stack between any two bases, is shown by the solid curve in Fig. 1.8. The yield of long-lived excited states varies with n in a manner that precisely matches the variation in the fraction of bases found in stacked domains in the noncooperative stacking model. This demonstrates that base stacking disorder controls the population of stacked bases and explains the length-dependent signal amplitudes in Fig. 1.6 without the need to invoke excited-state absorption cross sections that depend on the length of the stacked domains. The similar dynamics seen for $(dA)_2$ and for the much longer $(dA)_{18}$ further suggests that the long-lived excited states are delocalized over no more than a pair of nucleobases, regardless of the stacked domain length. Although such states could be compact excitons, recent femtosecond time-resolved IR experiments strongly argue for CT states resulting from photoinduced electron transfer between neighboring bases on the same strand (see the next section).

The value of $P = 0.47$ [75] is lower than room-temperature estimates of between 0.7 and 0.9 that can be derived from the experimental K values in Table 1.1 of Section 1.4.1. A possible explanation is that the solution temperature exceeded room temperature due to laser heating in the spinning cell used in these measurements. At elevated temperature, the degree of base stacking is reduced and K and P decrease.

Future measurements with a flow cell can test this hypothesis, but another possibility is that some excitations in stacked domains can undergo ultrafast nonradiative decay to the electronic ground state. Experiments on duplexes in which the vast majority of bases are believed to be stacked have revealed an ultrafast decay component consistent with this possibility [54, 112]. Clearly, a more complex kinetic scheme could be introduced that includes the possibility of both short- and long-lived excited states in base stacks. However, because experiments on duplex DNA include both base stacking and base pairing interactions, it is possible that ultrafast nonradiative decay in these systems is a consequence of a decay channel enabled by base pairing. Additional experiments

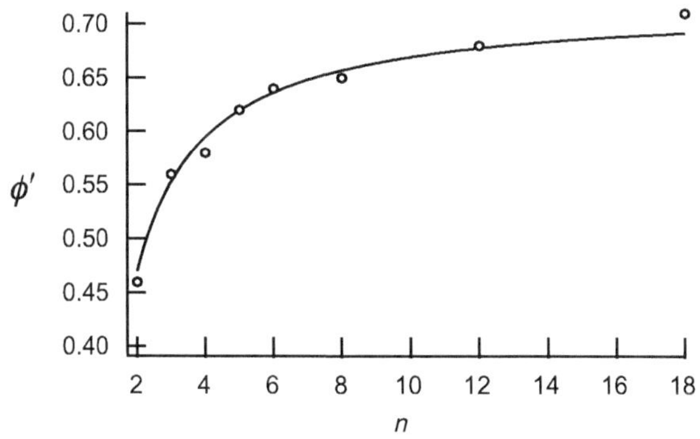

Figure 1.8 The fraction of nucleotides in $(dA)_n$ single strands that can form long-lived excited states versus n. The solid curve is a fit to Eq. 1.17.

and calculations are needed to explore the complex deactivation pathways arising in stacked base pairs.

1.4.4 Exciton Dynamics

The previous sections have summarized evidence from UV pump/UV probe TA experiments that the long-lived excited states of $(dA)_n$ oligonucleotides are localized on a pair of stacked nucleobases. Very recent UV pump/mid-IR probe experiments have shown that the long-lived excited states have vibrational difference spectra that match those of nucleobase radical ions formed by interbase electron transfer [113–115]. These new measurements thus support assigning the long-lived states seen in TA experiments to CT excited states.

Because the oscillator strength for reaching CT states is expected to be very low, the high yields seen experimentally indicate that these states are populated indirectly from the bright, Frenkel-type excitons believed to form when a DNA strand absorbs UV radiation. Experiments conducted to date have been unable to time resolve the formation of CT states, suggesting that they form faster than the experimental time resolution, which is on the order of several hundred femtoseconds. This suggests that Frenkel-type excitons in DNA may be extremely short lived.

There does not appear to be an unequivocal answer to the question about the size of excitons. The absorption spectrum of DNA is very similar to the sum of absorption spectra of a mixture of the constituent nucleotides [116, 117]. There is a reduction in intensity due to hypochromism [118], but the shape of the spectrum is sufficiently similar to that of the individual bases that Eisinger and Schulman concluded that excited states of DNA arise from absorption by single bases [119].

Of course, the coupling between nucleobases cannot be zero as demonstrated by the existence of exciton-coupled circular dichroism spectra of nucleic acids [120]. Dipolar coupling between nearby bases is responsible for the positive and negative Cotton effects seen in the CD spectra. Markovitsi and coworkers have demonstrated that even though the couplings are small, and do not produce large excitonic splittings in the UV/vis absorption spectra, they are nevertheless large enough to produce excitonic states that are spatially delocalized over several bases [3]. As already pointed out earlier, the relatively small size of stacked domains (<7 bases at room temperature) will likely limit the maximum extent of delocalization. Fluorescence anisotropy experiments provide evidence for rapid internal conversion within the manifold of excitonic states [105], but it has not yet been possible to reach unambiguous conclusions about delocalization lengths.

A challenge facing any experimental measurement of exciton size is the fact that delocalization is generally time dependent, causing the size of an exciton to depend on how much time has elapsed since it was created [121, 122]. Phase and energy relaxation are expected to cause an initially delocalized exciton to become increasingly localized. Calculations suggest that lower-energy exciton states are more localized than higher-energy ones, consistent with dynamic localization [3]. In contrast, Kwok et al. [73] argued that excitations in single-stranded $(dA)_{20}$ are initially localized on single adenines, but then delocalize over two or more bases to form excimer and excimer-like states.

Whether or not excitons undergo dynamic localization or delocalization, it is challenging to resolve the very rapid dynamics that occur in other multichromophoric systems in hundreds of femtoseconds or faster [122, 123]. Additionally, caution is warranted when comparing calculations that may predict exciton delocalization lengths for the initial optical transition (i.e., at time zero in a pump–probe experiment) [124] with experimental conclusions based on observables recorded at later times [74].

Finally, it is interesting to note that the TA bleach signals shown in Fig. 1.6 and those from other laboratories have been performed on relatively short oligomers containing 20 or fewer bases. If no more than 5% of all bases are excited (a typical criterion for avoiding saturation, see Section 1.2), then there will be no more than one excitation on average in every 20-mer DNA strand. The same fluence will therefore produce more than one excitation in longer strands. Although a few experiments have been performed in homopolymers having hundreds of nucleotides [64], it is unknown whether the simultaneous presence of multiple excitons in a single DNA strand can measurably affect dynamics. Very rapid trapping or localization of initially delocalized excited states will quite possibly frustrate the exciton–exciton interactions that are important in semiconductor quantum dots [125, 126] and conjugated organic polymers [127, 128].

1.5 Summary

This chapter has described both concepts and practical considerations behind TA experiments and introduced a rate equation model for predicting signal strengths from multiple one-photon absorbers. Appropriate kinetic schemes were used with the model to interpret magic angle TA bleach signals measured for variable-length homoadenine oligonucleotides in aqueous solution at room temperature. Excellent agreement with experiment is obtained using a model that recognizes that bases are found in stacked and unstacked contexts or domains. This base-stacking disorder gives rise to two principal relaxation pathways: one that is monomer-like and a second slower one in which CT states are formed on a subpicosecond timescale, which then decay via charge recombination in tens to hundreds of picoseconds.

It was shown further that the amplitudes obtained by fitting TA bleach signals can be used to estimate the quantum yield that describes the branching between the short- and long-lived excited-state populations. The results can be applied to bleach signals from other, non-DNA systems that exhibit excited-state heterogeneity.

The conclusion that UV excitation induces interbase electron transfer in DNA single strands raises many questions about how these events are modified in double-stranded DNA in which base stacking and base pairing interactions are present simultaneously. Much ongoing work is directed at this understanding. The proposal discussed here that excitons trap to long-lived CT states also warrants further exploration.

Appendix A. Rate Equation Analysis

A rate equation analysis is presented here for quantitatively predicting TA signals in the limit of small population transfer. In the model, a pump pulse with center wavelength λ_{pu} can create more than one kind of initial excited state either because the sample contains a mixture of absorbers (chemical species) or because excitation of a single chemical species by the pump pulse can stimulate transitions to more than one excited state. Each such excited state is the gateway to further excited states and other species that may contribute to the TA signal at later times. A kinetic mechanism connects the various states by first-order rate processes. In every mechanism, the initial excited state is denoted as state B and this state is reached by absorption from ground state A. States C, D, etc. represent further transient states or photoproducts that are populated from state B at later times. The states B, C, D, etc., and the kinetic scheme that describes transitions between these states may be different for each absorbing species. Figure A.1 summarizes these concepts.

Rate equations are solved to obtain the number densities (number of molecules per unit volume) of all states present in solution. The number density of state X of species i is denoted $n_{X,i}$. For simplicity, it will be assumed that the sum of the number densities over all states is conserved and equal to N_i, the initial

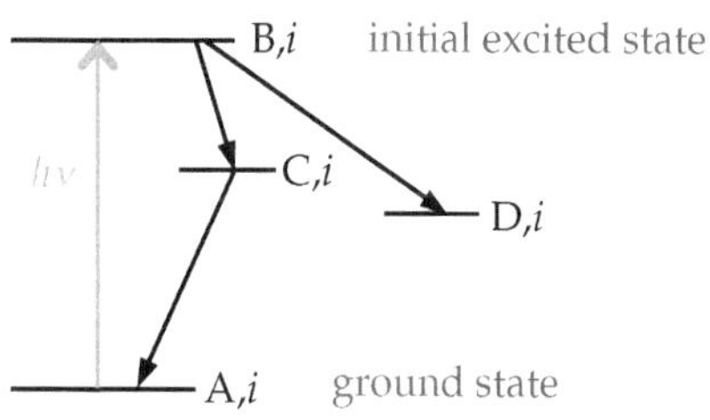

Figure A.1 Schematic showing an example mechanism linking various states of absorbing species i. The pump pulse induces transitions from ground state A to the initial excited state B. First-order rate equations are then solved to obtain the time-dependent populations of intermediate excited states and any permanent photoproducts.

(ground-state) number density of that species,

$$N_i = n_{\mathrm{A},i} + n_{\mathrm{B},i} + n_{\mathrm{C},i} + \cdots . \tag{A.1}$$

This will be the case when only a single state is created in each kinetic step, but it is not difficult to modify Eq. A.1 to account for photodissociation steps that create two or more states from a single precursor. The absorption rate for state X of species i is equal to $\sigma_{X,i} I$, where $\sigma_{X,i}$ is the absorption cross section (units of area) and I is the photon flux of the pump pulse measured in photons per unit area per unit time. The rate equation for the rate of change of the population of state B of species i is written as

$$\frac{\partial n_{\mathrm{B},i}}{\partial t} = n_{\mathrm{A},i}\sigma_{\mathrm{A},i} I - n_{\mathrm{B},i}/\tau_{\mathrm{B},i} + n_{\mathrm{A},i}\tilde{\sigma}_{\mathrm{A},i}\tilde{I} - n_{\mathrm{B},i}\tilde{\sigma}_{\mathrm{B},i}\tilde{I}, \tag{A.2}$$

where I and $\tilde{I}$ are the intensities of pump and probe pulses, respectively, and $\tau_{\mathrm{B},i}$ is the lifetime of state B of species i. Due to the small excited-state population, excited-state absorption by the pump pulse is neglected in Eq. A.2. Terms describing stimulated emission by the pump and probe pulses are also omitted from Eq. A.2. Stimulated emission by the pump pulse is insignificant when only a small fraction (<5%) of absorbers is excited. Because the probe pulse is much weaker than the pump pulse, any stimulated emission induced by the probe pulse will lead to a negligible change in the population of state B.

Tildes are used consistently to denote probe pulse quantities. Furthermore, spectral quantities with a tilde are to be evaluated at the probe wavelength, while those without are to be evaluated at the pump wavelength. For example, $\tilde{\sigma}_{\mathrm{A},i}$ is the absorption cross section of state A of species i at the probe wavelength, λ_{pr}. Both I and $\tilde{I}$ are functions of time, t, and of the spatial coordinate that measures distance through the sample, z. Their variation in space and time causes the number density of excited states of each absorber to depend on t and z.

In Eq. A.2, ground state A of each species i can only undergo a single transition to excited state B. When there are overlapping absorption bands, as in the case of the two lowest energy $^1\pi\pi^*$ states of the purine bases (adenine and guanine), the pump pulse causes a single chemical species to transition to two or more excited states. This can be properly described, while retaining the simple form of Eq. A.2, by treating any such species i as if it were actually several species, each of which undergoes a single transition to each of the various excited states. The cross section for a transition from ground state A of species i to any of several possible excited states indexed by j can be written $\sigma_{\mathrm{A},i}^{(j)}$. The sum of these cross sections is equal to the total absorption cross section determined with a spectrophotometer at the pump wavelength,

$$\sigma_{\mathrm{A},i} = \sum_j \sigma_{\mathrm{A},i}^{(j)}. \tag{A.3}$$

The number density of species i is then subdivided using Eq. A.4,

$$n_{\mathrm{A},i}^{(j)} = n_{\mathrm{A},i} \frac{\sigma_{\mathrm{A},i}^{(j)}}{\sigma_{\mathrm{A},i}}, \tag{A.4}$$

and used in Eq. A.2 with the total absorption cross section $\sigma_{\mathrm{A},i}$ whenever multiple excited states can be formed from species i.

Pump and probe beams are collimated in the region of space where they interact with the sample such that the pulse intensities in the absence of the sample are independent of distance z. The beams in a typical TA experiment are focused gently and the beam divergence θ is often 0.02 or smaller (corresponding to a spot of diameter 5 mm incident on a lens with focal length $f = 250$ mm). In this case, the variation in the beam spot size with distance through the sample is no greater than $\theta/2$ or 10 μm mm^{-1} when θ equals

0.02. The intensity variation through a sample path length of 1 mm is thus minor for a spot size of 100 μm at the front face of the sample. This justifies treating the intensity of each pulse as independent of z in the absence of the sample.

The intensity of a laser beam varies with the transverse distance r measured perpendicular to the direction of propagation. The spatial profile is usually assumed to be Gaussian (Appendix B). Because the molecules are motionless on the timescale of the pump–probe experiment, the signal can be computed separately at each transverse coordinate from pump and probe pulses of fixed intensity. This section presents solutions for pulses of constant spatial intensity (plane wave pulses), but the results can be easily averaged over r, if desired.

The probe pulse is sufficiently weak that it induces negligible population changes compared to the much stronger pump pulse, allowing the last two terms in Eq. A.2 to be neglected,

$$\frac{\partial n_{\mathrm{B},i}}{\partial t} = n_{\mathrm{A},i}\sigma_{\mathrm{A},i} I - n_{\mathrm{B},i}/\tau_{\mathrm{B},i}. \tag{A.5}$$

The sample path length, L, is chosen to be thin enough that temporal walk-off due to GVM (Eq. 1.7) can be neglected. For this reason and to simplify the analysis, a common group velocity, c, specifies the propagation speed of the intensity envelope of each pulse in the sample. In this case, the wave equations for pump and probe pulses are written as Eqs. A.6 and A.7, respectively,

$$\frac{\partial I}{\partial t} + c\frac{\partial I}{\partial z} = -cI\sum_i n_{\mathrm{A},i}\sigma_{\mathrm{A},i}, \tag{A.6}$$

$$\frac{\partial \tilde{I}}{\partial t} + c\frac{\partial \tilde{I}}{\partial z} = -c\tilde{I}\sum_i [n_{\mathrm{A},i}\tilde{\sigma}_{\mathrm{A},i} + n_{\mathrm{B},i}\tilde{\sigma}_{\mathrm{B},i} + \cdots]. \tag{A.7}$$

The right-hand side of Eq. A.6 is the negative of the absorption rate per unit volume due to the pump pulse multiplied by its group velocity. The right-hand side of Eq. A.7 is interpreted similarly for the probe pulse. Stimulated emission due to the probe pulse is neglected, although it would be straightforward to include it in the model by adjusting the cross sections in Eq. A.7.

Equations A.5 through A.7 are solved by making a change of variables $\tau = t - z/c$ and $z = \zeta$. Reduced time, τ, measures time in a moving reference frame that travels with each pulse [129]. With this change of variables Eqs. A.5–A.7 become

$$\frac{\partial n_{\mathrm{B},i}}{\partial t} = n_{\mathrm{A},i}\sigma_{\mathrm{A},i} I - n_{\mathrm{B},i}/\tau_{\mathrm{B},i}, \tag{A.8}$$

$$-\frac{1}{I}\frac{\partial I}{\partial \zeta} = \sum_i n_{\mathrm{A},i}\sigma_{\mathrm{A},i}, \tag{A.9}$$

$$-\frac{1}{\tilde{I}}\frac{\partial \tilde{I}}{\partial \zeta} = \sum_i \left[n_{\mathrm{A},i}\tilde{\sigma}_{\mathrm{A},i} + n_{\mathrm{B},i}\tilde{\sigma}_{\mathrm{B},i} + \cdots\right]. \tag{A.10}$$

Equations A.8–A.10 can be solved analytically in the small excitation limit for pump and probe pulses that have Gaussian intensity envelopes. The pump pulse incident on the front edge of the sample is written as,

$$I(\tau, \zeta = 0) = I_0 \exp\left(-\tau^2/\tau_\mathrm{P}^2\right), \tag{A.11}$$

where I_0 is the peak intensity and parameter τ_P is proportional to the pulse width. The pulse width is conventionally understood to be the FWHM of the intensity envelope function and is equal to $\tau_\mathrm{P}\sqrt{\ln 16}$. The peak of the probe pulse arrives at the front edge of the sample a time d after the peak of the pump pulse,

$$\tilde{I}(\tau, \zeta = 0) = \tilde{I}_0 \exp\left[-(\tau - d)^2/\tilde{\tau}_\mathrm{P}^2\right]. \tag{A.12}$$

The tildes in Eq. A.12 indicate probe pulse quantities. In this model, the pump pulse is only absorbed by molecules in state A, and any excited-state absorption by the pump pulse is neglected. Equations A.11 and A.12 are boundary conditions for the first-order differential equations (Eqs. A.9 and A.10) that describe the propagation of each pulse through the sample. Using the initial condition $n_{\mathrm{B},i}(\tau = -\infty, \zeta) = 0$, the general solution to Eq. A.8 is,

$$n_{\mathrm{B},i}(\tau, \zeta) = N_i\sigma_{\mathrm{A},i}e^{-\tau/\tau_{\mathrm{B},i}} \int_{-\infty}^{\tau} e^{\tau'/\tau_{\mathrm{B},i}} I(\tau', \zeta)\,d\tau'. \tag{A.13}$$

The pump pulse intensity is assumed to be sufficiently weak that the system is far from saturation and $n_{\mathrm{A},i} \approx N_i$. Equation A.9 can then be integrated immediately,

$$I(\tau, \zeta) = I(\tau, 0)\exp\left(-\zeta\sum_i N_i\sigma_{\mathrm{A},i}\right). \tag{A.14}$$

The pump is thus attenuated exponentially with distance through the sample in agreement with the Beer–Lambert law that holds in the absence of absorption saturation. Substituting Eqs. A.14 and A.11 into Eq. A.13 yields,

$$\begin{aligned} n_{\mathrm{B},i}(\tau,\zeta)/N &= \sigma_{\mathrm{A},i} I_0 e^{-\zeta\sum_i N_i\sigma_{\mathrm{A},i}} e^{-\tau/\tau_{\mathrm{B},i}} \int_{-\infty}^{\tau} e^{\tau'/\tau_{\mathrm{B},i}} e^{-\tau'^2/\tau_\mathrm{P}^2} d\tau' \\ &= \sigma_{\mathrm{A},i} F(0) e^{-\zeta\sum_i N_i\sigma_{\mathrm{A},i}} e^{-\tau/\tau_{\mathrm{B},i}} e^{\tau_\mathrm{P}^2/(4\tau_{\mathrm{B},i}^2)} \\ &\quad\times\frac{1}{2}\left[1+\mathrm{erf}\left(\frac{\tau}{\tau_\mathrm{P}}-\frac{\tau_\mathrm{P}}{2\tau_{\mathrm{B},i}}\right)\right] \end{aligned} \quad (A.15)$$

where $F(0)$ is the fluence of the pump pulse incident at $\zeta = 0$,

$$F(0) = \int_{-\infty}^{\infty} I(\tau,0)d\tau = I_0\tau_\mathrm{P}\sqrt{\pi}, \quad (A.16)$$

and erf is the error function, $\mathrm{erf}(x) = \frac{2}{\sqrt{\pi}}\int_0^x e^{-z^2}dz$. Equation A.15 describes a traveling wave of excited states. At each spatial position through the sample, the excited states build up with a rise time proportional to τ_P and then decay away again with a lifetime τ_B.

Equation A.15 for the fractional excited-state population can be simplified by introducing a new function $G(\tau)$, defined as the pump pulse intensity before the sample divided by its fluence before the sample,

$$G(\tau) = I(\tau,0)\Big/\int_{-\infty}^{+\infty} I(\tau,0)d\tau = \frac{1}{\tau_\mathrm{P}\sqrt{\pi}}\exp\left(-\tau^2/\tau_\mathrm{P}^2\right). \quad (A.17)$$

In this case, $G(\tau)$ is a normalized Gaussian function, that is, one whose integral over all values of τ equals unity. Let response function R describe the normalized population of state B in response to an infinitely short excitation pulse,

$$R_{\mathrm{B},i}(\tau) = \begin{cases} e^{-\tau/\tau_{\mathrm{B},i}} & \tau \geq 0 \\ 0 & \tau < 0 \end{cases}. \quad (A.18)$$

With Eqs. A.17 and A.18, Eq. A.15 can be written more succinctly as,

$$n_{\mathrm{B},i}(\tau,\zeta)/N_i = \sigma_{\mathrm{A},i} F(0) e^{-\zeta\sum_i N_i\sigma_{\mathrm{A},i}} (G * R_{\mathrm{B},i})(\tau), \quad (A.19)$$

where $G * R_{\mathrm{B},i}$ is the convolution of G and $R_{\mathrm{B},i}$,

$$(G * R_{\mathrm{B},i})(\tau) = \int_{-\infty}^{\infty} G(\tau - \tau') R_{\mathrm{B},i}(\tau')\, d\tau'. \tag{A.20}$$

After substituting $n_{\mathrm{A},i} = N_i - n_{\mathrm{B},i} - n_{\mathrm{C},i} - \cdots$ from Eq. A.1, Eq. A.10 becomes,

$$-\frac{1}{\tilde{I}}\frac{\partial \tilde{I}}{\partial \zeta} = \sum_i N_i \tilde{\sigma}_{\mathrm{A},i} + n_{\mathrm{B},i} \Delta\tilde{\sigma}_{\mathrm{B},i} + n_{\mathrm{C},i} \Delta\tilde{\sigma}_{\mathrm{C},i} + \cdots, \tag{A.21}$$

where $\Delta\sigma_{X,i} = \tilde{\sigma}_{X,i} - \tilde{\sigma}_{\mathrm{A},i}$ (X = B, C, ...) is the difference in absorption cross sections of state X and A at the probe wavelength. Equation A.21 is easily integrated,

$$\tilde{I}(\tau, L) = \tilde{I}(\tau, 0) \exp\left(-\sum_i N_i \tilde{\sigma}_{\mathrm{A},i} L\right) \tag{A.22}$$

$$\times \exp\left(-\sum_i \Delta\tilde{\sigma}_{\mathrm{B},i} \int_0^L n_{\mathrm{B},i}(\tau, \zeta)\, d\zeta - \Delta\tilde{\sigma}_{\mathrm{C},i} \int_0^L n_{\mathrm{C},i}(\tau, \zeta)\, d\zeta - \cdots\right).$$

The argument of the second exponential on the right-hand side of Eq. A.22 is small and can be replaced by the first two terms of its Taylor series expansion,

$$\tilde{I}(\tau, L) = \tilde{I}(\tau, 0) \exp\left(-L \sum_i N_i \tilde{\sigma}_{\mathrm{A},i}\right) \tag{A.23}$$

$$\times \left(1 - \sum_i \Delta\tilde{\sigma}_{\mathrm{B},i} \int_0^L n_{\mathrm{B},i}(\tau, \zeta)\, d\zeta - \Delta\tilde{\sigma}_{\mathrm{C},i} \int_0^L n_{\mathrm{C},i}(\tau, \zeta)\, d\zeta - \cdots\right).$$

The populations of all transient states B, C, ... are needed in Eq. A.23. Equation A.19 gives the population in the initial excited state B. The populations of states C, D, ... are not affected by optical pumping and are determined solely by the kinetic mechanism. These populations are determined from first-order differential equations in time and are therefore also given by Eq. A.19 after replacing $R_{\mathrm{B},i}$ by $R_{\mathrm{C},i}$, $R_{\mathrm{D},i}$, etc. The functions $R_{X,i}(t)$ (X = A, B, C, ...) give the fraction of all molecules initially excited to state B of species i by an infinitely short pulse that are found in state X a time t later. They are obtained by solving the system of differential equations for the kinetic scheme appropriate for species i subject to the initial conditions,

$$R_{\mathrm{A},i}(t=0)=0,\ R_{\mathrm{B},i}(0)=1,\ R_{\mathrm{C},i}(0)=R_{\mathrm{D},i}(0)=R_{\mathrm{E},i}(0)=\ldots=0. \tag{A.24}$$

The population in all states is constant at all times for first-order kinetics, so

$$R_{\mathrm{A},i}(t)+R_{\mathrm{B},i}(t)+R_{\mathrm{C},i}(t)+\cdots=1. \tag{A.25}$$

Substituting Eq. A.19 into Eq. A.23 and evaluating the integral yields,

$$\tilde{I}(\tau,L)=\tilde{I}(\tau,0)\exp\left(-L\sum_i N\tilde{\sigma}_{\mathrm{A},i}\right) \tag{A.26}$$
$$\times\left[1-F_{\mathrm{ave}}L\sum_i N_i\sigma_{\mathrm{A},i}\left(\left[R_{\mathrm{B},i}\Delta\tilde{\sigma}_{\mathrm{B},i}+R_{\mathrm{C},i}\Delta\tilde{\sigma}_{\mathrm{C},i}+\cdots\right]*G\right)(\tau)\right],$$

where F_{ave} is the average pump pulse fluence in the sample,

$$F_{\mathrm{ave}}=\frac{1}{L}\int_0^L F(0)e^{-\zeta\sum_i N_i\sigma_{\mathrm{A},i}}d\zeta. \tag{A.27}$$

Integrating Eq. A.26 over all values of τ gives the fluence of the probe pulse when it exits the sample,

$$\frac{\tilde{F}(L)}{\tilde{F}_{\mathrm{off}}(L)}=1-\frac{F_{\mathrm{ave}}L}{\tilde{F}(0)}\sum_i N_i\sigma_{\mathrm{A},i} \tag{A.28}$$
$$\times\int_{-\infty}^{+\infty}\tilde{I}(\tau,0)\left(\left[R_{\mathrm{B},i}\Delta\tilde{\sigma}_{\mathrm{B},i}+R_{\mathrm{C},i}\Delta\tilde{\sigma}_{\mathrm{C},i}+\cdots\right]*G\right)(\tau)d\tau.$$

With the pump pulse blocked, the fluence of the probe pulse decays exponentially with distance according to the Beer–Lambert law, and the fluence at $z=L$ is denoted $\tilde{F}_{\mathrm{off}}(L)=\tilde{F}(0)\,e^{-L\sum_i N_i\tilde{\sigma}_{\mathrm{A},i}}$ in Eq. A.28. The pump–probe signal, ΔA, is given by,

$$\Delta\tilde{A}=\tilde{A}-\tilde{A}_{\mathrm{off}}=-\ln\left[\frac{\tilde{F}(L)}{\tilde{F}_{\mathrm{off}}(L)}\right]/\ln 10. \tag{A.29}$$

Because the right-hand side of Eq. A.26 is close to unity, the approximation $-\ln(1-x)\approx x$ is well satisfied, allowing Eq. A.29

to be written as,

$$\Delta\tilde{A}(d) = \frac{F_{\text{ave}}L}{\ln(10)\,\tilde{F}(0)} \sum_i N_i \sigma_{\text{A},i} \tag{A.30}$$

$$\times \int_{-\infty}^{+\infty} \tilde{I}(\tau, 0)\left(\left[R_{\text{B},i}\Delta\tilde{\sigma}_{\text{B},i} + R_{\text{C},i}\Delta\tilde{\sigma}_{\text{C},i} + \cdots\right] * G\right)(\tau)d\tau.$$

After substituting in Eq. A.12 for the probe intensity, the integral on the right-hand side of A.30 can be written as a double convolution,

$$\Delta\tilde{A}(d) = \frac{F_{\text{ave}}L}{\ln(10)} \sum_i N_i \sigma_{\text{A},i}$$

$$\left(\left(\left[R_{\text{B},i}\Delta\tilde{\sigma}_{\text{B},i} + R_{\text{C},i}\Delta\tilde{\sigma}_{\text{C},i} + \cdots\right] * G\right) * \tilde{G}\right)(d), \tag{A.31}$$

where $\tilde{G}$ is the normalized intensity of the probe pulse defined analogously to Eq. A.17. Finally, using the associative property of convolution, $(R * G) * \tilde{G} = R * \left(G * \tilde{G}\right)$, and the fact that the convolution of two Gaussians yields a Gaussian, Eq. A.31 can be written,

$$\Delta\tilde{A}(d) = \frac{F_{\text{ave}}L}{\ln 10} \sum_i N_i \sigma_{\text{A},i} \left(\left[R_{\text{B},i}\Delta\tilde{\sigma}_{\text{B},i} + R_{\text{C},i}\Delta\tilde{\sigma}_{\text{C},i} + \cdots\right] * G'\right)(d), \tag{A.32}$$

where G' is the Gaussian function,

$$G' = \frac{1}{\tau_{\text{P}}\tilde{\tau}_{\text{P}}\pi} \exp\left[-d^2/\left(\tau_{\text{P}}^2 + \tilde{\tau}_{\text{P}}^2\right)\right]. \tag{A.33}$$

In other words, for Gaussian pump and probe pulses, the TA signal evolution is determined by the convolution of the kinetic response with a Gaussian with an effective pulse width of $\sqrt{\tau_{\text{P}}^2 + \tilde{\tau}_{\text{P}}^2}$.

Equation A.32 is the main result for the TA signal when multiple absorbing species can form multiple excited states that evolve in time according to first-order kinetics. The TA signal dynamics are determined by the kinetic response functions R convoluted with the instrument response function, which, as shown by Eq. A.33, is given by the convolution of pump and probe pulses [109]. Equation A.32 indicates that the TA signal increases linearly with the incident pump fluence provided that the fraction of population transferred to excited states is small (i.e., the system is far from saturation). The signal is also independent of the probe fluence, as expected.

Appendix B. Pulse Intensity and Fluence

In a rate equation analysis, the intensity envelope function, $I(r, t)$, of a laser pulse determines the rate of optical transitions. $I(r, t)$ varies slowly with respect to time, t, and a spatial coordinate, r, transverse to the direction of beam propagation. The intensity (energy flux) of the pulse is the energy that passes through unit area per unit time. The intensity envelope of a laser pulse that is Gaussian in space and time is written,

$$I(r, t) = \frac{2E_{\mathrm{P}}}{\tau\pi w^2}\left(\frac{\ln 16}{\pi}\right)^{1/2} \exp\left(-\ln 16 \cdot t^2/\tau^2\right) \cdot \exp\left(-2r^2/w^2\right), \tag{B.1}$$

where w is the radius at which the intensity has decreased by $1/e^2$ compared to its axial value, E_{P} is the pulse energy, and the pulse duration, τ, is defined to be the temporal FWHM. The on-axis peak intensity of the pulse is,

$$I(r=0, t=0) \equiv I_0 = \frac{2E_{\mathrm{P}}}{\tau\pi w^2}\left(\frac{\ln 16}{\pi}\right)^{1/2}. \tag{B.2}$$

This can be conveniently calculated in the commonly used units of GW cm^{-2} using Eq. B.3,

$$I_0 = \frac{59.81\ \frac{\mathrm{GW}}{\mathrm{cm}^2} \cdot (E_{\mathrm{P}}/\mu\mathrm{J})}{(\tau/\mathrm{fs}) \cdot (w/\mathrm{mm})^2}. \tag{B.3}$$

Integrating Eq. B.1 over r gives the instantaneous pulse power, $P(t)$,

$$P(t) = \int_0^{\infty} I(r, t) \cdot 2\pi r dr = \frac{E_{\mathrm{P}}}{\tau}\left(\frac{\ln 16}{\pi}\right)^{1/2} \exp\left(-\ln 16 \cdot t^2/\tau^2\right). \tag{B.4}$$

Integrating intensity (also called energy flux, Eq. B.1) versus time gives the energy fluence, $F(r)$ (SI units: J m^{-2}),

$$\begin{aligned} F(r) &= \int_{-\infty}^{\infty} I(r, t)dt = \frac{2E_{\mathrm{P}}}{\pi w^2} \exp\left(-2r^2/w^2\right) \\ &= I_0\tau\left(\frac{\pi}{\ln 16}\right)^{1/2} \exp\left(-2r^2/w^2\right). \end{aligned} \tag{B.5}$$

The peak energy fluence occurs on axis at $r = 0$ and is equal to,

$$F(0) = \frac{2E_{\mathrm{P}}}{\pi w^2} = I_0\tau\left(\frac{\pi}{\ln 16}\right)^{1/2}. \tag{B.6}$$

The integral relationships relating intensity, power and fluence are,

$$E_P = \int_{-\infty}^{\infty}\int_{0}^{\infty} I(r,t)2\pi r dr dt = \int_{-\infty}^{\infty} P(t)dt = \int_{0}^{\infty} F(r)2\pi r dr. \quad \text{(B.7)}$$

Each of the above quantities that depend on E_P can be converted to a photon quantity in the quasi-monochromatic approximation by dividing E_P by $h\nu$, where h is Planck's constant and ν is the pulse center frequency. In this way, intensity (energy flux) can be expressed as a photon flux, energy fluence can be expressed as a photon fluence, etc.

Acknowledgments

This contribution was made possible by grants from the US National Science Foundation (CHE-1112560 and CHE-1465277).

References

1. Pecourt, J.-M. L., Peon, J., and Kohler, B. (2000). Ultrafast internal conversion of electronically excited RNA and DNA nucleosides in water, *J. Am. Chem. Soc.*, **122**, pp. 9348–9349.
2. Crespo-Hernández, C. E., Cohen, B., Hare, P. M., and Kohler, B. (2004). Ultrafast excited-state dynamics in nucleic acids, *Chem. Rev.*, **104**, pp. 1977–2019.
3. Markovitsi, D., Gustavsson, T., and Talbot, F. (2007). Excited states and energy transfer among DNA bases in double helices, *Photochem. Photobiol. Sci.*, **6**, pp. 717–724.
4. Middleton, C. T., de La Harpe, K., Su, C., Law, Y. K., Crespo-Hernández, C. E., and Kohler, B. (2009). DNA excited-state dynamics: from single bases to the double helix, *Annu. Rev. Phys. Chem.*, **60**, pp. 217–239.
5. Markovitsi, D., Gustavsson, T., and Vayá, I. (2010). Fluorescence of DNA duplexes: from model helices to natural DNA, *J. Phys. Chem. Lett.*, **1**, pp. 3271–3276.
6. Kohler, B. (2010). Nonradiative decay mechanisms in DNA model systems, *J. Phys. Chem. Lett.*, **1**, pp. 2047–2053.

7. Kleinermanns, K., Nachtigallova, D., and de Vries, M. S. (2013). Excited state dynamics of DNA bases, *Int. Rev. Phys. Chem.*, **32**, pp. 308–342.
8. Mahr, H., and Hirsch, M. D. (1975). Optical up-conversion light gate with picosecond resolution, *Opt. Commun.*, **13**, pp. 96–99.
9. Block, D., Shah, J., and Gossard, A. C. (1986). Femtosecond luminescence measurements in GaAs, *Solid State Commun.*, **59**, pp. 527–531.
10. Shah, J., Damen, T. C., Deveaud, B., and Block, D. (1987). Subpicosecond luminescence spectroscopy using sum frequency generation, *Appl. Phys. Lett.*, **50**, pp. 1307–1309.
11. Shah, J. (1988). Ultrafast luminescence spectroscopy using sum frequency generation, *IEEE J. Quantum Electron.*, **24**, pp. 276–288.
12. Kahlow, M. A., Jarzeba, W., Dubruil, T. P., and Barbara, P. F. (1988). Ultrafast emission-spectroscopy in the ultraviolet by time-gated upconversion, *Rev. Sci. Instrum.*, **59**, pp. 1098–1109.
13. Peon, J., and Zewail, A. H. (2001). DNA/RNA nucleotides and nucleosides: direct measurement of excited-state lifetimes by femtosecond fluorescence up-conversion, *Chem. Phys. Lett.*, **348**, pp. 255–262.
14. Gustavsson, T., Sharonov, A., Onidas, D., and Markovitsi, D. (2002). Adenine, deoxyadenosine and deoxyadenosine 5′-monophosphate studied by femtosecond fluorescence upconversion spectroscopy, *Chem. Phys. Lett.*, **356**, pp. 49–54.
15. Gustavsson, T., Sharonov, A., and Markovitsi, D. (2002). Thymine, thymidine and thymidine 5′-monophosphate studied by femtosecond fluorescence upconversion spectroscopy, *Chem. Phys. Lett.*, **351**, pp. 195–200.
16. Norrish, R. G. W., and Porter, G. (1949). Chemical reactions produced by very high light intensities, *Nature,* **164**, pp. 658–658.
17. Porter, G., and Norrish, R. G. W. (1950). Flash photolysis and spectroscopy. A new method for the study of free radical reactions, *Proc. R. Soc. Lond., Ser. A,* **200**, pp. 284–300.
18. Porter, G., and Windsor, M. W. (1953). Triplet states in solution, *J. Chem. Phys.*, **21**, pp. 2088–2088.
19. Bloembergen, N. (1999). From nanosecond to femtosecond science, *Rev. Mod. Phys.*, **71**, pp. S283–S287.
20. Shank, C. V., and Ippen, E. P. (1974). Subpicosecond kilowatt pulses from a mode-locked CW dye laser, *Appl. Phys. Lett.*, **24**, pp. 373–375.
21. Ippen, E. P., and Shank, C. V. (1975). Dynamic spectroscopy and subpicosecond pulse-compression, *Appl. Phys. Lett.*, **27**, pp. 488–490.

22. Ippen, E. P., Shank, C. V., and Bergman, A. (1976). Picosecond recovery dynamics of malachite green, *Chem. Phys. Lett.,* **38**, pp. 611–614.
23. Shank, C. V., Ippen, E. P., and Teschke, O. (1977). Sub-picosecond relaxation of large organic-molecules in solution, *Chem. Phys. Lett.,* **45**, pp. 291–294.
24. Fork, R. L., Greene, B. I., and Shank, C. V. (1981). Generation of optical pulses shorter than 0.1 psec by colliding pulse mode-locking, *Appl. Phys. Lett.,* **38**, pp. 671–672.
25. Shank, C. V. (1983). Measurement of ultrafast phenomena in the femtosecond time domain, *Science,* **219**, pp. 1027–1031.
26. Pessot, M., Maine, P., and Mourou, G. (1987). 1000 times expansion compression of optical pulses for chirped pulse amplification, *Opt. Commun.,* **62**, pp. 419–421.
27. Maine, P., Strickland, D., Bado, P., Pessot, M., and Mourou, G. (1988). Generation of ultrahigh peak power pulses by chirped pulse amplification, *IEEE J. Quantum Electron.,* **24**, pp. 398–403.
28. Backus, S., Durfee, C. G., Murnane, M. M., and Kapteyn, H. C. (1998). High power ultrafast lasers, *Rev. Sci. Instrum.,* **69**, pp. 1207–1223.
29. Jonas, D. M., Lang, M. J., Nagasawa, Y., Joo, T., and Fleming, G. R. (1996). Pump-probe polarization anisotropy study of femtosecond energy transfer within the photosynthetic reaction center of *Rhodobacter sphaeroides* R26, *J. Phys. Chem.,* **100**, pp. 12660–12673.
30. Berera, R., van Grondelle, R., and Kennis, J. T. M. (2009). Ultrafast transient absorption spectroscopy: principles and application to photosynthetic systems, *Photosynth. Res.,* **101**, pp. 105–118.
31. Lessing, H. E., Jena, A. V., and Reichert, M. (1975). Orientational aspect of transient absorption in solutions, *Chem. Phys. Lett.,* **36**, pp. 517–522.
32. Lessing, H. E., and Jena, A. V. (1976). Separation of rotational diffusion and level kinetics in transient absorption spectroscopy, *Chem. Phys. Lett.,* **42**, pp. 213–217.
33. Megerle, U., Pugliesi, I., Schriever, C., Sailer, C. F., and Riedle, E. (2009). Sub-50 fs broadband absorption spectroscopy with tunable excitation: putting the analysis of ultrafast molecular dynamics on solid ground, *Appl. Phys. B: Lasers Opt.,* **96**, pp. 215–231.
34. Dobryakov, A. L., Kovalenko, S. A., Weigel, A., Perez-Lustres, J. L., Lange, J., Muller, A., and Ernsting, N. P. (2010). Femtosecond pump/supercontinuum-probe spectroscopy: optimized setup and signal analysis for single-shot spectral referencing, *Rev. Sci. Instrum.,* **81**, 113106.

35. Alfano, R. R., and Shapiro, S. L. (1970). Emission in region 4000 to 7000 Å via 4-photon coupling in glass, *Phys. Rev. Lett.*, **24**, pp. 584–587.
36. Heidt, A. M., Hartung, A., Bosman, G. W., Krok, P., Rohwer, E. G., Schwoerer, H., and Bartelt, H. (2011). Coherent octave spanning near-infrared and visible supercontinuum generation in all-normal dispersion photonic crystal fibers, *Opt. Express,* **19**, pp. 3775–3787.
37. Corkum, P. B., Ho, P. P., Alfano, R. R., and Manassah, J. T. (1985). Generation of infrared supercontinuum covering 3–14 μm in dielectrics and semiconductors, *Opt. Lett.,* **10**, pp. 624–626.
38. Ashihara, S., and Kawahara, Y. (2009). Spectral broadening of mid-infrared femtosecond pulses in GaAs, *Opt. Lett.,* **34**, pp. 3839–3841.
39. Domachuk, P., Wolchover, N. A., Cronin-Golomb, M., Wang, A., George, A. K., Cordeiro, C. M. B., Knight, J. C., and Omenetto, F. G. (2008). Over 4000 nm bandwidth of mid-IR supercontinuum generation in sub-centimeter segments of highly nonlinear tellurite PCFs, *Opt. Express,* **16**, pp. 7161–7168.
40. Bache, M., Guo, H. R., and Zhou, B. B. (2013). Generating mid-IR octave-spanning supercontinua and few-cycle pulses with solitons in phase-mismatched quadratic nonlinear crystals, *Opt. Mater. Express,* **3**, pp. 1647–1657.
41. Tauber, M. J., Mathies, R. A., Chen, X. Y., and Bradforth, S. E. (2003). Flowing liquid sample jet for resonance Raman and ultrafast optical spectroscopy, *Rev. Sci. Instrum.,***74**, pp. 4958–4960.
42. Treacy, E. B. (1969). Optical pulse compression with diffraction gratings, *IEEE J. Quantum Electron.,* **5**, pp. 454–458.
43. Fork, R. L., Martinez, O. E., and Gordon, J. P. (1984). Negative dispersion using pairs of prisms, *Opt. Lett.,* **9**, pp. 150–152.
44. Martinez, O. E., Gordon, J. P., and Fork, R. L. (1984). Negative group-velocity dispersion using refraction, *J. Opt. Soc. Am. A,* **1**, pp. 1003–1006.
45. Giuliano, C. R., and Hess, L. D. (1967). Nonlinear absorption of light: optical saturation of electronic transitions in organic molecules with high intensity laser radiation, *IEEE J. Quantum Electron.,* **3**, pp. 358–367.
46. Daimon, M., and Masumura, A. (2007). Measurement of the refractive index of distilled water from the near-infrared region to the ultraviolet region, *Appl. Opt.,* **46**, pp. 3811–3820.

47. Cho, B., Tiwari, V., Hill, R. J., Peters, W. K., Courtney, T. L., Spencer, A. P., and Jonas, D. M. (2013). Absolute measurement of femtosecond pump-probe signal strength, *J. Phys. Chem. A,* **117**, pp. 6332–6345.
48. Crisp, M. D. (1970). Propagation of small-area pulses of coherent light through a resonant medium, *Phys. Rev. A,* **1**, pp. 1604–1611.
49. Lamb, G. L. (1971). Analytical descriptions of ultrashort optical pulse propagation in a resonant medium, *Rev. Mod. Phys.,* **43**, pp. 99–124.
50. Sweetser, J. N., and Walmsley, I. A. (1996). Linear pulse propagation in stationary and nonstationary multilevel media in the transient regime, *J. Opt. Soc. B: Opt. Phys.,* **13**, pp. 601–612.
51. Bouvier, B., Gustavsson, T., Markovitsi, D., and Millié, P. (2002). Dipolar coupling between electronic transitions of the DNA bases and its relevance to exciton states in double helices, *Chem. Phys.,* **275**, pp. 75–92.
52. Hare, P. M., Crespo-Hernández, C. E., and Kohler, B. (2007). Internal conversion to the electronic ground state occurs via two distinct pathways for pyrimidine bases in aqueous solution, *Proc. Natl. Acad. Sci. U S A,* **104**, pp. 435–440.
53. de La Harpe, K. (2011). *Femtosecond UV and Infrared Time-Resolved Spectroscopy of DNA: From Well-Ordered Sequences to Genomic DNA,* PhD thesis, The Ohio State University.
54. Chen, J., Thazhathveetil, A. K., Lewis, F. D., and Kohler, B. (2013). Ultrafast excited-state dynamics in hexaethyleneglycol-linked DNA homoduplexes made of AT base pairs, *J. Am. Chem. Soc.,* **135**, pp. 10290–10293.
55. Pecourt, J.-M. L., Peon, J., and Kohler, B. (2001). DNA excited-state dynamics: ultrafast internal conversion and vibrational cooling in a series of nucleosides, *J. Am. Chem. Soc.,* **123**, pp. 10370–10378.
56. Nikogosyan, D. N., Oraevsky, A. A., and Rupasov, V. I. (1983). Two-photon ionization and dissociation of liquid water by powerful laser radiation, *Chem. Phys.,* **77**, pp. 131–143.
57. Reuther, A., Laubereau, A., and Nikogosyan, D. N. (1997). A simple method for the in situ analysis of femtosecond UV pulses in the pump-probe spectroscopy of solutions, *Opt. Commun.,* **141**, pp. 180–184.
58. Dragomir, A., McInerney, J. G., Nikogosyan, D. N., and Ruth, A. A. (2002). Two-photon absorption coefficients of several liquids at 264 nm, *IEEE J. Quantum Electron.,* **38**, pp. 31–36.
59. Gut, I. G., Hefetz, Y., Kochevar, I. E., and Hillenkamp, F. (1993). Two-photon absorption cross sections of guanosine 5′-monophosphate and

uridine 5′-monophosphate at 532 nm, *J. Phys. Chem.*, **97**, pp. 5171–5176.

60. Tycon, M. A., Chakraborty, A., and Fecko, C. J. (2011). Generation of DNA photolesions by two-photon absorption of a frequency-doubled Ti:Sapphire laser, *J. Photochem. Photobiol., B*, **102**, pp. 161–168.
61. Elles, C. G., Shkrob, I. A., Crowell, R. A., and Bradforth, S. E. (2007). Excited state dynamics of liquid water: insight from the dissociation reaction following two-photon excitation, *J. Chem. Phys.*, **126**, pp. 164503-1-8.
62. Hare, P. M., Price, E. A., and Bartels, D. M. (2008). Hydrated electron extinction coefficient revisited, *J. Phys. Chem. A*, **112**, pp. 6800–6802.
63. Jou, F.-Y., and Freeman, G. R. (1979). Temperature and isotope effects on the shape of the optical absorption spectrum of solvated electrons in water, *J. Phys. Chem.*, **83**, pp. 2383–2387.
64. Crespo-Hernández, C. E., and Kohler, B. (2004). Influence of secondary structure on electronic energy relaxation in adenine homopolymers, *J. Phys. Chem. B*, **108**, pp. 11182–11188.
65. Middleton, C. T. (2008). *Vibrational and Excited-State Dynamics of DNA Bases Revealed by UV and Infrared Femtosecond Time-Resolved Spectroscopy*, PhD thesis, Ohio State University.
66. Chen, J., Zhang, Y., and Kohler, B. (2014). Excited states in DNA strands investigated by ultrafast laser spectroscopy, *Top. Curr. Chem.*, **356**, pp. 39–88.
67. Crespo-Hernández, C. E., Cohen, B., and Kohler, B. (2005). Base stacking controls excited-state dynamics in A·T DNA, *Nature*, **436**, pp. 1141–1144.
68. Schwalb, N. K., and Temps, F. (2008). Base sequence and higher-order structure induce the complex excited-state dynamics in DNA, *Science*, **322**, pp. 243–245.
69. Lange, A. W., and Herbert, J. M. (2009). Both intra- and interstrand charge-transfer excited states in aqueous B-DNA are present at energies comparable to, or just above, the $^1\pi\pi^*$ excitonic bright states, *J. Am. Chem. Soc.*, **131**, pp. 3913–3922.
70. Kozak, C. R., Kistler, K. A., Lu, Z., and Matsika, S. (2010). Excited-state energies and electronic couplings of DNA base dimers, *J. Phys. Chem. B*, **114**, pp. 1674–1683.
71. Plasser, F., and Lischka, H. (2013). Electronic excitation and structural relaxation of the adenine dinucleotide in gas phase and solution, *Photochem. Photobiol. Sci.*, **12**, pp. 1440–1452.

72. Chen, J., and Kohler, B. (2014). Base stacking in adenosine dimers revealed by femtosecond transient absorption spectroscopy, *J. Am. Chem. Soc.,* **136**, pp. 6362–6372.
73. Kwok, W.-M., Ma, C., and Phillips, D. L. (2006). Femtosecond time- and wavelength-resolved fluorescence and absorption spectroscopic study of the excited states of adenosine and an adenine oligomer, *J. Am. Chem. Soc.,* **128**, pp. 11894–11905.
74. Buchvarov, I., Wang, Q., Raytchev, M., Trifonov, A., and Fiebig, T. (2007). Electronic energy delocalization and dissipation in single- and double-stranded DNA, *Proc. Natl. Acad. Sci. U S A,* **104**, pp. 4794–4797.
75. Su, C., Middleton, C. T., and Kohler, B. (2012). Base-stacking disorder and excited-state dynamics in single-stranded adenine homo-oligonucleotides, *J. Phys. Chem. B,* **116**, pp. 10266–10274.
76. Guckian, K. M., Schweitzer, B. A., Ren, R. X. F., Sheils, C. J., Tahmassebi, D. C., and Kool, E. T. (2000). Factors contributing to aromatic stacking in water: evaluation in the context of DNA, *J. Am. Chem. Soc.,* **122**, pp. 2213–2222.
77. Tso, P. O. P., Kondo, N. S., Schweizer, M. P., and Hollis, D. P. (1969). Studies of conformation and interaction in dinucleoside mono- and diphosphates by proton magnetic resonance, *Biochemistry,* **8**, pp. 997–1029.
78. Holcomb, D. N., and Tinoco, I., Jr. (1965). Conformation of polyriboadenylic acid: pH and temperature dependence, *Biopolymers,* **3**, pp. 121–133.
79. Hatters, D. M., Wilson, L., Atcliffe, B. W., Mulhern, T. D., Guzzo-Pernell, N., and Howlett, G. J. (2001). Sedimentation analysis of novel DNA structures formed by homo-oligonucleotides, *Biophys. J.,* **81**, pp. 371–381.
80. Mills, J. B., Vacano, E., and Hagerman, P. J. (1999). Flexibility of single-stranded DNA: use of gapped duplex helices to determine the persistence lengths of poly(dT) and poly(dA), *J. Mol. Biol.,* **285**, pp. 245–257.
81. Ke, C., Humeniuk, M., S-Gracz, H., and Marszalek, P. E. (2007). Direct measurements of base stacking interactions in DNA by single-molecule atomic-force spectroscopy, *Phys. Rev. Lett.,* **99**, p. 018302.
82. Goddard, N. L., Bonnet, G., Krichevsky, O., and Libchaber, A. (2000). Sequence dependent rigidity of single stranded DNA, *Phys. Rev. Lett.,* **85**, pp. 2400–2403.

83. Tyagi, S., and Kramer, F. R. (1996). Molecular beacons: probes that fluoresce upon hybridization, *Nat. Biotechnol.,* **14**, pp. 303–308.

84. Applequist, J., and Damle, V. (1966). Thermodynamics of one-stranded helix-coil equilibrium in polyadenylic acid, *J. Am. Chem. Soc.,* **88**, pp. 3895–3900.

85. Eisenberg, H., and Felsenfeld, G. (1967). Studies of the temperature-dependent conformation and phase separation of polyriboadenylic acid solutions at neutral pH, *J. Mol. Biol.,* **30**, pp. 17–37.

86. Nonin, S., Leroy, J.-L., and Gueron, M. (1995). Terminal base pairs of oligodeoxynucleotides: imino proton exchange and fraying, *Biochemistry,* **34**, pp. 10652–10659.

87. Singh, S., Patel, P. K., and Hosur, R. V. (1997). Structural polymorphism and dynamism in the DNA segment GATCTTCCCCCCGGAA: NMR investigations of hairpin, dumbbell, nicked duplex, parallel strands., and i-motif, *Biochemistry,* **36**, pp. 13214–13222.

88. Andreatta, D., Sen, S., Pérez Lustres, J. L., Kovalenko, S. A., Ernsting, N. P., Murphy, C. J., Coleman, R. S., and Berg, M. A. (2006). Ultrafast dynamics in DNA: "fraying" at the end of the helix, *J. Am. Chem. Soc.,* **128**, pp. 6885–6892.

89. Wong, K. Y., and Pettitt, B. M. (2008). The pathway of oligomeric DNA melting investigated by molecular dynamics simulations, *Biophys. J.,* **95**, pp. 5618–5626.

90. Breslauer, K. J., and Sturtevant, J. M. (1977). Calorimetric investigation of single stranded base stacking in ribo-oligonucleotide A_7, *Biophys. Chem.,* **7**, pp. 205–209.

91. Brahms, J., Michelson, A. M., and van Holde, K. E. (1966). Adenylate oligomers in single- and double-strand conformation, *J. Mol. Biol.,* **15**, pp. 467–488.

92. Buhot, A., and Halperin, A. (2004). Effects of stacking on the configurations and elasticity of single-stranded nucleic acids, *Phys. Rev. E,* **70**, 020902.

93. Olsthoorn, C. S., Bostelaar, L. J., Van Boom, J. H., and Altona, C. (1980). Conformational characteristics of the trinucleoside diphosphate dAp-dApdA and its constituents from nuclear magnetic resonance and circular dichroism studies. Extrapolation to the stacked conformers, *Eur. J. Biochem.,* **112**, pp. 95–110.

94. Fornasiero, D., and Kurucsev, T. (1984). Vibronic exciton interactions. Resolution and interpretation of the temperature-dependent circulardichroism and absorptionspectra of ApA and of dApdA, *Eur. J. Biochem.,* **143**, pp. 1–7.

95. Chen, J., and Kohler, B., unpublished work.
96. Ramreddy, T., Rao, B. J., and Krishnamoorthy, G. (2007). Site-specific dynamics of strands in ss- and dsDNA as revealed by time-domain fluorescence of 2-aminopurine, *J. Phys. Chem. B*, **111**, pp. 5757–5766.
97. Nordlund, T. M., Xu, D., and Evans, K. O. (1993). Excitation energy transfer in DNA: duplex melting and transfer from normal bases to 2-aminopurine, *Biochemistry*, **32**, pp. 12090–12095.
98. Xu, D.-G., and Nordlund, T. M. (2000). Sequence dependence of energy transfer in DNA oligonucleotides, *Biophys. J.*, **78**, pp. 1042–1058.
99. Takaya, T., Su, C., de La Harpe, K., Crespo-Hernández, C. E., and Kohler, B. (2008). UV excitation of single DNA and RNA strands produces high yields of exciplex states between two stacked bases, *Proc. Natl. Acad. Sci. U S A*, **105**, pp. 10285–10290.
100. Smith, V. R., Samoylova, E., Ritze, H. H., Radloff, W., and Schultz, T. (2010). Excimer states in microhydrated adenine clusters, *Phys. Chem. Chem. Phys.*, **12**, pp. 9632–9636.
101. Kabeláč, M., and Hobza, P. (2001). At nonzero temperatures, stacked structures of methylated nucleic acid base pairs and microhydrated nonmethylated nucleic acid base pairs are favored over planar hydrogen-bonded structures: a molecular dynamics simulations study, *Chem. Eur. J.*, **7**, pp. 2067–2074.
102. Conti, I., Altoè, P., Stenta, M., Garavelli, M., and Orlandi, G. (2010). Adenine deactivation in DNA resolved at the CASPT2//CASSCF/AMBER level, *Phys. Chem. Chem. Phys.*, **12**, pp. 5016–5023.
103. Eisinger, J., Guéron, M., Shulman, R. G., and Yamane, T. (1966). Excimer fluorescence of dinucleotides, polynucleotides., and DNA, *Proc. Natl. Acad. Sci. U S A*, **55**, pp. 1015–1020.
104. Nielsen, L. M., Hoffmann, S. V., and Nielsen, S. B. (2013). Electronic coupling between photo-excited stacked bases in DNA and RNA strands with emphasis on the bright states initially populated, *Photochem. Photobiol. Sci.*, **12**, pp. 1273–1285.
105. Markovitsi, D., Onidas, D., Gustavsson, T., Talbot, F., and Lazzarotto, E. (2005). Collective behavior of Franck-Condon excited states and energy transfer in DNA double helices, *J. Am. Chem. Soc.*, **127**, pp. 17130–17131.
106. West, B. A., Womick, J. M., and Morant, A. M. (2013). Interplay between vibrational energy transfer and excited state deactivation in DNA components, *J. Phys. Chem. A*, **117**, pp. 5865–5874.

107. Lu, Y., Lan, Z. G., and Thiel, W. (2011). Hydrogen bonding regulates the monomeric nonradiative decay of adenine in DNA strands, *Angew. Chem. Int. Ed.*, **50**, pp. 6864–6867.

108. Nachtigallova, D., Zeleny, T., Ruckenbauer, M., Muller, T., Barbatti, M., Hobza, P., and Lischka, H. (2010). Does stacking restrain the photodynamics of individual nucleobases?, *J. Am. Chem. Soc.*, **132**, pp. 8261–8263.

109. van Stokkum, I. H. M., Larsen, D. S., and van Grondelle, R. (2004). Global and target analysis of time-resolved spectra, *Biochim. Biophys. Acta*, **1657**, pp. 82–104.

110. Rich, A., and Tinoco, I., Jr. (1960). Effect of chain length upon hypochromism in nucleic acids and polynucleotides, *J. Am. Chem. Soc.*, **82**, pp. 6409–6410.

111. Warshaw, M. M., and Tinoco, I., Jr. (1966). Optical properties of sixteen dinucleoside phosphates, *J. Mol. Biol.*, **20**, pp. 29–38.

112. de La Harpe, K., and Kohler, B. (2011). Observation of long-lived excited states in DNA oligonucleotides with significant base sequence disorder, *J. Phys. Chem. Lett.*, **2**, pp. 133–138.

113. Doorley, G. W., Wojdyla, M., Watson, G. W., Towrie, M., Parker, A. W., Kelly, J. M., and Quinn, S. J. (2013). Tracking DNA excited states by picosecond-time-resolved infrared spectroscopy: signature band for a charge-transfer excited state in stacked adenine–thymine systems, *J. Phys. Chem. Lett.*, **4**, pp. 2739–2744.

114. Bucher, D. B., Pilles, B. M., Carell, T., and Zinth, W. (2014). Charge separation and charge delocalization identified in long-living states of photoexcited DNA, *Proc. Natl. Acad. Sci. U S A*, **111**, pp. 4369–4374.

115. Zhang, Y., Dood, J., Beckstead, A. A., Li, X.-B., Nguyen, K. V., Burrows, C. J., Improta, R., and Kohler, B. (2014). Efficient UV-induced charge separation and recombination in an 8-oxoguanine-containing dinucleotide, *Proc. Natl. Acad. Sci. U S A*, **111**, pp. 11612–11617.

116. Loofbourow, J. R. (1940). Borderland problems in biology and physics, *Rev. Mod. Phys.*, **12**, pp. 267–358.

117. Guéron, M., Eisinger, J., and Shulman, R. G. (1967). Excited states of nucleotides and singlet energy transfer in polynucleotides, *J. Chem. Phys.*, **47**, pp. 4077–4091.

118. Tinoco, I., Jr. (1960). Hypochromism in polynucleotides, *J. Am. Chem. Soc.*, **82**, pp. 4785–4790.

119. Eisinger, J., and Shulman, R. G. (1968). Excited electronic states of DNA, *Science*, **161**, pp. 1311–1319.

120. Johnson, W. C., Jr. (1985). Circular dichroism and its empirical application to biopolymers, *Methods Biochem. Anal.*, **31**, pp. 61–163.

121. Kühn, O., and Sundström, V. (1997). Pump-probe spectroscopy of dissipative energy transfer dynamics in photosynthetic antenna complexes: a density matrix approach, *J. Chem. Phys.*, **107**, pp. 4154–4164.

122. Book, L. D., Ostafin, A. E., Ponomarenko, N., Norris, J. R., and Scherer, N. F. (2000). Exciton delocalization and initial dephasing dynamics of purple bacterial LH2, *J. Phys. Chem. B*, **104**, pp. 8295–8307.

123. Edington, M. D., Riter, R. E., and Beck, W. F. (1996). Interexciton state relaxation and exciton localization in allophycocyanin trimers, *J. Phys. Chem.*, **100**, pp. 14206–14217.

124. Tonzani, S., and Schatz, G. C. (2008). Electronic excitations and spectra in single-stranded DNA, *J. Am. Chem. Soc.*, **130**, pp. 7607–7612.

125. Colombo, D. P., Roussel, K. A., Saeh, J., Skinner, D. E., Cavaleri, J. J., and Bowman, R. M. (1995). Femtosecond study of the intensity dependence of electron-hole dynamics in TiO_2 nanoclusters, *Chem. Phys. Lett.*, **232**, pp. 207–214.

126. Huang, J., Huang, Z. Q., Yang, Y., Zhu, H. M., and Lian, T. Q. (2010). Multiple exciton dissociation in CdSe quantum dots by ultrafast electron transfer to adsorbed methylene blue, *J. Am. Chem. Soc.*, **132**, pp. 4858–4864.

127. Herz, L. M., Daniel, C., Silva, C., Hoeben, F. J. M., Schenning, A. P. H. J., Meijer, E. W., Friend, R. H., and Phillips, R. T. (2003). Exciton dynamics in supramolecular assemblies of *p*-phenylenevinylene oligomers, *Synth. Met.*, **139**, pp. 839–842.

128. Martini, I. B., Smith, A. D., and Schwartz, B. J. (2004). Exciton-exciton annihilation and the production of interchain species in conjugated polymer films: comparing the ultrafast stimulated emission and photoluminescence dynamics of MEH-PPV, *Phys. Rev. B*, **69**, 035204.

129. Agrawal, G. P., and Olsson, N. A. (1989). Self-phase modulation and spectral broadening of optical pulses in semiconductor laser amplifiers, *IEEE J. Quantum Electron.*, **25**, pp. 2297–2306.

Chapter 2

Ultrafast Light-Induced Processes in DNA Photolyase and Its Substrate-Bound Complex

Klaus Brettel,[a] Martin Byrdin,[b] and Marten H. Vos[c]

[a] *Institute for Integrative Biology of the Cell (I2BC), IBITECS, CEA, CNRS, Univ. Paris Sud, Université Paris-Saclay, F-91198 Gif-sur-Yvette cedex, France*

[b] *Institut de Biologie Structurale Jean-Pierre Ebel, CEA, CNRS, UJF, 41 rue Jules Horowitz, 38027 Grenoble, France*

[c] *LOB, Ecole Polytechnique, CNRS, INSERM, Université Paris-Saclay, 91128 Palaiseau, France*

klaus.brettel@cea.fr, martin.byrdin@cea.fr, marten.vos@polytechnique.edu

2.1 Introduction

DNA photolyase is a fascinating multifaceted protein that, depending on the background or viewpoint, can be classified as flavoprotein, DNA repair enzyme, photoactivated enzyme, or bifunctional enzyme. These multiple features and its considerable history (first described more than 50 years ago [1]) make it a well-studied object and also a suitable model system for the study of a wide variety of biological and biophysical chemistry problems. Before focusing on the state

Ultrafast Dynamics at the Nanoscale: Biomolecules and Supramolecular Assemblies
Edited by Irene Burghardt and Stefan Haacke

ISBN 978-981-4745-33-8 (Hardcover), 978-981-4745-34-5 (eBook)
www.panstanford.com

of the art of the primary photoprocesses at the time of submission (2012), we shortly introduce the enzyme. A comprehensive review was written by Sancar in 2003 [2], and more recent work has been reviewed in Ref. [3].

The UV portion of sunlight is not completely absorbed by the ozone layer. The remainder, if absorbed by DNA, can trigger harmful photochemistry between two neighboring pyrimidines, the most frequently formed lesion being a cyclobutane pyrimidine dimer (CPD)—two pyrimidine bases covalently linked between their respective C6 and C5 atoms (see Fig. 2.3). Less frequent types of UV damage include the so-called (6-4) photoproduct and its Dewar isomer. If left unrepaired, all of these lesions can induce mutagenesis, cancer, and cell death [4]. While for the repair of the (6-4) lesion there exists a specialized enzyme, named (6-4) photolyase [5], the term "DNA photolyase" traditionally refers to the enzyme capable of light-induced repair of CPDs.

DNA photolyase is a soluble protein of medium size (ca. 55 kDa) found in species from all kingdoms of living beings, with the remarkable exception of placental mammals. Crystal structures have been solved both for the isolated protein [6, 7] and the complex [8] with its substrate: DNA containing a CPD lesion (Fig. 2.1).

DNA photolyase possesses an antenna cofactor (depending on the organism, a pterin or a deazaflavin) absorbing light in the blue/near-UV region and a second, catalytic cofactor, $FADH^-$ (flavin adenine dinucleotide [FAD] in its fully reduced form)[a] that harvests the antenna excitation by Förster transfer and injects a photoelectron into the prebound CPD. Upon electron uptake by the CPD, the two intradimer bonds split and the electron returns to the enzyme. Light-induced electron transfer (ET) between flavin and DNA is thus at the heart of the catalytic cycle of DNA repair by photolyase. This high-fidelity repair mechanism—as opposed to

[a]FAD can, in principle, exist in three redox states (oxidized, one-electron [or semi-] reduced, and two-electron (or fully) reduced) and three protonation states (unprotonated, monoprotonated, and biprotonated). For DNA photolyase, of these states the one-electron reduced monoprotonated radical $FADH^{\circ}$ and the two-electron reduced monoprotonated state $FADH^-$ are relevant; the oxidized unprotonated state FAD is also alluded to in the framework of comparison with cryptochromes.

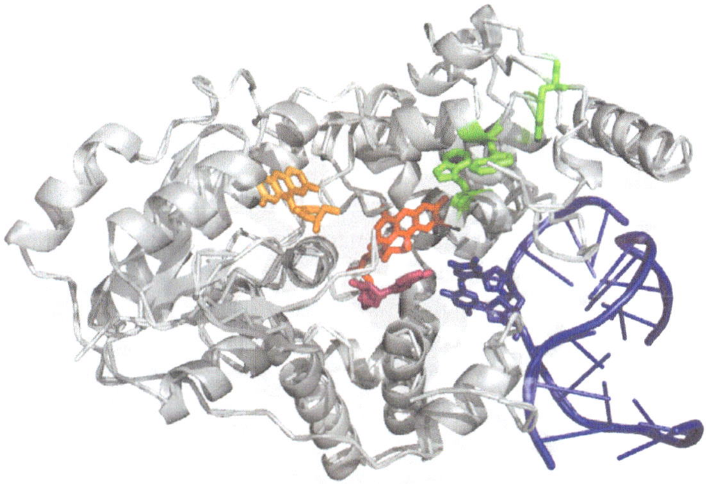

Figure 2.1 Overall structure of DNA photolyase from *A. nidulans* with a bound substrate: a short piece of double-stranded DNA containing a CPD (which was repaired during data collection) 1tez.pdb. Red: flavin; magenta: adenine; orange: antenna; green: triple-tryptophan electron transfer chain; blue: substrate. For visibility, part of the protein was rendered transparent. The figure was prepared using PyMOL.

evolutionary younger and more flexible machineries as, for example, excision repair—involves just a single enzyme, making it remarkably rapid [9] and effective [2, 10].

Analysis of photolyase primary sequences and crystal structures reveals a triad of evolutionary conserved tryptophan residues, linking the flavin deep inside the protein to the surface by a chain of "stepping stones" ~4–5 Å apart (green in Fig. 2.1). Remarkably, this chain is also conserved in most cryptochromes, constituting the other class of the so-called cryptochrome/photolyase protein family. These are structurally homologous to photolyases but they don't show DNA repair activity; instead they have been found to function, among others, as blue light receptors [11]. For completeness, we mention that several representatives of the cryptochrome/photolyase family have been identified very recently that contain differently arranged tryptophan chains [12–14]. These will further not be considered here.

Transient absorption spectroscopy has established a role of the triple tryptophan chain in the photoreduction of the flavin cofactor: light-induced ET from external sources to the flavin. In photolyase where the flavin is in the resting neutral radical FADH° state [15, 16] this reaction had been termed "preillumination effect" but is nowadays established under the name of photoactivation and can be considered as a built-in autorepair function at the disposal of photolyase upon oxidation "damage" to its catalytic flavin cofactor. In cryptochromes, the triple tryptophan chain has been linked to function, but this is still debated [17, 18].

ET along the triple tryptophan chain is coupled to de-protonation of the surface exposed tryptophan [15, 19], thus providing a possibility to study the interaction of these two processes of outmost biological significance. Remarkably, in photolyase, the electron and the proton follow different pathways and are given to different acceptors. It was possible to separate the two processes spectroscopically and to show that in this case they evolve on different timescales (picoseconds for ET versus nanoseconds for proton transfer). For the much slower recombination reaction (outside the scope of this review), coupling of ET and proton uptake is relevant [19]. In this context the issue of concertation between the two reactions has recently been raised [20]. This illustrates nicely the role that study of charge transfer processes in photolyase can play for elucidating general biophysical-chemistry mechanisms.

This paradigmatic role is conferred to photolyase by the fact that its chemistry is triggered by light and can thus be easily synchronized over a macroscopic ensemble of molecules, facilitating the study of various processes with time-resolved spectroscopic techniques, in particular the high-time-resolution femtosecond pump–probe transient absorption technique (cf. Chapter 1 for more details). As signals obtained from light-triggered enzymes are often small, extensive averaging may be required in order to obtain reasonable signal-to-noise (S/N) ratios. This can become challenging with delicate biological material or if enzymes need to be provided with fresh substrate for every subsequent turnover. These complications are easier to overcome using classical continuous-probe techniques, the time resolution of which has been pushed recently to the subnanosecond range [21]. In certain cases where averaging becomes limiting or slower processes are involved, such a picosecond classical approach may be considered a worthwhile complement/alternative.

In the following sections, we first give a general overview of the relatively well-established processes of excitation energy transfer (Section 2.2) and photoactivation (Section 2.3) and will then concentrate in more detail mainly on results and controversies from emerging time-resolved studies on the very CPD repair mechanism (Section 2.4).

2.2 Energy Transfer

The single nonflavin cofactor, termed "antenna" in Fig. 2.1, has a substantially higher extinction than flavin in the near UV and enhances the cross section of photoactivity of the enzyme by allowing transfer of excitation energy to the redox-active flavin via Förster resonance energy transfer (FRET). A system with a single antenna chromophore as present in the cryptochrome/photolyase family is unique in nature (antenna pigments in photosynthesis are invariably present in large excess of photochemically active pigments) and also provides a relatively simple model system for the mechanism of energy transfer.

The experimentally best characterized system is that of the *Escherichia coli* photolyase, that contains a methenyltetrahydrofolate (MTHF) chromophore (absorption and emission maximum at ~380 nm and ~470 nm), located ~17 Å (center to center) from the flavin. Here, $MTHF^* \rightarrow FADH^{\circ *}$ and $MTHF^* \rightarrow FADH^{-*}$ transfer occur in ~18 and ~160 ps, respectively, yielding, with the intrinsic MTHF* lifetime of ~350 ps, quantum yields (QYs) of ~95% and ~55% [22, 23]. These values can be reasonably well understood in the framework of FRET (cf. Chapter 3); the difference in rates being largely due to the difference in the absorption of $FADH^-$ and $FADH^\circ$ in the MTHF emission region. A similar energy transfer time (~160 ps) was reported [24] and computed [25] from the deazaflavin derivative antenna in the catalytically active ($FADH^-$) form of *Anacystis nidulans* photolyase. However, here the QY is higher (>90%) due to the longer intrinsic excited-state lifetime of this chromophore.

The excitation hopping rates in photolyase are much lower than those encountered in specialized photosynthetic antenna proteins, where generally more closely coupled and higher extinction (chlorin) chromophores are employed. However the transfer occurs with fair yield and on similar timescales as the subsequent photochemical processes that are the main subjects of this chapter. Experimentally, those processes are therefore generally studied under conditions where the flavin is directly excited and complications due to energy transfer can be avoided.

2.3 Photoactivation

The purified enzyme is typically in the inactive semireduced blue $FADH^{\circ}$ state. This state can be converted to the catalytically active light yellow $FADH^{-}$ state by photoreduction of the flavin. The mechanism by which this happens has been unraveled in fair detail in the first decade of this century, as will be summarized below. It makes use of the well-known property of protein-bound flavins that their fluorescence can be quenched by ET from nearby aromatic amino acids to the flavin excited state. This property has been used extensively to study the close environment of flavin-containing active sites in proteins by ultrafast time-resolved methods [26, 27]. The resulting charge pair is unstable in these cases. Charge recombination occurs on the femtosecond or picosecond timescale and the initial ground state is rapidly recovered. The situation in photolyase differs in two main points: (a) The flavin initially is in the neutral radical state $FADH^{\circ}$ and not in the oxidized state (FAD), the resting state of the vast majority of flavoproteins, and (b) the initially created charge pair is partly stabilized by further electron transport, leading to a sizeable yield of long-lived $FADH^{-}$ state.

Early on, it has been recognized that the state $FADH^{-}/W306^{\circ +}$ is an intermediate in this process [28]. W306 (numbering for the *E. coli* enzyme) is a surface-exposed tryptophan residue that is located 15 Å (edge to edge) from the flavin isoalloxazine ring. This distance is too long to allow direct ET within the flavin excited-state lifetime

(<100 ps, see below). Two other tryptophan residues (W382 and W359) have been noted to bridge the flavin and W306 in the X-ray structure [6]. For generality, they will be termed "Wn," "Wm," and "Wf" for the tryptophans near, middle, and far with respect to the flavin, respectively (Fig. 2.2). This triad of tryptophans was proposed to act as an ET chain between the flavin and the aqueous environment on the basis of theoretical studies [29] and time-resolved studies on the wild-type enzyme [15]. In the latter work, FADH°* was shown to decay in ~30 ps and the surface-exposed Wf oxidized within 10 ns before releasing a proton into the aqueous phase in ~200 ns.

Optical absorption spectroscopy can be used to investigate the oxidation and subsequent deprotonation of tryptophan, as, unlike tryptophan in its usual state, both forms of the oxidized tryptophan (cation radical and neutral, deprotonated radical) absorb in the visible region. Their absorptions are weak (extinction ~2000–3000 $M^{-1}cm^{-1}$) but spectrally distinguishable. Reversely, the fact that the photoactive flavin signals are also relatively weak (extinction of FADH° ~5000 $M^{-1}cm^{-1}$) makes this system suitable for exploration of light-induced tryptophan radical dynamics. However, discrimination between the three tryptophans in the ET chain of photolyase is challenging, as their spectra in the oxidized state are expected to be very similar. Furthermore, in view of the range of timescales involved in the processes, results from different experimental techniques have to be connected. From a series of studies using wild-type *E. coli* photolyase and three mutant forms (in which the tryptophans were replaced one by one with redox-inactive phenylalanines [30–33]), where polarization photoselection was exploited to discriminate between Wf and the other two tryptophans [33, 34] (Wm and Wn unfortunately have near-parallel optical transitions, cf. Fig. 2.2), and in which absolute QY references were used [35], the reaction scheme of Fig. 2.2 could be deduced.

The striking overall results are that (a) the three tryptophans are indeed functional in the ET chain and (b) the overall oxidation of the surface-exposed W306 is rate limited by the first 30 ps FADH°* decay, that is, the two-step oxidation of Wf by Wn occurs faster than 30 ps. Thus the tryptophan chain functions as an effective nanowire connecting distant parts of the protein, using very fast

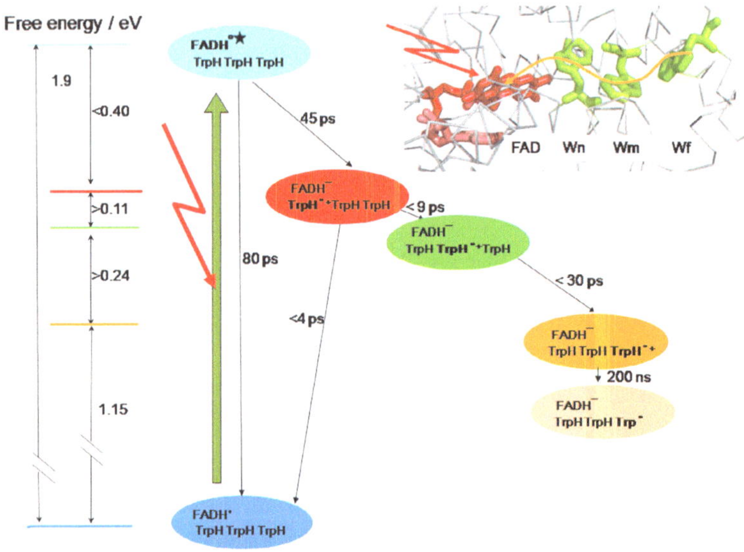

Figure 2.2 Simplified scheme of electron transfer through the tryptophan triad to photoexcited FADH°. Only picosecond electron transfer reactions and the final deprotonation reaction discussed here are included. The scheme is based on a body of data from wild-type and mutant photolyases from *E. coli*, as discussed in Ref. [35], and reprinted with minimal adaptation from Ref. [3], Copyright (2010), with permission from Elsevier. The H in TrpH indicates the N1 proton of tryptophan. The orange arrow in the inset indicates the pathway and direction of light-induced electron transfer.

electron hopping along the intermediate states. Indeed, a more than 4 orders of magnitude slower reaction was predicted from an early calculation in the framework of a superexchange mechanism (a mechanism where the coupling between remote redox partners is enhanced by an intervening component without the latter acting as a real intermediate) [29].

The photochemical QY of $FADH^-/Wf^{\circ+}$ formation was found to be ~20% [35]. The QY losses essentially take place via processes competing with the first two steps. First, whereas the intrinsic rate of the $FADH^{\circ*} \rightarrow FADH^-/Wn^{\circ+}$ reaction is $(45\ ps)^{-1}$, competing $FADH^{\circ*}$ decay processes (presumably predominantly due to intrinsic $FADH^{\circ*}$ decay [36]) occur in ~80 ps [30], leading to net $FADH^{\circ*}$

decay in 30 ps. More important QY losses arise from recombination of the primary $FADH^-/Wn^{\circ+}$ charge pair (<4 ps). The latter process highlights the parallel with the common ET-quenching phenomenon in flavoproteins, where direct recombination is often the unique decay route of the primary charge pair. Indeed such phenomena have also been observed not only in the ubiquitous oxidized flavoproteins, but also in flavoproteins brought into the semireduced radical state [36].

Considering the scheme in Fig. 2.2 allows an interesting comparison with the reaction scheme of photosynthetic reaction centers (RCs). In photolyase, a long-range ET can take place in a distance (~15 Å edge to edge) and time range (picoseconds) comparable to those of RCs [37]. As in RCs, the charge pair is stabilized by proceeding along a range of intermediates, yet without making use of specialized cofactors (chlorins and quinones in RCs) other than the flavin. On the other hand, the QY of the net charge separation is far lower than the near unity of RCs.

The energy of a photon corresponding to the lowest $FADH^\circ \rightarrow FADH^{\circ*}$ transition (1.9 eV) is largely sufficient to drive the charge separation to the final $FADH^-/Wf^{\circ+}$ state, at a free energy of 1.15 eV above the neutral ground state, as deduced from measured (flavin) and calculated (Wf) redox potentials [35]. Yet the two intermediate states involving oxidized Wm and Wn are located at higher free energy, leaving less than about half of the driving force (<0.40 eV) for the initial charge separation.

The free energies of the three different $FADH^-/W^{\circ+}$ states vary over more than 0.35 eV, implying a striking variation in the effective redox potentials of the tryptophan residues. This assessment supports the idea that tryptophan redox properties can be substantially modulated by their close environment [38]. The lower free energy of the final state, where the positive charge resides on Wf, can be rationalized in terms of the fact that it is surface exposed and has a more polar environment. This feature was also predicted from an early computational study of the energetics of the intermediate states based on the crystal structure [39]. By contrast, in the same study [39], the free energy of the intermediate state involving oxidized Wm was predicted to lie ~60 meV above that of the primary charge pair, whereas experiments show that it lies >110 meV below (Fig. 2.2). In very recent studies combining classical molecular dynamics simulations of the protein with a quantum mechanical treatment of the tryptophan redox partners, sequentially stabilizing effective energy levels that agree with experiment were computed [40, 41]. In the most extensive of these studies, the time constants of both intertryptophan ET steps were computed at ~70 ps [41], within only 1 order of magnitude of the 9 ps and 30 ps upper limits set by the experimental results (Fig. 2.2). Interestingly, these simulations also indicated that the protein structural reorganization following charge transfer extends to timescales far beyond 30 ps, thus lowering the effective reorganization energy with respect to those corresponding to fully relaxed states. In combination with the low driving forces, these low effective reorganization energies allow fast low-barrier ET.

Altogether, the photochemical triggering of ET through the aromatic residue chain possible in DNA photolyase gives valuable insight into the mechanism of efficient through-protein oxidoreduction reactions in a regime where protein motions occur on the same timescale as ET. In view of the timescales of the intertryptophan transfer rates (a few picoseconds or less) one might even speculate that coherent off-equilibrium vibrational dynamics modulate or drive these reactions, as has been observed in photosynthetic RCs [42, 43]. Yet, the slower speed of the initial charge separation in photolyase, that should drive such dynamics, complicates experimental investigations in this direction.

The protein electron wire described here highlights tryptophan as a conduction medium. Yet other aromatic residues, in particular tyrosines that have a slightly lower redox potential than tryptophan [44], are located within ET distance from this wire and could be involved. Tyrosine oxidation by the surface-exposed tryptophan has indeed been observed but on the microsecond timescale and along with proton transfer and uniquely in *A. nidulans* photolyase [45]. Furthermore, to explain a heterogeneity observed in the $FADH^{\circ *}$ decay of *E. coli* photolyase, an alternative photoreduction pathway involving transient oxidation of a phenylalanine residue (F366) and presumably excluding Wn has been invoked [46]. In the absence of clear examples of phenylalanine photo-oxidation by flavin, the available mutagenesis data, showing that each of the tryptophan replacements by phenylalanine inhibits the electron donor capacity of the chain, plead against such a possibility.

As mentioned, the tryptophan triad is also present and largely conserved in the more recently discovered cryptochrome blue light receptor family. These proteins contain oxidized—yellow—FAD in the resting state, and the excited state is quenched by flavin reduction by the close-lying tryptophan more than an order of magnitude faster (~1 ps [47–49]) than in—blue—$FADH^{\circ-}$ containing photolyase, presumably due to the higher driving force. The faster formation of the primary charge pair, in combination with a potentially higher QY of full ET, should allow a more direct dissection of the electron hopping than possible in photolyase using spectrally and polarization resolved transient absorption spectroscopy. A first such study on wild-type algae cryptochromes indicated a similar overall mechanism as in photolyase, yet with somewhat slower individual steps [47], indicating fine-tuning properties of the protein environment. When combined with site-directed mutagenesis these systems will allow even more detailed insight into the mechanism of electron hopping along aromatic residues.

2.4 Photorepair

In this section, we focus on two recent fast and ultrafast spectroscopic studies that advanced the understanding of the reaction mechanism and dynamics of CPD repair by photolyase [50, 51]. As a framework, we use the general reaction scheme that is shown in Fig. 2.3 for a thymine–thymine dimer, T=T.

It has been known for more than 20 years (see, for example, Refs. [2, 3] for reviews) that only the enzyme with fully reduced flavin cofactor, $FADH^-$, is active in photorepair. The enzyme specifically recognizes CPD lesions in DNA. The CPD is flipped out of the double helix and enters into the binding pocket of the enzyme, closely approaching the FAD cofactor (Figs. 2.1 and 2.3, inset). In this configuration, photoexcitation of the $FADH^-$ (either directly or via energy transfer upon excitation of the antenna pigment; see Section 2.2) triggers a chemical reaction sequence that restores two intact pyrimidines. These are finally released from the binding pocket, enabling binding and repair of another CPD by the same enzyme. The QY of CPD repair by photolyase is as high as 0.5 to 1 [2, 10], which is 1 order of magnitude higher than that of the repair of the (6-4) photoproduct by its specific photolyase [5].

The first light-dependent step in the CPD repair sequence is ET from the excited $FADH^-$ to the CPD in ~200 ps (k_{FET} in Fig. 2.3). This was originally suggested on the basis of the observation that the presence of CPDs accelerated the decay of fluorescence of $FADH^{-*}$ to ~160 ps (from ~1.4 ns in the absence of a substrate) [22, 52]. ET as the origin of the acceleration was established by the detection of $FADH^\circ$ as a transient reaction product by ultrafast absorption spectroscopy [53]. The kinetics of $FADH^\circ$ formation could be described either biexponentially (see last line of Table 2.1) with time constants of 60 and 355 ps or by a stretched exponential, $\exp(-t/\tau)^{\beta}$, with $\tau = 170$ ps and $\beta = 0.71$. The deviation from a monoexponential decay was assigned to a solvation process [53]. The driving force for the k_{ET} process can be estimated on the basis of the 0–0 transition energy of a photon absorption by $FADH^-$ (2.48 eV at 500 nm), the one-electron reduction potential of $FADH^\circ$ in photolyase (~0.05 V versus a standard hydrogen electrode, when the substrate is bound [54, 55]) and numerical estimates for the one-electron reduction potential of T=T (−1.96 to −1.65 V, depending on the assumptions made for the effects of the protein environment [56, 57]. Hence, this ET is downhill by 0.47 to 0.78 eV. With respect to the ET pathway (see Fig. 2.3, inset), it has been discussed (partly reviewed in Ref. [3]) whether the transfer is direct (along the shortest distance of 4.3 Å from C8′ of the flavin to O4′ of the 3′ thymine) or via the adenine moiety of the FAD (from C10′ of the flavin via the adenine to O4 of the 5′ thymine), either as a real intermediate or a virtual one in superexchange coupling. A very recent study [56] analyzing the ET rate for different CPDs (T=T, T=U, U=T, and U=U; U stands for (deoxy)uracil, which is slightly easier to reduce than thymine) in the framework of semiclassical ET theories, favored the pathway, including the adenine as a virtual intermediate (see also Chapter 4).

On the basis of theoretical studies and studies of model compounds (see references in Ref. [3]) it had been anticipated that the two intradimer bonds of T=T split rapidly upon electron uptake

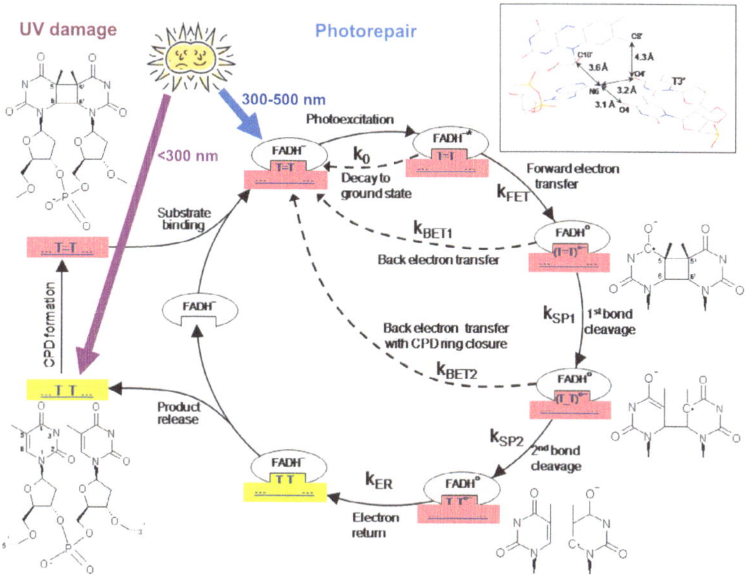

Figure 2.3 Reaction scheme of light-induced CPD formation and repair by DNA photolyase for the example of a thymine–thymine dimer. Gray ellipses represent the enzyme with the flavin cofactor in the indicated redox state; green and red rectangles represent intact and damaged DNA. Dashed arrows indicate possible side reactions leading to loss in repair yield. The drawn structures of $(T{=}T)^{\circ -}$, $(T\_T)^{\circ -}$, and T $T^{\circ -}$ do not imply definitive knowledge of the predominant mesomeric structures and localizations of the excess electron between the two rings. The inset shows the geometry of the FAD cofactor and the CPD (in situ repaired during structure determination, PDB entry 1TEZ) with relevant electron transfer distances indicated.

(within 400 fs according to a recent simulation [58]), leaving one intact thymine and one thymine anion radical, $T^{\circ -}$. The latter should return its excess electron to FADH°, restoring an active photolyase. This electron return rate k_{ER} was for the first time observed by Kao et al. [53] in 2005 by ultrafast transient absorption in a study of the kinetics of the FADH° intermediate. The decay of FADH° could be described either biexponentially with time constants of 460 and 1240 ps or by a stretched exponential with $\tau = 560$ ps and $\beta = 0.71$ (Table 2.1). Again, the deviation from a monoexponential decay was assigned to a solvation process. As an alternative explanation,

Brettel and Byrdin [3] suggested that the faster phase might have reflected back ET (k_{BET}) prior to complete CPD splitting that would result in re-formation of the CPD lesion and hence diminish the repair QY. Unfortunately, the repair QY had not been quantified in the experiments reported in [53]. The driving force of electron return from $T^{\circ -}$ to FADH° can be estimated from the respective one electron reduction potentials of −2.18 V [56] to −1.62 V [57] for T and ~0.05 V for FADH° [54, 55], yielding 2.23 to 1.67 eV, much higher than for forward ET from $FADH^{-*}$ to T=T (see above). On the basis of modifications of the electron return kinetics upon T-to-U replacements, Liu et al. [56] suggested that the excess electron was located on the 3′ base and returned to FADH° by superexchange via the adenine.

The product state, T T, has its unique spectroscopic signature around 265 nm. A first attempt to monitor DNA-based intermediates of the repair reaction in the UV was reported by MacFarlane and Stanley in 2003 [59]. They used ultrafast transient absorption at 265 nm and observed an instantaneous bleaching (attributed to formation of $FADH^{-*}$), a subsequent absorption increase (described biexponentially with time constants of 32 ± 20 and 600 ± 300 ps) and accumulation of FADH° during the experiment. The two kinetic phases were tentatively assigned to ET from $FADH^{-*}$ to the CPD and the second to bond cleavage. However, both the very fast ET from $FADH^{-*}$ to the CPD and the apparent lack of electron return to FADH° are in contrast to studies monitoring the flavin kinetics [22, 52]. Reviewing the state of the art in 2010, Brettel and Byrdin [3] pointed out that a more conclusive study would have to monitor the kinetics of the DNA based intermediates and the kinetics of the flavin species in the same experimental system.

In 2011, the first two studies of this type were published [50, 51]. They developed rather different approaches to face three specific major technical challenges:

- The difficulty to provide a sufficiently large sample volume containing photolyase with a fully reduced flavin cofactor and an (unrepaired) CPD in the binding pocket at each excitation flash while averaging over a large number of shots in order to obtain transient absorption data at a satisfying S/N ratio

- The strong background absorption of intact nucleotides flanking the CPD (which are required for efficient binding of the substrate to the enzyme) hampers transient absorption studies in the 265 nm band
- The need to disentangle absorption changes due to formation of intact bases in the UV from large contributions due to the flavin excited state and radical

In the first study, Thiagarajan et al. [50] used a specially conceived extension [21] of the classical real-time transient absorption technique: detection of continuous probe light from ultrastable continuous-wave lasers with GHz bandwidth electronics, capturing a complete kinetic trace upon a single shot of the pump pulse. Thus, a reasonable S/N ratio was obtained by averaging not more than a few hundred shots (compared to $\sim 10^6$ in a typical ultrafast pump–probe experiment). Because of the relatively modest number of shots to be accumulated, it was possible to strictly control the state of the sample (redox state of the flavin cofactor and amount of restored thymines/repair QY) by measuring the UV/vis absorption spectrum, and to replace the sample when necessary. An obvious disadvantage of the real-time approach was the limited time resolution (rise time $t_{20-80} \approx 280$ ps); by deconvolution, time constants down to ~ 100 ps could be extracted.

To address the second challenge, Thiagarajan et al. [50] used a specially conceived substrate (a modified thymidine 10-mer with a central CPD [10]) that is repaired as well as conventional substrates, but absorbs only weakly around 265 nm. The enzyme in this study was the photolyase from the cyanobacterium *A. nidulans* that lacked the antenna pigment because of heterologous expression in *E. coli*. Transient absorption changes upon excitation of $FADH^-$ were recorded at 266 nm and at key wavelengths in the visible for the enzyme–substrate complex and the enzyme without substrate. In addition, the well-characterized actinometer $[Ru(bpy)_3]^{2+}$ [21, 60] was used as a reference and for QY determination in identical geometry. From these data, Thiagarajan et al. systematically constructed the population kinetics of $FADH^{\circ}$ and of restored thymines (Fig. 2.4;

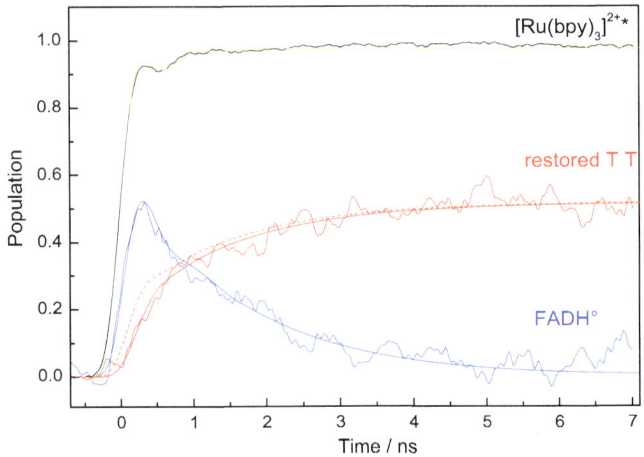

Figure 2.4 Time resolution of T=T repair by DNA photolyase from *A. nidulans* as measured (noisy traces) and modeled (smooth traces) in Ref. [50]. The green/black curves show the reaction of the detection system to the instantaneous absorption change of $[Ru(bpy)_3]^{2+}$ upon a 100 ps excitation flash. The blue and red curves show the reconstructed kinetics of FADH° formation/decay and of the thymine pair restoration, respectively, upon $FADH^-$ excitation with such a flash. Modeling assumed the reaction scheme as in Fig. 2.3 (without distinction between the $(T{=}T)^{\circ -}$ and $(T\_T)^{\circ -}$ intermediates) and rate constants as listed in Table 2.1. Dashed red line: same with $k_{sp} = 1\ ps^{-1}$ (and amplitude downscaled by 0.58).

note that the red curve refers to restoration of a pair of thymines, that is, complete repair of one T=T). The procedure required precise knowledge of the FADH° *minus* $FADH^-$ and of the $FADH^{-*}$ *minus* $FADH^-$ difference spectra that were independently measured in photolyase [50].

The results show that FADH° (blue curve in Fig. 2.4) rises to a maximum at ~300 ps and then decays with apparent biphasic kinetics with time constants of a few hundred picoseconds and ~1.5 ns, similar to the previous report that used the ultrafast pump–probe technique [53]. Thymine restoration (red curve in Fig. 2.4) shows a slight lag followed by a biphasic rise with time constants of ~200 ps and 1.5 ns. The final level corresponds to a repair QY of ~0.5.

These population kinetics were modeled using a simplification of the reaction scheme in Fig. 2.3, where the two bond cleavage steps are merged into one single step, resulting also in a single back-ET channel. The best description (smooth solid lines in Fig. 2.4) was obtained with the parameters listed in Table 2.1. Remarkably, dimer splitting was best described with a time constant of 260 ps, much slower than the predicted subpicosecond cleavage of both intradimer bonds (simulated by the dashed red line in Fig. 2.4) [58].

This study [50] provided direct experimental demonstration of the hitherto hypothetical sequence of photorepair events (Fig. 2.3): forward ET is followed by dimer splitting (forming one intact thymine) and subsequent electron return, restoring the other thymine and $FADH^-$.

The subsequent study by Liu et al. [51] used the pump–probe technique with an instrument response time of 250 fs. The repetition rate was 1 kHz and it was mentioned that the reaction mixtures were continuously mixed and contained an at least 60-fold excess of the substrate over enzyme and could be used for several hours (corresponding to $\sim 10^7$ shots). However, no control over sample state and repair QY was documented.

The second challenge was addressed by using a T=T substrate without any flanking bases, which does virtually not absorb around 265 nm (as long as it is not repaired). To compensate for the poor binding by photolyase of CPDs without flanking bases [61], a high (18 mM) substrate concentration was used. The enzyme was photolyase from *E. coli* with its antenna pigment removed by photodecomposition.

Transient absorption kinetics of enzyme–substrate complexes upon excitation of $FADH^-$ were measured at 15 wavelengths between 260 and 430 nm and at 620 and 710 nm. As the data were presented with normalized amplitudes, spectral information and information on yields are not available to the reader. The kinetic traces were decomposed into kinetics of species and associated spectra according to the reaction scheme shown in Fig. 2.3, with considerable absorption in the UV attributed to the CPD repair intermediates $(T\_T)^{\circ -}$ and T $T^{\circ -}$. The reaction rate constants are listed in the second line of Table 2.1. For most of the rate constants there is a rough agreement with the preceding study [50] that had a lower time resolution.

In detail, according to Liu et al. [51] cleavage of the second (C6-C6′) bond in 90 ps would be the rate limiting step of dimer splitting, with an activation barrier estimated to 0.174 eV from transition state theory [56]. For the cleavage of the first (C5-C5′) bond, an upper limit of 10 ps was derived from the absence of a detectable contribution of the precursor species $(T{=}T)^{\circ-}$ [51], so that the first bond cleavage may in fact be barrierless, as predicted in Ref. [58].

Apart from the moderate difference in the dimer splitting kinetics, there is a very pronounced difference between the back-ET kinetics (k_{BET}) derived from the two experimental studies: 2.4 ns [51] versus 0.35 ns [50] (Table 2.1). Possible origins of this difference were discussed in a letter to PNAS [62] and a reply [63] focusing on the precise value of the repair QY (see Table 2.1) under the experimental conditions, on which the assessment of this rate in Ref. [51] very critically depends. For instance, with a repair QY of 0.7, within the range referred to from the older literature in the reply [63], an estimation from competition with bond cleavage (as applied in Ref. [51] with an assumed QY of 0.82) would yield a time constant of 500 ps for back ET. Further clarification will have to await experimental determination of the repair QY during the pump–probe experiments. More generally, it is difficult to assess the quantitative description of the photorepair reaction sequence in Ref. [51], because essential experimental details, controls, and results (e.g., geometry of the cell and pump and probe beams, sample replacement method, absorption spectra before and after data collection, repair QY during the pump–probe experiments, amplitudes of the absorption changes) were not documented and because the uniqueness of the decomposition of the transient absorption data was not demonstrated.

Comparing the *experimental data* (i.e., transient absorption kinetics presented in the two studies), the one obvious difference is in the ratio between the long-lived absorption increase at 266 nm

Table 2.1 Rate constants for elementary steps of enzymatic CPD repair by photolyase as determined in relevant studies

Study	$1/k_{FET}$	$1/k_{sp1}$	$1/k_{sp2}$	$1/k_{ER}$	$1/k_0$	$1/k_{BET1}$	$1/k_{BET2}$	Repair quantum yield
Thiagarajan et al., 2011 [50][a]	~100 ps	260 ps		1.55 ns	1.25 ns	0.35 ns		0.52
Liu et al., 2011 [51][b]	250 ps[c]	<10 ps[d]	90 ps[c]	0.7 ns[c]	1.3 ns		2.4 ns[e]	0.82[f]
Kao et al., 2005 [53][b]	Observed $FADH^{-*}$ decay[g] 60 ps (45%) 335 ps (55%)	n.d.		Observed $FADH^{\circ}$ decay[g] 0.46 ns 1.24 ns	1.3 ns	n.d.		0.89[h]

n.d.: Not determined

[a]Enzyme from *A. nidulans*, no antenna cofactor due to heterologous expression, modified thymine 10-mer with central T=T and all bases except the one at the 3′ end replaced by dihydrothymine, measured by real-time transient absorption

[b]Enzyme from *E. coli*, antenna cofactor photodecomposed, T=T without flanking bases, measured by pump–probe transient absorption

[c]Average time constant of a stretched exponential with $\beta = 0.71$ [56]

[d]Concluded from nonobservation of the precursor $(T{=}T)^{\circ -}$

[e]Obtained from quantum yield considerations

[f]No method or reference given

[g]Biexponential fit

[h]Cited from Ref. [51]

due to restoration of intact thymine and the initial bleaching due to formation of $FADH^{-*}$ (~1:1 in Ref. [51] versus ~2.5:1 in Ref. [50]).[b]

This suggests that the effective repair QY during the pump–probe experiments in Ref. [51] was considerably lower than during the real-time experiments probing a fresh sample volume in each shot [50]. Alternatively, excitation of $FADH^{-}$ would have to be accompanied by a much stronger bleaching in the *E. coli* photolyase used in Ref. [51] than in the *A. nidulans* photolyase used in Ref. [50] which appears unlikely.

Analysis of the DNA-based contributions to the absorption changes critically depends on reliable information on the $FADH^{\circ}$ *minus* $FADH^{-}$ difference spectrum. In the critical UV region, the difference of the two species' spectra presented in [51] (obtained by apoprotein subtraction) substantially deviates from that obtained by photoreduction of $FADH^{\circ}$ [50]. It thus appears that the diverging quantitative descriptions of bond cleavage and competing back ET in the two studies [50, 51] are at least in part due to discrepancies in the estimation of flavin contributions to the absorption changes in the UV and to the unresolved issue of the effective repair QY during the pump–probe experiments. Future precisions on the flavin contributions and the possible availability of suitable reference spectra for the CPD repair intermediates $(T{=}T)^{\circ-}$, $(T\_T)^{\circ-}$, and T $T^{\circ-}$ may contribute to the firm establishment of these intermediates' kinetics.

In conclusion, experimental data on the kinetics of forward ET from $FADH^{-*}$ to the T=T lesion and on the electron return to $FADH^{\circ}$ after dimer splitting (with restoration of the second thymine) converge to time constants of 100–250 ps and ~1 ns, respectively. The kinetics of cleavage of the two intradimer bonds (<1 ps from simulation [58], 260 and 90 ps from the two experimental studies [50, 51] and of potential back ET in competition with bond cleavage are yet to be convincingly established. High time resolution, as in the study by Liu et al. [51], should be combined with rigorous control of the state of the sample and the repair QY and careful

[b]Note that the amplitude of the initial bleaching in Ref. [50] was measured in the absence of a substrate ($FADH^{-*}$ lifetime of 1.25 ns) to avoid underestimation due to the limited time resolution; cf. the controversy in Refs. [62, 63].

measurement of contributing flavin spectra in the UV, as in the study by Thiagarajan et al. [50], to allow (a) establishing definitively how much the formation of the first intact thymine is delayed with respect to the forward ET (which is a measure of the overall bond cleavage time) and (b) searching for kinetic and spectral signatures of the repair intermediates $(T{=}T)^{\circ-}$, $(T\_T)^{\circ-}$, and $T\ T^{\circ-}$.

As a final perspective, temperature-dependent studies should help to estimate the activation barriers of the two bond cleavage steps and may allow stabilizing the intermediates $(T{=}T)^{\circ-}$ and/or $(T\_T)^{\circ-}$ and their precise spectroscopic and kinetic characterization.

Acknowledgments

Our research was supported by ANR grants 05-BLAN-0304 (K.B., M.B., and M.H.V.), 09-BLAN-0248-03 (K.B. and M.B.) and 09-PIRI-0019-01 (M.H.V.) and by FRISBI (ANR-10-INSB-05-01).

References

1. Rupert, C. S., Goodgal, S. H., and Herriott, R. M. (1958). Photoreactivation in vitro of ultraviolet inactivated, *J. Gen. Physiol.*, **41**, pp. 451–471.
2. Sancar, A. (2003). Structure and function of DNA photolyase and cryptochrome blue-light photoreceptors, *Chem. Rev.*, **103**, pp. 2203–2238.
3. Brettel, K., and Byrdin, M. (2010). Reaction mechanisms of DNA photolyase, *Curr. Opin. Struct. Biol.*, **20**, pp. 693–701.
4. Vink, A. A., and Roza, L. (2001). Biological consequences of cyclobutane pyrimidine dimers, *J. Photochem. Photobiol. B*, **65**, pp. 101–104.
5. Hitomi, K., Kim, S.-T., Iwai, S., Harima, N., Otoshi, E., et al. (1997). Binding and catalytic properties of *Xenopus* (6-4) photolyase, *J. Biol. Chem.*, **272**, pp. 32591–32598.
6. Park, H.-W., Kim, S.-T., Sancar, A., and Deisenhofer, J. (1995). Crystal-structure of DNA photolyase from *Escherichia coli*, *Science*, **268**, pp. 1866–1872.
7. Tamada, T., Kitadokoro, K., Higuchi, Y., Inaka, K., Yasui, A., et al. (1997). Crystal structure of DNA photolyase from *Anacystis nidulans*, *Nat. Struct. Mol. Biol.*, **4**, pp. 887–891.

8. Mees, A., Klar, T., Gnau, P., Hennecke, U., Eker, A. P. M., et al. (2004). Crystal structure of a photolyase bound to a CPD-like DNA lesion after in situ repair, *Science,* **306**, pp. 1789–1793.
9. Espagne, A., Byrdin, M., Eker, A. P. M., and Brettel, K. (2009). Very fast product release and catalytic turnover of DNA photolyase, *ChemBioChem,* **10**, pp. 1777–1780.
10. Thiagarajan, V., Villette, S., Espagne, A., Eker, A. P. M., Brettel, K., and Byrdin, M. (2010). DNA repair by photolyase: a novel substrate with low background absorption around 265 nm for transient absorption studies in the UV, *Biochemistry,* **49**, pp. 297–303.
11. Chaves, I., Pokorny, R., Byrdin, M., Hoang, N., Ritz, T., et al. (2011). The cryptochromes: blue light photoreceptors in plants and animals, *Ann. Rev. Plant Biol.,* **62**, pp. 335–364.
12. Okafuji, A., Biskup, T., Hitomi, K., Getzoff, E. D., Kaiser, G., et al. (2010). Light-induced activation of class II cyclobutane pyrimidine dimer photolyases, *DNA Repair,* **9**, pp. 495–505.
13. Biskup, T., Hitomi, K., Getzoff, E. D., Krapf, S., Koslowski, T., et al. (2011). Unexpected electron transfer in cryptochrome identified by time-resolved EPR spectroscopy, *Angew. Chem. Int. Ed.,* **50**, pp. 12647–12651.
14. Kiontke, S., Geisselbrecht, Y., Pokorny, R., Carell, T., Batschauer, A., and Essen, L. O. (2011). Crystal structures of an archael class II DNA photolyase and its complex with UV-damaged duplex DNA, *EMBO J.,* **30**, pp. 4437–4449.
15. Aubert, C., Vos, M. H., Mathis, P., Eker, A. P. M., and Brettel, K. (2000). Intra-protein radical transfer during photoactivation of DNA photolyase, *Nature,* **405**, pp. 586–590.
16. Giovani, B., Byrdin, M., Ahmad, M., and Brettel, K. (2003). Light-induced electron transfer in a cryptochrome blue-light photoreceptor, *Nat. Struct. Biol.,* **10**, pp. 489–490.
17. Liu, B., Liu, H., Zhong, D., and Lin, C. (2010). Searching for a photocycle of the cryptochrome photoreceptors, *Curr. Opin. Plant Biol.,* **13**, pp. 578–586.
18. Zeugner, A., Byrdin, M., Bouly, J.-P., Bakrim, N., Giovani, B., et al. (2005). Light-induced electron transfer in *Arabidopsis* cryptochrome-1 correlates with in vivo function, *J. Biol. Chem.,* **280**, pp. 19437–19440.
19. Byrdin, M., Sartor, V., Eker, A. P. M., Vos, M. H., Aubert, C., et al. (2004). Intraprotein electron transfer and proton dynamics during photoactivation of DNA photolyase from E. coli: review and new insights from an "inverse" deuterium isotope effect, *Biochim. Biophys. Acta,* **1655**, pp. 64–70.

20. Zieba, A. A., Richardson, C., Lucero, C., Dieng, S. D., Gindt, Y. M., and Schelvis, J. P. M. (2011). Evidence for concerted electron proton transfer in charge recombination between FADH(−) and (306)Trp(•) in *Escherichia coli* photolyase, *J. Am. Chem. Soc.,* **133**, pp. 7824–7836.

21. Byrdin, M., Thiagarajan, V., Villette, S., Espagne, A., and Brettel, K. (2009). Use of ruthenium dyes for subnanosecond detector fidelity testing in real time transient absorption, *Rev. Sci. Instr.,* **80**, 043102.

22. Kim, S. T., Heelis, P. F., Okamura, T., Hirata, Y., Mataga, N., and Sancar, A. (1991). Determination of rates and yields of interchromophore (folate $\rightarrow$ flavin) energy transfer and intermolecular (flavin $\rightarrow$ DNA) electron transfer in *Escherichia coli* photolyase by time-resolved fluorescence and absorption spectroscopy, *Biochemistry,* **30**, pp. 11262–11270.

23. Saxena, C., Sancar, A., and Zhong, D. (2004). Femtosecond dynamics of DNA photolyase: energy transfer of antenna initiation and electron transfer of cofactor reduction, *J. Phys. Chem. B,* **108**, pp. 18026–18033.

24. Kim, S. T., Heelis, P. F., and Sancar, A. (1992). Energy transfer (deazaflavin $\rightarrow$ FADH2) and electron transfer (FADH2 $\rightarrow$ T<>T) kinetics in *Anacystis nidulans* photolyase, *Biochemistry,* **31**, pp. 11244–11248.

25. Zheng, X., Garcia, J., and Stuchebrukhov, A. A. (2008). Theoretical study of excitation energy transfer in DNA photolyase, *J. Phys. Chem. B,* **112**, pp. 8724–8729.

26. Nunthaboot, N., Tanaka, F., Kokpol, S., Chosrowjan, H., Taniguchi, S., and Mataga, N. (2008). Simultaneous analysis of ultrafast fluorescence decays of FMN binding protein and its mutated proteins by molecular dynamic simulation and electron transfer theory, *J. Phys. Chem. B,* **112**, pp. 13121–13127.

27. Zhong, D., and Zewail, A. H. (2001). Femtosecond dynamics of flavoproteins: charge separation and recombination in rioboflavin (vitamin B2)-binding protein and in glucose oxidase enzyme, *Proc. Natl. Acad. Sci. U S A,* **98**, pp. 11867–11872.

28. Li, Y. F., Heelis, P. F., and Sancar, A. (1991). Active site of DNA photolyase: tryptophan-306 is the intrinsic hydrogen atom donor essential for flavin radical photoreduction and DNA repair in vitro, *Biochemistry,* **30**, pp. 6322–6329.

29. Cheung, M. S., Daizadeh, I., Stuchebrukhov, A. A., and Heelis, P. F. (1999). Pathways of electron transfer in Escherichia coli DNA photolyase: Trp306 to FADH, *Biophys. J.,* **76**, pp. 1241–1249.

30. Byrdin, M., Eker, A. P. M., Vos, M. H., and Brettel, K. (2003). Dissection of the triple tryptophan electron transfer chain in Escherichia coli DNA

photolyase: Trp382 is the primary donor in photoactivation, *Proc. Natl. Acad. Sci. U S A,* **100**, pp. 8676–8681.

31. Byrdin, M., Villette, S., Eker, A. P. M., and Brettel, K. (2007). Observation of an intermediate tryptophanyl radical in W306F mutant DNA photolyase from Escherichia coli supports electron hopping along the triple tryptophan chain, *Biochemistry,* **46**, pp. 10072–10077.
32. Lukacs, A., Eker, A. P. M., Byrdin, M., Villette, S., Pan, J., et al. (2006). Role of the middle residue in the triple tryptophan electron transfer chain of DNA photolyase: ultrafast spectroscopy of a Trp>Phe mutant, *J. Phys. Chem. B,* **110**, pp. 15654–15658.
33. Lukacs, A., Eker, A. P. M., Byrdin, M., Brettel, K., and Vos, M. H. (2008). Electron hopping through the 15 Å triple tryptophan molecular wire in DNA photolyase occurs within 30 ps, *J. Am. Chem. Soc.,* **130**, pp. 14394–14395.
34. Byrdin, M., Villette, S., Espagne, A., Eker, A. P. M., and Brettel, K. (2008). Polarized transient absorption spectroscopy to resolve electron transfer between tryptophans in DNA photolyase, *J. Phys. Chem. B,* **112**, pp. 6866–6871.
35. Byrdin, M., Lukacs, A., Thiagarajan, V., Eker, A. P. M., Brettel, K., and Vos, M. H. (2010). Quantum yield measurements of short-lived photoactivation intermediates in DNA photolyase: toward a detailed understanding of the triple tryptophan electron transfer chain, *J. Phys. Chem. A,* **114**, pp. 3207–3214.
36. Pan, J., Byrdin, M., Aubert, C., Eker, A. P. M., Brettel, K., and Vos, M. H. (2004). Excited-state properties of flavin radicals in flavoproteins: femtosecond spectroscopy of DNA photolyase, glucose oxidase and flavodoxin, *J. Phys. Chem. B,* **108**, pp. 10160–10167.
37. Hoff, A. J., and Deisenhofer, J. (1997). Photophysics of photosynthesis. Structure and spectroscopy of reaction centers of purple bacteria, *Phys. Rep.,* **287**, pp. 1–247.
38. Chen, X., Dai, H., Li, J., Huang, X., and Wei, Z. (2012). The effects of biological environments on the electron-relay functionality of tryptophan residues in proteins, *ChemPhysChem,* **13**, pp. 183–192.
39. Popovic, D. M., Zmiric, A., Zaric, S. D., and Knapp, E.-W. (2002). Energetics of radical transfer in DNA photolyase, *J. Am. Chem. Soc.,* **124**, pp. 3775–3782.
40. Krapf, S., Koslowski, T., and Steinbrecher, T. (2010). The thermodynamics of charge transfer in DNA photolyase: using thermodynamic integration calculations to analyse the kinetics of electron transfer reactions, *Phys. Chem. Chem. Phys.,* **12**, pp. 9516–9525.

41. Woiczikowski, P. B., Steinbrecher, T., Kubar, T., and Elstner, M. (2011). Nonadiabatic QM/MM simulations of fast charge transfer in *Escherichia coli* DNA photolyase, *J. Phys. Chem. B*, **115**, pp. 9846–9863.

42. Vos, M. H., Rischel, C., Jones, M. R., and Martin, J.-L. (2000). Electrochromic detection of a coherent component in the formation of the charge pair P+HL- in bacterial reaction centers, *Biochemistry*, **39**, pp. 8353–8361.

43. Shuvalov, V. A., and Yakovlev, A. G. (2003). Coupling of nuclear wavepacket motion and charge separation in bacterial reaction centers, *FEBS Lett.*, **540**, pp. 26–34.

44. DeFelippis, M. R., Murthy, C. P., Faraggi, M., and Klapper, M. H. (1989). Pulse radiolytic measurement of redox potentials: the tyrosine and tryptophan radicals, *Biochemistry*, **28**, pp. 4847–4853.

45. Aubert, C., Mathis, P., Eker, A. P. M., and Brettel, K. (1999). Intraprotein electron transfer between tyrosine and tryptophan in DNA photolyase from *Anacystis nidulans*, *Proc. Natl. Acad. Sci. U S A*, **96**, pp. 5423–5427.

46. Wang, H., Saxena, C., Quan, D., Sancar, A., and Zhong, D. (2005). Femtosecond dynamics of flavin cofactor in DNA photolyase: radical reduction, local solvation, and charge recombination, *J. Phys. Chem. B*, **109**, pp. 1329–1333.

47. Brazard, J., Usman. A., Lacombat, F., Ley, C., Martin, M. M., et al. (2010). Spectro-temporal characterization of the photoactivation mechanism of two new oxidized cryptochrome/photolyase photoreceptors, *J. Am. Chem. Soc.*, **132**, pp. 4935–4945.

48. Kao, Y.-T., Tan, C., Song, S.-H., Öztürk, N., Li, J., et al. (2008). Ultrafast dynamics and anionic active states of the flavin cofactor in cryptochrome and photolyase, *J. Am. Chem. Soc.*, **130**, pp. 7695–7701.

49. Shirdel, J., Zirak, P., Penzkofer, A., Breitkreuz, H., and Wolf, E. (2008). Absorption and fluorescence spectroscopic characterisation of the circadian blue-light photoreceptor cryptochrome from *Drosophila melanogaster* (dCry), *Chem. Phys.*, **352**, pp. 35–47.

50. Thiagarajan, V., Byrdin, M., Eker, A. P. M., Müller. P., and Brettel, K. (2011). Kinetics of cyclobutane thymine dimer splitting by DNA photolyase directly monitored in the UV, *Proc. Natl. Acad. Sci. U S A*, **108**, pp. 9402–9407.

51. Liu, Z., Tan, C., Guo, X., Kao, Y.-T., Li, J., et al. (2011). Dynamics and mechanism of cyclobutane pyrimidine dimer repair by DNA photolyase, *Proc. Natl. Acad. Sci. U S A*, **108**, pp. 14831–14836.

52. Okamura, T., Sancar, A., Heelis, P. F., Begley, T. P., Hirata, Y., and Mataga, N. (1991). Picosecond laser photolysis studies on the photorepair

of pyrimidine dimers by DNA photolyase .1. Laser photolysis of photolyase-2-deoxyuridine dinucleotide photodimer complex, *J. Am. Chem. Soc.,* **113**, pp. 3143–3145.

53. Kao, Y.-T., Saxena, C., Wang, L., Sancar, A., and Zhong, D. (2005). Direct observation of thymine dimer repair in DNA by photolyase, *Proc. Natl. Acad. Sci. U S A,* **102**, pp. 16128–16132.
54. Balland, V., Byrdin, M., Eker, A. P. M., Ahmad, M., and Brettel, K. (2009). What makes the difference between a cryptochrome and DNA photolyase? A spectroelectrochemical comparison of the flavin redox transitions, *J. Am. Chem. Soc.,* **131**, pp. 426–427.
55. Gindt, Y. M., Schelvis, J. P. M., Thoren, K. L., and Huang, T. H. (2005). Substrate binding modulates the reduction potential of DNA photolyase, *J. Am. Chem. Soc.,* **127**, pp. 10472–10473.
56. Liu, Z., Guo, X., Tan, C., Li, J., Kao, Y.-T., et al. (2012). Electron tunneling pathways and role of adenine in repair of cyclobutane pyrimidine dimer by DNA photolyase, *J. Am. Chem. Soc.,* **134**, pp. 8104–8114.
57. Sokolowsky, K., Newton, M., Lucero, C., Wertheim, B., Freedman, J., et al. (2010). Spectroscopic and thermodynamic comparisons of Escherichia coli DNA photolyase and *Vibrio cholerae* cryptochrome 1, *J. Phys. Chem. B,* **114**, pp. 7121–7130.
58. Masson, F., Laino, T., Rothlisberger, U., and Hutter, J. (2009). A QM/MM investigation of thymine dimer radical anion splitting catalyzed by DNA photolyase, *ChemPhysChem,* **10**, pp. 400–410.
59. MacFarlane, A. W., and Stanley, R. J. (2003). *Cis-Syn* thymidine dimer repair by DNA photolyase in real time, *Biochemistry,* **42**, pp. 8558–8568.
60. Müller, P., and Brettel K. (2012). $[Ru(bpy)_3]^{2+}$ as a reference in transient absorption spectroscopy: differential absorption coefficients for formation of the long-lived ^{3}MLCT excited state, *Photochem. Photobiol. Sci.,* **11**, pp. 632–636.
61. Kim, S. T., and Sancar, A. (1991). Effect of base, pentose, and phosphodiester backbone structures on binding and repair of pyrimidine dimers by *Escherichia coli* DNA photolyase, *Biochemistry,* **30**, pp. 8623–8630.
62. Brettel, K., and Byrdin, M. (2012). DNA photolyase: is the nonproductive back electron transfer really much slower than forward transfer? *Proc. Natl. Acad. Sci. U S A,* **109**, E1462.
63. Zhong, D., Sancar, A., and Stuchebrukhov, A. (2012). Reply to Brettel and Byrdin: on the efficiency of DNA repair by photolyase, *Proc. Natl. Acad. Sci. U S A,* **109**, E1463.

Chapter 3

Dynamics and Mechanisms of Ultraviolet-Damaged DNA Repair by Photolyases

Zheyun Liu, Lijuan Wang, and Dongping Zhong

Department of Physics, The Ohio State University, Columbus, OH 43210, USA
Department of Chemistry and Biochemistry, The Ohio State University, Columbus, OH 43210, USA
Programs of Biophysics, Chemical Physics, and Biochemistry, The Ohio State University, Columbus, OH 43210, USA
zhong.z8@osu.edu

The ultraviolet (UV) component of sunlight irradiation causes DNA damage by inducing the formation of cyclobutane pyrimidine dimer (CPD) and pyrimidine-pyrimidone (6-4) photoproduct (6-4PP), which are mutagenic and finally lead to mutations and cell death. Photolyase, which exists in many organisms, is a flavoprotein that can restore the damaged DNA. With femtosecond spectroscopy, the dynamics and mechanisms of repairs of CPD and 6-4PP by photolyases were revealed. After electron injection, the CPD anion radical splits in two sequential steps within 90 picoseconds before the electron returns to the cofactor to complete the photocycle. For 6-4PP repair, the anionic 6-4PP proceeds to a critical cyclic proton transfer with a nearby histidine residue to close the photocycle.

Ultrafast Dynamics at the Nanoscale: Biomolecules and Supramolecular Assemblies
Edited by Irene Burghardt and Stefan Haacke

ISBN 978-981-4745-33-8 (Hardcover), 978-981-4745-34-5 (eBook)
www.panstanford.com

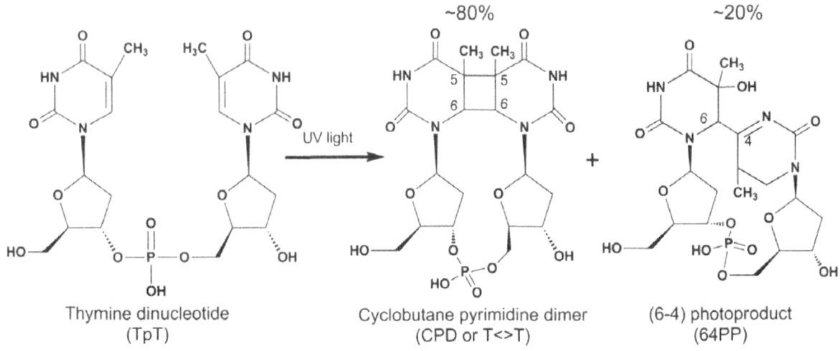

Figure 3.1 Structures of undamaged thymine bases and two major DNA photolesions.

Significantly, the intervening adenine moiety of the cofactor in both CPD and 6-4PP repair is critical to mediating the electron tunneling between the enzyme and substrate with a superexchange mechanism. The unique electron-tunneling pathway ensures the synergy of all elementary steps in the photorepair machinery.

3.1 Introduction

The ultraviolet (UV) component in sunlight damages DNA by inducing cyclobutane pyrimidine dimer (CPD, 80%) and a less frequently pyrimidine-pyrimidone (6-4) photoproduct (6-4PP, 20%) (Fig. 3.1) [1]. In CPD, the two nearby thymine bases in the same strand connect with each other to form a cyclobutane ring. On the other side, the 6-4PP is a relatively complicated chemical structure while the C6 of 5′ thymine and C4 of 3′ thymine are linked and the hydroxyl group of 5′ base is transferred from 3′ base. Both photoproducts lead to mutagenesis and even skin cancer [2–4]. Photolyase, a special class of flavoprotein, restores the damaged DNA through the absorption of blue light [1, 5–7]. Two kinds of photolyases share similar primary sequences and folding structures but one of the two photolyases repairs one photoproduct and cannot repair another. Therefore, these photolyases are usually classified as CPD photolyase and (6-4) photolyase on the basis of their functions.

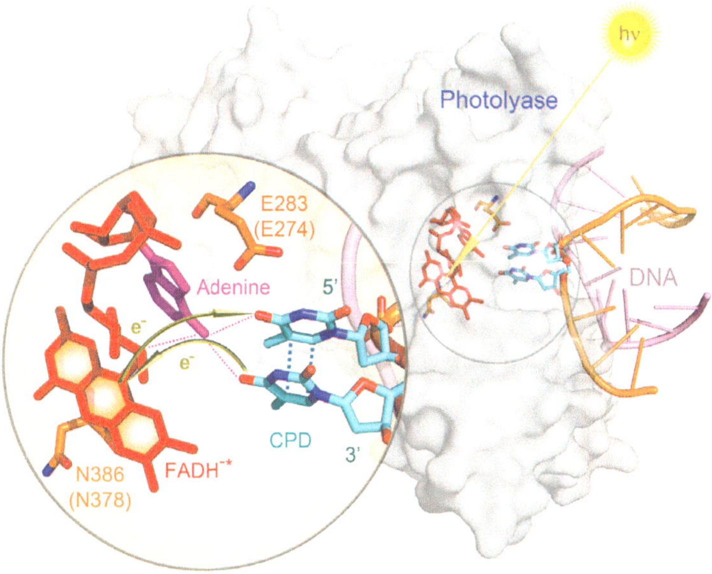

Figure 3.2 X-ray complex structure of *Anacystis nidulans* photolyase with DNA containing a repaired photoproduct of a thymine dimer. The thymine dimer is flipped out of DNA and inserted into the active site. A close-up view shows the relative positions of the catalytic cofactor $FADH^-$ and the repaired substrate with the electron-tunneling pathways in repair. *Escherichia coli* photolyase adopts a similar structure.

Both photolyases contain a noncovalently bound, fully reduced flavin adenine dinucleotide ($FADH^-$) molecule as the active cofactor. During the past two decades, the crystal structures of photolyases [8–13], especially of the enzyme–substrate (CPD or 6-4PP) complexes, were solved (Figs. 3.2 and 3.3) [9, 10], in which the $FADH^-$ cofactor takes an unusual U-shaped folding configuration in both enzymes. The adenine moiety of the bent cofactor lies between the isoalloxazine ring of flavin and the damaged photolesions, which flip out of the double-stranded DNA and intrude into the active site of the enzymes [9, 10].

The mechanism of CPD photorepair has been proposed and tested in the past 30 years [14–17] until we reported the cyclic electron transfer (ET) reaction in repair in 2005 with femtosecond-

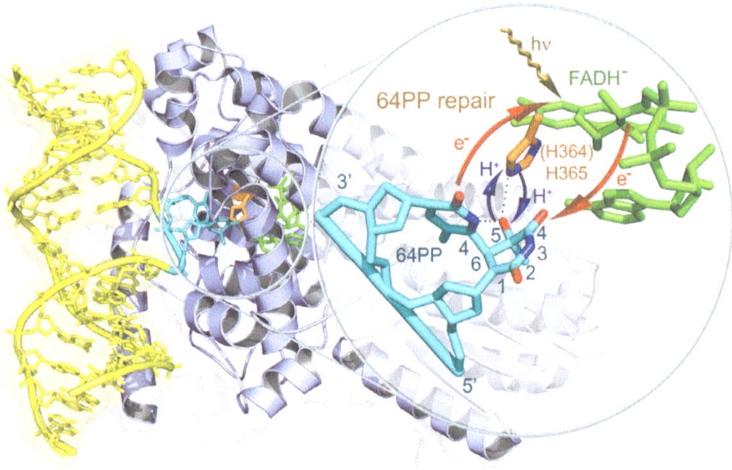

Figure 3.3 X-ray structure of *Drosophila melanogaster* (6-4) photolyase (blue) bound to DNA (yellow) containing a (6-4) photoproduct. *Arabidopsis thaliana* photolyase has a similar structure and conserved histidine residue in the active site (His 364 in *Arabidopsis thaliana* and His 365 in *Drosophila melanogaster*). The 6-4PP is flipped out of DNA and inserted into the active site. The close-up view shows the relative positions of the catalytic cofactor $FADH^-$, conserved His 364 (His 365) residue, and 6-4PP substrate.

resolved spectroscopy [18]. On the other side, several hypothesis of 6-4PP repair were proposed [19–26] but the detailed mechanism remained very elusive. In this chapter, two major issues are covered: what the detailed repair mechanisms of CPD and 6-4PP photolesions are, and how the photolyase delivers one electron into damaged DNA with blue light excitation. Here, we review the repair mechanisms and electron-tunneling pathways of both CPD and 6-4PP repairs by photolyases using ultrafast spectroscopy combined with molecular biology methods.

3.2 Reaction Mechanism of CPD Repair by Photolyase

Figure 3.4 shows the proposed repair mechanism of CPD with a series of elementary reactions of ET and bond-breaking/bond-making

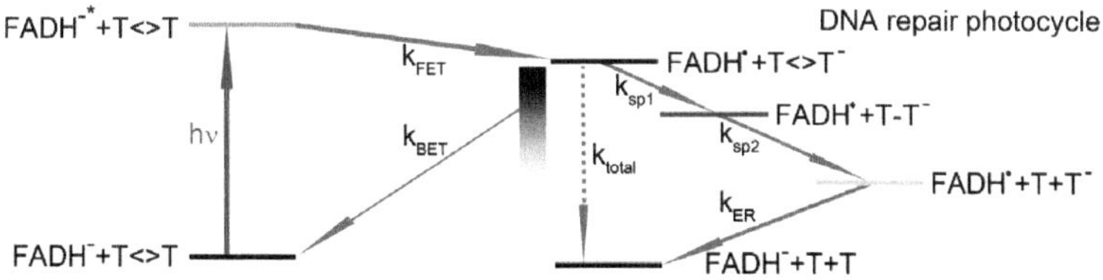

Figure 3.4 A proposed sequential repair scheme is shown. After the forward electron transfer (reaction rate k_{FET}) from $FADH^{-*}$ to a thymine dimer upon light excitation, the repair channel includes splitting of two bonds of C5-C5′ (k_{sp1}) and C6-C6′ (k_{sp2}) in the thymine dimer, accompanied with unproductive back electron transfer (k_{BET}) without repair. In the productive pathway, the subsequent electron return (k_{ER}) after complete ring splitting finishes the reaction photocycle. k_{total} is the overall decay rate of intermediate-state $FADH^{\bullet}$ after the initial charge separation.

steps. These reactions are ultrafast and involve various intermediate species. Using femtosecond spectroscopy with tunable wavelengths, we can follow the dynamics of the entire reaction evolution. We first used femtosecond-resolved fluorescence spectroscopy [27] and solved the dynamics of the initial forward-ET step.

We measured both the radiative lifetime of $FADH^{-*}$ in the absence and presence of CPD. While the former decays in 1.3 nanoseconds (ns), the latter becomes much faster, which can be best represented by a stretched single exponential decay at 170 picoseconds (ps) with $\beta = 0.71$ [18]. This stretched behavior reflects a heterogeneous ET dynamics continuously modulated by the active-site solvation, which has been observed in photolyase [18, 28, 29], and even in light-harvesting complexes [30]. With the derived average decay time at 212 ps [31] and taking the effective lifetime of $FADH^{-*}$ at 1.3 ns ($1/k_d$), the forward ET was determined to finish in 250 ps. The initial charge separation has a very high branching of 0.85 [$B_{FET} = k_{FET}/(k_{FET} + k_d)$]. Next, we captured the signal of radical $FADH^{\bullet}$ using the transient absorption technique in the visible region and proved that the reaction photocycle finishes in 700 ps with a cyclic ET process.

To resolve the detailed mechanism of CPD splitting after electron injection and to probe the signals of DNA intermediates and repaired thymine products, the detection wavelengths were extended from visible to deep UV light [32]. Figure 3.5 shows a striking pattern

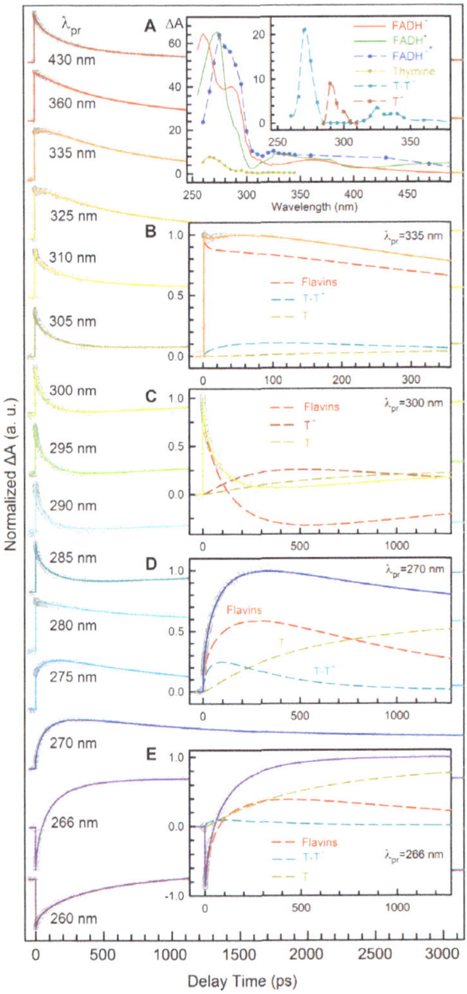

Figure 3.5 Femtosecond-resolved transient absorption dynamics of reactants, various intermediates, and products involved in repair of a thymine dimer. The repair dynamics are probed systematically from 800 to 260 nm, and shown are the typical results in the UV region with a distinct pattern. Inset A: The absorption coefficients of all species involved in repair. Insets B–E: Transient absorption signals probed at 335, 300, 270, and 266 nm. These dynamics are systematically fitted by total flavin-related species ($FADH^{-*}+ FADH^{\bullet}+ FADH^{-}$, dashed red), thymine dimer intermediate T-T^{-} (dashed cyan), thymine anion (dashed dark red), and thymine products (dashed dark yellow).

of the transient absorption signals of the complex of photolyase with substrate T<>T, probed at fifteen wavelengths. At 430 nm, the signal is only the summation of all three flavin species ($FADH^{-*}$, $FADH^{\bullet}$ and $FADH^{-}$) and the thymine-related species do not have absorption in the visible range. Thus, the signal returns to zero with a time constant of 700 ps. Knowing the flavin dynamics and the timescale for completion of the cyclic reaction, we tuned the probe wavelength to the UV region to detect the thymine-related species. At 335 nm, we captured the formation and decay of thymine-related intermediates, that is, $(T\text{-}T)^{-}$, and at 300 nm we clearly observed the long-component formation of final repaired thymine molecules (insets B–E in Fig. 3.5). By knowing the dynamics of $FADH^{-*}$ and the absorption coefficients of $FADH^{\bullet}$ and $FADH^{-}$ (inset A in Fig. 3.5), we systematically and successfully fit all the absorption transients only with the sequential model shown in Fig. 3.4, but not any other concerted schemes of thymine splitting and electron return, resulting in deconvolution of various species in insets B–E and the related absorption coefficients in inset A of Fig. 3.5, and thus obtain the entire dynamics of thymine dimer splitting.

Our data indicate that in contrast to the computational reaction scheme models [33, 34] the thymine dimer splits by a sequential pathway. The first-bond C5-C5′ breakage ($1/k_{sp1}$) finishes in much less than 10 ps, consistent with the theoretical prediction of a nearly barrierless process [33–41]. Also, the driving force of direct back ET between $T{<>}T^{-}$ and $FADH^{\bullet}$ is quite large (~2.0 eV) which falls in the Marcus inverted ET region and should occur in hundreds of picoseconds. Thus, the slow formation (k_{FET}) and ultrafast decay (k_{sp1}) result in negligible accumulation of $T{<>}T^{-}$ population and a close-to-unity efficiency of C5-C5′ splitting. In the following, we observed the formation and decay of $(T\text{-}T)^{-}$ intermediate after the first-bond C5-C5′ breakage (Fig. 3.4 and insets B–E in Fig. 3.5). The decay dynamics in 87 ps mainly represents the second-bond C6-C6′ splitting (k_{sp2}). Given the total repair quantum yield as 0.82 [15, 42, 43], the forward-ET branching as 0.85, and the yield of C5-C5′ breaking as 1.0, the branching of C6-C6′ splitting is 0.96. Therefore, the second-bond breakage finishes in 90 ps ($1/k_{sp2}$) which is much longer than that of theoretical calculations [18], and the back ET without the second-bond splitting occurs in 2.4 ns ($1/k_{BET}$), which

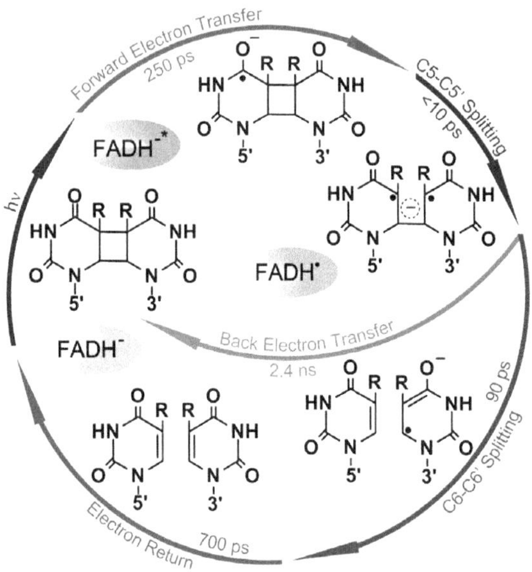

Figure 3.6 Complete photocycle of CPD repair by photolyase. All resolved elementary steps of CPD (thymine dimer) repair, showing the complete repair photocycle on the ultrafast timescales and the elucidated molecular mechanism.

is much slower than that reported by a recent time-resolved study with subnanosecond resolution [44]. After the second-bond C6-C6′ cleavage, the signal of T^- forms and decays in 700 ps ($1/k_{ER}$ and inset C of Fig. 3.5), reflecting that the electron return from T^- to $FADH^\bullet$ is completely decoupled from the second-bond breaking. This is confirmed by the observation of final products of two repaired thymine molecules, which are formed in two sequential steps in 90 and 700 ps upon the initial electron injection (see insets D and E). The electron return in 700 ps falls in the inverted Marcus ET region as well with a driving force of −2.26 eV [32].

The dynamics of each elementary step are summarized in Fig. 3.6. The latter highlights our direct observation of the ultrafast sequential splitting dynamics of the cyclobutane ring and the complete reaction photocycle in 700 ps.

To test how photolyase achieves a high repair efficiency about 0.8, we also characterized the repair processes in a biomimetic flavin–thymine dimer adduct by following the temporal evolution of reactants and intermediates using femtosecond spectroscopy [45]. We first studied the excited $FMNH^{-*}$ dynamics using femtosecond-resolved fluorescence spectroscopy. The fluorescence transient of $FMNH^{-*}$ in the absence of dimer exhibited multiple decay dynamics with lifetimes ($1/k_d$) of 5.8 ps (82%), 35 ps (16%), and 1.5 ns (2%), as shown in Fig. 3.7. Such multiple-decay dynamics reflects the ultrafast deactivation of the excited $FMNH^{-*}$ with a butterfly bending of the isoalloxazine ring [46]. By considering the ET reaction in each deactivation process of excited LFH^{-*} (lumiflavin moiety) we obtained an ET dynamics lifetime ($1/k_{FET}$) of 79 ps. Thus, the first-step ET quantum branching is only 0.124, considering all three deactivation processes, and it is much lower than that in photolyase (0.85). In addition, the ring splitting of CPD in aqueous solution has a similar sequential mechanism and we obtained dimer splitting in a sequential manner of a few picoseconds and 435 ps ($1/k_{sp2}$) and unproductive back ET in 95 ps ($1/k_{BET}$) by probing the absorption of various intermediates. The fast back ET of 95 ps in solution indicates a relatively unstable charge-separated intermediate ($LFH^{\bullet}$–T-T^-), leading to a significant competition between ring splitting (435 ps) and futile back ET. The branching of the dimer ring splitting with electron return is 0.179, which is also smaller than that in photolyase (0.96). Considering the branching of forward ET as 0.124 acquired from the fluorescence experiment, we obtained the total repair efficiency of 0.022, which is close to steady-state measurement but more than one order smaller than that in the enzyme.

Compared with excited $FMNH^{-*}$ in solution, the dynamics of the excited $FADH^{-*}$ in photolyase has a dominant lifetime of 1.3 ns [18] that provides a time window long enough for initial forward ET. Also compared with the biomimetic system, the active site of photolyase contains the appropriate functional groups to optimize redox potentials and active-site mobility for electron tunneling [18, 29, 40], leading to maximum of the repair quantum yield and balance of the four elementary processes between the forward-ET and deactivation process, and between the second-bond splitting and back ET.

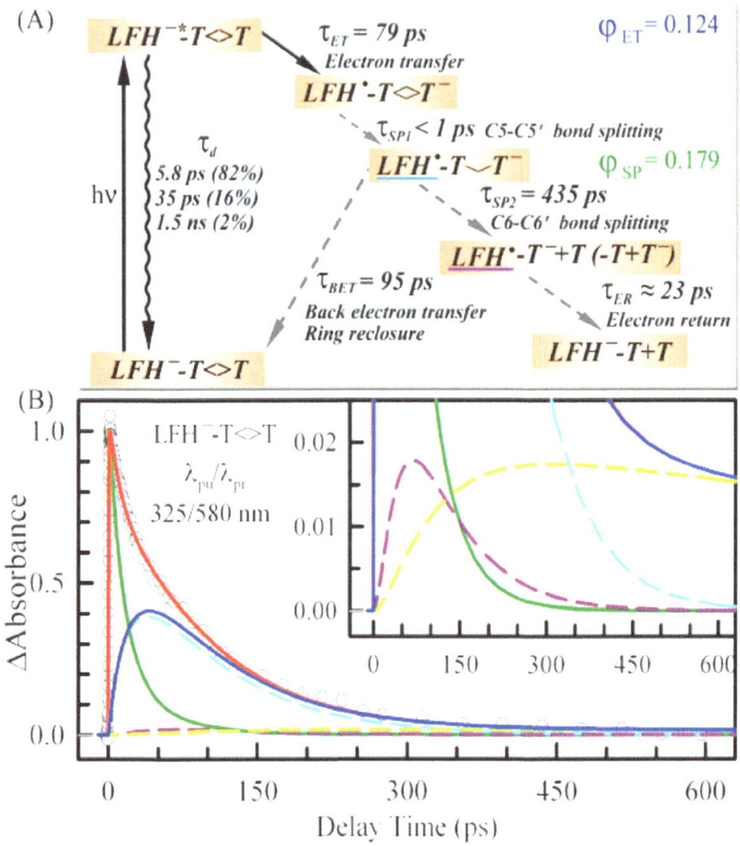

Figure 3.7 (A) Repair scheme of a model system of a flavin–thymine dimer (LfH$^-$–T<>T) with forward ET (τ_{ET}) after light excitation, ultrafast first C5-C5′ bond splitting (τ_{sp1}), back ET (τ_{BET}) and ring reclosure without repair, and the repair channel including C6-C6′ bond splitting (τ_{sp2}) and electron return (τ_{ER}). (B) Femtosecond-resolved absorption signal of LFH^{-}*–T<>T probed at 580 nm upon 325 nm excitation, with detection of both LFH^{-}*–T<>T (green curve; mainly probed at 710 nm) and LFH$^\bullet$ (blue curve). The total LFH$^\bullet$ signal is from the one dominant contribution of the initially formed LFH$^\bullet$ (dashed cyan curve) and two minor contributions of the branched LFH$^\bullet$ in the repair channel (≥85%, dashed purple; <15%, dashed lime green). A minor plateau from the deactivation channels in the 710 nm transient was removed for clarity. The inset shows the dynamics of the two fitted minor channels. Reprinted with permission from Ref. [45]. Copyright (2012) American Chemical Society.

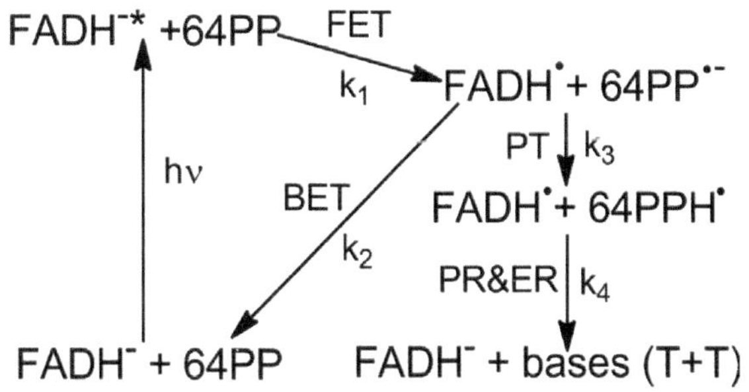

Figure 3.8 The proposed repair scheme includes forward electron transfer (FET, reaction rate k_1) after light excitation, back electron transfer (BET, reaction rate k_2) without repair, and the repair channel, comprising probable initial proton transfer (PT, reaction rate k_3) and late proton and electron return (PR and ER, reaction rate k_4) after repair.

3.3 Reaction Mechanism of 6-4PP Repair by Photolyase

For 6-4PP repair, various hypothetical repair models [10, 19–25], including formation of an oxetane intermediate in the ground state before the photochemical reaction [19, 21] and, more recently, a model that includes formation of a water molecule following the primary photochemical reaction [10, 20], have been proposed to rationalize the bond breakages and atom arrangements necessary for repair. Most models invoke proton transfer from a neighboring histidine residue, leading to a plausible scheme (Fig. 3.8).

To identify the reaction intermediates and test the proposed repair models of 6-4PP, we performed ultrafast spectroscopy and followed the similar strategy as CPD repair. We first used femtosecond fluorescence spectroscopy to solve the dynamics of forward ET [28]. By capturing emission from excited $FADH^-$ at 550 nm (Fig. 3.9A, inset), the femtosecond-resolved transient in the complex can be best represented by a stretched single exponential decay with a time constant of 225 ps and $\beta = 0.8$ (Fig. 3.9B). Considering the effective lifetime of $FADH^{-*}$ in 3.0 ns, the forward ET finishes in 280 ps. When the potential active-site proton donor, His 364, was replaced by charged (Lys and Asp), polar (Asn and Tyr), and hydrophobic (Ala and Met) residues, we observed similar ET

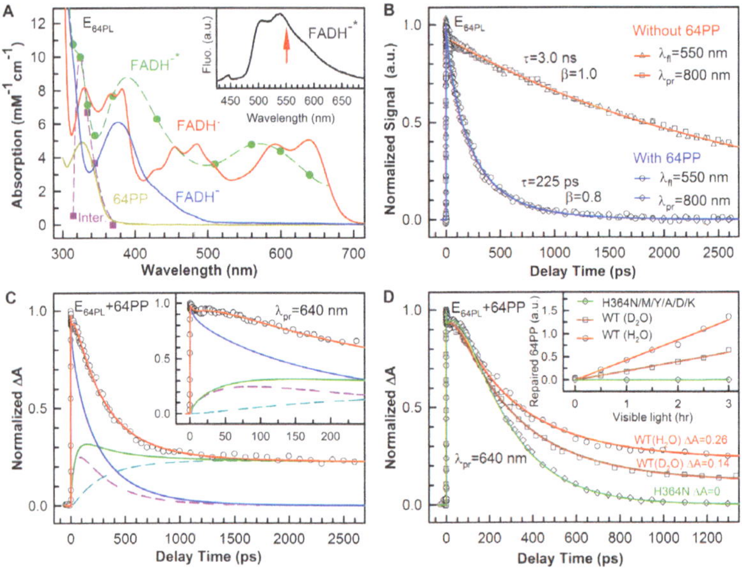

Figure 3.9 Femtosecond-resolved dynamics of flavin species involved in the repair of damaged DNA by (6-4) photolyase enzyme. (A) Absorption spectra and coefficients of purified protein with $FADH^\bullet$ (red), converted active form $FADH^-$ (blue), damaged 6-4PP (yellow), and $FADH^{-*}$ (green) determined by this study. Inset: Fluorescence emission of $FADH^{-*}$, with an arrow indicating the gated wavelength. (B) Normalized signals detected by both fluorescence (gated around the emission peak of 550 nm) and absorption (probed at 800 nm) methods without (red) and with (blue) the substrate in the active site show the same lifetime and forward-electron-transfer decays. λ_{fl}, wavelength of fluorescence; λ_{pr}, probing wavelength; a.u., arbitrary units. (C) Transient absorption signal probed at 640 nm with both $FADH^{-*}$ (blue) and $FADH^\bullet$ detection (green). The total $FADH^\bullet$ signal is from the two contributions of the initially formed $FADH^\bullet$ signal and the branched $FADH^\bullet$ signal in the repair channel. Inset: The flat signal in tens of picoseconds reflects an apparent fast signal rise. ΔA, absorption change. (D) Transient absorption signals probed at 640 nm of the His 364 Asn mutant (green line) and wild-type (WT) enzymes in D_2O (dark-red line) compared with the WT in H_2O (light-red line). The corresponding relative steady-state quantum yield measurements are shown in the inset.

dynamics with time constants $\tau = 165\text{–}358$ ps. This finding, along with the observation that the 6-4PP seems to be in its standard form in the structure of enzyme–substrate complex [10, 20], excludes the early repair model that requires a mixture of oxetane precursor and ground-state 6-4PP before the photochemical excitation [19, 21].

We conclude that after one electron injection 6-4PP repair takes place completely in the anionic ground state of 6-4PP. After the photoinduced charge separation (FADH$^{\bullet}$+6-4PP$^{\bullet-}$), the reaction can evolve along two pathways: back ET (reaction rate constant k_2) or 6-4PP repair (k_3) (Fig. 3.8). Knowing the forward-ET dynamics of FADH^{-*} we can map out the temporal evolution of FADH$^{\bullet}$ by probing at wavelengths from 500 to 700 nm (Fig. 3.9A) to follow 6-4PP repair. In Fig. 3.9C, the transient probed at 640 nm (red curve) shows drastically different behavior from that probed at 800 nm (blue curve), owing to the capture of the radical FADH$^{\bullet}$ (green curve). We observed an apparent rise signal of FADH$^{\bullet}$ in 51 ps (initial flat part of Fig. 3.9C, inset) and a long plateau, indicating that complete 6-4PP repair takes longer than several nanoseconds. Because forward ET takes about 280 ps, the 51 ps is the overall decay time of the initially formed FADH$^{\bullet}$ in Fig. 3.8 (dashed purple curve in Fig. 3.9C) but appears as an apparent rise. Slower formation and faster decay result in apparent reverse kinetics and less FADH$^{\bullet}$ accumulation. The dynamics of the branched FADH$^{\bullet}$ in the repair channel (dashed light-blue curve in Fig. 3.9C) exhibit a complex formation (largely determined by the k_1 process) and a slow decay (k_4), but the amplitude is mainly determined by the k_3 rate. By deconvolution, we measured back ET in 57 ps (k_2) and repair in 481 ps (k_3). From these rate constants we obtained a repair branching constant of 0.1, a value which is in excellent agreement with the reported steady-state repair quantum yield, suggesting that after the k_3 step, all subsequent reaction steps proceed to the final 6-4PP repair without any back ET that would lead to a futile cycle. These findings reveal that the underlying reason for the low repair quantum yield of (6-4) photolyase (0.1), compared to CPD photolyase (0.82), is the fast rate of back ET ($k_2 = 57$ ps) from 6-4PP$^{\bullet-}$ to FADH$^{\bullet}$, relative to the parallel repair channel ($k_3 = 481$ ps), although the branching of forward ET is quite high in both cases.

For the series of mutants (His 364 Asn/Met/Tyr/Ala/Asp/Lys) used to test the proposed reaction mechanism, we observed that all transients probed in the 500–700 nm region show similar back-ET dynamics in the range of 80–298 ps [28], but decay to zero without any long plateaus (shown in Fig. 3.9D for the His 364 Asn mutant). This observation is critical to the proposed reaction scheme, as it indicates that in the mutants, even though ET from $FADH^{-*}$ to 6-4PP is essentially normal, the repair channel is completely shut off, and all $FADH^{\bullet}$ formed by initial charge separation follows a futile cycle back to $FADH^{-}$ by charge recombination (Fig. 3.8). These results are also consistent with our steady-state quantum yield measurements, which revealed a total lack of repair in any of the mutants (inset in Fig. 3.9D). Collectively, our data, in agreement with an earlier report, indicate that His 364 in the active site is a functional residue that is irreplaceable in the repair channel, and proton transfer from His 364 to 6-4PP is conceivably the rate-limiting step (k_3) in our reaction scheme (Fig. 3.8). To test for proton transfer from His 364 during the repair reaction, the reaction was carried out in D_2O. As shown in Fig. 3.9D for wild-type photolyase, we observed a different transient with an obviously lower plateau, reflecting a slower repair process (1250 ps), which nevertheless had similar forward (260 ps) and back (68 ps) ET dynamics to those seen in H_2O. The X-ray structure shows a hydrogen bond distance of 2.7 Å between the amine group of His 364 and the hydroxyl group at the C5 position of the 5′ base (Fig. 3.3) [10]; hence, proton transfer from His 364 to the hydroxyl group is quite feasible. The lower plateau in the transient in D_2O is about half of that in H_2O (Fig. 3.9D), and thus corresponds to a halving in the repair branching, in good agreement with the steady-state measured quantum yield ratio of 1:2 for D_2O:H_2O (Fig. 3.9D, inset). We also studied the repair dynamics and the steady-state enzyme activity over a pH range of 7 to 9 and did not note any changes, consistent with the observation that His 364 remains protonated over even a wide basic pH range. All these results are consistent with proton transfer from His 364 to anionic 6-4PP, to generate a protonated neutral radical $6\text{-}4PPH^{\bullet}$, as a key step in the repair pathway (Fig. 3.3). This critical proton transfer, facilitated by the initial photoinduced ET, completely blocks the futile back ET from 6-4PP and allows the reaction to proceed to repair with 100% efficiency after this step.

On the basis of these findings and previous data, including the crystal structure of the enzyme–substrate complex, a catalytic photocycle for the repair of thymine (6-4) photoproduct was proposed (Fig. 3.10). In this scheme, the primary reactions are the initial ET (I to II in Fig. 3.10) and the subsequent proton transfers (II to III in Fig. 3.10). The proton transfer from a His residue in photolyase to the 6-4PP induced by the initial ET is a key step

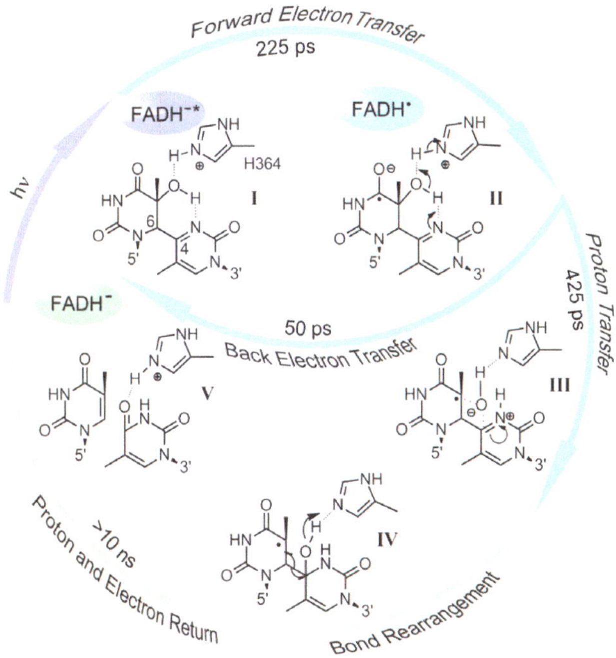

Figure 3.10 Repair photocycle of (6-4) thymine photoproduct by (6-4) photolyase. The resolved elementary steps include a forward electron transfer in 225 ps upon excitation (I to II), a back electron transfer in 50 ps without any repair (II to I), and a parallel, catalytic proton transfer between the enzyme (His 364) and the substrate (II to III), induced by the initial electron transfer, in 425 ps. This proton transfer is the critical step in repair and determines the overall repair quantum yield. The subsequent repair reactions involve a series of atom arrangements with bond breaking and making (III to IV), and final proton and electron returns (to His 364 residue and flavin cofactor) to convert the 6-4PP to two thymine bases on timescales of longer than 10 ns (IV to V).

in the repair photocycle, like the "dividing line" in the transition state, and makes the subsequent reactions "downhill" without the possibility of back reaction. This critical step competes with the back ET, resulting in an overall repair quantum yield of about 0.1, which is probably the maximum value that could be achieved for such a structurally and chemically challenging reaction. The successive elementary steps naturally proceed to an intramolecular proton transfer from the hydroxyl group on the C5 of the 5′ base to the N3 at the 3′ base [10, 47] to form a transient zwitterion, and then an oxygen atom attack of the C4 position at the 3′ base to form a transient oxetane-type structure (III in Fig. 3.10). In our scheme, in which a simple transient oxetane formation facilitates the oxygen atom transfer from the 5′ to 3′ base followed by a C6-C4 bond split (IV in Fig. 3.10), would be less prone to mutagenic side reactions. This is because, after oxygen atom transfer and C-C bond cleavage, the proton returns to the essential His 364 residue, and the electron returns to $FADH^{\bullet}$, restoring the enzyme to its active form and the 6-4PP to two thymine bases (V in Fig. 3.10).

3.4 Electron-Tunneling Pathways in DNA Restoration

In the previous sections, we provided detailed reaction photocycles for both CPD and 6-4PP repairs. However, a key question remains unresolved: How the electron tunnels between the excited $FADH^{-}$ and T<>T in the forward ET and between $FADH^{\bullet}$ and T^{-} in the electron return following splitting of the cyclobutane ring. Two main tunneling schemes have been proposed based on theoretical calculations. One model suggests that, in CPD photolyase, the electron tunnels through the intervening adenine moiety at a longer distance of about 8 Å with the unusual U-shaped configuration of the cofactor [48, 49]. The other model concludes that the electron would directly travel through space at a "shorter" distance of 4.3 Å (Fig. 3.2) [50, 51].

To answer this question, we prepared a series of (deoxy)uracil-modified dimers (Fig. 3.11) (T<>T, U<>T, T<>U and U<>U) [16,

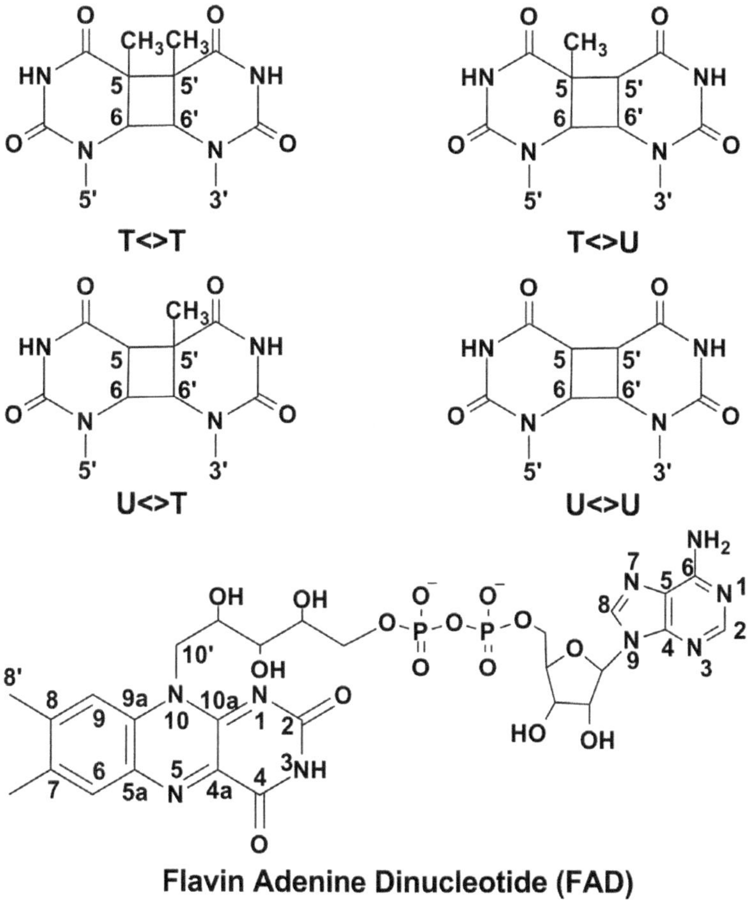

Figure 3.11 Chemical structures of various damaged substrates and flavin adenine dinucleotide studied for investigating the electron-tunneling pathway (Section 3.4).

32, 52]. We quantitatively measured the dissociation constants of enzyme–CPD complexes and thus determined the enzyme activities of various substrate (T<>U, U<>T, U<>U and T<>T) repair by photolyase. With the steady-state information and the solved reaction mechanism (Fig. 3.6), a complete series of femtosecond-resolved studies from visible to UV light region were performed to study the dynamics of each step for these substrates [52].

Table 3.1 Results of reaction times, efficiencies of the elementary steps and overall repair quantum yields of various damaged substrates[a]

Substrate	QY	β	$< \tau_{\mathrm{FET}} >$	B_{FET}	$< \tau_{\mathrm{BET}} >$	$< \tau_{\mathrm{sp}} >$	B_{sp}	$< \tau_{\mathrm{ER}} >$
T<>T	0.82	0.71	250	0.85	2400	90	0.96	700
T<>U	0.88	0.87	85	0.94	1175	75	0.94	185
U<>T	0.86	0.89	63	0.95	315	35	0.90	1220
U<>U	0.84	0.92	73	0.95	260	35	0.88	210

[a]All times are in units of picosecond. Here β is the stretched parameter and QY is the overall repair quantum yield of damaged substrates, calculated by $QY = B_{\mathrm{FET}} \times B_{\mathrm{sp}}$ or by direct measurement.

At 710 nm, only signal of $FADH^{-*}$ were probed, while the dynamics were observed to follow a stretched single exponential decay, $Ae^{-(t/\tau)^{\beta}}$, in 75, 57, and 66 ps (τ) with the stretched parameters of 0.87, 0.89, and 0.92 (β) for T<>U, U<>T, and U<>U, respectively (Figs. 3.12–3.14), which are much faster than that of T<>T (170 ps, see above). We obtained the average timescales of 212, 80, 60, and 69 ps and taking the $FADH^{-*}$ lifetime (1.3 ns) contributions into account, and final ET times of 250, 85, 63, and 73 ps (Table 3.1) for T<>T, T<>U, U<>T, and U<>U, respectively. Thus, the ET dynamics (rates) from the excited $FADH^-$ to four CPD substrates vary in a decreasing order of 5′-U<>T-3′ >U<>U >T<>U >T<>T. This observation is significant and indicates that the electron tunnels toward ending at the 5′ side, not the 3′ side, of the dimers because U is at the 5′ side and obtains an electron easier than T. U<>T and T<>U have the similar reduction potentials, but the ET time lengthens by more than 35% from 63 ps of U<>T to 85 ps of T<>U, that is, the rate decreases to 74%.

Theoretical studies showed the similar binding configurations for U<>T and T<>U and thus the observed differences in ET rates should not come from the electronic couplings because of the similar donor–acceptor separation [9, 12, 48].

The uracil base has a reduction potential more positive than thymine by ~0.11 V [53]. Although the electronic interactions between T and U in the covalent CPD species delocalize the CPD states and result in similar redox properties of U<>T and T<>U, the U moiety without a methyl group at the C5 position at the 5′ side should have a larger electron affinity than the T moiety. Thus, if the adenine moiety of $FADH^-$ is involved in electron tunneling,

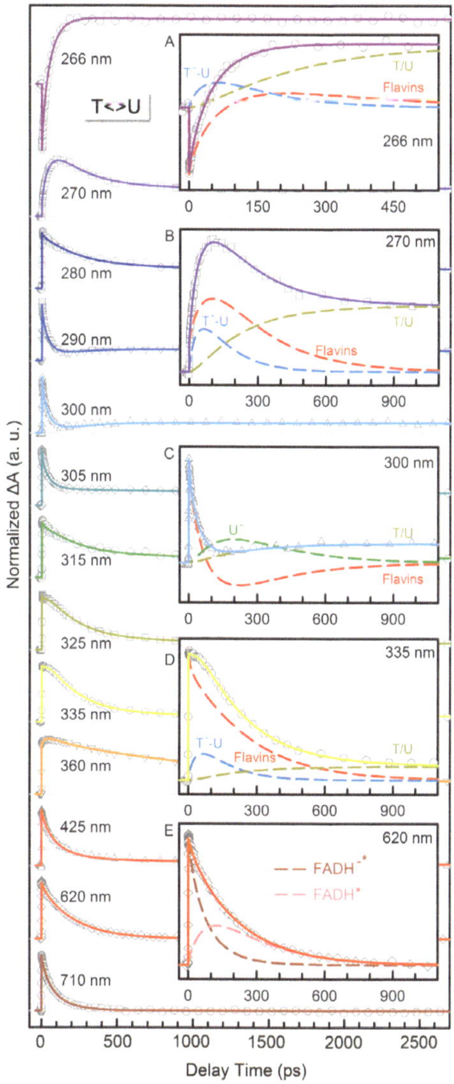

Figure 3.12 Femtosecond-resolved transient absorption dynamics of T<>U repair by photolyase at probe wavelengths from 710 to 266 nm. The transients probed at 266, 270, 300, 335, and 620 nm are deconvoluted in insets A–E, respectively. All transients were fitted by considering flavin-related species (dashed red in insets A–D and dashed pink and dark red in inset E), substrate intermediates of T^--U (dashed blue) and U^- (dashed dark green), and thymine/uracil products (dashed dark yellow).

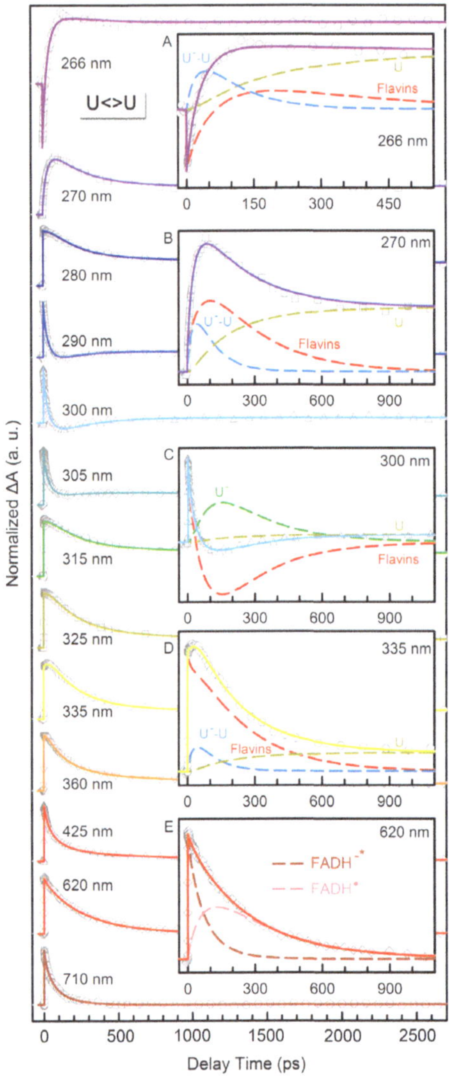

Figure 3.13 Femtosecond-resolved transient absorption dynamics of U<>U repair by photolyase at probe wavelengths from 710 to 266 nm. The transients probed at 266, 270, 300, 335, and 620 nm are deconvoluted in insets A–E, respectively. All transients were fitted by considering flavin-related species (dashed red in insets A–D and dashed pink and dark red in inset E), substrate intermediates of U^--U (dashed blue) and U^- (dashed dark green), and uracil products (dashed dark yellow).

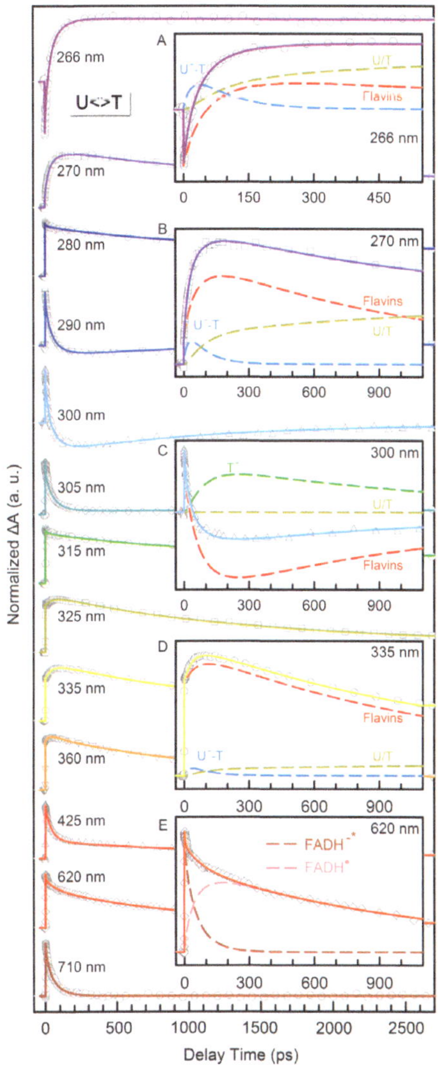

Figure 3.14 Femtosecond-resolved transient absorption dynamics of U<>T repair by photolyase at probe wavelengths from 710 to 266 nm. The transients probed at 266, 270, 300, 335, and 620 nm are deconvoluted in insets A–E, respectively. All transients were fitted by considering flavin-related species (dashed red in insets A–D and dashed pink and dark red in inset E), substrate intermediates of U^--T (dashed blue) and T^- (dashed dark green), and uracil/thymine products (dashed dark yellow).

our results would indicate that the local configuration and orientation favors the electron tunneling to the 5′ side, rather than the 3′ side (Fig. 3.2), although the distances from the adenine to the two sides are nearly equal, 3.1 and 3.2 Å, respectively. The larger ET rate of U<>T versus T<>U also excludes the electron-tunneling pathway directly from the oxylene ring of the flavin to the 3′ side of the dimer (Fig. 3.2) [50, 51]. Thus, we conclude that the electron tunnels through the adenine moiety and this conclusion will also be separately obtained from the ring splitting dynamics as shown below.

Given that the reduction potentials of adenine and $FADH^-$ are -2.52 V [53] and $+0.08$ V [54] versus the normal hydrogen electrode (NHE), respectively, and assuming the $S_1 \leftarrow S_0$ transition of $FADH^-$ near the end of absorption at 480–500 nm (2.58–2.48 eV), the intramolecular electron hopping from the isoalloxazine ring to adenine in $FADH^{-*}$ is unfavorable due to the slightly positive free energy $\Delta G^0 \approx +$ (0.0–0.12) eV. Also, we did not observe any fast decay of $FADH^{-*}$ and the excited-state dynamics is in nanoseconds in the absence of substrate [18, 32]. Thus, we can rule out the possibility of sequential two-step electron hopping between the isoalloxazine ring and the substrates bridged by adenine in a total of 63–85 ps. Thus, we conclude that the electron tunneling from the isoalloxazine ring to substrates must be mediated by adenine through a superexchange mechanism due to the long distance of ~8 Å.

We can estimate the ET times for the two proposed ET pathways, direct tunneling to the 3′ side through space or mediated by adenine to the 5′ side. Although the position of the electron in $FADH^{-*}$ has discrepancy in recent theoretical calculations [48–51, 55], the recent experiments show that the difference of dipole moments between excited state and ground state is quite small [56, 57] and thus the electron should reside at the side of pyrimidine and pyrazine rings. Here we assume the N10 (Fig. 3.11) position in the flavin cofactor as the initial starting point for the calculation of two potential ET pathways. Therefore the total tunneling length for the adenine-mediated route is 8.2 Å, while the total tunneling length in the second through-space pathway could be 10 Å, although the distance between the isoalloxazine ring and the 3′ side of the dimer is the shortest at 4.3 Å. Thus, using the empirical ET formula below [58,

59], we can estimate the ET times for the two hybrid tunneling pathways.

$$\log k = 13 - A - 3.1\frac{(\Delta G^0 + \lambda)^2}{\lambda}, \quad (3.1)$$

$$A = \beta(r - r_0)\log e = 0.434\beta(r - r_0)$$

where k is the ET rate in s^{-1}, A is related to the electronic coupling term, ΔG^0 is the total free energy of the reaction in electron volts, λ is the reorganization energy in eV, β here is the empirical ET parameter in $Å^{-1}$, and r is the separation distance in Å. r_0 is the van der Waals distance at 3.0 Å for calculating the long tunneling distance.

For different tunneling configurations, the β parameters are different for covalent bond, through protein architecture, or even through space [59–61]. For the T<>T dimer, the free energy ΔG^0 is -0.44 eV and the reorganization energy λ is 1.21 eV (see below). Thus, we obtained ET times of 0.39 to 1.7 ns for the first hybrid pathway and of 8.1–34.7 ns for the second hybrid route (for A detailed estimation, see Ref. [52]). The measured ET time for T<>T is 250 ps, closer to the estimated value of 389 ps through the first hybrid tunneling pathway. Thus, the electron tunnels through the covalent bond N10-C10′ at 1.5 Å and then from C10′ to the 5′ side of the dimer in 6.7 Å mediated by adenine nearly in the middle pathway. Using the measured ET time of 250 ps, we obtain $\beta = 0.83$ $Å^{-1}$ for the tunneling from C10′ to the 5′-side dimer due to the mediation by the adenine moiety. Thus, from the C10′ position to the 5′-side dimer, the nature of tunneling is more likely toward the covalent bond tunneling ($\beta = 0.71$ $Å^{-1}$) [59, 62] by 3.6 Å from C10′ to adenine (N atom near C6 position in Fig. 3.11) and 3.1 Å from adenine to the 5′-side dimer. Such a hybrid tunneling pathway mediated by the adenine moiety is also observed in 6-4PP repair by (6-4) photolyase. In (6-4) photolyase, the first pathway to the 5′ side of the 6-4PP through adenine is nearly the same as in the dimer repair, but the separation between C8′ to the nearest 3′ side of the 6-4PP in the second route increases to 6.3 Å. However, the forward ET has a similar dynamics of 280 ps due to the first unique hybrid tunneling pathway through the adenine moiety in the unusual U-shaped configuration of $FADH^-$ at the active site. Thus, the adenine moiety plays a critical role in mediating electron tunneling in repair of damaged DNA through a superexchange mechanism.

With the measured forward-ET dynamics, we can determine the electronic coupling constant (J) and related free energy for different substrates. We can rewrite Eq. 3.1 in a semiclassical ET expression.

$$k_{\mathrm{ET}} = \sqrt{\frac{4\pi^3}{h^2 \lambda k_{\mathrm{B}} T}} J^2 x 10^{\frac{3.1(\Delta G^0+\lambda)^2}{\lambda}} \tag{3.2}$$

where h and k_{B} are Planck and Boltzmann constants, respectively, and T is the temperature in Kelvin. For the four different substrates, the coupling constant (J) and the reorganization energy (λ) can be considered as constants. Considering the free energy ΔG^0 forward ET between FADH^{-*} and T<>T as -0.44 eV and the reported reorganization energy of λ to be 1.2–1.4 eV in flavoprotein ETs [63–65] as a fitting reference value and taking into account a larger J and a smaller λ in the forward ET than in the electron return, we obtain the free energy for ET with U<>U to be -0.59 eV, $\lambda = 1.21$ eV and $J = 3.0$ meV [52]. The coupling constant of 3.0 meV is in a good agreement with the theoretical calculation which assumed electron tunneling through adenine instead of space [48]. Using the obtained λ and J values, we calculate a free energy (ΔG^0) of -0.61 eV for U<>T and -0.57 eV for T<>U (Table 3.2). The difference of 40 meV for U<>T and T<>U probably results from the preferred

Table 3.2 The derived free energies of all elementary steps in repair of various substrates[a]

Substrate	$\Delta G^0_{\mathrm{FET}}$ [b]	$\Delta G^0_{\mathrm{BET}}$	$\Delta G^{\ddagger 0}_{\mathrm{sp}}$ [c]	ΔG^0_{ER}
T<>T	-0.44	-0.22	0.174	-2.26
T<>U	-0.57	-0.29	0.170	-2.10
U<>T	-0.61	-0.42	0.152	-2.32
U<>U	-0.59	-0.44	0.152	-2.12

[a]All free energies are in units of electron volts. The reduction potentials of several involved species are $E(\mathrm{FADH^{\bullet}/FADH^-}) = +0.08$ V versus the NHE in the presence of substrate binding to photolyase, $E(\mathrm{T/T^-}) = -2.18$ V, $E(\mathrm{U/U^-}) = -2.07$ V, and E(T<>T/T<>T$^-$) $= -1.96$ V.

[b]The 2.48 eV for the $S_1 \leftarrow S_0$ transition at 500 nm is used in calculation of $\Delta G^0_{\mathrm{FET}}$.

[c]The pre-exponential factor A in calculation of $\Delta G^{\ddagger 0}_{\mathrm{sp}}$ from $k = A \exp[-\Delta G^{\ddagger 0}_{\mathrm{sp}}/(k_{\mathrm{B}} T)]$ is assumed to be 10^{13} s^{-1}.

electron-tunneling directionality to the 5′ side rather than the 3′ side of the dimer, assuming the same electronic coupling constants for both ETs.

With understanding of the forward-ET dynamics probed at 710 nm, we tuned the probe wavelength to 620 nm to detect the flavin intermediate $FADH^{\bullet}$. To detect the substrate-related intermediates and products, we extended the probe wavelengths to UV region from 360 to 266 nm and observed a series of striking features of transients. All these transients can be fit systematically only with the sequential model shown in Fig. 3.4 and the deconvoluted transients probed at 270, 300, and 335 nm are shown in Figs. 3.12–3.14B–D. The resulting reaction times of all elementary steps are listed in Table 3.1.

At wavelengths shorter than 360 nm, we observed the anionic intermediates of T^--U, U^--U, and U^--T. Following the C5-C5′ bond breaking within a few picoseconds, the dynamics of T^--U, U^--U, and U^--T decay in 64, 31, and 30 ps, representing the total dynamics (rates) of both the C6-C6′ bond breakage of anionic CPD intermediates and futile back ET after the C5-C5′ splitting. Given the splitting branching of T^--U, U^--U and U^--T as 0.94, 0.88, and 0.90 (Table 3.1), respectively, the T^--U takes the similar bond splitting time of 75 ps as T^--T in 90 ps, but the C6-C6′ bond in U^--U and U^--T splits on a much faster timescale of 35 ps. These splitting times are much longer than our recent theoretical values (less than 1 ps) of the C-C bond splitting in a CPD model system in bulk water. Clearly, the second C6-C6′ bond breaking has a longer time with T and a shorter time with U at the 5′ side. Thus, after the C5-C5′ bond breakage, the excess electron mainly remains at the 5′ side because the T moiety at the 5′ side with a methyl group at the C5 position can significantly stabilize the anionic radical and thus has a longer time of the C6-C6′ splitting. This observation also supports the electron tunneling toward the 5′ side of the dimer, which also must pass through the adenine moiety. From the measured quantum yields (Table 3.1), we obtained the back-ET timescales of 1175, 315, and 260 ps for T^--U, U^--T and U^--U, respectively, much shorter than that for T^--T in 2.4 ns. Thus, the back-ET dynamics change in a decreasing order of 5′-U^--U-3′ >U^--T >T^--U >T^--T. Using the same λ and J values of 1.21 eV and 3 meV, respectively, we obtained the free energy of back ET to be −0.22 eV for T^--T, −0.29 eV for T^--U, −0.42 eV for U^--T, and −0.44 eV for U^--U, as given in Table 3.2. Clearly, these reactions are in the Marcus normal ET region.

The total repair quantum yield is the product of two-step yields: forward-ET branching and complete splitting branching. As shown in Table 3.1, the forward ET with T<>T has the longest timescale of 250 ps and thus has the smallest first-step branching of 0.85. However, due to the longest back-ET time of 2.4 ns, even with the longest C6-C6′ bond splitting time of 90 ps, the second-step splitting branching is highest, 0.96, among all four substrates. Thus, the highest second-step splitting branching compensates for the lowest first-step forward-ET branching, still leading to a high total quantum yield of 0.82. Similarly, for the three uracil-substituted substrates, the forward ET is faster and results in a larger first-step branching of 0.94–0.95, but the splitting branching becomes relatively smaller, 0.88–0.94, due to the faster back-ET process. However, the total quantum yields are similar in the range of 0.84–0.88, which seems to be optimized for all four CPD substrates.

After a sequential breakage of both C-C bonds, the cyclobutane ring is repaired but the electron must return to the semiquinone $FADH^{\bullet}$ to complete the photocycle and restore the catalytic state of $FADH^-$. In the UV region, we did observe the signal of T^- and U^- intermediates around 300 nm and the products of T and U in various substrates. The $(T+U)^-$ and $(U+U)^-$ have faster return dynamics of 185 and 210 ps than $(T+T)^-$ with 700 ps, while $(U+T)^-$ has the longest electron return time of 1220 ps. The electron return dynamics are much slower than the ring splitting and thus both processes are decoupled. Also, the electron return dynamics for the four substrates vary in a decreasing order of 5′-T+U^--3′ >U+U^- >T+T^- >U+T^-. From the four measured electron return times, the U at the 3′ side has faster charge recombination and the T at the 3′ side has slower electron return dynamics. After the cyclobutane ring splits, the electron mostly stays at the 3′ side. From the X-ray structure [9, 12, 13], several water molecules, polar/charged residues, and the highly reducing $FADH^{\bullet}$ are all around the 3′ side, which probably stabilizes and solvates the electron. The final reaction profiles are shown in Fig. 3.15.

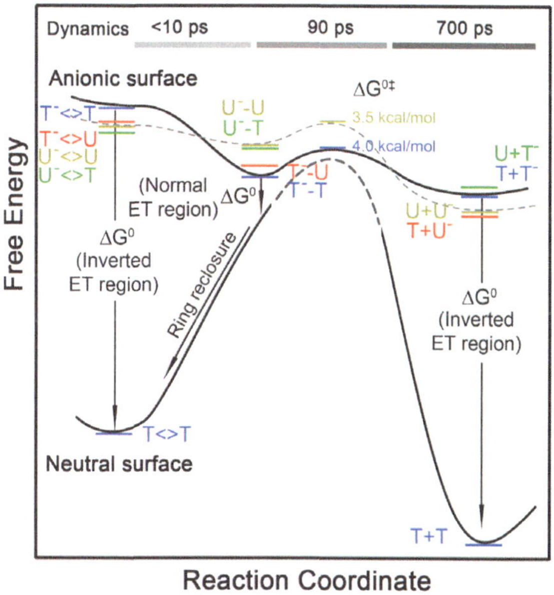

Figure 3.15 The free-energy profiles along the reaction coordinate after the forward ET in repair of CPD substrates with the timescales of the dynamics shown at the top. On the anionic surface, the solid curve represents the splitting of $T<>T^-$, while the dashed curve for $U<>U^-$. On the neutral surface, the bond-breaking activation barrier (dashed curve) is very high according to theoretical calculations [15]. Note the different regions, normal or inverted, of three ET processes and the ring reclosure after the futile back ET.

3.5 Concluding Remarks

Two of the major issues in the DNA photorepair field are the reaction mechanism and the related repair efficiency. To solve these important questions, we used femtosecond laser spectroscopy to synchronize the reactions by exciting the enzyme–substrate complexes and following the subsequent reactions [28, 32, 52]. Not surprisingly, the identical $FADH^-$ cofactor in both photolyases donates one electron to damaged photolesions upon excitation, which take similar time constants of about 250 ps and achieves a very high branching of forward ET compared with the biomimetic system [45]. The following repair mechanisms vary due to the very different chemical nature of these damaged DNA, while there are only two C-C bonds connecting thymine bases in CPD but there are both cross-base linkage and atom misplacement in 6-4PP. In CPD repair the initial charge separation destabilizes the pyrimidine dimer so that the CPD anion experiences a sequential bond splitting in a few and 90 ps. During the repair, the back ET without the second-bond splitting tremendously slows down to 2.4 ns to maximize the repair channel and the electron return after the repair is completed in 700 ps, completely decoupled from the ring splitting. The branching of ring splitting is larger than 90% so that the overall efficiency of CPD repair reaches 0.82, which is much more efficient than the mimetic system. On the other side, the unproductive back ET between 6-4PP anion and $FADH^{\bullet}$ is much faster (57 ps) than that of CPD, while it competes with a repair channel of proton transfer from the functionally critical His 364 residue to 6-4PP anion in 480 ps. The proton transfer neutralizes the negative charge of the 6-4PP radical and induces the consequent bond rearrangement in tens of nanoseconds. The extra flexibility of sidechains would be necessary to search for the favorable configuration for ultrafast proton tunneling but it also facilitates the back ET, which decreases the repair efficiency. Thus, to maximize the repair quantum yield of damaged DNA lesions, the elementary reaction dynamics, including the critical ET and proton transfer, have to be balanced. The resulting repair quantum yield of 6-4PP as 0.1 is probably the highest value it can reach over the course of evolution. Besides that, the pathway of electron tunneling is also a pivotal issue since it does not follow the general case of ET within the protein backbone. With femtosecond resolution and (deoxy)uracil-substituted substrates of CPDs, we completely mapped out the entire dynamic evolution of DNA repair processes for four different base combinations. The dynamics of all elementary steps significantly depend on the locations of thymine and uracil bases at the 5′ or 3′ side of DNA and thus provide unique and strong evidence for determination of the electron-tunneling pathways for the forward ET and final electron return. Similar as (6-4) photolyase (Fig. 3.16), the forward ET ends on the 5′ side of DNA with a hybrid of tunneling configurations and the electron passes through the critical intervening adenine moiety of the U-shaped cofactor

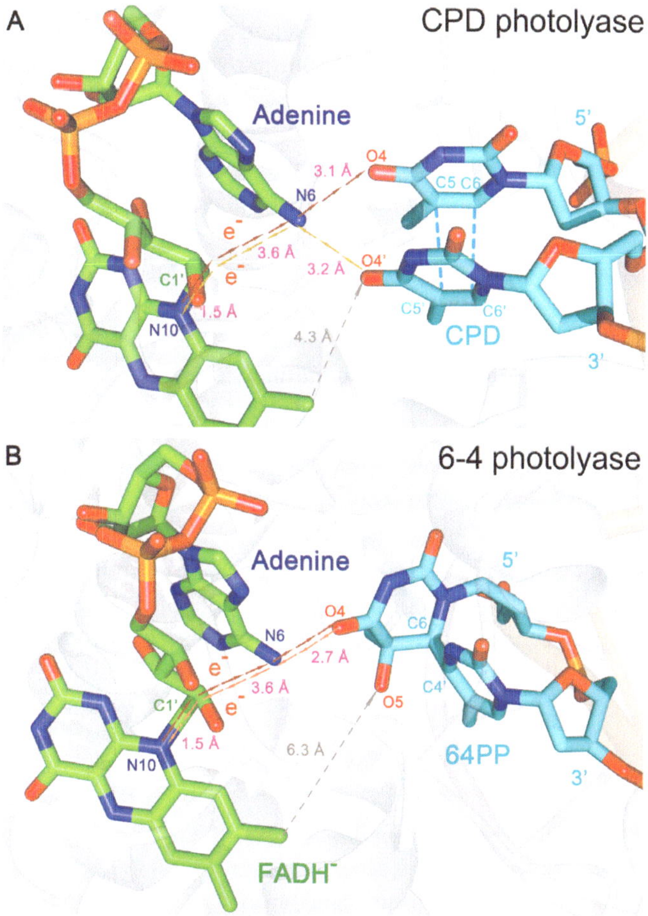

Figure 3.16 Active-site structures of enzyme–substrate complexes. (A) The relative positions of active cofactor $FADH^-$ and repaired substrate at CPD photolyase (*Anacystis nidulans*). The red and golden arrows represent two electron-tunneling pathways of the forward ET and electron return after repair, respectively. The gray arrows show the unfavorable through-space electron-tunneling pathway. (B) The relative positions of $FADH^-$ and substrate 6-4 photoproduct of (6-4) photolyase (*Arabidopsis thaliana*). Similarly, the red and dark-red arrows show the tunneling pathways of the forward and futile back ET between flavin and 6-4 photoproduct, similar to CPD photolyase. The through-space tunneling distance is 6.3 Å in a gray arrow line.

flavin through a superexchange mechanism. On the other side, the electron return after repair in CPD photolyase starts at the 3′ side, shifting from the 5′ side during splitting, and follows the reverse direction through the mediated adenine (Fig. 3.16). Thus, the unique bent configuration of the flavin cofactor assures the critical adenine position to mediate all forward and back-ET processes in tens of picoseconds to nanoseconds with a delicate balance of timescales for both repair reactions. These ET times are not too slow to decrease the efficiency of the forward ET and not too fast to increase the futile back-ET channel. All electron-tunneling pathways have unique directionalities and dynamics and these ET reactions are critical to determining the optimal repair quantum yield.

Acknowledgments

We thank Prof. Aziz Sancar (University of North Carolina at Chapel Hill) for his long-time collaboration and helpful discussion and all group members involved in the DNA repair project. The work is supported in part by the National Institutes of Health (grant GM118332).

References

1. Sancar, A. (2003). Structure and function of DNA photolyase and cryptochrome blue-light photoreceptors, *Chem. Rev.*, **103**(6), pp. 2203–2237.
2. Taylor, J. S. (1994). Unraveling the molecular pathway from sunlight to skin-cancer, *Acc. Chem. Res.*, **27**(3), pp. 76–82.
3. Daya-Grosjean, L., Dumaz, N., and Sarasin, A. (1995). The specificity of p53 mutation spectra in sunlight-induced human cancers, *J. Photochem. Photobiol. B*, **28**(2), pp. 115–124.
4. Lima-Bessa, K. M., and Menck, C. F. M. (2005). Skin cancer: lights on genome lesions, *Curr. Biol.*, **15**(2), pp. R58–R61.
5. Dulbecco, R. (1949). Reactivation of ultra-violet-inactivated bacteriophage by visible light, *Nature*,**163**(4155), pp. 949–950.

6. Husain, I., et al. (1988). Photoreactivation of killing in E. coli K-12 phr- cells is not caused by pyrimidine dimer reversal, *Photochem. Photobiol.*, **48**(2), pp. 233–234.
7. Todo, T., et al. (1993). A new photoreactivating enzyme that specifically repairs ultraviolet light-induced (6-4) photoproducts, *Nature*, **361**(6410), pp. 371–374.
8. Park, H. W., et al. (1995). Crystal structure of DNA photolyase from Escherichia coli, *Science*, **268**(5219), pp. 1866–1872.
9. Mees, A., et al. (2004). Crystal structure of a photolyase bound to a CPD-like DNA lesion after in situ repair, *Science*, **306**(5702), pp. 1789–1793.
10. Maul, M. J., et al. (2008). Crystal structure and mechanism of a DNA (6-4) photolyase, *Angew. Chem. Int. Ed. Engl.*, **47**(52), pp. 10076–10080.
11. Hitomi, K., et al. (2009). Functional motifs in the (6-4) photolyase crystal structure make a comparative framework for DNA repair photolyases and clock cryptochromes, *Proc. Natl. Acad. Sci. U S A*, **106**(17), pp. 6962–6967.
12. Kiontke, S., et al. (2011). Crystal structures of an archaeal class II DNA photolyase and its complex with UV-damaged duplex DNA, *EMBO J.*, **30**(21), pp. 4437–4449.
13. Pokorny, R., et al. (2008). Recognition and repair of UV lesions in loop structures of duplex DNA by DASH-type cryptochrome, *Proc. Natl. Acad. Sci. U S A*, **105**(52), pp. 21023–21027.
14. Sancar, G. B., et al. (1987). Action mechanism of Escherichia coli DNA photolyase. 3. Photolysis of the enzyme-substrate complex and the absolute action spectrum, *J. Biol. Chem.*, **262**(1), pp. 492–498.
15. Kim, S. T., and Sancar, A. (1991). Effect of base, pentose, and phosphodiester backbone structures on binding and repair of pyrimidine dimers by Escherichia coli DNA photolyase, *Biochemistry*, **30**(35), pp. 8623–8630.
16. Langenbacher, T., et al. (1997). Substrate and temperature dependence of DNA photolyase repair activity examined with ultrafast spectroscopy, *J. Am. Chem. Soc.*, **119**(43), pp. 10532–10536.
17. MacFarlane, A. W., and Stanley, R. J. (2003). Cis-Syn thymidine dimer repair by DNA photolyase in real time, *Biochemistry*, **42**(28), pp. 8558–8568.
18. Kao, Y.-T., et al. (2005). Direct observation of thymine dimer repair in DNA by photolyase, *Proc. Natl. Acad. Sci. U S A*, **102**(45), pp. 16128–16132.

19. Zhao, X. D., et al. (1997). Reaction mechanism of (6-4) photolyase, *J. Biol. Chem.*, **272**(51), pp. 32580–32590.
20. Glas, A. F., et al. (2009). Crystal structure of the T(6-4)C lesion in complex with a (6-4) DNA photolyase and repair of UV-induced (6-4) and dewar photolesions, *Chem. Eur. J.*, **15**(40), pp. 10387–10396.
21. Hitomi, K., et al. (2001). Role of two histidines in the (6-4) photolyase reaction, *J. Biol. Chem.*, **276**(13), pp. 10103–10109.
22. Joseph, A., Prakash, G., and Falvey, D. E. (2000). Model studies of the (6-4) photoproduct photolyase enzyme: laser flash photolysis experiments confirm radical ion intermediates in the sensitized repair of thymine oxetane adducts, *J. Am. Chem. Soc.*, **122**(45),pp. 11219–11225.
23. Borg, O. A., Eriksson, L. A., and Durbeej, B. (2007). Electron-transfer induced repair of 6-4 photoproducts in DNA: a computational study, *J. Phys. Chem. A*, **111**(12), pp. 2351–2361.
24. Yamamoto, J., et al. (2009). Role of the carbonyl group of the (6-4) photoproduct in the (6-4) photolyase reaction, *Biochemistry*, **48**(39), pp. 9306–9312.
25. Domratcheva, T., and Schlichting, I. (2009). Electronic structure of (6-4) DNA photoproduct repair involving a non-oxetane pathway, *J. Am. Chem. Soc.*, **131**(49), pp. 17793–17799.
26. Sadeghian, K., et al. (2010). Theoretical study on the repair mechanism of the (6-4) photolesion by the (6-4) photolyase, *J. Am. Chem. Soc.*, **132**(45), pp. 16285–16295.
27. Saxena, C., Sancar, A., and Zhong, D. (2004). Femtosecond dynamics of DNA photolyase: energy transfer of antenna initiation and electron transfer of cofactor reduction, *J. Phys. Chem. B*, **108**(46), pp. 18026–18033.
28. Li, J., et al. (2010). Dynamics and mechanism of repair of ultraviolet-induced (6-4) photoproduct by photolyase, *Nature*, **466**(7308), pp. 887–890.
29. Chang, C.-W., et al. (2010). Ultrafast solvation dynamics at binding and active sites of photolyases, *Proc. Natl. Acad. Sci. U S A*, **107**(7), pp. 2914–2919.
30. Wang, H. Y., et al. (2007). Protein dynamics control the kinetics of initial electron transfer in photosynthesis, *Science*, **316**(5825), pp. 747–750.
31. Kao, Y. T., et al. (2007). Femtochemistry in enzyme catalysis: DNA photolyase, *Cell Biochem. Biophys.*, **48**(1), pp. 32–44.

32. Liu, Z., et al. (2011). Dynamics and mechanism of cyclobutane pyrimidine dimer repair by DNA photolyase, *Proc. Natl. Acad. Sci. U S A,* **108**(36), pp. 14831–14836.

33. Hassanali, A. A., Zhong, D., and Singer, S. J. (2011). An AIMD study of the CPD repair mechanism in water: reaction free energy surface and mechanistic implications, *J. Phys. Chem. B,* **115**(14), pp. 3848–3859.

34. Masson, F., et al. (2009). A QM/MM investigation of thymine dimer radical anion splitting catalyzed by DNA photolyase, *ChemPhysChem,* **10**(2), pp. 400–410.

35. Voityuk, A. A., and Rosch, N. (1997). Ab initio study on the structure and splitting of the uracil dimer anion radical, *J. Phys. Chem. A,* **101**(44), pp. 8335–8338.

36. Durbeej, B., and Eriksson, L. A. (2000). Thermodynamics of the photoenzymic repair mechanism studied by density functional theory, *J. Am. Chem. Soc.,* **122**(41), pp. 10126–10132.

37. Harrison, C. B., O'Neil, L. L., and Wiest, O. (2005). Computational studies of DNA photolyase, *J. Phys. Chem. A,* **109**(32), pp. 7001–7012.

38. Masson, F., et al. (2008). Computational study of thymine dimer radical anion splitting in the self-repair process of duplex DNA, *J. Am. Chem. Soc.,* **130**(11), pp. 3443–3450.

39. Saettel, N. J., and O. Wiest, O. (2001). DFT study of the [2+2] cycloreversion of uracil dimer anion radical: waters matter, *J. Am. Chem. Soc.,* **123**(11), pp. 2693–2694.

40. Hassanali, A. A., Zhong, D., and Singer, S. J. (2011). An AIMD study of CPD repair mechanism in water: role of solvent in ring splitting, *J. Phys. Chem. B,* **115**(14), pp. 3860–3871.

41. Tachikawa, H., and Kawabata, H. (2008). Interaction between thymine dimer and flavin-adenine dinucleotide: a DFT and direct ab initio molecular dynamics study, *J. Phys. Chem. B,* **112**(24), pp. 7315–7319.

42. Payne, G., and A. Sancar, A. (1990). Absolute action spectrum of E-FADH2 and E-FADH2-MTHF forms of Escherichia coli DNA photolyase, *Biochemistry,* **29**(33), pp. 7715–7727.

43. Ramsey, A. J., Alderfer, J. L., and Jorns, M. S. (1992). Energy transduction during catalysis by Escherichia coli DNA photolyase, *Biochemistry,* **31**(31), pp. 7134–7142.

44. Thiagarajan, V., et al. (2011). Kinetics of cyclobutane thymine dimer splitting by DNA photolyase directly monitored in the UV, *Proc. Natl. Acad. Sci. U S A,* **108**(23), pp. 9402–9407.

45. Kao, Y. T., et al. (2012). Dynamics and mechanism of DNA repair in a biomimetic system: flavin-thymine dimer adduct, *J. Am. Chem. Soc.*, **134**(3), pp. 1501–1503.
46. Kao, Y.-T., et al. (2008). Ultrafast dynamics of flavins in five redox states, *J. Am. Chem. Soc.*, **130**(39), pp. 13132–13139.
47. Yamamoto, J., Tanaka, Y., and Iwai, S. (2009). Spectroscopic analysis of the pyrimidine(6-4)pyrimidone photoproduct: insights into the (6-4) photolyase reaction, *Org. Biomol. Chem.*, **7**(1), pp. 161–166.
48. Antony, J., Medvedev, D. M., and Stuchebrukhov, A. A. (2000). Theoretical study of electron transfer between the photolyase catalytic cofactor FADH(-) and DNA thymine dimer, *J. Am. Chem. Soc.*, **122**(6), pp. 1057–1065.
49. Medvedev, D., and Stuchebrukhov, A. A. (2001). DNA repair mechanism by photolyase: electron transfer path from the photolyase catalytic cofactor FADH(-) to DNA thymine dimer, *J. Theo. Biol.*, **210**(2), pp. 237–248.
50. Prytkova, T. R., Beratan, D. N., and Skourtis, S. S. (2007). Photoselected electron transfer pathways in DNA photolyase, *Proc. Natl. Acad. Sci. U S A*, **104**(3), pp. 802–807.
51. Acocella, A., Jones, G. A., and Zerbetto, F. (2010). What is adenine doing in photolyase?, *J. Phys. Chem. B*, **114**(11), pp. 4101–4106.
52. Liu, Z., et al. (2012). Electron tunneling pathways and role of adenine in repair of cyclobutane pyrimidine dimer by DNA photolyase, *J. Am. Chem. Soc.*, **134**(19), pp. 8104–8114.
53. Seidel, C. A. M., Schulz, A., and Sauer, M. H. M. (1996). Nucleobase-specific quenching of fluorescent dyes .1. Nucleobase one-electron redox potentials and their correlation with static and dynamic quenching efficiencies, *J. Phys. Chem.*, **100**(13), pp. 5541–5553.
54. Gindt, Y. M., et al. (2005). Substrate binding modulates the reduction potential of DNA photolyase, *J. Am. Chem. Soc.*, **127**(30), pp. 10472–10473.
55. Zheng, X. H., Garcia, J., and Stuchebrukhov, A. A. (2008). Theoretical study of excitation energy transfer in DNA photolyase, *J. Phys. Chem. B*, **112**(29), pp. 8724–8729.
56. Kodali, G. (2009). Excited state electronic properties of DNA photolyase and fluorescent nucleobase analogues (FBA): an experimental and theoretical study, *PhD Thesis*, Temple University.
57. Chang, C.-W., et al. (2010). Mapping solvation dynamics at the function site of flavodoxin in three redox states, *J. Am. Chem. Soc.*, **132**(36), pp. 12741–12747.

58. Hopfield, J. J. (1974). Electron-transfer between biological molecules by thermally activated tunneling, *Proc. Natl. Acad. Sci. U S A*, **71**(9), pp. 3640–3644.

59. Page, C. C., et al. (1999). Natural engineering principles of electron tunnelling in biological oxidation-reduction, *Nature*, **402**(6757), pp. 47–52.

60. Gray, H. B., and Winkler, J. R. (2005). Long-range electron transfer, *Proc. Natl. Acad. Sci. U S A*, **102**(10), pp. 3534–3539.

61. Oevering, H., et al. (1987). Long-range photoinduced through-bond electron-transfer and radiative recombination via rigid nonconjugated bridges - distance and solvent dependence, *J. Am. Chem. Soc.*, **109**(11), pp. 3258–3269.

62. Gray, H. B., and Winkler, J. R. (2003). Electron tunneling through proteins, *Q. Rev. Biophys.*, **36**(3), pp. 341–372.

63. Wilson, E. K., et al. (1997). An exposed tyrosine on the surface of trimethylamine dehydrogenase facilitates electron transfer to electron transferring flavoprotein: kinetics of transfer in wild-type and mutant complexes, *Biochemistry*, **36**(1), pp. 41–48.

64. Twitchett, M. B., et al. (1997). Intramolecular electron transfer kinetics of a synthetic flavocytochrome c, *J. Am. Chem. Soc.*, **119**(2), pp. 435–436.

65. Roth, J. P., et al. (2004). Oxygen isotope effects on electron transfer to O_2 probed using chemically modified flavins bound to glucose oxidase, *J. Am. Chem. Soc.*, **126**(46), pp. 15120–15131.

Chapter 4

Photoactive Yellow Protein: Converting Light into a Metastable Structural Change

Marie Louise Groot[a] and Klaas J. Hellingwerf[a,b]

[a] *VU University Amsterdam, Department of Sciences, LaserLab Amsterdam, De Boelelaan 1081, 1081 HV Amsterdam, The Netherlands*
[b] *University of Amsterdam, Swammerdam Institute for Life Sciences, Science Park 904,1098 WS Amsterdam, The Netherlands*
m.l.groot@vu.nl

Two questions at the forefront of biophysical sciences are biological sensing and energy conversion. Photoactive yellow protein is at the crossing point of these two topics as it converts light energy into a structural change in the process of biological light sensing. This bacterial photosensor is an excellent model system to study how a protein achieves such a function as it is relatively small and very stable. Over the years crystallography, spectroscopy, and multiscale modeling techniques have been applied to study the first step in the signal transduction process that it catalyzes—the ultrafast isomerization of the *p*-coumaric acid chromophore intrinsic to PYP. This has culminated in an ever-better understanding of the mechanism of its isomerization and the role of the protein

Ultrafast Dynamics at the Nanoscale: Biomolecules and Supramolecular Assemblies
Edited by Irene Burghardt and Stefan Haacke

ISBN 978-981-4745-33-8 (Hardcover), 978-981-4745-34-5 (eBook)
www.panstanford.com

in this process. Here we provide a review of the current state of knowledge on this issue.

4.1 Introduction

Signal transduction proteins provide an important role in the communication of a biological cell with its environment. These cells sense external stimuli, such as light or small molecules, and signal their presence to other proteins and are thus an important element in their functioning. A full understanding, with atomic details, of how these proteins achieve such a result and to what extent quantum effects or protein conformational changes play a role in these processes is for most proteins still elusive. However, with the advancement of ultrafast spectroscopic and structural techniques and developments in multiscale modeling, our knowledge of enzymology has greatly advanced over the past few years and will undoubtedly continue to do so. Here, we will discuss how such a combined theoretical and experimental approach by researchers in the field has advanced the understanding of the signal transduction process in photoactive yellow protein (PYP) [1].

PYP is a photosensory receptor, that is, a member of one of a set of protein families that can generate a biological signal upon the absorption of a photon of visible electromagnetic radiation [2]. Several different families of these photosensory receptors have been characterized, each with their own specific (chromophore) structure and mechanism of primary photochemistry. Best known are the rhodopsins, in which the absorption of green light leads to Z/E isomerization of their retinal chromophore, and the LOV domains that show blue-light-induced covalent adduct formation between their flavin chromophore and a nearby cysteine. PYP is a member of the xanthopsin family of photoreceptors and acts as a photosensory receptor for a photophobic response in *Halorhodospira halophila* [3].

This protein displays an exceptional chemical and photochemical stability and is therefore an excellent model system to study how a light absorption event gives rise to the structural change that is necessary to form a signaling state. PYP is a relatively small protein (125 amino acids), containing a 4-hydroxy cinnamic acid

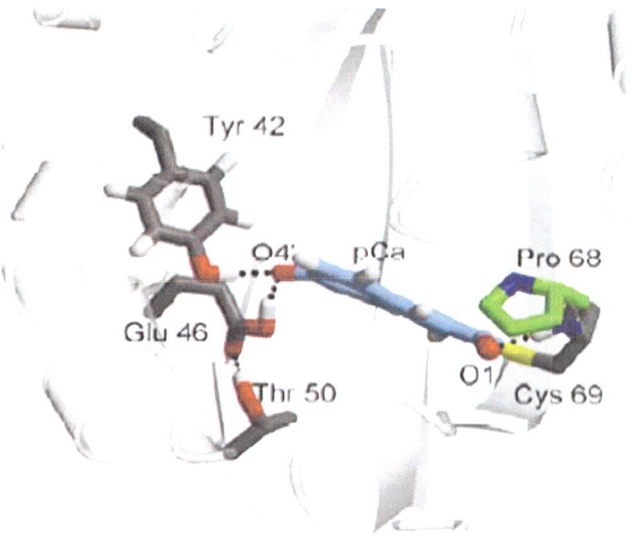

Figure 4.1 The chromophore-binding pocket of wt-PYP. The *p*-coumaric acid (pCa) and the neighboring hydrogen-bonded residues are represented as sticks. The phenolate oxygen atom is labeled as O4′ and the carbonyl oxygen as O1. Proline 68 is indicated in green color. The phenolate ring of pCa is restrained by a hydrogen bond network including residues Glu46 and Tyr42 and indirectly Thr50 and Arg52 (not shown).

(or *p*-coumaric acid [pCa]) chromophore, which is thiol-esterified to a cysteine residue, C69 of apo-PYP [4]. The chromophore is buried inside the main hydrophobic core of this α-/β-fold protein as a phenolate anion [5], stabilized by hydrogen bonding with E46 and Y42 [6] in the *trans*-configuration [7]. The pCa carbonyl group is hydrogen-bonded with the nitrogen atom of the peptide bond between P68 and C69. The chromophore is therefore part of an intricate hydrogen bond network also involving multiple amino acids that keeps pCa in a fixed conformation [8, 9] (Fig. 4.1).

Light absorption by the chromophore of PYP induces a photocycle of which the first step is the formation of a *cis*-configuration of the pCa chromophore within a few picoseconds [10]. This first step in the photocycle is followed by the formation of various red-shifted intermediates and, with a rate of approximately 10^3 s^{-1}, the transfer

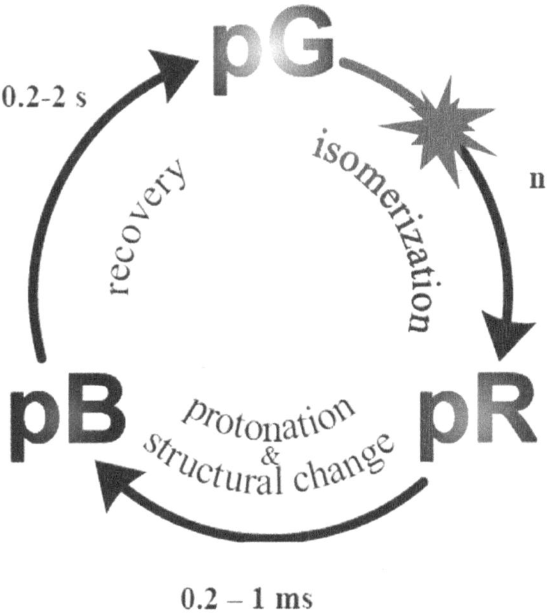

Figure 4.2 Photocycle of photoactive yellow protein. Upon light absorption by the chromophore in the ground state, a state denoted as pG, the protein enters a photocycle that consists of isomerization of the chromophore into the *cis*-conformer, pR, followed by protonation of the chromophore and a structural change of the protein, pB. After about a second the ground state is recovered. pR is also denoted as I_1, and intermediate earlier than I_1, I_0, has been identified (see text) [34].

of a proton from E46 to the chromophore [11, 12], to form the first of a series of blue-shifted intermediates that is accompanied by the (partial) unfolding of the protein forming the signaling state of PYP [13]. A complex thermal recovery leads to reformation of the ground state within 1 s [14–18] (see Fig. 4.2).

Only inside the protein a long-lived isomerized state is formed: In solution or in the denatured protein pCa photoionizes or forms twisted *trans*-conformations that relax back to the *trans*-ground state, but it does not form stable *cis*-photoproducts at all [19–22], demonstrating the role of the protein in the formation of a long-lived photoproduct. In agreement with this, recently a model system was described in which it is shown that a cyclodextrin ring

PHYSICAL CONSTANTS

Note. All entries are given to five significant figures where these are known. Otherwise the error is given in brackets.

Quantity	*Symbol*	*Value*
Speed of light in vacuum	c	2.9979×10^{8} m/s
Planck's constant	h	4.1356×10^{-24} GeV s
Planck's constant reduced	$\hbar \equiv h/2\pi$	6.5821×10^{-25} GeV s
	$\hbar c$	1.9733×10^{-16} GeV m
	$(\hbar c)^2$	3.8938×10^{-32} GeV^2 m^2
Electron charge (magnitude)	e	1.6022×10^{-19} C
Fine structure constant	$\alpha \equiv e^2/4\pi\varepsilon_0\,\hbar c$	$7.2974 \times 10^{-3} = 1/137.04$
Electron mass	m_e	0.51100 MeV/c^2
Proton mass	m_p	0.93827 GeV/c^2
W boson mass	M_W	80.385(0.015) GeV/c^2
Z boson mass	M_Z	91.188(0.002) GeV/c^2
Higgs boson mass	M_H	125.7(0.4) GeV/c^2
Fermi coupling constant	$G_F/(\hbar c)^3$	1.1664×10^{-5} GeV^{-2}
Strong coupling constant	$\alpha_s(M_Z c^2)$	0.1185(0.0006)

CONVERSION FACTORS

$1\ \mathrm{eV} = 1.6022 \times 10^{-19}\ \mathrm{J}$	$1\ \mathrm{eV/c^2} = 1.7827 \times 10^{-36}\ \mathrm{kg}$
$1\ \mathrm{fermi} = 1\ \mathrm{fm} \equiv 10^{-15}\ \mathrm{m}$	$1\ \mathrm{barn} = 1\ \mathrm{b} \equiv 10^{-28}\ \mathrm{m}^2$

NATURAL UNITS

Throughout the book we have used natural units, $\hbar = c = 1$. They can be converted back to mks units by the method discussed in Section 1.5. The following conversion factors are often useful.

Time	$1\ \mathrm{GeV}^{-1} = 6.5821 \times 10^{-25}\ \mathrm{s}$
Length	$1\ \mathrm{GeV}^{-1} = 0.19733\ \mathrm{fm}$
Cross-section	$1\ \mathrm{GeV}^{-2} = 0.38938\ \mathrm{mb}$

system can—but only partially—replace the apoprotein [23]. Site-directed mutagenesis has been used in several studies to reveal the role of individual amino acids in the isomerization process [24–33]. None of these mutants displayed an increased quantum yield of photoisomerization: so far it appears that evolution has optimized the protein environment of pCa in PYP to achieve the highest isomerization quantum yield. These studies do, however, shed light on the question, How do the amino acid residues in the wild-type (wt) protein promote efficient isomerization of the PYP chromophore?

Earlier reviews on the structure and function of PYP have been published over the years [34–44]. This review is focused on the initial ultrafast events in PYP: The formation of a stable isomerized ground state of the pCa chromophore, the role of the amino acid residues near pCa in this process, and how they modulate the reaction via hydrogen bond interactions and localized charges, as revealed by spectroscopic and X-ray diffraction studies. Also, we will discuss some forms in which this understanding of the initial events in PYP can be exploited for practical application.

In doing so, we will review the literature and discuss results from multiscale modeling, (ultrafast) crystallography, and ultrafast spectroscopy. To generate more insight into one of these experimental tools, we first provide here a description of pump–probe spectroscopy protocols, including a few PYP-specific issues.

4.2 Ultrafast Experiments: How to . . . ?

To follow fast processes in time, a high time resolution and a method that is sensitive to these processes are needed. In ultrafast spectroscopy a coupled pair of short laser pulses is used to induce a reaction, by promoting a chromophore from its ground state to an electronically excited state, and to probe a reaction, by measuring changes in the chromophore absorption spectrum, respectively (cf. Chapter 1). Short laser pulses are nowadays usually obtained from Ti:sapphire oscillators in combination with a Ti:sapphire regenerative amplifier (see Refs. [45] for a general description of pump–probe experiments and [46] for a detailed review of the pump–probe setup used in most of the PYP studies published by our group). Pulses of typically around 100 fs, but also as short as 10 fs, can routinely be produced. These pulses are centered around 800 nm due to the Ti:sapphire lasing medium. The absorption spectrum of pCa in PYP peaks at 440 nm, extending from ~380–480 nm (see Fig. 4.2). pCa can therefore be excited at 400 nm by frequency doubling of the 800 nm laser pulses in a nonlinear crystal (usually beta barium borate [BBO], BaB_2O_4). However, at this (and more blue) excitation wavelengths the probability of a two-photon excited electron detachment process, that is, photoionization of the chromophore leading to a free electron in the protein cavity [47–49], is larger than when excitation wavelengths of 450 nm or longer are used. The latter wavelengths can be generated with more elaborate nonlinear mixing schemes in a noncollinear parametric oscillator [46]. Induced absorption changes can be followed by converting a second 800 nm pulse into a pulse with a broad white light spectrum by focusing either the fundamental or a wavelength-converted pulse on a CaF_2 or sapphire plate. The high pulse energy in the plate induces self-phase modulation, leading to the appearance of new frequencies to the high- and low-energy sides of the original pulse and thus to a broadening of its spectrum. Using a 800 nm pulse and a CaF_2 plate, a usable spectrum from ~380 nm up to 750 nm can be generated. Around 800 nm the fundamental interferes; therefore, when a smooth spectrum at longer wavelength is required, a 1000 nm pulse can be used to generate a smooth spectrum between 500 and 900 nm [48, 50].

Experiments in the midinfrared (mid-IR) spectral range (between 3 and 10 μm) can be performed to probe both chromophore structural changes and protein signals via changes in their vibrational mode transitions. This requires the generation of a probe tunable in the 3–10 μm region via nonlinear mixing schemes that involve difference frequency generation of two pulses in the 1–2 μm range (see Ref. [46]).

In all cases, the pump and probe beams are spatially and temporally overlapped in the sample. After interrogation of the sample, the probe light is dispersed in a spectrograph and detected on a detector, for example, a 256-element diode array for experiments in the visible or a HgCdT array for detection in the mid-IR. The polarization of the excitation pulse must be set at the magic angle with respect to the probe pulse (54.7°) to avoid interference of anisotropy effects with the population dynamics. The repetition rate of the laser depends on the type of laser system and is in our case 1 kHz, which allows for the readout of all detectors at every shot. The time delay between a pump and probe pulse pair, which are derived from one single laser pulse, can be varied by sending one of the pulses over a variable delay line. By increasing the difference in optical path length between the two pulses in steps of, for example, 3 μm, the probe pulse is delayed each time by 10 fs. Correspondingly larger steps are taken for longer delay times, and typically a time window in the order of 6 ns is covered. The excitation intensity must be kept as low as possible to avoid nonlinear excitation processes [48] and is usually constrained in PYP experiments by a ≤50 nJ energy per pulse focused on a 100-μm-diameter spot. The photocycle lifetime of PYP is ~1 s; therefore, at a 1 kHz repetition frequency of the activating pulses, care must be taken to place a fresh part of the sample in the laser focus for every laser pulse. As these pulses arrive every 1 ms, devices like a Lissajous scanner, which move the sample perpendicular to the laser beam, have to be used [46]. To ensure a linear illumination of the sample volume, the concentration of the PYP sample has to be adjusted to an optical density (OD) of ~0.3–0.5 at the absorption maximum of 445 nm by adjusting the concentration and/or the path length of the sample cell. The integrity of the sample can be determined by measuring its absorption spectrum before and after experiments. However, a better practice is to measure several complete datasets consecutively and inspect them on consistency, instead of averaging each time delay point during an extended period. One dataset will then consist of a matrix of absorbance values, with a size of X time points and Y wavelengths, where Y is determined by the number of detectors and X by an appropriate sampling of the time region of interest.

The two-dimensional datasets can be analyzed by global analysis, in which it is assumed that the data over the full matrix can be described by a minimal model that consists of a sum of exponential decays, or is of a more complex nature [51]. The public domain software package Glotaran allows for the interactive inspection and fitting of data to superposition models [52].

4.3 Timescale and Mechanism of Isomerization

4.3.1 Timescales

Upon excitation of the pCa chromophore with light, an excited state is formed. The resulting pCa absorption difference spectra are characterized by absorption to higher-lying electronic states on the blue side and stimulated emission on the red side of the bleached ground-state absorption. The stimulated emission shifts to 495 nm in less than 0.25 ps [53] and shows—apart from a modest narrowing on a picosecond timescale—hardly any spectral dynamics [54–56]. The narrowing has been ascribed to small-scale excited-state reorganization and vibrational relaxation of the molecule [54–56]; however, it could also reflect evolution along the twisting coordinate, as locked variants of the chromophore lacked this spectral dynamics [56]. The emission decay is multiphasic, and depending on the quality of the fluorescence decay or pump–probe data, up to 4 time constants are required to fit the time traces: 0.25, 1.7, 7.7, and 49 ps for fluorescence emission decay recorded with a streak camera [53]; 0.1, 0.75, 3.3, and 40 ps for fluorescence decay recorded via upconversion [56]; and 0.74, 1.6, and 12 for pump–probe signals [24], to give a few examples. The major part, >90%, of the decay occurs with the <10 ps time constants.

Part of the decay is to the ground state, part to the first ground-state intermediate (GSI) of the photocycle, denoted as I_0, and part to a GSI that decays in a few picoseconds to the ground state, pG (see Fig. 4.2), discussed in more detail later). The I_0 minus pG absorption difference spectrum is characterized by a red-shifted product absorption peaking at ~510 nm (see Fig. 4.2). The relative yield of I_0 formation associated with each of the time constants varies, with the process with the fastest rate of ~0.6–0.7 ps having a ~twofold higher I_0 yield than the 2.8 ps rate [24, 47]. The longer-lived component of 20–40 ps is small in amplitude (<10%) and therefore its relative yield is difficult to estimate in global analysis.

Both a sequential model, in which each short-lived excited state relaxes into the next longer-living state, and a parallel model, in which the excited states decay independently, gives an equally good fit to the data. It is therefore up to now unclear whether the different time constants and their different reactivity are representative of a relaxation process in the excited state (i.e., an evolution from a more toward a less productive minimum on the excited-state surface) or whether they are due to heterogeneity or inhomogeneity due to small structural differences between proteins [57].

With a time constant of 0.8–1.5 ns [24, 47] a next photocycle intermediate is formed with an absorption difference maximum located at 480 nm, in between the absorption maximum of the ground state (pG) and of I_0. Originally, the partial overlap between I_0, I_1, and pG made it difficult to estimate the quantum yield of the I_0 to I_1 transition. However, femtosecond mid-IR absorption difference spectra showed a ~30% decay of the bleached ground-state bands with the ES to I_0 transition, which remained constant upon formation of I_1. This clearly showed that the quantum yield of I_0 formation is ~30% and that of the I_0 to I_1 transition ~100%. The 30% I_1 yield is in good agreement with estimates of the quantum yield made on intermediates in the later part of the photocycle [58]. Nearly all studies on the quantum yield of PYP converge on this value (except for some very early papers, for example, Ref. [59]) and a recent paper from the group of van Thor reports a significantly lower value based on results obtained with femtosecond spectroscopy [49]. Therefore, the consensus view is that the quantum efficiency of signaling state formation is determined by the success of the first step in the photocycle, the formation of I_0, occurring on a picosecond timescale.

The quantum yield of 30% of the I_0 and I_1 states implies that 70% of the excited pCa chromophores do not enter the photocycle by making a stable *cis*-isomerized pCa ground state but instead return to the pCa *trans*-ground state. A large fraction of these molecules (50%–100%) does this via a transient ground state that has a lifetime of about 6 ps, denoted as GSI. This state was first identified by overpopulating this state via a pump-dump-probe sequence of laser pulses [47, 60]. Subsequently, it was shown to be inherent to the deactivation pathway [61]. Its visible absorption spectrum is slightly red-shifted to that of the pG state and its IR difference spectrum resembles that of the I_0 and I_1 intermediates but without a $+1667/-1650\ cm^{-1}$ band shift, signaling breaking of the hydrogen bond between the pCa carbonyl and the nitrogen atom of the peptide bond between P68 and C69 [61]. The nature of the GSI state is a matter of debate: It could either represent a vibrationally hot pCa ground state, consistent with a red-shifted absorption spectrum, or represent a pCa with a distorted spatial configuration due to evolution along the twisting coordinate that occurred in the excited state. The similarity between the IR difference spectra of GSI, I_0, and I_1 provide support for the latter interpretation [61]. Interestingly, the hydrogen bond to the carbonyl of pCa is intact in GSI, suggesting a key role for the breaking of this bond in the successful entry of the protein into the photocycle [61], in line with quantum mechanics/molecular mechanics (QM/MM) simulations of Groenhof et al. [62, 63].

4.3.2 Isomerization

The structural changes in chromophores and proteins in the excited state and in the I_0 and I_1 states cannot be determined by optical spectroscopic methods. Especially in the I_1 state the chromophore is characterized optically by only a modest shift from its ground-state absorption maximum. A spectral decomposition of the absorption difference spectra showed this to be only ~10 nm [64]. Nevertheless, the chromophore has been shown to have isomerized from *trans* to *cis* in both I_0 and I_1 states by a combination of structure sensitive techniques, mainly X-ray diffraction experiments on crystals of PYP and IR spectroscopy. These experiments showed that whereas in the ground state the chromophore is in a deprotonated *trans*-form, negatively charged, and possibly stabilized by the positive Arg52 residue [65], in I_1 it is in a *cis*-conformation created via isomerization about the C-2=C-3 double bond of the chromophore tail. The carbonyl oxygen of the chromophore had rotated away from its original position in which it forms a hydrogen bond to the amide nitrogen of Cys69, to the other side of the chromophore [11, 66, 67]. The structure of I_0 proved harder to determine, as the short lifetime of this state makes it difficult to cryotrap, for crystallographic and

Fourier transform infrared spectroscopy (FTIR) studies, or time-resolve in X-ray Laue diffraction experiments. However, several cryotrapped X-ray studies, as well as time-resolved X-ray studies, have been published, claiming to have resolved the I_0 state [68–71]. The excited-state structural changes have so far only been accessible via analysis of vibrational mode spectra obtained via femtosecond IR spectroscopy [61, 72, 73], impulsive excitation of vibrational wave packets [74], or femtosecond Raman spectroscopy [75].

4.3.2.1 Excited state

The hydrogen bond of Glu46 with the phenolic oxygen was shown to upshift in the excited state and thus be slightly weaker than in the ground state and in the subsequent I_0 and I_1 states [61, 72, 73], possibly providing the chromophore with the necessary freedom to start twisting along an isomerization coordinate.

Downshift of a phenol ring bend mode at $\sim$1155 cm^{-1} was observed [61, 72–75] and discussed to be in line with a change in bond length as a consequence of reduction of the C-2=C-3 bond order, presumed to lead up to isomerization [74]. In the E46Q mutant this downshift had a similar magnitude in the IR absorption difference spectra [61], though it was not observed in stimulated Raman experiments [75]. In E46Q the hydrogen bond between the phenolic oxygen and Gln is weaker; X-ray crystallography has indicated the difference in the position of the hydrogen atom in wt-PYP and E46Q in the ground state [76, 77]. Another phenol ring mode at 1555 cm^{-1} was observed in the Raman measurements and shown to be sensitive to the hydrogen-bonding network around the phenolic oxygen of the chromophore. It was more strongly coupled to the S_0-S_1 transition in wt-PYP than in E46Q-PYP, but upon excitation of the wt-PYP chromophore, the hydrogen-bonding network immediately rearranged to a structure similar to that in E46Q PYP within the time resolution of 150 fs [75].

4.3.2.2 I_0 state

The discussion with regard to the structure of the I_0 state focuses on the angle of the C-2,C-3 bonds (whether the C=C bond is distorted, *trans*, or *cis*) and the rotation angle of the carbonyl group of the chromophore. Initially, a 80° double-bond torsion angle between C(2) and C(3) was reported, based on crystallographic analysis of a photocycle intermediate trapped at 150 K [68], indicative of an incomplete isomerization reaction around this double bond in the first step of the photocycle. In a subsequent study a complete isomerization of this bond in the chromophore was reported [69]. Both studies reported a flipping of the thioester linkage in conjunction with a 166° or full 180° rotation of the thioester carbonyl relative to the plane of the aromatic ring [68, 69]. In two recent time-resolved X-ray crystallographic studies this discussion continued, as one reported I_0 to have a torsion angle of 85° around its C(2), C(3) bond, right in between the *cis*- and *trans*-values (169° in pG and 1° in I_1) [70], and the other study reported I_0, based on density functional theory (DFT)-optimized structures and kinetic arguments, to be in a strained *cis*-conformation with angles only 30° displaced from the classic *cis*-conformation [71]. In both studies the pCa carbonyl was found to be rotated away from its position in pG, but now only by 90°, while the hydrogen bond between the carbonyl and Cys69 remained intact. This 90° rotation of the carbonyl is therefore only half of that observed in cryotrapped intermediates [68, 69]. Apparently, either in the cryotrapped experiments a later intermediate was trapped, or under the conditions of the time-resolved experiments rotation of the carbonyl along the chromophore axis is slower or hindered. QM/MM simulations of the early steps in isomerization predicted a full rotation of the carbonyl with I_0 formation [62]. Upon returning from the excited state to the S_0 state, the chromophore was found to relax to a *cis*-conformation, which also involved flipping of the thioester linkage in which the carbonyl oxygen atom rotated 180° clockwise [62]. In time-resolved mid-IR studies [61, 72] one of the few positive (i.e., product) bands in the absorption difference spectra, a band at 1667 cm^{-1} with a negative counterpart at 1650 cm^{-1}, was assigned [67, 78] to an upshift of the pCa carbonyl stretch mode due to the loss of a hydrogen bond in I_0 and I_1. As no, or hardly any, evolution in the frequency of this product band occurred with the I_0 to I_1 transition, it appears that these experiments support a sequence of events where the hydrogen bond of the carbonyl and Cys69 is broken with I_0 formation and where the carbonyl assumes a position similar to that in I_1. Polarized time-resolved mid-IR spectra were also shown to be consistent with a *cis*-structure with a flipped-over carbonyl group [73]. The assignment of bands in an IR difference spectrum to modes of particular molecular groups is a critical issue. The assignment of the $+1667/-1650\ cm^{-1}$ feature to the pCa carbonyl mode was strengthened by a femtosecond IR study on proteins in which mutations to residue P68 were made. This yielded proteins in which the hydrogen bond(s) to the carbonyl varied in strength, resulting in consistent shifts of the bands at 1667 cm^{-1} and 1650 cm^{-1} [24]. However, considering the importance of this discussion it appears relevant to strengthen the IR assignment of these bands to the carbonyl of pCa by, for example, IR experiments on isotope-labeled PYP samples.

4.3.2.3 I_1 state

The new time-resolved X-ray studies also reveal more issues regarding the I_1 state. They show the I_0 intermediate to transition into a state denoted as pR_1 in the Schotte study [71] and into a mixture of two states denoted as pR_1 and I_{CT} in the Jung study [70]. According to the former, pCa in pR_1 assumes a more planar configuration, with the carbonyl–hydrogen bond with Cys69 broken and the carbonyl rotating further to the other side of the chromophore [71]. In the Jung study, it is the state denoted I_{CT} in which the carbonyl completes its excursion to the opposite side of the chromophore. In parallel, ~30% of the proteins follow another path in which the carbonyl returns back to the same side of the chromophore and where the hydrogen bond between the phenolate oxygen and Glu46 is broken, forming the state denoted as pR_1 [70]. Time-resolved IR spectra of the Glu carbonyl band showed only a minor downshift in I_0 and I_1 and thus indicated that the hydrogen bond between Glu46 and the chromophore was still intact on this timescale and was even strengthened [61, 72, 73]. This suggests that this intermediate is not formed on this timescale under the conditions of the IR experiments. In E46Q, however, where the hydrogen bond with residue 46 is slightly weaker, this bond was shown to break with the I_0 to I_1 transition, in agreement with the crystallographic data on E46Q [70]. On a 16 ns timescale, in the Schotte study a pR1-to-pR2 transition occurs in which sterical strain of the *syn* conformation of the C(2)-C(1)-S-C(b) dihedral angle is relieved via a dihedral rotation to an *anti* conformation [71]. On the same timescale a similar movement is observed in Ref. [70].

We conclude that despite several coincident observations in the two new crystallographic studies, major differences were also reported, both between each other and with earlier studies. Clearly, the level of signal to noise has evolved greatly in X-ray experiments but not yet to the level that data can be interpreted without resorting to modeling, with DFT structures in Schotte [71] and with extrapolated refinement in Jung et al. [70]. It is unclear to what extent this influences the results and how dissimilar the raw data of both studies are. Differences in the data may arise from differences in illumination conditions or in sample preparations used in the two studies [70, 71]. Further, it may be that the complicated PYP photocycle itself plays a role and leads to "unexpected" interference of other transients with the photocycle intermediates in crystals: As I_0 and I_1 have a quantum yield of 30%, 70% of the chromophores return to the ground state after excitation. A large fraction of these (~50%–100%) does this via a GSI [47] in which, according to its IR difference spectrum, the chromophore backbone is of a similar structure as in I_0 but in which the hydrogen bond between the pCa carbonyl and Cys69 is intact [61]. For PYP in solution, this GSI decays to the ground-state pG in about 6 ps [47, 61]. If this decay would be slower in crystals than in solution, it might coexist with the earliest intermediate resolved in the time-resolved X-ray studies, which shows a similar pCa structure, with the hydrogen bond of the carbonyl and Cys69 intact.

The following picture on the isomerization process in PYP emerges from the available data: Light absorption by the chromophore induces primary photochemistry in a photocycle that leads to the *cis*-configuration of this chromophore within a few picoseconds, with a quantum yield of ~0.35. The structural transition necessary for this *trans-/cis*-isomerization can only be completed in a complex reaction, in which also the hydrogen bond between the C=O group of the chromophore and the nitrogen atom of C69 is disrupted, after which the C=O group starts to rotate around the long axis of the chromophore to stabilize the isomerization of the $C_7{=}C_8$ ethylenic bond [11, 63, 79]. In parallel, a transient GSI is formed, which may have an isomerized $C_7{=}C_8$ ethylenic bond as well but has its carbonyl group still hydrogen-bonded to the nitrogen atom of the proteins' backbone [61]. This GSI then rapidly (i.e., with a half-life of 6 ps) thermally relaxes back to the stable GSI.

4.3.3 Mechanism

In principle, isomerization may occur either around the single bond or around the double bond of pCa. When embedded in the protein, the available space for the chromophore is severely restricted [80, 81]. Restrictions on the space available for the moving atoms during isomerization may necessitate a concerted movement along both single and double bond(s) in a volume-conserving mechanism, such as a bicycle-pedal (BP) movement or a hula twist (HT).

According to QM/MM simulations the double-bond mechanism is favored over the single-bond isomerization [62, 79, 82–84], whereas most crystallographic studies [66, 68, 69, 85, 86] suggest the HT mechanism. Jung et al. [70] proposed I_0 to be of a transition state nature and to bifurcate into two pathways, one in which isomerization occurs via the HT mechanism and in one via the BP mechanism, occurring in a 2:1 ratio in favor of the BP pathway. BP processes involve the simultaneous rotation of two adjacent double bonds and the HT mechanism involves the simultaneous rotation of an adjacent double and single bond.

We addressed the issue to what extent single-bond rotation is involved in the isomerization of pCa in the protein, in addition to double-bond rotation, by reconstituting apo-PYP with a chromophore analogue in which rotation across the C_4–C_7 single bond was locked with an ethane bridge, thus creating a cyclopentane ring [87]. The primary photochemistry and subsequent photocycle transitions of this protein were characterized with UV/visible transient absorption spectroscopy in the time domain from picoseconds to seconds and compared to that of protein reconstituted with pCa. The result of reconstitution with the modified chromophore was that the quantum yield of photochemistry was ~60% reduced in the locked variant and the excited-state decay dynamics of the chromophore accelerated. Furthermore, the yield of the transient GSI in the early phase of the photocycle in the rotation-locked samples was considerably higher than in the corresponding samples reconstituted with pCa. This revealed that the C_4–C_7 single-bond rotation is not critically required for the primary photochemistry of PYP, but nevertheless, rotational flexibility of the C_4–C_7 bond does contribute to increase the quantum yield of signaling state formation in PYP. Restriction of this flexibility decreases the quantum yield of product formation in PYP and dramatically increases the yield of the GSI [87].

4.4 Role of Charges in the Protein Cavity

Apart from restricting the space around the chromophore, the protein provides an environment with local charges, due to the presence of charged and polar amino acids. pCa in the ground state is in its anionic form, with a negative charge on the phenolic oxygen [11, 88]. In general, an isolated charge buried in a protein's interior will destabilize the protein structure; to be electrically neutralized, it must be coupled with a counterion. The counterion against pCa's negative charge was thought to be the cationic form of R52 at a distance of 6.34 Å from the phenolic oxygen of pCa [65]. Formation of the electronically excited singlet state through absorption of a blue photon is accompanied by a profound charge redistribution in the cinnamyl chromophore of PYP, as demonstrated with Stark spectroscopy [89]. This charge redistribution has been suggested to consist of migration of the negative charge on the phenolic oxygen toward the thioester linkage, giving the double bond of the hydrogen-bonded carbonyl group more of a single-bond character, thereby lowering the barrier for isomerization [89]. The large charge redistribution upon photoexcitation in PYP has been suggested to be stabilized by the amino acid residues around the chromophore, with a particular role for the positively charged guanidinium group of Arg52. QM/MM calculations predicted that the positive charge of R52 is crucial in favoring double-bond over single-bond isomerization via electrostatic interactions with the chromophore, which lower the energy of both the twisted DB minimum and the barrier to reach that minimum [62, 79, 82–84]. Experimental observations made on site-directed mutant derivatives of PYP in which R52 had been replaced by a noncharged isosteric amino acid [90], however, have so far not provided indications for such a stabilizing role of R52. We further investigated this [87] by experiments on the chromophore analogue, in which rotation across the C_4–C_7 single bond was locked with an ethane bridge reconstituted in the R52A mutant, and by altering the charge of R52 by varying the buffer pH. The obtained results revealed that the net charge of R52 does not determine the primary photochemistry of PYP. Indeed, this is in good agreement with the finding that Arg52 is neutral rather than charged in recent neutron diffraction experiments [91]. It appears therefore that the negative charge on pCa is stabilized solely by the hydrogen bonds with Glu46 and Tyr42.

4.5 Role of Hydrogen Bonds to the Phenol Ring and the Carbonyl Group of the Chromophore

The hydrogen bond network, which connects amino acid residues Glu46 and Tyr42 to the phenolate oxygen of the chromophore in the active site and the hydrogen bond from the carbonyl to the nitrogen atom of C69, in combination with the geometry of the protein nanospace, imposes strong steric constraints on the chromophore [68, 92, 93]. In addition, it shows a dynamic response in the excited state that possibly favors the decay into the isomerized ground state.

4.5.1 Hydrogen Bonds to the Phenol Ring

The hydrogen bonds that the phenolic oxygen accepts from Glu46 and Tyr42 are very short (cf. Fig. 4.1). X-ray crystallographic studies of PYP with better than 1.0 Å resolution [9, 68, 94] revealed 2.5–2.6 Å oxygen–oxygen distances for the hydrogen bonds donated by Tyr42 and Glu46. Normal hydrogen bonds display bond lengths of 0.95 Å for the donor OH and about 2.07 Å for hydrogen acceptor bond lengths, resulting in an average oxygen–oxygen distance of

3.02 Å [95]. QM/MM simulations [82] suggest the hydrogen bonds of the phenolic oxygen to Tyr42 and Glu46 to be the crucial factor for the ultrafast decay from the twisted double-bond minimum. The stabilization of the double-bond torsion was found to depend on the length of the Tyr42/Glu46 hydrogen bonds [96]. Shorter hydrogen bonds were suggested to significantly enhance the stabilization and again proposed to be an important factor for the successful *trans–cis*-isomerization of the chromophore in the native protein [96]. The phenolic oxygen hydrogen bonds were even proposed to have a special nature: With high-resolution neutron crystallography Yamaguchi et al. [95] managed to identify 87% of the hydrogen positions and proposed that the short hydrogen bond between pCa and E46 is a low-barrier hydrogen bond (LBHB), whereas the hydrogen bond between pCa and Y42 should be called a short ionic hydrogen bond (SIHB). In LBHBs the proton is equally shared by the donor and acceptor atoms, with respect to distance and bond strength, and thus involves migration of the bridging proton toward the acceptor oxygen. This arrangement has an effect on the electron distribution of the pCa chromophore, as it implies the OH bond to be of a partial covalent nature [97]. Furthermore, the short hydrogen bond arrangement was shown to create a hydrogen bond framework that effectively transmits conformational changes along the network and therefore may help in the isomerization and signal transduction process [98]. The LBHB nature of the short hydrogen bonds of the phenolic oxygen has been strongly refuted by Saito et al., who used QM/MM calculations and energetic consideration of the relative pKa values of Glu and pCa to argue that the short distance between the oxygen atoms of Glu46 and pCa does not imply that the proton is located precisely in between but more closely to the Glu46-O on a ~1 Å distance [97]. Possibly, the use of deuterons in Ref. [95] has interfered with analysis of the neutron crystallography data, as deuterons were shown to shift relative to hydrogens between donor and acceptor molecules [98].

The rigidity of the hydrogen bonds and of the protein nanospace has been modified by point mutations to residues Glu46, Y42, and Thr50, leading to alterations of the hydrogen bond network around the chromophore [60, 92]. The hydrogen bond coordination of the chromophore phenolate group is directly weakened (E46Q) or deleted (E46A, Y42F) by mutations to E46 and Y42, while the slightly more distant hydrogen bond between Thr50 and Tyr42 is weakened in T50V. Weakening or deletion of these hydrogen bond interactions increased the excited-state lifetime [92, 99]. Pump-probe experiments confirmed the longer excited-state lifetime to be correlated with a lower quantum yield for the isomerized I_0/I_1 states [60]. A weaker hydrogen bond between residue 46 and the phenol ring, as in mutant E46Q, does not alter the isomerization yield [60, 61]. Examination of the effect of temperature on the fluorescence dynamics of wt-PYP and several mutants revealed that the fastest excited-state decay, which is in the femtosecond to 1 ps regime in wt-PYP, is temperature independent and therefore that the reaction responsible for it, presumably a twisting around the C=C bond, could be of a coherent or barrier-less origin, whereas thermal activation is necessary for the reaction in the few picosecond to 10 ps regime [92]. Apparently, the barrier-less reaction in the case of wt-PYP effectively promotes isomerization, while modification of the potential energy surface by the introduction of point mutations introduces barriers and local minima.

In a study on a set of mutants of PYP containing all 20 side chains at the position of the active site residue 46 spectral tuning of pCa was shown to be affected by a broadening of the excited-state potential with respect to the ground-state potential, rather than by a modulating of the energy gap between the two potentials [32]. Partly, the broader excited-state potential energy surface in the wt is brought about by a weakening of the hydrogen bond(s) to the phenolic oxygen, as the hydrogen bond between E46 and the phenolic oxygen was shown to be weaker in the excited state [61, 72], but also changes in the excited-state electron redistribution in pCa stabilized or favored by the amino acid at position 46 must play a role [32]. The wt-PYP appears in the high range of "wider" excited-state potentials. Replacement of the wt residue Glu by Arg, Lys, Asp, and Pro results in even wider potential energy surfaces [32], but most likely not in a higher isomerization yield: These mutants have a significantly higher fluorescence quantum yield [32], and in light of the anticorrelation between a long excited-state lifetime (i.e., higher fluorescence quantum yield) and a higher isomerization quantum yield in the E46X mutants (see before), a low isomerization quantum yield is expected. As discussed in Philip et al. [32], Arg, Lys, and Asp would have an effect on the charge distribution in the pCa chromophore—and thus on the C–C bond order—in the S_1 state opposite to that of weakening of the hydrogen bond between residue 46 and the pCa: The Arg and Lys substitutions could reduce conjugation in the chromophore by localizing electrons on the phenolic oxygen of the pCa, while the Asp substitution would result in charge localization on its carbonyl oxygen. Therefore, it appeared that the hydrogen bonds with E46 and Y42 indeed do play a very specific role in promoting isomerization by restricting the chromophore in the ground state, allowing for more freedom in the excited state, and by affecting a specific charge distribution in pCa. Recent experiments on PYP and the mutant E46A, in protonated and deprotonated states to reduce the number of hydrogen bond interactions between the pCa phenolate oxygen and the protein, showed, however, that the negative charge on pCa stabilized by the short hydrogen bonds with E46 and Y42 affects the rate of isomerization but not the quantum yield of isomerization [64]. The longer-lived excited state in protonated and deprotonated E46A is in agreement with the higher fluorescence quantum yield, but notably, this does not correlate with a lower isomerization quantum yield, suggesting that other chromophore/protein motions control the quantum yield of isomerization.

4.5.2 Hydrogen Bonds to the Carbonyl Group

As discussed in Section 4.3.2 breaking of the hydrogen bond to the carbonyl is a key event for the isomerization, as it enables the carbonyl group to rotate along the long axis of the chromophore. Control over the isomerization yield via a modulation of the strength of this hydrogen bond would be extremely interesting and provide a solid support for breaking of the hydrogen bond as a key event in the isomerization. In Rupenyan et al. [24] we prepared a set of mutants to modulate the strength of this bond. This included mutant P68A for which a higher than wt quantum yield had been reported [100]. However, analysis of the mutants by femtosecond mid-IR spectroscopy and molecular dynamics (MD) simulations revealed a strengthening of the hydrogen bond(s) to the carbonyl, as new bonds to water molecules or other amino acids were formed, and a lower quantum yield of photochemistry. A *weakening* of hydrogen bonds in combination with an *increase* in quantum yield has not yet been observed.

In the proline 68 mutants the excited-state lifetime was not found to correlate inversely with the quantum yield. The quantum yields were observed to decrease, but the excited-state lifetimes remained almost the same, apart from the longest-lived fraction: In two of the mutants (P68A and P68G) the relative amplitude of the long lifetime doubled, while this longest time constant increased to ~40 ps [24]. This enabled the observation of significant I_0 formation on this timescale, comparable to that of the fastest phase [24], which had previously escaped analysis in the wt [15]. Combined with the information on the excited potential discussed earlier it appears that in these mutants the evolution in the excited-state potential, presumably along a twisting coordinate, is as fast and as barrier-less as in the wt, and the crossing point of the excited-state and product potential energy surfaces is reached in the same time. Then, however, the "decision" along which ground-state potential surface, the *trans* or *cis*, the evolution is continued is modified in the mutant by the stronger hydrogen bond to the carbonyl, and more chromophores decay to the *trans*-ground state.

Clearly, more residues than the ones discussed earlier play a role in facilitating isomerization, albeit not via the dynamic modulation of hydrogen bonds but by providing a fluidic, dynamic environment that provides the space necessary for the chromophore to isomerize. It is interesting to speculate whether such motions are triggered by the excitation of the chromophore and/or whether they are intrinsic to the protein structure. The latter does at least seem to be the case as concerted motions of Pro68, Tyr98, and the thioester linkage of the chromophore were observed in high-resolution pG, that is, ground state, structures and were suggested to facilitate the C=O flip during isomerization [101]. In fact this concerted motion was suggested to be essential for the functionality of the protein [101].

4.6 Role of Water Molecules

At neutral pH in aqueous solution, and in the denatured protein, pCa photoionizes. Upon illumination, it forms twisted *trans*-

conformations that relax back to the *trans*-ground state, but it does not form stable *cis*-photoproducts at all [19–22]. In the protein, where a stable isomerization product is formed at a reasonably high yield, it may be expected that the number of water molecules interacting with pCa is actively restricted. In the crystal structure of PYP at least one water molecule (water 200) is observed, which is far from the active site [102]. In a combined subpicosecond mid-IR spectroscopy and MD simulation study, we investigated the effect of replacing proline-68, positioned near but not in direct contact with the chromophore, with the neutral amino acids alanine, glycine, and valine. In the wt protein in solution, the MD simulations indicated a higher number of water molecules near the chromophore than in the crystal structure: on average one molecule near the alkylic tail of the chromophore and around three near the phenol ring up to the carbonyl group [24]. It appeared that P68 actively restrains the entrance of water molecules into the chromophore-binding pocket, close to the alkylic part of pCa, as there were one or two more water molecules in that area in all mutants. This resulted in additional hydrogen bonds to the chromophore, directly to the new residue and to the extra water molecules, and had a negative effect on the isomerization quantum yield [24]. Interestingly, the chromophore active site was found to resemble a water-like dielectric environment, as the diffusion behavior of a solvated electron inside PYP was similar to that in aqueous solvent environment [48]. This could be either because of direct interaction of the electron with the water molecules or because hydrogen bond solvation by residues in the chromophore pocket leads to electron diffusion similar as in water [48].

4.7 Emerging Picture and Open Questions

The studies reviewed earlier have revealed much detail of how PYP works. They are excellent examples of the detailed insight that can be obtained by the combined application of theoretical and computational multiscale modeling, spectroscopy, crystallography, site-directed mutagenesis, and chemical engineering. The application of all these different techniques to one specific problem leads to a synergy in which hypotheses can be formulated on the basis of one study and tested with the next. We expect that many of the still existing controversies and open questions regarding the isomerization in PYP, discussed in this review, will be resolved by applying these techniques in an even more coordinated and synergistic manner. An important example is a full account of all spectroscopically identified intermediates (like the GSI) and their dynamics, in structural studies like time-resolved diffraction and time-resolved scattering [103]. Technical progress continues to be made and will enable novel studies. For example, in our laboratory we are currently preparing for spectroscopic studies in which "inactive" and "active" proteins, that is, those ~70% that do not enter the photocycle after excitation and those ~30% that do, are being compared to see which structural details are required to switch a protein from one state to the other and on which timescale this happens. We expect that finally the efforts made by the different labs will result in a validated model with atomic detail of the light-driven events in PYP. This achievement implies better and more accurate models for protein dynamics and function in general, with applications beyond that of PYP, and it will enable the ability to control and modify the function of PYP for different applications.

However, already with the currently available knowledge, light-triggered proteins such as PYP are being used in several applications. The fact that it is possible to trigger a photoactive protein with a single photon, the generation and application of which is under highly precise special and temporal control, to let that protein make a transition from one structural state to another, has not escaped the attention of those who are interested to transfer knowledge from basic science into concrete applications. Initially these applications were in the field of advanced, optically addressable materials, in which the aspect of photoactivation could be exploited for information storage [104, 105]. These developments were largely driven by the excellent physicochemical stability of bacteriorhodopsin from the purple membranes of *Halobacterium salinarum* [106].

An important innovation in this field, however, was achieved with the introduction, and subsequent ample demonstration of the feasibility, of an approach nowadays referred to as optogenetics [107], which recently even further developed into optopharmacology [108]. In optogenetics one modulates a biological response of choice with light, via the construction and heterologous expression, of a fusion protein composed of a light-sensitive, photoactive, input domain, fused to a key regulator (domain) of the pathway of choice. This approach was developed with the application of the channel rhodopsins from *Chlamydomonas reinhartii* as very selective light-dependent triggers of neural circuits [109] in 2005 but has expanded exponentially during the past seven years, while it now also includes examples of the regulation of gene expression [110] and enzyme activity [111]. In the final part of this review, we will discuss general aspects of importance when one wants to use optically activated proteins in such applications, and discuss examples of specific applications of PYP.

4.8 Applications of PYP: General Aspects

A rich variety of light-sensitive proteins/domains is available for these applications, although not all families of photosensory receptor proteins [112] have had their light-sensitive input domains tested yet. Most popular so far have been the rhodopsins, the LOV domains (e.g., of phototropins), PYP, and the phytochromes (e.g., Refs. [113–118]). The rhodopsins, particularly channelrhodopsin, have the key advantage that they can directly generate or dissipate action potentials, but they can also be used to modulate the input signal for a bacterial kinase, controlling, for example, chemotaxis. LOV domains have the advantage that they use a chromophore that is naturally abundant in nearly all cell types, the oxidized form of flavin-adenine dinucleotide.

A limited number of key structural and functional aspects of the light-sensitive input domains of the chimeric modulator proteins are important for the design of optogenetically active proteins. First is of course their color, which in the case of phytochromes comes with such selectivity that even multichrome experiments can be envisaged [116]. Most other domains, however, have rather broad UV/visible absorbance bands with a full-width at half-maximum (FWHM) of about 50 nm. Second, the structure of the linker that connects the two domains of the chimeras is subject to the constraint of efficient transfer of structural change. For LOV domains this can be achieved either via the (de)-docking of its Jα-helix on the back surface of its central β sheet or via a rotary movement in a coiled/coil structure with the LOV domain in synchrony with its characteristic hepta-helical repeat [113]; nevertheless, the brute force approach of just screening all possible linkers may also work [114].

The third important aspect is the photochemical properties of the input domain, that is, its photochemical quantum yield and the rate of thermal recovery of the signaling state back to the ground state (Fig. 4.3). In a simple model representation of such a domain

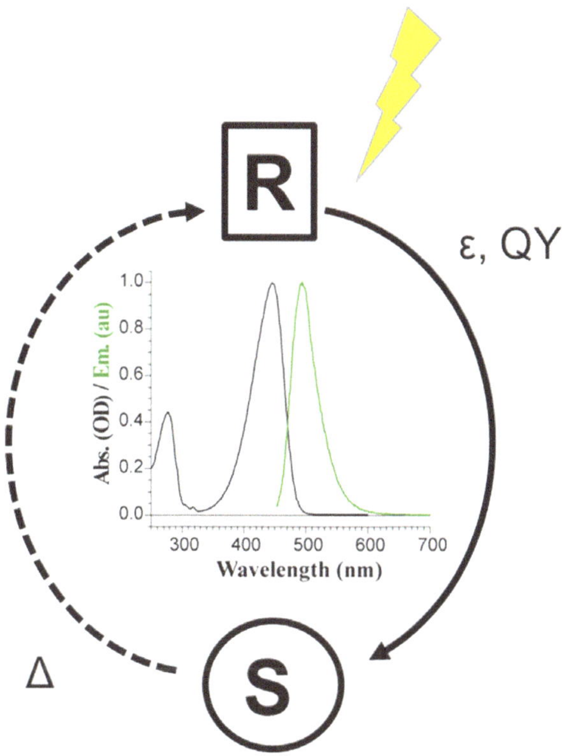

Figure 4.3 Simplified photocycle of a photoactive protein (domain), relevant for optogenetics applications. R, receptor state; S, signaling state; ε, extinction coefficient; QY, quantum yield; Δ, rate of thermal recovery. The insert shows the absorption and emission spectrum of PYP.

these two factors are the main determinants, next to the light intensity—which is under direct control of the experimenter—of the steady-state distribution over the two relevant protein states, that is, receptor and signaling states. A high quantum yield and low recovery rate lower the light intensity that is required to reach a significant population of the signaling state, whereas a too low recovery rate generates the requirement for absolute darkness in order to be able to de-populate the signaling state (which, for example, is an issue for the blue-light receptor protein YtvA from *Bacillus subtilis*).

4.9 Applications of PYP: Examples

PYP can be synthesized in living cells as well as fully in vitro [119]. This allows preparation of selected derivatives that facilitate testing of new hypotheses regarding the molecular mechanism of PYP [119].

The coumaryl-binding PAS domain that forms the structural basis of PYP has been used to create (i) a light-sensitive α-hemolysin and (ii) a light-sensitive transcriptional activator. Actually this latter work arose from a nonbiological optogenetic approach in which a light-sensitive azo-compound was attached to the helix of a leucine zipper in order to modulate its α-helical content.

The chimeric protein, which is obtained when the carboxy terminus of PYP is fused to the N-terminus of α-hemolysin from *Staphylococcus aureus,* has been used to study the rate of hemolysis of red blood cells from sheep [120]. It was observed that this rate increases by 50% when, and only when, PYP is functionally reconstituted with its coumaryl chromophore and is illuminated with 450 nm light. Woolley and collaborators [121–123] fused an N-terminally-truncated form of PYP to the carboxy-terminus of the zipper region of GCN4, the prototypical leucine-zipper-type DNA-binding protein, which is found in many eukaryotic DNA-binding proteins. The resulting fusion protein, GCN4Δ25PYP-v2, is monomeric in the dark, while blue-light irradiation causes a reversible twofold increase in its specific DNA-binding affinity [121].

Significantly, DNA binding also causes a significant (sixfold) slowing down of the recovery rate of the ground state of this fusion protein, which even further lowers the requirement for light saturation of the dark/light transition as compared to the N-terminally-truncated derivative of PYP itself [124]. Via a combination of directed and random mutagenesis [123] the performance of this fusion protein was enhanced up to 12 times in the fraction of bound DNA upon irradiation, and a tenfold slower recovery rate was achieved.

In none of these publications was the holo-PYP derivative synthesized in vivo (compare Refs. [25, 124]), even though this was reported for a related light-switchable derivative of the leucine zipper domain of GCN4 [125]. In the related but divergent **p**hotoactive yellow protein turn **o**ff/**o**n **l**abel (POOL) approach [126] expression of PYP-fusion proteins is exploited for quantitation of heterologous protein expression.

In anticipation of further applications of PYP, and to maximize transmission of structural changes between light-input and output domains, these authors recently reported construction of a circularly permutated variant of PYP [127], such that the insertion/linker site for the output domain can be selected closest to the region of the PYP protein in which the largest structural change is generated by isomerization of the coumaryl chromophore. This was achieved by inserting a GGSGGSGG linker between its amino- and carboxy-terminus and making new termini at G115 and S114. The resulting protein does show large structural changes near the introduced linker. Furthermore, to make it even more suitable for optogenetic applications, the authors tuned down the thermal recovery rate of the protein (from the millisecond to the second timescale), so the necessity of the use of too high light intensities can be avoided.

The most recently proposed application of PYP is that of a fluorogenic probe for quick, no-wash live-cell imaging of intracellular proteins [128]. To achieve this, a coumarin-derived chromophore analogue was synthesized that allows selective in vivo labeling, such that the label becomes fluorescent only after this selective labelling.

References

1. Meyer, T. E. (1985). Isolation and characterization of soluble cytochromes, ferredoxins and other chromophoric proteins from the halophilic phototrophic bacterium *Ectothiorhodospira halophila, Biochim. Biophys. Acta,* **806**, pp. 175–183.
2. van der Horst, M. A., and Hellingwerf, K. J. (2004). Photoreceptor proteins, "star actors of modern times": a review of the functional dynamics in the structure of representative members of six different photoreceptor families, *Acc. Chem. Res.*, **37**, pp. 13–20.
3. Sprenger, W. W., Hoff, W. D., Armitage, J. P., and Hellingwerf, K. J. (1993). The eubacterium Ectothiorhodospira halophila is negatively phototactic, with a wavelength dependence that fits the absorption spectrum of the photoactive yellow protein, *J. Bacteriol.*, **175**, pp. 3096–3104.
4. Van Beeumen, J. J., Devreese, B. V., Van Bun, S. M., Hoff, W. D., Hellingwerf, K. J., Meyer, T. E., McRee, D. E., and Cusanovich, M. A. (1993). Primary structure of a photoactive yellow protein from the phototrophic bacterium Ectothiorhodospira halophila, with evidence for the mass and the binding site of the chromophore, *Protein Sci.*, **2**, pp. 1114–1125.
5. Kim, M., Mathies, R. A., Hoff, W. D., and Hellingwerf, K. J. (1995). Resonance Raman evidence that the thioester-linked 4-hydroxycinnamyl chromophore of photoactive yellow protein is deprotonated, *Biochemistry,* **34**, pp. 12669–12672.
6. Borgstahl, G. E. O., Williams, D. R., and Getzoff, E. D. (1995). 1.4 Angstrom structure of photoactive yellow protein, a cytosolic photoreceptor: unusual fold, active-site, and chromophore, *Biochemistry,* **34**, pp. 6278–6287.
7. Kort, R., Vonk, H., Xu, X., Hoff, W. D., Crielaard, W., and Hellingwerf, K. J. (1996). Evidence for trans-cis isomerization of the p-coumaric acid chromophore as the photochemical basis of the photocycle of photoactive yellow protein, *FEBS Lett.*, **382**, pp. 73–78.
8. Hoff, W. D., Devreese, B., Fokkens, R., Nugteren-Roodzant, I. M., Van Beeumen, J., Nibbering, N., and Hellingwerf, K. J. (1996). Chemical reactivity and spectroscopy of the thiol ester-linked p-coumaric acid chromophore in the photoactive yellow protein from Ectothiorhodospira halophila, *Biochemistry,* **35**, pp. 1274–1281.

9. Getzoff, E. D., Gutwin, K. N., and Genick, U. K. (2003). Anticipatory active-site motions and chromophore distortion prime photoreceptor PYP for light activation, *Nat. Struct. Biol.*, **10**, pp. 663–668.
10. Gensch, T., Gradinaru, C. C., van Stokkum, I. H. M., Hendriks, J., Hellingwerf, K. J., and van Grondelle, R. (2002). The primary photoreaction of photoactive yellow protein (PYP): anisotropy changes and excitation wavelength dependence, *Chem. Phys. Lett.*, **356**, pp. 347–354.
11. Xie, A., Hoff, W. D., Kroon, A. R., and Hellingwerf, K. J. (1996). Glu46 donates a proton to the 4-hydroxycinnamate anion chromophore during the photocycle of photoactive yellow protein, *Biochemistry*, **35**, pp. 14671–14678.
12. Xie, A., Kelemen, L., Hendriks, J., White, B. J., Hellingwerf, K. J., and Hoff, W. D. (2001). Formation of a new buried charge drives a large-amplitude protein quake in photoreceptor activation, *Biochemistry*, **40**, pp. 1510–1517.
13. van Brederode, M. E., Hoff, W. D., Van Stokkum, I. H., Groot, M. L., and Hellingwerf, K. J. (1996). Protein folding thermodynamics applied to the photocycle of the photoactive yellow protein, *Biophys. J.*, **71**, pp. 365–380.
14. Meyer, T. E., Yakali, E., Cusanovich, M. A., and Tollin, G. (1987). Properties of a water-soluble, yellow protein isolated from a halophilic phototrophic bacterium that has photochemical activity analogous to sensory rhodopsin, *Biochemistry*, **26**, pp. 418–423.
15. Hoff, W. D., van Stokkum, I. H., van Ramesdonk, H. J., van Brederode, M. E., Brouwer, A. M., Fitch, J. C., Meyer, T. E., van Grondelle, R., and Hellingwerf, K. J. (1994). Measurement and global analysis of the absorbance changes in the photocycle of the photoactive yellow protein from Ectothiorhodospira halophila, *Biophys. J.*, **67**, pp. 1691–1705.
16. Hendriks, J., van Stokkum, I. H. M., and Hellingwerf, K. J. (2003). Deuterium isotope effects in the photocycle transitions of the photoactive yellow protein, *Biophys. J.*, **84**, pp. 1180–1191.
17. Hoersch, D., Otto, H., Joshi, C. P., Borucki, B., Cusanovich, M. A., and Heyn, M. P. (2007). Role of a conserved salt bridge between the PAS core and the N-terminal domain in the activation of the photoreceptor photoactive yellow protein, *Biophys. J.*, **93**, pp. 1687–1699.
18. Hendriks, J., and Hellingwerf, K. J. (2009). pH dependence of the photoactive yellow protein photocycle recovery reaction reveals a new

late photocycle intermediate with a deprotonated chromophore, *J. Biol. Chem.*, **284**, pp. 5277–5288.

19. Larsen, D. S., Vengris, M., van stokkum, I. H. M., van der Horst, M., Cordfunke, R., Hellingwerf, K. J., and van Grondelle, R. (2003). Initial photo-induced dynamics of the photoactive yellow protein chromophore in solution, *Chem. Phys. Lett.*, **369**, pp. 563–569.
20. Mataga, N., Chosrowjan, H., Shibata, Y., Imamoto, Y., Kataoka, M., and Tokunaga, F. (2002). Ultrafast photoinduced reaction dynamics of photoactive yellow protein (PYP): observation of coherent oscillations in the femtosecond fluorescence decay dynamics, *Chem. Phys. Lett.*, **352**, pp. 220–225.
21. Chosrowjan, H., Mataga, N., Shibata, Y., Imamoto, Y., and Tokunaga, F. (1998). Environmental effects on the femtosecond-picosecond fluorescence dynamics of photoactive yellow protein: chromophores in aqueous solutions and in protein nanospaces modified by site-directed mutagenesis, *J. Phys. Chem. B*, **102**, pp. 7695–7698.
22. Changenet-Barret, P., Espagne, A., Charier, S., Baudin, J. B., Jullien, L., Plaza, P., Hellingwerf, K. J., and Martin, M. M. (2004). Early molecular events in the photoactive yellow protein: role of the chromophore photophysics, *Photochem. Photobiol. Sci.*, **3**, pp. 823–829.
23. Changenet-Barret, P., Loukou, C., Ley, C., Lacombat, F., Plaza, P., Mallet, J.-M., and Martin, M. M. (2010). Primary photodynamics of a biomimetic model of photoactive yellow protein (PYP), *Phys. Chem. Chem. Phys.*, **12**, pp. 13715–13723.
24. Rupenyan, A. B., Vreede, J., van Stokkum, I. H. M., Hospes, M., Kennis, J. T. M., Hellingwerf, K. J., and Groot, M. L. (2011). Proline 68 enhances photoisomerization yield in photoactive yellow protein, *J. Phys. Chem. B*, **115**, pp. 6668–6677.
25. Joshi, C. P., Otto, H., Hoersch, D., Meyer, T. E., Cusanovich, M. A., and Heyn, M. P. (2009). Strong hydrogen bond between glutamic acid 46 and chromophore leads to the intermediate spectral form and excited state proton transfer in the Y42F mutant of the photoreceptor photoactive yellow protein, *Biochemistry*, **48**, pp. 9980–9993.
26. Imamoto, Y., Harigai, M., Morimoto, T., and Kataoka, M. (2008). Low-temperature spectroscopy of Met100Ala mutant of photoactive yellow protein, *Photochem. Photobiol.*, **84**, pp. 970–976.
27. Shimizu, N., Kamikubo, H., Yamazaki, Y., Imamoto, Y., and Kataoka, M. (2006). The crystal structure of the R52Q mutant demonstrates a role for R52 in chromophore pK(a) regulation in photoactive yellow protein, *Biochemistry*, **45**, pp. 3542–3547.

28. Rajagopal, S., Anderson, S., Srajer, V., Schmidt, M., Pahl, R., and Moffat, K. (2005). A structural pathway for signaling in the E46Q mutant of photoactive yellow protein, *Structure*, **13**, pp. 55–63.
29. Derix, N. M., Wechselberger, R. W., van der Horst, M. A., Hellingwerf, K. J., Boelens, R., Kaptein, R., and van Nuland, N. A. J. (2003). Lack of negative charge in the E46Q mutant of photoactive yellow protein prevents partial unfolding of the blue-shifted intermediate, *Biochemistry*, **42**, pp. 14501–14506.
30. Sasaki, J., Kumauchi, M., Hamada, N., Oka, T., and Tokunaga, F. (2002). Light-induced unfolding of photoactive yellow protein mutant M100L, *Biochemistry*, **41**, pp. 1915–1922.
31. Devanathan, S., Lin, S., Cusanovich, M. A., Woodbury, N., and Tollin, G. (2000). Early intermediates in the photocycle of the Glu46Gln mutant of photoactive yellow protein: femtosecond spectroscopy, *Biophys. J.*, **79**, pp. 2132–2137.
32. Philip, A. F., Nome, R. A., Papadantonakis, G. A., Scherer, N. F., and Hoff, W. D. (2010). Spectral tuning in photoactive yellow protein by modulation of the shape of the excited state energy surface, *Proc. Natl. Acad. Sci. U S A*, **107**, pp. 5821–5826.
33. Philip, A. F., Eisenman, K. T., Papadantonakis, G. A., and Hoff, W. D. (2008). Functional tuning of photoactive yellow protein by active site residue 46, *Biochemistry*, **47**, pp. 13800–13810.
34. Hellingwerf, K. J., Hendriks, J., and Gensch, T. (2003). Photoactive yellow protein, a new type of photoreceptor protein: will this "yellow lab" bring us where we want to go?, *J. Phys. Chem. A*, **107**, pp. 1082–1094.
35. Cusanovich, M. A., and Meyer, T. E. (2003). Photoactive yellow protein: a prototypic PAS domain sensory protein and development of a common signaling mechanism, *Biochemistry*, **42**, pp. 4759–4770.
36. Imamoto, Y., and Kataoka, M. (2007). Structure and photoreaction of photoactive yellow protein, a structural prototype of the PAS domain superfamily, *Photochem. Photobiol.*, **83**, pp. 40–49.
37. Larsen, D. S., and van Grondelle, R. (2005). Initial photoinduced dynamics of the photoactive yellow protein, *ChemPhysChem*, **6**, pp. 828–837.
38. Changenet-Barret, P., Espagne, A., Plaza, P., Hellingwerf, K. J., and Martin, M. M. (2005). Investigations of the primary events in a bacterial photoreceptor for photomotility: photoactive yellow protein (PYP), *New J. Chem.*, **29**, pp. 527–534.

39. Gomelsky, M., and Hoff, W. D. (2011). Light helps bacteria make important lifestyle decisions, *Trends Microbiol.*, **19**, pp. 441–448.
40. Kennis, J. T. M., and Groot, M.-L. (2007). Ultrafast spectroscopy of biological photoreceptors, *Curr. Opin. Struct. Biol.*, **17**, pp. 623–630.
41. Kyndt, J. A., Meyer, T. E., and Cusanovich, M. A. (2004). Photoactive yellow protein, bacteriophytochrome, and sensory rhodopsin in purple phototrophic bacteria, *Photochem. Photobiol. Sci.*, **3**, pp. 519–530.
42. Martin, M. M., Plaza, P., and Changenet-Barret, P. (2008). Real-time probing of fast photoinduced charge-transfer in electron donor-acceptor model compounds and in biological photosensors, *Pol. J. Chem.*, **82**, pp. 759–771.
43. Meyer, T. E., Kyndt, J. A., Memmi, S., Moser, T., Colon-Acevedo, B., Devreese, B., and Van Beeumen, J. J. (2012). The growing family of photoactive yellow proteins and their presumed functional roles, *Photochem. Photobiol. Sci.*, **11**, pp. 1495–1514.
44. van der Horst, M. A., Laan, W., Yeremenko, S., Wende, A., Palm, P., Oesterhelt, D., and Hellingwerf, K. J. (2005). From primary photochemistry to biological function in the blue-light photoreceptors PYP and AppA, *Photochem. Photobiol. Sci.*, **4**, pp. 688–693.
45. Maciejewski, A., Naskrecki, R., Lorenc, M., Ziolek, M., Karolczak, J., Kubicki, J., Matysiak, M., and Szymanski, M. (2000). Transient absorption experimental set-up with femtosecond time resolution. Femto- and picosecond study of DCM molecule in cyclohexane and methanol solution, *J. Mol. Struct.*, **555**, pp. 1–13.
46. Groot, M. L., van Wilderen, L., and Di Donato, M. (2007). Time-resolved methods in biophysics. 5. Femtosecond time-resolved and dispersed infrared spectroscopy on proteins, *Photochem. Photobiol. Sci.*, **6**, pp. 501–507.
47. Larsen, D. S., van Stokkum, I. H. M., Vengris, M., van der Horst, M. A., de Weerd, F. L., Hellingwerf, K. J., and van Grondelle, R. (2004). Incoherent manipulation of the photoactive yellow protein photocycle with dispersed pump-dump-probe spectroscopy, *Biophys. J.*, **87**, pp. 1858–1872.
48. Zhu, J., Paparelli, L., Hospes, M., Arents, J., Kennis, J. T. M., van Stokkum, I. H. M., Hellingwerf, K. J., and Groot, M. L. (2013). Photoionization and electron radical recombination dynamics in photoactive yellow protein investigated by ultrafast spectroscopy in the visible and near-infrared spectral region, *J. Phys. Chem. B*, **117**, pp. 11042–11048.

49. Lincoln, C. N., Fitzpatrick, A. E., and van Thor, J. J. (2012). Photoisomerisation quantum yield and non-linear cross-sections with femtosecond excitation of the photoactive yellow protein, *Phys. Chem. Chem. Phys.*, **14**, pp. 15752–15764.
50. Zhu, J., van Stokkum, Ivo H. M., Paparelli, L., Jones, M. R., and Groot, M. L. (2013). Early bacteriopheophytin reduction in charge separation in reaction centers of *Rhodobacter sphaeroides, Biophys. J.*, **104**, pp. 2493–2502.
51. Van Stokkum, I. H. M., Larsen, D. S., and Van Grondelle, R. (2004). Global and target analysis of time-resolved spectra, *Biochim. Biophys. Acta*, **1657**, pp. 82–104.
52. Snellenburg, J. J., Laptenok, S. P., Seger, R., Mullen, K. M., and van Stokkum, I. H. M. (2012). Glotaran: a Java-based graphical user interface for the R package TIMP, *J. Stat. Softw.*, **49**, pp. 1–22.
53. van Stokkum, I. H. M., Gobets, B., Gensch, T., van Mourik, F., Hellingwerf, K. J., van Grondelle, R., and Kennis, J. T. M. (2006). (Sub)-picosecond spectral evolution of fluorescence in photoactive proteins studied with a synchroscan streak camera system, *Photochem. Photobiol.*, **82**, pp. 380–388.
54. Hanada, H., Kanematsu, Y., Kinoshita, S., Kumauchi, M., Sasaki, J., and Tokunaga, F. (2001). Ultrafast fluorescence spectroscopy of photoactive yellow protein, *J. Lumin.*, **94**, pp. 593–596.
55. Changenet, P., Zhang, H., van der Meer, M. J., Hellingwerf, K. J., and Glasbeek, M. (1998). Subpicosecond fluorescence upconversion measurements of primary events in yellow proteins, *Chem. Phys. Lett.*, **282**, pp. 276–282.
56. Vengris, M., van der Horst, M. A., Zgrablic, G., van Stokkum, I. H. M., Haacke, S., Chergui, M., Hellingwerf, K. J., van Grondelle, R., and Larsen, D. S. (2004). Contrasting the excited-state dynamics of the photoactive yellow protein chromophore: protein versus solvent environments, *Biophys. J.*, **87**, pp. 1848–1857.
57. Kim, P. W., Rockwell, N. C., Martin, S. S., Lagarias, J. C., and Larsen, D. S. (2014). Dynamic inhomogeneity in the cyanobacterial phytochrome Cph1, *Proc. Natl. Acad. Sci. U S A*, **53**, pp. 2818–2826.
58. van Brederode, M. E., Gensch, T., Hoff, W. D., Hellingwerf, K. J., and Braslavsky, S. E. (1995). Photoinduced volume change and energy storage associated with the early transformations of the photoactive yellow protein from Ectothiorhodospira halophila, *Biophys. J.*, **68**, pp. 1101–1109.

59. Meyer, T. E., Tollin, G., Hazzard, J. H., and Cusanovich, M. A. (1989). Photoactive yellow protein from the purple phototrophic bacterium, *Ectothiorhodospira halophila*. Quantum yield of photobleaching and effects of temperature, alcohols, glycerol, and sucrose on kinetics of photobleaching and recovery, *Biophys. J.*, **56**, pp. 559–564.
60. Changenet-Barret, P., Plaza, P., Martin, M. M., Chosrowjan, H., Taniguchi, S., Mataga, N., Imamoto, Y., and Kataoka, M. (2009). Structural effects on the ultrafast photoisomerization of photoactive yellow protein. Transient absorption spectroscopy of two point mutants, *J. Phys. Chem. C*, **113**, pp. 11605–11613.
61. van Wilderen, L., Van der Horst, M. A., van Stokkum, I. H. M., Hellingwerf, K. J., van Grondelle, R., and Groot, M. L. (2006). Ultrafast infrared spectroscopy reveals a key step for successful entry into the photocycle for photoactive yellow protein, *Proc. Natl. Acad. Sci. U S A*, **103**, pp. 15050–15055.
62. Groenhof, G., Bouxin-Cademartory, M., Hess, B., De Visser, S. P., Berendsen, H. J. C., Olivucci, M., Mark, A. E., and Robb, M. A. (2004). Photoactivation of the photoactive yellow protein: why photon absorption triggers a trans-to-cis Isomerization of the chromophore in the protein, *J. Am. Chem. Soc.*, **126**, pp. 4228–4233.
63. Groenhof, G., Lensink, M. F., Berendsen, H. J. C., Snijders, J. G., and Mark, A. E. (2002). Signal transduction in the photoactive yellow protein. I. Photon absorption and the isomerization of the chromophore, *Proteins*, **48**, pp. 202–211.
64. Zhu, J. Y., Vreede, J., Hospes, M., Arents, J., Kennis, J. T. M., van Stokkum, I. H. M., Hellingwerf, K. J., and Groot, M. L. (2015). Short hydrogen bonds and negative charge in photoactive yellow protein promote fast isomerization but not high quantum yield, *J. Phys. Chem. B*, **119**, pp. 2372–2383.
65. Borgstahl, G. E., Williams, D. R., and Getzoff, E. D. (1995). 1.4 Å structure of photoactive yellow protein, a cytosolic photoreceptor: unusual fold, active site, and chromophore, *Biochemistry*, **34**, pp. 6278–6287.
66. Ren, Z., Perman, B., Srajer, V., Teng, T.-V., Pradervand, C., Bourgeois, D., Schotte, F., Ursby, T., Kort, R., Wulff, M., and Moffat, K. (2001). A molecular movie at 1.8 Å resolution displays the photocycle of photoactive yellow protein, a eubacterial blue-light receptor, from nanoseconds to seconds, *Biochemistry*, **40**, pp. 13788–13801.
67. Brudler, R., Rammelsberg, R., Woo, T. T., Getzoff, E. D., and Gerwert, K. (2001). Structure of the I1 early intermediate of photoactive yellow protein by FTIR spectroscopy, *Nat. Struct. Biol.*, **8**, pp. 265–270.

68. Genick, U. K., Soltis, S. M., Kuhn, P., Canestrelli, I. L., and Getzoff, E. D. (1998). Structure at 0.85 Å resolution of an early protein photocycle intermediate [see comments], *Nature*, **392**, pp. 206–209.

69. Kort, R., Hellingwerf, K. J., and Ravelli, R. B. G. (2004). Initial events in the photocycle of photoactive yellow protein, *J. Biol. Chem.*, **279**, pp. 26417–26424.

70. Jung, Y. O., Lee, J. H., Kim, J., Schmidt, M., Moffat, K., Šrajer, V., and Ihee, H. (2013). Volume-conserving trans–cis isomerization pathways in photoactive yellow protein visualized by picosecond X-ray crystallography, *Nat. Chem.*, **5**, pp. 212–220.

71. Schotte, F., Cho, H. S., Kaila, V. R. I., Kamikubo, H., Dashdorj, N., Henry, E. R., Graber, T. J., Henning, R., Wulff, M., Hummer, G., Kataoka, M., and Anfinrud, P. A. (2012). Watching a signaling protein function in real time via 100-ps time-resolved Laue crystallography, *Proc. Natl. Acad. Sci. U S A*, **109**, pp. 19256–19261.

72. Groot, M. L., van Wilderen, L. J. G. W., Larsen, D. S., van der Horst, M. A., van Stokkum, I. H. M., Hellingwerf, K. J., and van Grondelle, R. (2003). Initial steps of signal generation in photoactive yellow protein revealed with femtosecond mid-infrared spectroscopy, *Biochemistry*, **42**, pp. 10054–10059.

73. Heyne, K., Mohammed, O. F., Usman, A., Dreyer, J., Nibbering, E. T. J., and Cusanovich, M. A. (2005). Structural evolution of the chromophore in the primary stages of trans/cis isomerization in photoactive yellow protein, *J. Am. Chem. Soc.*, **127**, pp. 18100–18106.

74. Liu, J., Yabushita, A., Taniguchi, S., Chosrowjan, H., Imamoto, Y., Sueda, K., Miyanaga, N., and Kobayashi, T. (2013). Ultrafast time-resolved pump-probe spectroscopy of PYP by a sub-8 fs pulse laser at 400 nm, *J. Phys. Chem. B*, **117**, pp. 4818–4826.

75. Nakamura, R., Hamada, N., Abe, K., and Yoshizawa, M. (2012). Ultrafast hydrogen-bonding dynamics in the electronic excited state of photoactive yellow protein revealed by femtosecond stimulated Raman spectroscopy, *J. Phys. Chem. B*, **116**, pp. 14768–14775.

76. Anderson, S., Crosson, S., and Moffat, K. (2004). Short hydrogen bonds in photoactive yellow protein, *Acta Crystallogr. D*, **60**, pp. 1008–1016.

77. Sugishima, M., Tanimoto, N., Soda, K., Hamada, N., Tokunaga, F., and Fukuyama, K. (2004). Structure of photoactive yellow protein (PYP) E46Q mutant at 1.2 Å resolution suggests how Glu46 controls the spectroscopic and kinetic characteristics of PYP, *Acta Crystallogr. D*, **60**, pp. 2305–2309.

78. Unno, M., Kumauchi, M., Sasaki, J., Tokunaga, F., and Yamauchi, S. (2002). Resonance Raman spectroscopy and quantum chemical calculations reveal structural changes in the active site of photoactive yellow protein, *Biochemistry,* **41**, pp. 5668–5674.
79. Groenhof, G., Schafer, L. V., Boggio-Pasqua, M., Grubmuller, H., and Robb, M. A. (2008). Arginine52 controls the photoisomerization process in photoactive yellow protein, *J. Am. Chem. Soc.*, **130**, pp. 3250–3251.
80. Liu, R. S. H., and Asato, A. E. (1985). The primary process of vision and the structure of bathorhodopsin: a mechanism for photoisomerization of polyenes, *Proc. Natl. Acad. Sci. U S A*, **82**, pp. 259–263.
81. Warshel, A., and Barboy, N. (1982). Energy-storage and reaction pathways in the 1st step of the vision process, *J. Am. Chem. Soc.*, **104**, pp. 1469–1476.
82. Boggio-Pasqua, M., Robb, M. A., and Groenhof, G. (2009). Hydrogen bonding controls excited-state decay of the photoactive yellow protein chromophore, *J. Am. Chem. Soc.*, **131**, pp. 13580–13581.
83. Virshup, A. M., Punwong, C., Pogorelov, T. V., Lindquist, B. A., Ko, C., and Martinez, T. J. (2009). Photodynamics in complex environments: ab initio multiple spawning quantum mechanical/molecular mechanical dynamics, *J. Phys. Chem. B*, **113**, pp. 3280–3291.
84. Ko, C., Virshup, A. M., and Martinez, T. J. (2008). Electrostatic control of photoisomerization in the photoactive yellow protein chromophore: ab initio multiple spawning dynamics, *Chem. Phys. Lett.*, **460**, pp. 272–277.
85. Imamoto, Y., Kataoka, M., and Liu, R. S. H. (2002). Mechanistic pathways for the photoisomerization reaction of the anchored, tethered chromophore of the photoactive yellow protein and its mutants, *Photochem. Photobiol.*, **76**, pp. 584–589.
86. Anderson, S., Srajer, V., and Moffat, K. (2004). Structural heterogeneity of cryotrapped intermediates in the bacterial blue light photoreceptor, photoactive yellow protein, *Photochem. Photobiol.*, **80**, pp. 7–14.
87. Stahl, A. D., Hospes, M., Singhal, K., van Stokkum, I., van Grondelle, R., Groot, M. L., and Hellingwerf, K. J. (2011). On the involvement of single-bond rotation in the primary photochemistry of photoactive yellow protein, *Biophys. J.*, **101**, pp. 1184–1192.
88. Imamoto, Y., Mihara, K., Hisatomi, O., Kataoka, M., Tokunaga, F., Bojkova, N., and Yoshihara, K. (1997). Evidence for proton transfer from

Glu-46 to the chromophore during the photocycle of photoactive yellow protein, *J. Biol. Chem.*, **272**, pp. 12905–12908.

89. Premvardhan, L. L., van der Horst, M. A., Hellingwerf, K. J., and van Grondelle, R. (2003). Stark spectroscopy on photoactive yellow protein, E46Q, and a nonisomerizing derivative, probes photo-induced charge motion, *Biophys. J.*, **84**, pp. 3226–3239.
90. Changenet-Barret, P., Plaza, P., Martin, M. M., Chosrowjan, H., Taniguchi, S., Mataga, N., Imamoto, Y., and Kataoka, M. (2007). Role of arginine 52 on the primary photoinduced events in the PYP photocycle, *Chem. Phys. Lett.*, **434**, pp. 320–325.
91. Yamaguchi, S., Kamikubo, H., Kurihara, K., Kuroki, R., Niimura, N., Shimizu, N., Yamazaki, Y., and Kataoka, M. (2009). Low-barrier hydrogen bond in photoactive yellow protein, *Proc. Natl. Acad. Sci. U S A*, **106**, pp. 440–444.
92. Mataga, N., Chosrowjan, H., Shibata, Y., Imamoto, Y., and Tokunaga, F. (2000). Effects of modification of protein nanospace structure and change of temperature on the femtosecond to picosecond fluorescence dynamics of photoactive yellow protein, *J. Phys. Chem. B*, **104**, pp. 5191–5199.
93. Mataga, N., Chosrowjan, H., and Taniguchi, S. (2004). Investigations into the dynamics and mechanisms of ultrafast photoinduced reactions taking place in photoresponsive protein nanospaces (PINS), *J. Photochem. Photobiol. C*, **5**, pp. 155–168.
94. Fisher, S. Z., Anderson, S., Henning, R., Moffat, K., Langan, P., Thiyagarajan, P., and Schultz, A. J. (2007). Neutron and X-ray structural studies of short hydrogen bonds in photoactive yellow protein (PYP), *Acta Crystallogr. D*, **63**, pp. 1178–1184.
95. Yamaguchi, S., Kamikubo, H., Kurihara, K., Kuroki, R., Niimura, N., Shimizu, N., Yamazaki, Y., and Kataoka, M. (2009). Low-barrier hydrogen bond in photoactive yellow protein, *Proc. Natl. Acad. Sci. U S A*, **106**, pp. 440–444.
96. Gromov, E. V., Burghardt, I., Koppel, H., and Cederbaum, L. S. (2012). Native hydrogen bonding network of the photoactive yellow protein (PYP) chromophore: impact on the electronic structure and photoinduced isomerization, *J. Photochem. Photobiol. A*, **234**, pp. 123–134.
97. Saito, K., and Ishikita, H. (2012). Energetics of short hydrogen bonds in photoactive yellow protein, *Proc. Natl. Acad. Sci. U S A*, **109**, pp. 167–172.

98. Sigala, P. A., Tsuchida, M. A., and Herschlag, D. (2009). Hydrogen bond dynamics in the active site of photoactive yellow protein, *Proc. Natl. Acad. Sci. U S A*, **106**, pp. 9232–9237.
99. Devanathan, S., Lin, S., Cusanovich, M. A., Woodbury, N., and Tollin, G. (2001). Early photocycle kinetic behavior of the E46A and Y42F mutants of photoactive yellow protein: femtosecond spectroscopy, *Biophys. J.*, **81**, pp. 2314–2319.
100. Takeshita, K., Imamoto, Y., Kataoka, M., Mihara, K., Tokunaga, F., and Terazima, M. (2002). Structural change of site-directed mutants of PYP: new dynamics during pR state, *Biophys. J.*, **83**, pp. 1567–1577.
101. Coureux, P. D., Fan, Z. P., Stojanoff, V., and Genick, U. K. (2008). Picometer-scale conformational heterogeneity separates functional from nonfunctional states of a photoreceptor protein, *Structure*, **16**, pp. 863–872.
102. Kandori, H., Iwata, T., Hendriks, J., Maeda, A., and Hellingwerf, K. J. (2000). Water structural changes involved in the activation process of photoactive yellow protein, *Biochemistry*, **39**, pp. 7902–7909.
103. Kim, T. W., Lee, J. H., Choi, J., Kim, K. H., van Wilderen, L. J., Guerin, L., Kim, Y., Jung, Y. O., Yang, C., Kim, J., Wulff, M., van Thor, J. J., and Ihee, H. (2012). Protein structural dynamics of photoactive yellow protein in solution revealed by pump-probe X-ray solution scattering, *J. Am. Chem. Soc.*, **134**, pp. 3145–3153.
104. Birge, R. R. (1992). Protein-based optical computing and memories, *Computer*, **25**, pp. 56–67.
105. Hellingwerf, K. J., van Thor, J. J., Hoff, W. D., and Hendriks, J. (1998). Nanodevices from photobiological meterials, In *Nanotechnology: Towards a Molecular Construction Kit*, Wolde, A. T., ed., The Hague, the Netherlands: STT Netherlands Study Center for Technology Trends, pp. 265–272.
106. Oesterhe, D., and Stoecken, W. (1971). Rhodopsin-like protein from purple membrane of Halobacterium-halobium, *Nat. New Biol.*, **233**, pp. 149–152.
107. Miller, G. (2006). Optogenetics: shining new light on neural circuits, *Science*, **314**, pp. 1674–1676.
108. Nevola, L., Martín-Quirós, A., Eckelt, K., Camarero, N., Tosi, S., Llobet, A., Giralt, E., and Gorostiza, P. (2013). Light-regulated stapled peptides to inhibit protein–protein interactions involved in clathrin-mediated endocytosis, *Angew. Chem. Int. Ed.*, **52**, pp. 7704–7708.

109. Boyden, E. S., Zhang, F., Bamberg, E., Nagel, G., and Deisseroth, K. (2005). Millisecond-timescale, genetically targeted optical control of neural activity, *Nat. Neurosci.*, **8**, pp. 1263–1268.

110. Levskaya, A., Chevalier, A. A., Tabor, J. J., Simpson, Z. B., Lavery, L. A., Levy, M., Davidson, E. A., Scouras, A., Ellington, A. D., Marcotte, E. M., and Voigt, C. A. (2005). Engineering *Escherichia coli* to see light: these smart bacteria "photograph" a light pattern as a high-definition chemical image, *Nature*, **438**, pp. 441–442.

111. Wu, Y. I., Frey, D., Lungu, O. I., Jaehrig, A., Schlichting, I., Kuhlman, B., and Hahn, K. M. (2009). A genetically encoded photoactivatable Rac controls the motility of living cells, *Nature*, **461**, pp. 104–108.

112. van der Horst, M. A., and Hellingwerf, K. J. (2004). Photoreceptor proteins, "star actors of modern times": a review of the functional dynamics in the structure of representative members of six different photoreceptor families, *Acc. Chem. Res.*, **37**, pp. 13–20.

113. Piatkevich, K. D., Subach, F. V., and Verkhusha, V. V. (2013). Engineering of bacterial phytochromes for near-infrared imaging, sensing, and light-control in mammals, *Chem. Soc. Rev.*, **42**, pp. 3441–3452.

114. Moglich, A., and Moffat, K. (2010). Engineered photoreceptors as novel optogenetic tools, *Photochem. Photobiol. Sci.*, **9**, pp. 1286–1300.

115. Shimizu-Sato, S., Huq, E., Tepperman, J. M., and Quail, P. H. (2002). A light-switchable gene promoter system, *Nat. Biotechnol.*, **20**, pp. 1041–1044.

116. Tabor, J. J., Levskaya, A., and Voigt, C. A. (2011). Multichromatic control of gene expression in *Escherichia coli*, *J. Mol. Biol.*, **405**, pp. 315–324.

117. Krauss, U., Lee, J., Benkovic, S. J., and Jaeger, K. E. (2010). LOVely enzymes: towards engineering light-controllable biocatalysts, *Microb. Biotechnol.*, **3**, pp. 15–23.

118. Hegemann, P., and Nagel, G. (2013). From channelrhodopsins to optogenetics, *EMBO Mol. Med.*, **5**, pp. 173–176.

119. Gordon, W. R., Bang, D., Hoff, W. D., and Kent, S. B. H. (2013). Total chemical synthesis of fully functional photoactive yellow protein, *Bioorgan. Med. Chem.*, **21**, pp. 3436–3442.

120. Ui, M., Tanaka, Y., Araki, Y., Wada, T., Takei, T., Tsumoto, K., Endo, S., and Kinbara, K. (2012). Application of photoactive yellow protein as a photoresponsive module for controlling hemolytic activity of staphylococcal alpha-hemolysin, *Chem. Commun.*, **48**, pp. 4737–4739.

121. Morgan, S. A., Al-Abdul-Wahid, S., and Woolley, G. A. (2010). Structure-based design of a photocontrolled DNA binding protein, *J. Mol. Biol.*, **399**, pp. 94–112.

122. Morgan, S. A., and Woolley, G. A. (2010). A photoswitchable DNA-binding protein based on a truncated GCN4-photoactive yellow protein chimera, *Photochem. Photobiol. Sci.*, **9**, pp. 1320–1326.

123. Fan, H. Y., Morgan, S. A., Brechun, K. E., Chen, Y. Y., Jaikaran, A. S. I., and Woolley, G. A. (2011). Improving a designed photocontrolled DNA-binding protein, *Biochemistry*, **50**, pp. 1226–1237.

124. Kyndt, J. A., Vanrobaeys, F., Fitch, J. C., Devreese, B. V., Meyer, T. E., Cusanovich, M. A., and Van Beeumen, J. J. (2003). Heterologous production of *Halorhodospira halophila* holo-photoactive yellow protein through tandem expression of the postulated biosynthetic genes, *Biochemistry*, **42**, pp. 965–970.

125. Zhang, F. Z., Timm, K. A., Arndt, K. M., and Woolley, G. A. (2010). Photocontrol of coiled-coil proteins in living cells, *Angew. Chem. Int. Ed.*, **49**, pp. 3943–3946.

126. Kim, Y., Ganesan, P., and Ihee, H. (2013). High-throughput instant quantification of protein expression and purity based on photoactive yellow protein turn off/on label, *Protein Sci.*, **22**, pp. 1109–1117.

127. Kumar, A., Burns, D. C., Al-Abdul-Wahid, M. S., and Woolley, G. A. (2013). A circularly permuted photoactive yellow protein as a scaffold for photoswitch design, *Biochemistry*, **52**, pp. 3320–3331.

128. Hori, Y., Norinobu, T., Sato, M., Arita, K., Shirakawa, M., and Kikuchi, K. (2013). Development of fluorogenic probes for quick no-wash live-cell imaging of intracellular proteins, *J. Am. Chem. Soc.*, **135**, pp. 12360–12365.

Chapter 5

Energy Transfer Mechanisms in Nanobiohybrid Structures Based on Quantum Dots and Photosensitive Membrane Proteins

Svetlana V. Sizova,[a,b] Vladimir A. Oleinikov,[a,b] Nicolas Bouchonville,[c] Michael Molinari,[c] Pavel Samokhvalov,[a] Alyona Sukhanova,[a,c] and Igor Nabiev[a,c]

[a]*Laboratory of Nano-Bioengineering, National Research Nuclear University MEPhI (Moscow Engineering Physics Institute), 31 Kashirskoe sh., 115409 Moscow, Russian Federation*
[b]*Shemyakin and Ovchinnikov Institute of Bioorganic Chemistry, Russian Academy of Sciences, 117871 Moscow, Russian Federation*
[c]*Laboratoire de Recherche en Nanosciences LRN-EA4682, Université de Reims, Champagne-Ardenne, 51100 Reims, France*
igor.nabiev@univ-reims.fr

The integration of novel nanomaterials with highly functional biological molecules has numerous advanced applications, including optoelectronics, biosensing, imaging, and energy harvesting. This review summarizes recent progress in understanding the mechanisms of energy transfer between semiconductor nanocrystal (so-called quantum dots [QDs]) and photosensitive proteins in heterostructures, such as hybrids of semiconductor nanocrystals

Ultrafast Dynamics at the Nanoscale: Biomolecules and Supramolecular Assemblies
Edited by Irene Burghardt and Stefan Haacke

ISBN 978-981-4745-33-8 (Hardcover), 978-981-4745-34-5 (eBook)
www.panstanford.com

with purple membranes containing bacteriorhodopsin (bR) or with photosynthetic reaction centers (RCs). Understanding of these mechanisms should enable prediction of the possible ways to improve the biological function of biomolecules by means of their assembling with QDs and develop novel functional materials with controlled photonic properties and applications. The possible mechanisms of energy transfer from QDs to photochromic biomolecules are discussed and correlated with experimental data. The principles of hybrid structures engineering, donor/acceptor parameters affecting both energy transfer efficiency and biological function, and functionality of these hybrid structures are described. New nanobiohybrid materials are shown to have advanced implications for optoelectronics, photonics, and photovoltaics due to the ability of nanocomponents of these materials for efficient energy harvesting, conversion, and transfer of additional energy to Biosystems, thus making them working more efficiently.

5.1 Introduction

One of the trends of modern nanobiotechnology is the use of natural and modified biological molecules in various devices, such as photonic and electronic ones. The photosynthetic membrane pigment–protein complexes of photosynthetic reaction centers (RCs) and bacteriorhodopsin (bR, an evolutionarily ancient protein) play a key role in natural photoconverting processes. Energy transfer between quantum dots (QDs) of different sizes and photochromic proteins in nanobiohybrid structures is not just of basic scientific interest; it also has implications for light-harvesting purposes [9, 61, 71].

The unique photochromic, electrochromic, and photoelectric properties of photosynthetic RCs and bR, as well as their relatively high resistance to environmental factors, make them promising objects of nanobiotechnology. High energy efficiency and quantum sensitivity of RCs and bR and the possibility to control their properties by means of chemical modification or genetic engineering make it possible to optimize the molecular characteristics of photovoltaic cells created on their basis [48]. Immobilization of these proteins in a polymer matrix to form thin oriented films, into which various semiconductors and other conductive components can also be included, allows one to assess the possible practical application of RC proteins and bR [5, 6, 32, 33, 44, 73]. However, we know that these light-sensitive structures cannot use the energy of UV photons (which they normally do not absorb), because these photons may destroy the chromophore. Changing the properties of these complexes by forming hybrid nanostructures that include QDs is a way to overcome this obstacle. This is currently one of the most important lines of research in this field.

QDs are an important class of materials that offer great promises for a wide range of applications from energy conversion to biomedicine [4, 64, 65, 69, 71]. With their high absorption capacity, QDs may serve as donors of energy for light-sensitive proteins. The use of QDs as additional antennas allows the use of light in the spectral region where RCs and bR do not absorb it (from deep UV to blue). QDs convert the nonusable incident light into photons with longer wavelengths that are efficiently absorbed by RCs and bR, which therefore improves the biological function of the proteins.

To date, the interaction of QDs with photochromic natural molecules has not been studied in detail. Better understanding the energy transfer mechanisms in heterostructures of, for example, QD–bR and QD–RC types, especially the mechanism of nonradiative transfer, will offer new possibilities for advanced applications of hybrid nanobiomaterials, including optical switches, transmission and storage elements, nanosensors, nanomechanical devices, light sources, solar cells, and biofuel elements.

5.2 Possible Energy Transfer Mechanisms in Nanobiohybrid Structures Based on Photosensitive Biomolecules and Quantum Dots: Nonradiative and Radiative Energy Transfer

Energy transfer from excited atoms, ions, and molecules to unexcited ones is a phenomenon that frequently occurs in nature. It has been established that QDs are ideal for devices based on fluorescence resonance energy transfer due to their narrow emission and broad excitation spectra. Furthermore, the large size of QDs compared to organic dyes makes it possible to design a configuration where multiple acceptors interact with a single donor, which enhances the efficiency and, hence, increases the sensitivity [63]. Energy transfer between QDs of different sizes in closely packed assemblies may also be applied to light harvesting [11, 71] similar to the natural photosynthetic process [45]. Hybrid organicinorganic nanomaterials have recently attracted much attention due to the appealing possibility of synergetic properties exploiting the best of both components [3, 4, 38, 58, 59].

Traditionally, two types of excitation energy transfer are distinguished, nonradiative and radiative [1].

Radiative transfer is a well-known physical process in which a photon is emitted by an excited particle and is then absorbed by an unexcited particle. Radiative transfer depends only on optical properties of the sample, including the size of the optical cell, the optical density of the sample at the excitation and fluorescence wavelengths, and the exact geometry of the exciting and emitted light beams.

Another type of energy transfer is the nonradiative one; it can be mediated by an electronic charge or a free electron–hole pair that has been created by an incident photon and migrates carrying both energy and charge. Nonradiative energy transfer causes de-excitation of the donor and simultaneous excitation of the acceptor. The resonant interaction in which the excitation energy is transferred without direct charge transfer is an example of a nonradiative energy transfer. The energy transfer takes place through absorption and emission of a virtual photon. Time-resolved photoluminescence (PL) measurements showing a reduction of the donor lifetime accompanied by an increase in the acceptor PL lifetime provide further evidence for nonradiative energy transfer. Since nonradiative energy transfer is faster than radiative one, the donor decay is increased and, hence, the donor PL lifetime is reduced [21].

It should be noted that the study of energy transfer yields intermolecular information that cannot be obtained from the parameters of solvent relaxation, the reaction in the excited state, fluorescence quenching and polarization, etc., from one component alone. All these spectral properties of fluorophores depend on their interactions with solvent molecules in the surrounding solvation shell. Except for the effect on the spectral properties of the donor and acceptor, these interactions are less important for energy transfer. Nonradiative energy transfer is effective at distances of up to 50 Å. Solvent molecules or a macromolecule located within this distance have only a little effect on the efficiency of energy transfer, which mainly depends on the distance between the donor and the acceptor.

As noted above, the interaction of QDs with photochromic molecules has been poorly studied to date. Therefore, the development of both experimental and theoretical methods for studying the mechanisms of energy transfer in photosystems is of primary importance.

The initial step of energy transfer from the donor to the acceptor is resonant or isoenergetic transfer, which means that only the transfer from the excited donor level (E_{D}) into the vibronically excited state of the acceptor ($E_{\mathrm{A}}^{1,*}$) is possible (Fig. 5.1a). The excess of the vibronic energy is rapidly dissipated, leaving the acceptor molecule in the state E_{A}^{1}. Förster theory holds for donor–acceptor distances where the pure dipole–dipole interaction prevails. For substantially smaller distances, interactions between the dipole and quadrupole moments eventually leading to the so-called Dexter transfer should be considered [61]. For even smaller distances (less than 1 nm), the wave functions begin to overlap noticeably, and strong coupling between the excitons, as well as the tunneling of electrons and holes, may take place.

Förster resonance energy transfer (FRET) involves the nonradiative transfer of the excitation energy from an excited donor fluorophore (D) (after the absorption of a photon) to a ground-state acceptor fluorophore (A) brought into close proximity, which can radiatively emit a lower-energy photon [15, 22]. FRET processes are driven by dipole–dipole interactions and depend on the degree of the spectral overlap between the donor PL and the acceptor absorption and on the sixth power of the separation distance between the donor and the acceptor (r). Within the Förster formalism, the rate of nonradiative energy transfer is given by the equation

$$k_{\mathrm{DA}} = \frac{B \times Q_{\mathrm{D}} I}{\tau_{\mathrm{D}} r^6} = \left(\frac{1}{\tau_{\mathrm{D}}}\right) \times \left(\frac{R_0}{r}\right)^6 \tag{5.1}$$

where τ_{D} is the excited-state radiative lifetime of the isolated donor and R_0 is the Förster radius corresponding to the rate of FRET equal to the rate of radiative decay ($k_{\mathrm{DA}} = 1/\tau_{\mathrm{D}}$). R_0 is a function of the refractive index of the medium (n_{D}), the Avogadro number (N_{A}), donor PL quantum yield (Q_{D}), the overlap integral (I), and the parameter $k\rho$ that depends on the relative orientation of the donor and acceptor dipoles; it is expressed as

$$R_0 = \left(\frac{9000\,(\ln 10)\,k_{\rho}^2\,Q_{\mathrm{D}}}{N_{\mathrm{A}} 128\pi^5 n_{\mathrm{D}}^4} I\right)^{1/6}. \tag{5.2}$$

The spectral overlap integral I defined as

$$I = \int J\,(\lambda) d\lambda = \int \mathrm{PL}_{\text{D-corr}}\,(\lambda) \times \lambda^4 \times \varepsilon_{\mathrm{A}}\,(\lambda)\,d\lambda \tag{5.3}$$

is a quantitative measure of the donor–acceptor spectral overlap over all wavelengths λ, where $\mathrm{PL}_{\text{D-corr}}$ and ε_{A} represent the donor emission (normalized dimensionless spectrum) and the spectrum of the acceptor absorption extinction coefficient, respectively. These equations imply that a PL quantum yield measurement is necessary for deriving an accurate estimate of the Förster distance. The FRET efficiency (E) defined as

$$E = \frac{k_{\mathrm{DA}}}{k_{\mathrm{DA}} + \tau_{\mathrm{D}}^{-1}} = \frac{R_0^6}{R_0^6 + r^6} \tag{5.4}$$

accounts for the fraction of excitons nonradiatively transferred from the donor to the acceptor, that is, the fraction of energy transfer per donor excitation event. This can be experimentally measured by monitoring the changes in the donor and/or acceptor fluorescence intensities or changes in the fluorescent lifetimes of the fluorophores as follows:

$$E = 1 - \tau'_{\mathrm{D}}/\tau_{\mathrm{D}} \tag{5.5}$$

where τ'_{D} and τ_{D} are the donor fluorescence lifetimes in the presence and absence of the acceptor, respectively, or as

$$E = 1 - \mathrm{PL}'_{\mathrm{D}}/\mathrm{PL}'_{\mathrm{D}} \tag{5.6}$$

where PL'_D and PL_D are the donor fluorescence intensities with and without the acceptor, respectively.

During FRET, energy transfer is expressed as a decrease in the donor fluorescence intensity and shortening of its exciton lifetime; this may, in turn, be coupled with an enhancement of the acceptor fluorescence and an increase in its effective exciton lifetime. For more detailed derivation of Förster's energy transfer formalism and the methods for monitoring FRET, see Refs. [15, 22]. For the configuration where a single donor interacts with multiple acceptors brought into close proximity, we have extended the above analysis upon the assumption that all acceptors are equally separated from a central donor and derived the following equation for the FRET efficiency:

$$E(n, r) = \frac{nR_0^6}{nR_0^6 + r^6} \tag{5.7}$$

where n is the number of acceptors surrounding a single donor [14, 50].

In the configuration where a single donor interacts with a flat monolayer of acceptors, the FRET efficiency is proportional to the fourth power of the inverse distance between the donor and the surface of the thin layer.

Finally, if there is a bulk layer of acceptors, the FRET efficiency is proportional to third power of the inverse distance between the donor and the thick layer surface (Fig. 5.1b) [61].

Förster theory agrees with experimental data where the conditions of its applicability are met. A common criterion is the condition of the applicability of the theory of weak interactions and rapid vibrational relaxation. At distances 5–10 times smaller than R_0, the weak interaction theory is inapplicable.

As noted above, the distance between the donor and the acceptor is an important factor in Förster's formalism. The nanoparticle (NP) shape may affect the energy transfer. Guyot-Sionnest and coworkers demonstrated that the dipole moment of CdSe nanocrystals scales linearly with the crystallite diameter [7]. Furthermore, Li and Alivisatos [46] showed that the permanent dipole moment of CdSe nanorods increases linearly with volume. The energy transfer depends on the inverse of the sixth power of r for a point acceptor

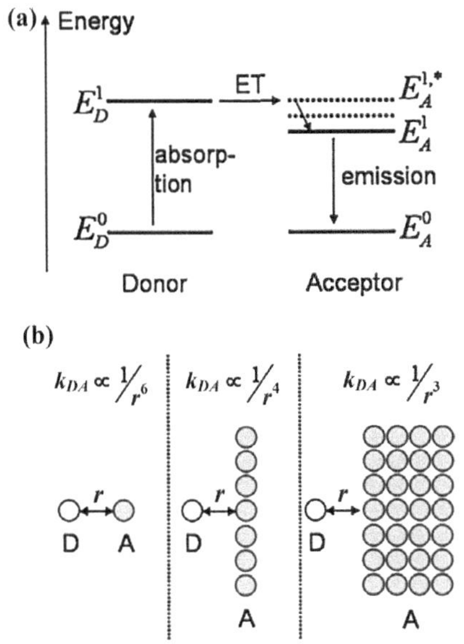

Figure 5.1 Schematic presentation of FRET. (a) After absorption of a photon by the donor, the energy is transferred resonantly, that is, isoenergetically, to a vibronically excited state $E_A^{1,*}$ of the acceptor. The excess vibronic energy is quickly dissipated, rendering FRET a one-way process. Finally, fluorescence is emitted from the acceptor. (b) Proportionality of the energy transfer rate to the 6th, 4th, or 3rd power of the inverse distance (r) between the donor (D) and a single acceptor (A), a thin layer of acceptors, or a bulk layer of acceptors, respectively. Reprinted from Ref. [61], Copyright (2011), with permission from Elsevier.

(Fig. 5.1b). If the donor fluorophore is in front of a thin plane of acceptors, the FRET rate becomes proportional to the inverse of the fourth power of r. If the donor is in front of a bulk layer of acceptors, the FRET rate is even less sensitive to the distance, decreasing proportionally to only the inverse of the third power of r (Fig. 5.1b) [61].

Recently, Schleifenbaum et al. [67] reported the configuration of one donor dye molecule interacting with multiple acceptor dye molecules. Medintz et al. [50] estimated the potential of luminescent semiconductor QDs for the development of hybrid inorganic-bio-receptor-sensing materials. They demonstrated the use of luminescent CdSe/ZnS QDs as energy donors in assays based on fluorescence resonance energy transfer, with organic dyes serving as energy acceptors in QD–dye-labeled protein conjugates. Dayal et al. [18] observed non–Förster energy transfer behavior in QD–phthalocyanine conjugates; they also reported surface effects on QD-based energy transfer.

Pons et al. [56] described the advantage of size heterogeneity in terms of the effects of the spectral overlap on the FRET between QD donors and proximal dye acceptors (in a QD–protein dye conjugate). They found that the same acceptor may have different donor–acceptor spectral overlaps due to the donor size heterogeneity.

A robust minimal model for FRET that captures the key features of Förster's level-broadening idea in a microscopic quantum mechanical setting has been developed by the research groups headed by Renugopalakrishnan and King [42, 60]. At short distances and times, this model predicts coherent quantum effects, while at long distances and times, it reproduces the standard FRET rates and distance scales (Fig. 5.2). The model is based on a master equation with Lindblad operators that takes into account the fluorescence

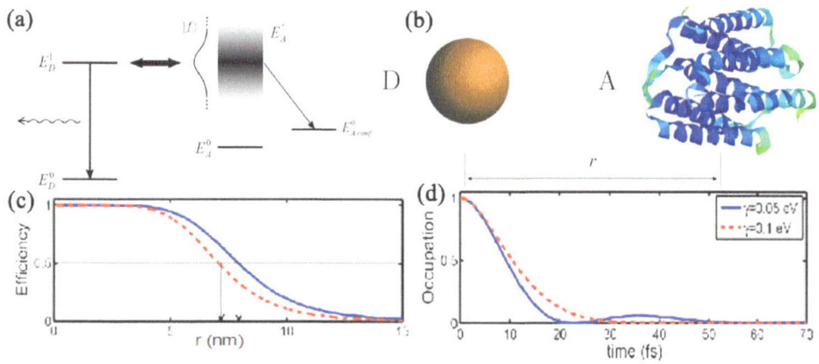

Figure 5.2 Basic principles of the continuum model of FRET. (a) Schematics of the continuum model of FRET. $E^0_{\text{A,conf}}$, the energy level of bR after the conformational change; $|f\rangle$, the normalized wave packet superposition. (b) Geometry of interaction: the donor is a quantum dot (QD), and the acceptor is bacteriorhodopsin (bR). (c) QD–bR efficiency as a function of the separation distance. The arrows show the positions of R_0 for two values of γ (the width of the wave packet). (d) Occupation probability of the initial excited state as a function of time. $r = 2$ nm. Coherent oscillations are apparent when $\gamma = 0.05$ eV. Reprinted (figure) with permission from Ref. [42]. Copyright (2012) by the American Physical Society.

and relaxation effects and uses a continuum of excited states in the acceptor to implement the level-broadening idea. The model can be easily extended to include other effects, for example, temperature dependence and phonon interactions. A specific application of the model is the analysis of FRET coupling between a QD and bR, where it makes realistic predictions of the FRET distance (Fig. 5.2). The authors use the following model: A donor (QD) passes from the ground state into the discrete excited state after absorption of a photon. The donor can transfer energy to the acceptor through FRET or emit a photon as a result of relaxation and return to the ground state (Fig. 5.2). The acceptor (bR) is a quantum system passing into an excited state with the energy level in a continuum after the transfer of the donor excitation. As a result of photoexcitation, the bR molecule changes its conformation (photoisomerization), which is reflected in an additional discrete level of $E^0_{\mathrm{A,conf}}$.

On the basis of mathematical calculations using Lindblad density matrix equations and operators and coherent interactions, a new mathematical model describing the efficiency of nonradiative energy transfer with account for the probability of vibrational relaxation of both the donor and the acceptor has been developed [42]. The example of calculation with the development model is demonstrated by Fig. 5.2. The curve of the efficiency (Fig. 5.2c) as a function of r almost exactly matches the phenomenological formula (Eq. 5.4). The results are robust against changes in wave function wave, γ. Figure 5.2d shows the occupation probability of the initial donor exited state as a function of time for $r = 2$ nm.

5.3 The Quantum Dot–Bacteriorhodopsin Nanobiohybrid Structure

5.3.1 Bacteriorhodopsin: Structure and Function

bR is a retinal-containing protein that has been intensely studied over the years due to its inherent ability to function as a light-driven proton pump. bR is structurally similar to the visual rhodopsin in mammalian retinal cells. It is found in the cell membrane of *Halobacterium salinarum* and is utilized for ATP synthesis under

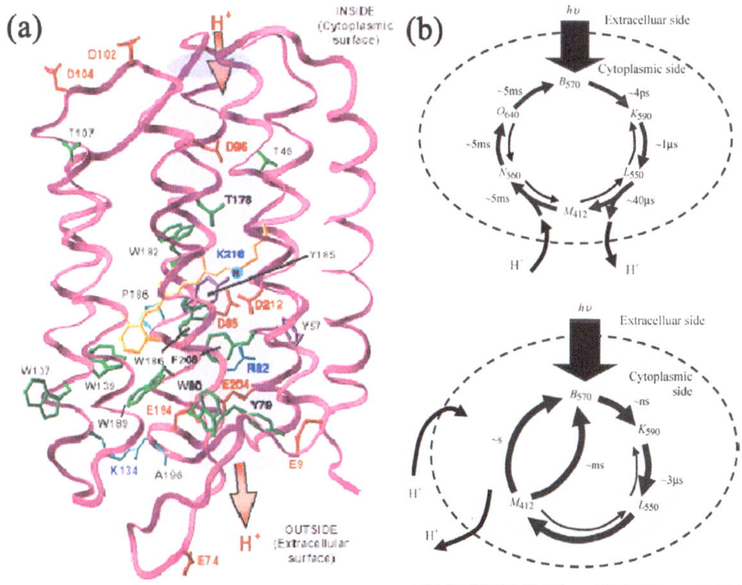

Figure 5.3 Bacteriorhodopsin (bR) structure and photocycle. (a) Schematic representation of the bR tertiary structure. Adapted with permission from Ref. [6]. Copyright (1999) American Chemical Society. (b) Basic photochemical cycles of bR exposed to visible light ($h\upsilon$) for both aqueous (top) and dry (bottom) forms. The subscripts indicate the peak wavelengths for the identified intermediates.

anaerobic conditions. It has a molecular weight of 26,784 Da and consists of 248 amino acid residues [34, 47] arranged in seven α-helical bundles inside the lipid membrane and that form a cage where a retinylidene residue attached to lysine 216 is located (Fig. 5.3a). The retinylidene residue is tilted about 20° toward the outer side of the cell membrane. Upon illumination with light in the range 500–650 nm (λ_{max} = 570 nm), the retinal chromophore absorbs a photon, which initiates a conformational change from the all-*trans* to the 13-*cis* retinal state and a series of structural transformations corresponding to a well-defined photocycle (Fig. 5.3b). The photocycle results in the translocation of a proton from the cytoplasmic side of the membrane to the extracellular side, a process in which the Schiff base, Asp85, and Asp96 play a key role. Within this photocycle, the protein moves through a number of spectrally distinguishable quasi-stable states (K, L, M, N, and O) [29]. The only step of this photocycle that requires energy is the first step, namely the formation of K. The remaining steps are a sequence of thermal relaxations that return the retinal to its all-*trans* conformation. A wild-type bR molecule needs about 10 ms for a complete catalytic cycle.

It should be noted that bR in the *H. salinarum* cell membrane (the so-called purple membrane [PM]) is ultrastable; in particular, it is photostable under a variety of conditions, such as high ionic strength, the presence of organic solvents, and extreme pH values. It can be dehydrated, arranged into multilayers, or embedded into a polymer matrix where it withstands temperatures as high as 150°C. In essence, bR acts as an exquisitely sensitive and robust, wavelength-specific, photoactive proton pump with excellent photochromic conversions. A lot of technical applications have been suggested for bR. Most of them are based on its unique photochromic and photoelectric properties. They include 2D and 3D optical storage systems, holographic media, light switchers, optical filtering, neural networks, spatial light modulators, pattern recognition, interferometry, motion detection, artificial retinas, etc. [32, 33]. bR–QD heterosystems with good energy coupling may provide a new material that will be more suitable for these applications.

5.3.2 Energy Coupling between Quantum Dots and Bacteriorhodopsin in Aqueous Media

QDs can serve as a FRET donor with a wide array of acceptor particles. This process is affected by a number of QD parameters, such as surface charge, diameter, and organic coating thickness [8]. QDs can be selected to provide the maximal overlap of the QD emission and bR absorbance peak. The effects of the surface charge and diameter of QDs on the efficiency of FRET from QDs to the retinal of bR have been analyzed [9]. Modification of CdSe/ZnS QDs with cysteine, cysteamine, and polyethylene glycol (PEG) derivatives containing terminal carboxylic (SH-PEG-COOH) or hydroxylic (SH-PEG-OH) groups yields QDs with strongly negative, strongly positive, negative, and slightly negative surface charges, respectively. Additionally, PEG-capped QDs are known to have larger hydrodynamic diameters (12–15 nm) than those of cysteine- or cysteamine-capped nanocrystals (5–7 nm). QDs functionalized with cysteine, PEG-COOH, or PEG-OH have been found to be negatively charged with ξ-potentials of –38, –30, and –5 mV, respectively, whereas cysteamine-capped QDs are positively charged with a ξ-potential of +36 mV [8]. QDs with such charged organic coatings adsorb spontaneously on PMs. It has been shown that the efficiency of FRET can be controlled by varying the surface charge and thickness of the QD organic coating. The FRET efficiency can also be tuned by varying the degree of the spectral overlap between the QD PL and bR absorption (Fig. 5.4). The same authors used QDs emitting PL at 570 nm to analyze how the QD surface charge and thickness of their organic capping layer affect the FRET efficiency and the morphology of the complexes formed. Figures 5.5a–5.5f

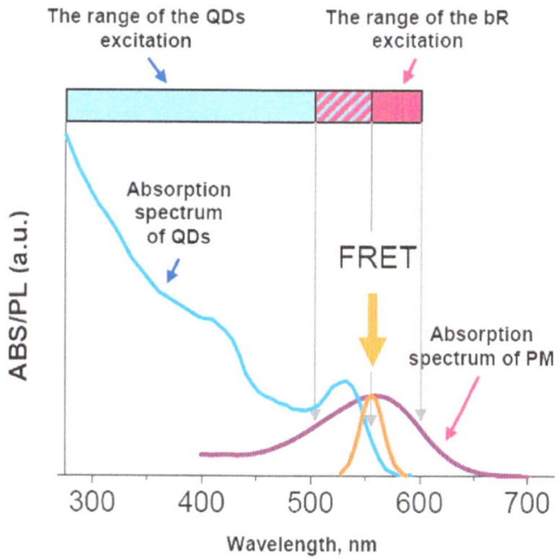

Figure 5.4 Optical properties of bacteriorhodopsin (bR) in purple membranes (PMs) and schematic diagram of energy transfer from quantum dots (QDs) to bR. The UV-Vis spectra of PMs (the purple line) and the typical adsorption spectra of QDs tuned in resonance with bR (the blue line), as well as the photoluminescence (PL) emission spectrum of these QDs (the orange line), are shown. The spectral regions of the bR and QD effective excitation are shown as purple and blue bars, respectively.

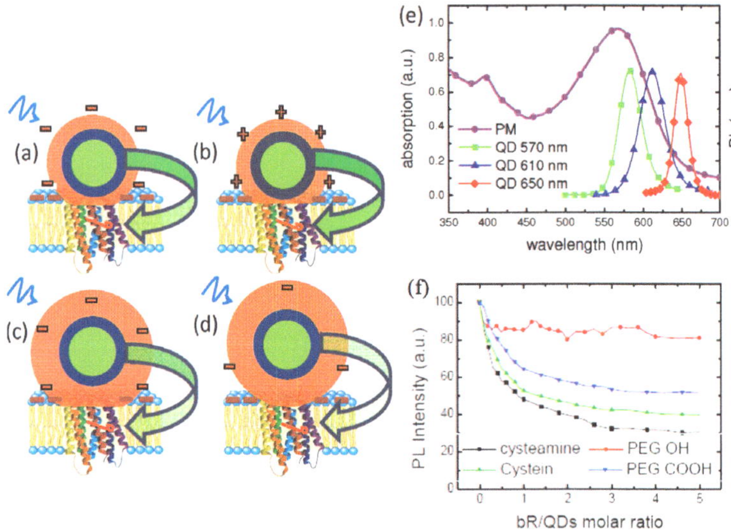

Figure 5.5 Schematics of hybrid materials consisting of quantum dots (QDs) with different surface charges and thicknesses of the organic coating and purple membranes (PMs) containing bacteriorhodopsin (bR). (a, b) CdSe/ZnS QDs coated with negatively charged cysteine or positively charged cysteamine self-assembled on the PM surface. These coatings are the thinnest, and the hybrid materials exhibit a very efficient FRET from QDs to bR. (c, d) CdSe/ZnS QDs coated with the SH-PEG-COOH and SH-PEG-OH trifunctional ligands, respectively, self-assembled on the PM surface. These QD coatings are thicker than the cysteine and cysteamine ones and exhibit a less intense FRET (in the case of SH-PEG-COOH) or no energy transfer at all (in the case of SH-PEG-OH) when QDs are assembled with PMs. (e) Energy absorption by bR (the energy acceptor, the purple line) and photoluminescence (PL) of QDs (the energy donors) of different diameters (PL colors). (f) Effects of charge and thickness of the QD coating layer on PL lifetimes of CdSe/ZnS QD570 as a function of the bR-to-QD molar ratio. The concentration of QDs was 1.4 μM in all cases; the dilution factor and internal filter effects were taken into account. The optical density in all these experiments was <0.1. Adapted with permission from Ref. [8]. Copyright 2011, AIP Publishing LLC. Adapted with permission from Ref. [57]. Copyright (2010) American Chemical Society.

show that the formation of complexes of PMs with QDs capped with the cysteine or cysteamine leads to quenching of QD PL as strong as 65%–70%. Otherwise, QDs solubilized with the trifunctional SH-PEG-COOH ligand show a PL quenching of about 40%, and QDs capped with the SH-PEG-OH ligand show a negligible PL quenching of $<15\%$, a value close to that obtained in the control experiment with the so-called white membranes. It was shown that the most FRET-efficient QD–PM complexes are characterized by the highest level of QD ordering; hence, atomic force microscopy (AFM) imaging of bR–PM hybrid material provides the basis for optimization of the assembling design in order to engineer nanobiohybrid materials with advanced optical and photovoltaic properties.

Rakovich et al. [58, 59] have demonstrated that QDs immobilized on the PM surface can play the role of a built-in light energy convertor by harvesting light in the spectral region where it cannot be absorbed efficiently by PMs alone (from UV to blue). The influence of QDs of two types (CdSe/ZnS, 570 nm, and CdTe, 650 nm) on the efficiency of energy transfer between the QDs and bR has been investigated by this group. It has been shown that in the case of CdTe QDs with a thinner shell, the FRET efficiency is much higher (quenching of QD fluorescence by PMs is as high as 64% in the case of a PM–QD complex with a molar ratio of 0.78) compared to CdSe/ZnS QDs (26.3% at a PM-to-QD molar ratio of 21.8). The authors attribute this fact to larger diameters of CdSe/ZnS QDs compared to CdTe QDs due to the thickness of the additional ZnS shell and of the additional organic shell formed during the water solubilization of CdSe/ZnS QDs, which reduces FRET.

The study of the kinetics of fluorescence decay confirms that the energy transfer in these heterostructures is accounted for by FRET: the fluorescence decay time of CdTe QDs is 9.94 ns in the absence of bR and 3.54 ns at a bR-to-QD molar ratio of 0.78; in the case of CdSe/ZnS QDs, it is 4.79 ns in the absence of bR and 3.53 ns at a bR-to-QD molar ratio of 21.8.

The effect of the separation distance between the bR film and deposited QDs has been found by another research group [26]. The bR–QD separation distance was varied by using CdSe (core) and CdSe/ZnS (core–shell) QDs. The ZnS shell on CdSe QDs provided an additional 1 to 1.5 nm layer between the CdSe QD and the bR retinal in the dried films. The addition of a ZnS shell on CdSe QDs was found to reduce the energy transfer to the bR molecules by 19.5% compared to the original CdSe QDs. Thus, an increase in the QD–bR separation distance by adding shells to the QDs reduces the nonradiative energy transfer from the QD to the bR retinal. This results in a decreased QD emission quenching compared to nonshell QDs.

In an aqueous suspension, a uniform bR–QD distance can be achieved by using streptavidin–biotin-linking schemes. The small biotin molecule has an extremely high binding affinity for the streptavidin protein, which ensures specific, tight binding at all locations. The estimated bR retinal–QD separation distance due to this bond is ~8.5 nm, assuming a streptavidin length of 5 nm [16], a negligible biotin size, a QD shell thickness of 1.0 nm, and retinal location in the center of bR, 2.0 nm from the biotin–streptavidin bond. It was estimated that upon addition of biotinylated bR to the streptavidin-coated QDs, the intensity of the QD optical emission decreased, and the FRET efficiency in this system was 32%. This result fits the theoretically calculated FRET efficiency of 34.17% for a bR–QD FRET coupling pair at this separation distance [24]. The same research group observed a reduction of the QD emission of up to 49% in the system created by coupling QDs and bR with a separation distance of approximately 3.5 nm; that is, a decrease in the distance from 8.5 to 3.5 nm increased the FRET efficiency by 17%.

5.3.3 Methods of Forming Heterostructures Containing Quantum Dot–Bacteriorhodopsin Complexes

A wide range of methods for immobilization of bR have been explored and published [54]. They include simple drop drying, which results in nonoriented multilayers [64]; immobilization of a bR-PM film onto a solid support using the Langmuir–Blodgett (LB) deposition approaches, in which films are initially compressed at an air–water interface [51, 74]; electrophoretic sedimentation (EPS) [41], where external electric fields affect the PM dipole moments and generate orientated mono- and multilayers;

host–guest-mediated immobilization [13]; self-assembly [10, 40, 49, 62]; electrostatic layer-by-layer (LbL) adsorption [35, 55]; antigen–antibody molecular recognition [43]; sol–gel encapsulation [17]; and the use of polymers as immobilizing matrices [12]. As outlined above, QDs can be applied directly onto the PM film as a suspension or bound with bR covalently or electrostatically.

As mentioned above, the potential applications of PMs in engineered devices are extensive. For many of the applications, such as photoelectric systems, it is important that PMs be oriented during deposition. This is necessary to facilitate unidirectional transport of protons within the layered system, producing a stronger electrical signal. A number of techniques, such as Langmuir–Blodgett (LB), immunochemical binding, and EPS, the simplest and most widely used technique for PM orientation, can be used to provide a high degree of PM orientation upon deposition. Although the antigen–antibody technique ensures good bR orientation [37, 43, 69], the procedure is very time consuming because it is necessary to synthesize antigens, monoclonal antibodies, and bi-antibodies. Another efficient approach suggested recently is genetic engineering of PMs containing cysteine residues on the surface in order to bind them to the gold substrate via thiols [2].

To investigate the effects of FRET and bR, optical absorption layers of bR were added to a single QD monolayer. It was shown that the first bR monolayer induced a 39% reduction of the QD emission. All subsequent layers resulted in QD quenching of less than 10%. Since FRET is primarily observed when the bR (acceptor) and QD (donor) are within 10 nm from each other, only the first bR layer was within the FRET distance; hence, resonance energy transfer was the probable cause of the additional quenching by the first layer of PM [25]. It turns out that the monomer form of bR is a more efficient energy acceptor than bR in the PM structure. By changing the separation distance between QDs and the bR monomer from 8.6 to 3.5 nm, one can increase the energy transfer from 51.1 to 88.2% [27].

5.3.4 Quantum Dot–Bacteriorhodopsin Hybrids in Dried Films as Media for Sensing and Optical Applications

It has been shown that at low humidity, the proton is actually shifted between the Schiff base and Asp85 over a distance of about 4 Å rather than pumped through the bR molecule. However, bR retains its main photochromic and electrogenic properties determined by the formation of the M intermediate, which occurs even in dry proteins. Therefore, the thin-film technology can be used in bR-based systems.

An ideal bR immobilization process would result in an ultrathin PM layer covering the entire substrate. However, experimental data have shown that monolayer fabrication methods do not yield a continuous uniform monolayer of closely packed membranes [69]. The sporadic distribution of PM fragments over the substrate may result from weak electrostatic repulsion between the constituent fragments accounted for by their net negative charge above the PM isoelectric point.

A positive polymer adhesion layer (0.5 nm in thickness) has been used to obtain multilayered films of bR, QDs, and QD–bR hybrids by the LbL technique. In QD–bR multilayered electrodes, distinct QD quenching determined by each bR layer has been monitored [26]. Analyzing the QD FRET coupling strength with different numbers of bR monolayers showed that the QD layer could directly couple to the nearest two bR monolayers. This agrees with the FRET theory, because a single bR monolayer has a thickness of 5.5 nm, and any bR that is more than two monolayers away from the QDs would be out of the FRET coupling range (less than 10 nm). Fluorescence lifetime decay studies confirm a decrease in QD lifetimes associated with bR film thickness, with effective lifetimes of 14.5 ns (the QD control), 11.4 ns (QDs + one bR monolayer), 9.7 ns (QDs + two bR monolayers), and 9.9 ns (QDs + three bR monolayers). The FRET data indicated that the optimal architecture to maximize the output with the minimal material is a three-monolayer bR electrode topped with a single-monolayer QD film (an ~18 nm thick hybrid electrode), which can utilize both FRET and photonic absorption energy processes to power the bR output. The three-monolayer bR electrode without QDs yielded a photovoltaic output of approximately 0.4 mV. The response increased to approximately 1.2 mV after electrostatic absorption of a single QD monolayer, which was due to the contributed QD energy. It has been shown that dried biotinylated bR electrodes created by EPS retain the capacity to effectively bind functional streptavidin groups of QD–streptavidin conjugates. This linkage scheme allows the binding of a single QD monolayer on top of the bR electrode, determining a QD–bR separation distance of approximately 6–8 nm. The estimation of the overall effect on bR showed that the binding of QDs to the surface of the PM electrode resulted in an increased bR photovoltaic response to white light illumination. In the absence of QDs, the bR electrode yielded a 0.249 V peak transient response upon illumination followed by a −0.158 V peak upon termination of the illumination. When a QD layer was then bound to the same sample and reactivated, there was an initial bR photovoltage peak of 0.306 V followed by a −0.2 V peak upon termination of the illumination. The addition of QDs to the PM electrode increased the bR photovoltaic output by approximately 57 mV or 23% (Fig. 5.6a,b) [26].

5.3.5 Enhancement of the Biological Functions of Bacteriorhodopsin by Means of Coupling with Quantum Dots

The first proof-of-principle evidence of an increase in the efficiency of the biological function of bR in the presence of QDs was reported in Ref. [57]. This demonstration was done using oriented proteoliposomes (Fig. 5.7a,b), that is, liposomes with bR molecules incorporated in the liposome walls vesicles in a highly oriented manner. Proteoliposomes with an average diameter of 240 nm were formed. These proteoliposomes had a surface density of 0.06 nm^{-2}, which corresponded to the density of hexagonal bR lattice package in native PMs. The orientation of bR in the prepared proteoliposomes was opposite to that in bacterial PMs, where bR pumps protons out of the cell; hence, protons were pumped into proteoliposomes. As a result, photoillumination provoked a photoinduced decrease in the concentration of protons (increase in pH) in the bulk suspension. Integration of CdTe QD590 into PMs even at a QD-to-bR molar ratio of 1:5 induced a pronounced (>25%)

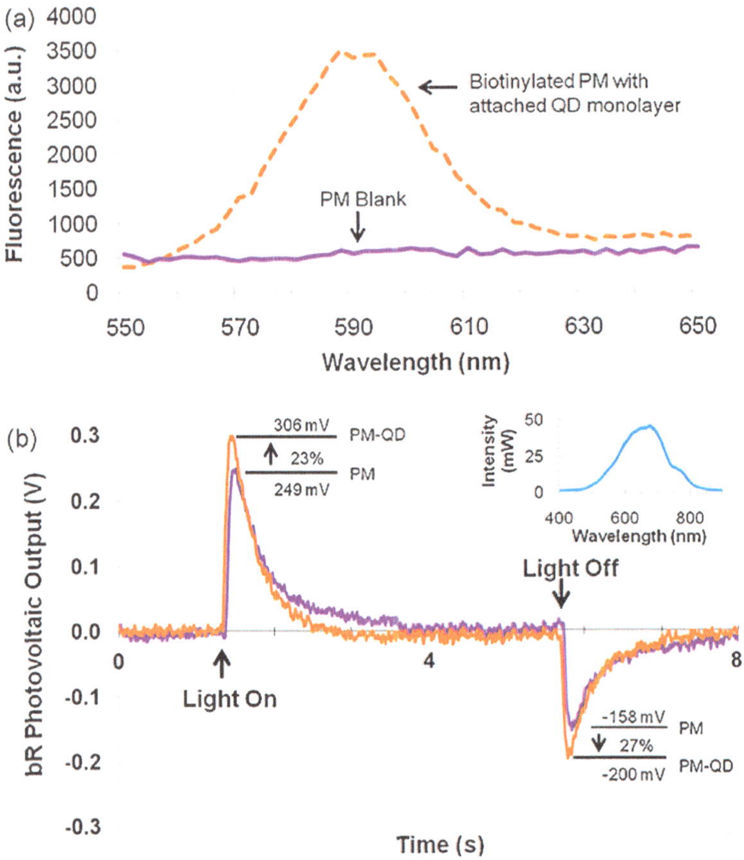

Figure 5.6 Improvement of the bacteriorhodopsin (bR) photovoltaic properties with the use of semiconductor quantum dots (QDs) in a photovoltaic cell. (a) Emission spectra confirming the attachment of a single QD monolayer to a biotinylated purple membrane (PM) electrode and the absence of attachment to a nonbiotinylated PM. (b) bR electrode activation with and without the attached QD monolayer. The inset shows the spectrum of incident white light. When a QD layer was then bound to the same sample and reactivated, there was an initial bR photovoltage peak of 0.306 V followed by a −0.2 V peak upon termination of the light. The addition of QDs to the PM electrode increased the bR photovoltaic output by approximately 57 mV, or 23%. The photovoltaic on/off peak ratios for the PM and PM–QD electrodes were close to 1.5 (1.42 and 1.55, respectively). Reproduced with permission from Ref. [8]. Copyright 2011, AIP Publishing LLC.

increase in the photoresponse in terms of bR proton pumping (Fig. 5.7c). It is noteworthy that the thickness of the organic shell covering the QD fluorescent core is essential both for the efficiency of FRET and for the enhancement of the biological function of bR. Therefore, the efficiency of FRET and an increase in the photoresponse were found to be maximum for CdTe QDs covered with a stabilizing TGA shell <0.3 nm in thickness [57].

The photoresponse of bR within the bR–QD hybrid increases exponentially with an increase in the photoillumination power density and reaches saturation at about 2 kW/m^2 (Fig. 5.7d). Note that the increase in photoresponse due to the presence of QDs is the largest at the highest illumination power densities and depends on the amount of QDs incorporated in the proteoliposomes. Binding of QDs in amounts exceeding one QD per bR molecule leads to a decrease in the efficiency of bR proton pumping. This may be explained by the internal filter effect due to the large quantity of QDs forming multilayers or unbound nanocrystals in the suspension. All these nanocrystals may not participate in FRET to bR, but their internal filter effect precludes the photoexcitation of bR.

Estimation based on the analysis of the excitation and absorption spectra of bR, as well as the excitation/emission spectra of QDs, shows a maximum increase in the proton transport efficiency of 8%–10%. The actual increase in the efficiency is even higher. Moreover, at the highest photoirradiation intensities in the region of saturation, where bR receives enough photons for the highest efficiency of proton transport, addition of QDs has a positive effect (Fig. 5.7d). Such an unexpected behavior may be explained by the shift of equilibrium between the intermediate states and shortening of the bR photocycle duration. These results in an increase in the number of photocycles per time unit and, consequently, the number of protons transported through the membrane.

We have to take into account that the photocycle of bR is a complicated process where the molecules pass through a sequence of photodistinguishable states, most of which are photosensitive. We believe that the use of QDs of different, carefully selected colors may specifically affect some selected intermediates of the bR photocycle, thus permitting the development of more complex hybrids with metastable or stable states and paving the way to the development of novel optoelectronic switches or memory elements.

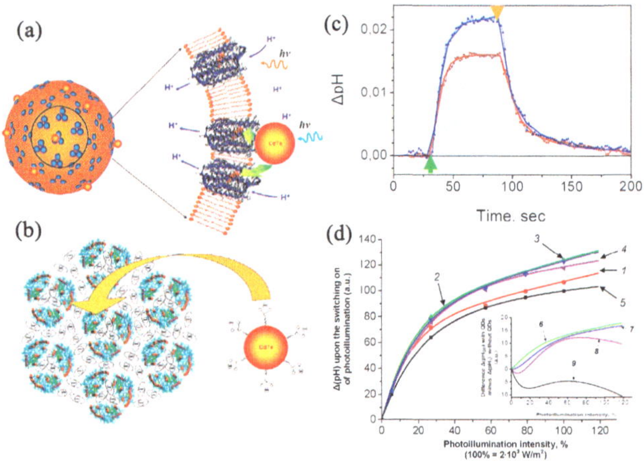

Figure 5.7 Improvement of the bacteriorhodopsin (bR) photovoltaic properties with the use of semiconductor quantum dots (QDs) in proteoliposomes containing highly oriented bR. (a) Organization and functionality of a complex composed from liposomes containing highly oriented bR (proteoliposomes) and QDs immobilized on their surface. All objects are shown to scale. The close-up shows a fragment of the liposome membrane where bR is shown in blue and retinal is shown in red. Excitation with white light was used. Illumination with photons in the orange region of the visible spectrum provides direct excitation of bR. Photons with higher energy (the blue region) excite only QDs, which transfer their energy to bR via FRET. Absorption of photons by bR induces the transfer of protons into proteoliposomes and the photoinduced increase in pH in the solution outside of proteoliposomes, which serves as a measure of the efficiency of the bR function. (b) bR is organized in trimers in the membrane of the proteoliposomes. The size of these trimers is very similar to the diameter of QDs used in the experiments on energy transfer. (c) Effects of QDs on the efficiency of the photoinduced proton pumping by bR reconstructed within the liposomes: variation of pH upon switching on (the green arrow) or switching off (the yellow arrow) photoillumination of the proteoliposome suspension with white light. The red curve shows the pH photoresponse of bR-containing liposomes (ΔpH_0). The blue curve shows the pH photoresponse for the same liposomes tagged with QDs ($\lambda_{em} = 590$ nm) immobilized on their surface at a QD-to-bR molar ratio (ΔpH_{QD}) of 1:5. Reproduced with permission from Ref. [57]. Copyright (2010) American Chemical Society. (d) The amplitude of the pH jump (ΔpH) versus the intensity of photoillumination for different contents of QDs solubilized with PEG–OH. Inset: The difference between ΔpH_{QD} and ΔpH_0; here, ΔpH_0 is the jump of pH for the liposome solution without QDs and ΔpH_{QD} is the jump of pH for proteoliposomes containing QDs. Curve 1 is without QDs; curves 2, 3, 4, and 5 are after the addition of 5, 10, 15, and 20 μL of a 7.7 mg/mL CdSe/ZnS QD dispersion, respectively, to the liposome solution; curves 6, 7, 8, and 9 have been obtained by subtracting curve 1 from curves 2, 3, 4, and 5, respectively.

5.4 The Quantum Dot–Photosynthetic Reaction Center Nanobiohybrid Structure

Research aimed at bridging photosynthetic complexes with inorganic nanostructures started only a few years ago, yet several important observations have already been made. The most important ones are plasmon-enhanced absorption and emission of light-harvesting complexes, tunability of the energy transfer due to plasmon coupling, and conjugation of photosystems with carbon nanotubes, as well as metal and semiconductor surfaces [48].

The development of artificial photosynthetic systems that utilize solar energy is one of the most challenging goals in material sciences. In natural photosynthetic organisms, light is initially absorbed by protein–pigment antenna complexes and transferred to specialized RCs, in which the captured light energy is converted into chemical energy for synthesis of high-energy molecules that fuel the body. Mimicking this process in artificial solar energy conversion requires, among other components, an efficient light-harvesting antenna capable of transferring the excitation energy to RCs. The use of organic chromophores is rather limited due to their narrow light absorption spectra and lack of photostability. However, inorganic nanocrystals may provide a significantly greater absorption than natural photosystems, thus enhancing the light harvesting. These nanocrystals are also very efficient in excitation energy transfer. We have already demonstrated a threefold increase in the rate of generation of excitons in RCs in specially designed QD–RC complexes, and theoretical estimates predict an even larger enhancement.

5.4.1 Bacterial Photosynthetic Reaction Centers: Structure and Properties

Many recent studies in the field of photosynthesis aimed at creating artificial photosynthetic systems based on the existing pigment–protein complexes. Changing the properties of these complexes through interaction with nanostructures is one of the most important trends of these studies. Theoretical considerations show that the efficiency of the photosynthetic apparatus can be increased several times through these interactions. Attempts have been made to use fullerenes [30], as well as various dendrimers (highly branched synthetic macromolecules) [53] as artificial photosynthetic RCs. A number of studies have demonstrated the efficiency of energy transfer in these systems; however, as mentioned above, the use of organic fluorophores in light-harvesting systems is very limited due to their narrow spectral range and low photostability [31].

Light-harvesting complexes and the photosynthesis mechanism of purple bacteria have been studied in most detail. The RCs of photosynthetic bacteria are chromophore–protein complexes that are capable of directional photoinduced electron transfer through biological membranes [19]. Its quantum yield (charges separated per photon absorbed) is close to unity because of the highly efficient transfer of excited-state energy through the antenna pigments of the photosystem to the RC and highly efficient charge separation inside the RC.

As a result, the issue of whether the high-yield photovoltaic properties of natural RCs can be employed in various devices has attracted considerable interest, and a number of laboratories have investigated photocurrent generation by RCs encapsulated in a solar cell or adhered to a functionalized metal electrode. Photocurrents have been reported for RC photosystem I and photosystem II from plants or cyanobacteria and from purple bacterium RCs isolated mainly from *Rhodobacter sphaeroides*. Other applications of RCs have also been explored, including photosensors, optoelectronic components, gating elements for phototransistors, and biosensors for the detection of herbicides and other environmental pollutants (these data are comprehensively summarized in Ref. [20]).

Although RCs differ from one to another in the precise composition and mechanism of action, the various types of RCs found in nature display common structural and mechanistic characteristics [36]. An RC is composed of domains arranged in two membrane-spanning branches (A and B), nevertheless only one of which (branch A) is active in catalyzing electron transfer across the RC. The active branch is responsible for electron–hole separation and sequential transmembrane electron transfer. Each branch (A and B,

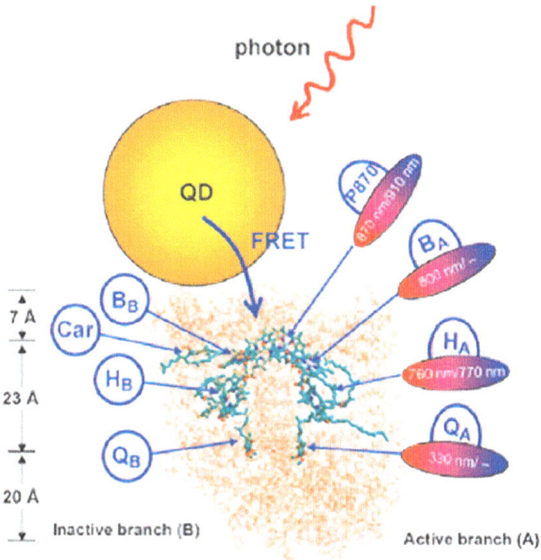

Figure 5.8 Organization and functionality of a complex composed of a reaction center (RC) and a QD. Branches A and B are active and inactive in electron transfer, respectively. The positions of the absorption/fluorescence maxima for bacteriochlorophyll (BChl) special pair (P), BChl monomer (B), bacteriopheophytine (H), and quinone (Q) are indicated only for the active branch (branch A). Photons are absorbed by both the RC and the QD. An exciton from the QD is transferred to the RC via FRET. Excitons inside the RC relax to the lowest energy level located at the special pair. In nature, excitons trapped in the special pair P are dissociated (i.e., photoinduced charge separation occurs), and an optically excited electron is transferred along the active branch, leaving to a positively charged hole at P. In the presence of QDs the energy transfer from QDs to RCs occurs. The effect causes the increases of RC fluorescence, which is due to a larger rate of exciton generation. Reproduced from Ref. [52], with permission from John Wiley & Sons, Inc.

Fig. 5.8) consists of one bacteriochlorophyll (BChl; B_A or B_B), one bacteriopheophytin (H_A or H_B), and one quinone (Q_A or Q_B) molecule. Both branches are connected to the key element of the RC, which is a dimer of BChl molecules, the so-called special pair (P or P870) [52]. In nature, a solar photon first creates an excitation in the RC, and then this excitation moves toward P870, where the electron separates from the hole within 2–3 ps. At the next step, this electron moves along the electron transfer chain toward QA. Afterward, this electron takes part in chemical transformations, eventually leading to the synthesis of high-energy molecules (ATP), which satisfy most energy needs of the bacterium [52].

5.4.2 Energy Transfer from Quantum Dots to Bacterial Reaction Centers

The key function of photosynthesis systems is to harvest sunlight energy and transfer the excitation for rapid charge separation of optically generated electron–hole pairs. Light is initially absorbed by antenna pigment–protein complexes, in which it induces an excited electronic state (exciton), and then excitons (or electron–hole pairs) are transferred by means of FRET to chlorophyll cofactors of specialized RCs. Recently, it was suggested that the rate of optical generation of electron–hole pairs within a photosynthetic system could be considerably increased through conjugation with semiconductor NPs [28]. QDs, due to their high absorption capacity, can act as complementary elements to the antenna complexes of the photosynthetic apparatus. Furthermore, it is easy to select QDs with the maximum fluorescence in the desired spectral range, which makes it possible to create a system with efficient energy migration. These properties make QDs promising candidates as donors of excitation energy for electron transfer in photosynthetic systems.

An example of the efficient transfer of excitation energy harvested by NPs (QDs) to a complex biological photosynthetic system (RC) is presented in Ref. [52]. The QD–RC complexes were prepared by electrostatic assembly of purified RCs from the purple bacteria *Rhodobacter sphaeroides* with photoluminescent QDs (Fig. 5.8). The diagram of energy relaxation inside the RC, which is of crucial importance for understanding the photoelectronic response of the QD–RC system, is shown in Fig. 5.9a. The photon energy absorbed by QDs is transferred to one of RC cofactors by FRET. The excitation in the RC quickly decays to the lowest excited level of P870 (Fig. 5.8a). In a natural environment, the exciton is split into an electron and a hole. In the opposite case, under the experimental conditions of Ref. [52], the exciton decays with a characteristic emission at 910 nm.

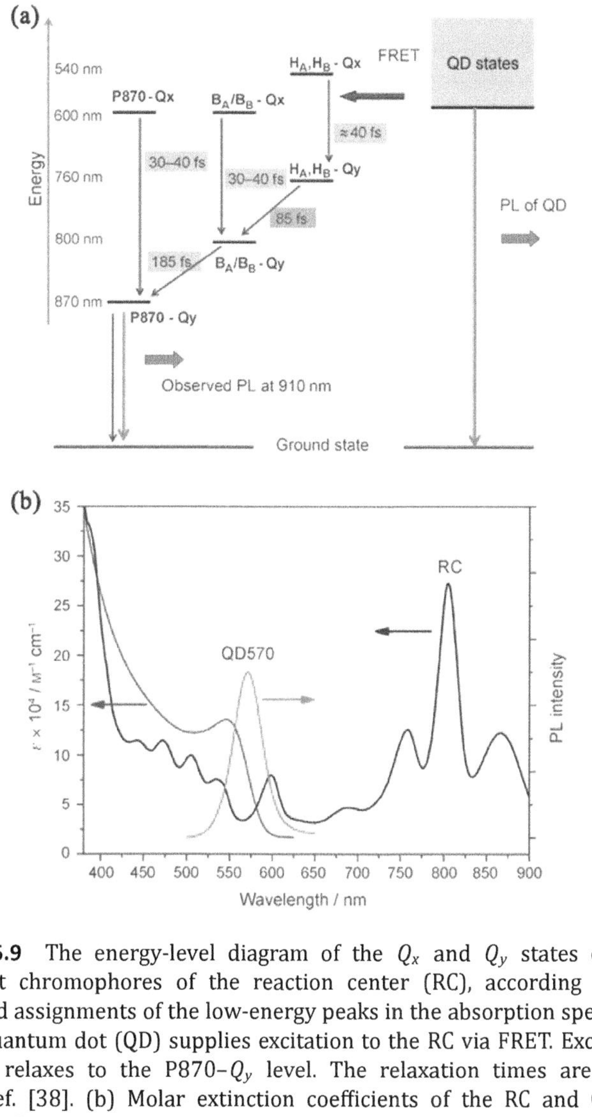

Figure 5.9 The energy-level diagram of the Q_x and Q_y states on the different chromophores of the reaction center (RC), according to the standard assignments of the low-energy peaks in the absorption spectrum. (a) A quantum dot (QD) supplies excitation to the RC via FRET. Excitation quickly relaxes to the P870–Q_y level. The relaxation times are taken from Ref. [38]. (b) Molar extinction coefficients of the RC and QD570 as dependent on the wavelength and the QD570 photoluminescence (PL) spectrum. Adapted from Ref. [52], with permission from John Wiley & Sons, Inc.

This energy structure reflects the RC absorption spectrum with discrete peaks assigned to the Q_x and Q_y absorption bands of the different chromophores bound to the RC (Fig. 5.9b). Interestingly, the energy transfer from the QD onto the special pair P870 makes also use of the bacteriochlorophylls (BA or BB), one bacteriopheophytins (HA or HB) as intermediate relay systems involving, for the former, the intramolecular electronic relation from the Q_x into the Q_y states.

The experimental extinction coefficient for the CdTe QDs emitting at 570 nm (QD570) used in this study is shown as an example [52]. Note that at comparable molar concentrations, the QD570 absorption at wavelengths from 400 to 570 nm is stronger than that of the RC; therefore, optical enhancement of the RC absorption is possible.

The complex displayed efficient quenching of QD exciton emission (up to factor 5 at the ratio of RC/QD $=$ 2), which suggested efficient FRET from QDs to RCs. The efficiency of FRET is conveniently described by the parameter E:

$$E = \frac{1 - \mathrm{PL}_{\mathrm{QD\text{-}RC}}}{\mathrm{PL}_{\mathrm{QD}}} \tag{5.8}$$

where $\mathrm{PL}_{\mathrm{QD\text{-}RC}}$ and $\mathrm{PL}_{\mathrm{QD}}$ are the peak intensities of QD exciton emission for QD–RC complexes and individual QDs, respectively.

It has found that FRET is more efficient for CdTe QDs than for CdSe QDs; this fact can be explained by the smaller radii and thinner ligand shells of CdTe QDs and the resulting smaller QD–RC separation distances. According to the standard theory, the FRET efficiency for a complex comprising one QD and one RC (the QD–RC complex) can be expressed as

$$E0 = \frac{1}{1 + \left(\frac{1}{R_{\mathrm{F}}}\right)^6} \tag{5.9}$$

where r is the QD–RC spatial separation and R_{F} is the Förster radius.

The QD luminescence quenching as a result of FRET was also confirmed by time-resolved PL measurements.

Energy transfer from QDs to RCs has also been monitored by measuring the enhancement of the PL emission of the BChl special pair in the RC (at 910 nm) determined as

$$A_{910\,\mathrm{nm}}(\lambda_{\mathrm{exc}}) = \frac{\mathrm{PL}_{\mathrm{QD\text{-}RC}}^{910\,\mathrm{nm}}}{\mathrm{PL}_{\mathrm{RC}}^{910\,\mathrm{nm}}} = \frac{n_{\mathrm{QD\text{-}RC}}}{n_{\mathrm{RC}}} \tag{5.10}$$

where λ_{exc} is the excitation wavelength,

$\mathrm{PL}_{\mathrm{QD\text{-}RC}}$ and $\mathrm{PL}_{\mathrm{RC}}$ are the PL peak intensities at 910 nm for QD–RC complexes and for individual RCs, respectively, and n_{RC} and $n_{\mathrm{QD\text{-}RC}}$ are the numbers of electron–hole pairs generated at the special pair in an RC and in a QD–RC hybrid complex, respectively.

The experimentally determined factors $E(\lambda_{\mathrm{exc}})$ have been found to be larger than unity for $\lambda_{\mathrm{exc}} < \lambda_{\mathrm{QD,exciton}}$, where $\lambda_{\mathrm{QD,exciton}}$ is the QD exciton peak wavelength. Therefore, the hybrid structures demonstrate optical enhancement; that is, the number of electron–hole pairs arriving at the special pair of an RC is increased. The physical reason for the optical enhancement of the RC function is that additional electron–hole pairs are transported to generated at the RCs from QDs by means of FRET coupling with QDs. Indeed, the QD570–RC complex shows a stronger enhancement (by a factor of up to three at $\lambda_{\mathrm{exc}} = 550$ nm) than QD530–RC (Fig. 5.10), which may be explained by a stronger absorption of QD570.

The FRET efficiency of the RC–QD complexes is an important factor of the enhancement effect. The authors have also shown that at smaller RC-to-QD molar ratios, the enhancement effect increases because more QDs interact with a single RC and, hence, more energy can be transferred (Fig. 5.10c,d).

Theoretical estimation of the influence of QDs on the optical generation of electron–hole pairs inside the photosynthetic system has been performed [28]. The model consisted of an RC of the cyanobacterium *Synechocystis* sp. PCC6803 and a single semiconductor QD. This RC contains 96 chlorophyll and 22 carotenoid molecules, all of which are involved in light harvesting. The central six chlorophylls also take part in the electron transfer. The electron transfer chain starts at the P700 chlorophyll molecules (the special pair or primary electron donor) and includes chlorophylls as primary acceptors, two phylloquinones as secondary acceptors, and three Fe4S4 clusters. In principle, the QD can be attached to either side (acceptor or donor), but the author assumed that the QD is bound to the RC on the electron donor side. Such conjugates can

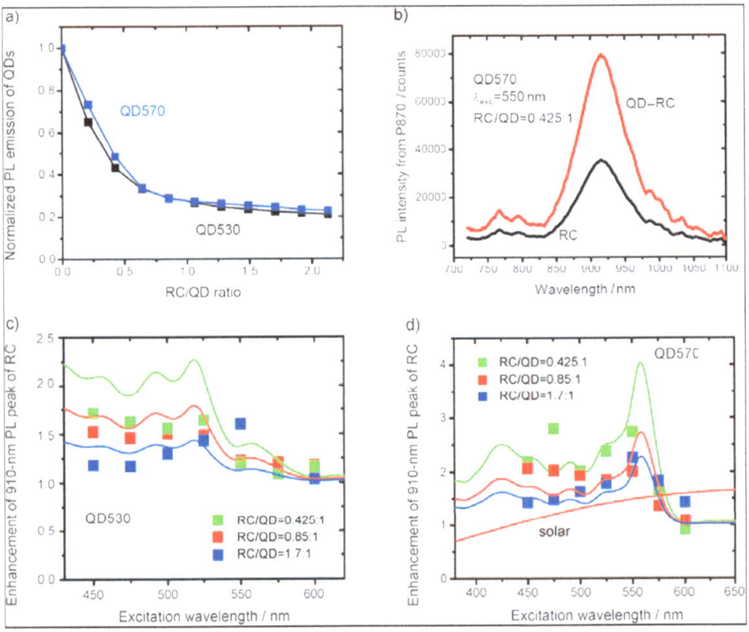

Figure 5.10 FRET and optical enhancement in complexes of CdTe quantum dots (QDs) and reaction centers (RCs). (a) Quenching of the QD photoluminescence (PL) as a function of the RC-to-QD molar ratio. Excitation wavelength, 450 nm; QD530 concentration, 0.5 mm; QD570 concentration, 0.25 mm. The volume of the RC solution added at each step was 4.8 and 2.4 mL, respectively. The dilution factors and inner filter and reabsorption effects were taken into account. (b) Increase in the PL peak at 910 nm associated with the PL emission from the bacteriochlorophyll (BChl) special pair (P870) of the RC. (c, d) Summary of the data of PL emission enhancement from RCs. The enhancement coefficients (colored squares) as a function of the excitation wavelength are given for the 910 nm PL band of P870 for complexes based on CdTe QD530 and QD570; sodium ascorbate (10 mM) was added to prevent the RCs from photo-oxidizing. The solar spectrum is shown in (d). Reproduced from Ref. [52], with permission from John Wiley & Sons, Inc.

be obtained using a sulfide bond with the cysteine residue of a mutant protein, as described in Ref. [23]. In the light-harvesting process, an optically excited electron–hole pair becomes trapped at the specatial pair. Then, the special pair loses an electron, which travels across the membrane along the electron transfer chain to

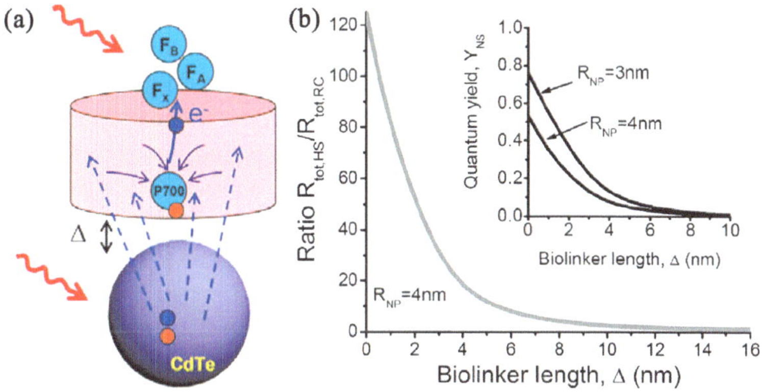

Figure 5.11 FRET efficiency within the hybrid systems engineered from CdTe quantum dots (QDs) and bacterial photosynthetic reaction centers (RCs). (a) Geometry of the hybrid system. Förster energy transfer couples the CdTe QD and the RC. (b) Calculated ratios of the integrated rates of generation of exited electrons for the RC and for the hybrid RC–QD system as a function of the biolinker length. Inset: Quantum yields for the RC and the hybrid complex as a function of the biolinker length. Reproduced from Ref. [28], with permission from John Wiley & Sons, Inc.

the Fe4S4 clusters. This electron transfer and subsequent excitation of the Fe4S4 clusters triggers a series of reactions that eventually results in the synthesis of ATP and reduction of NADP+.

An ensemble of absorbing chlorophylls has been modelled as a cylinder, which is 5 nm in radius and 6 nm in height (Fig. 5.11a). The RC contains many chlorophyll molecules with different orientations of dipoles and absorbs photons coming from all directions; that is, the absorption of light by the "chlorophyll cylinder" is isotropic.

Incident photons are absorbed by both the QD and the RC. Excitons optically generated inside the QD may recombine there or be transferred to the RC via the FRET mechanism. The absorption spectrum of the RC from the cyanobacterium *Synechocystis* sp. has two bands with the wavelengths $\lambda_{RC1} \approx 680$ nm and $\lambda_{RC2} \approx 430$ nm. The 96 chlorophyll inside the RC are responsible for the 680 nm band. FRET is effective if the overlap between the emission spectrum of the QD and the absorption spectrum of the RC is significant. The absorption of optical energy inside the QD occurs if $\lambda < \lambda_{exc}$, where λ and λ_{exc} are the wavelengths of incident light and exciton emission, respectively. Therefore, to ensure efficient FRET between QDs and RCs, λ_{RC1} should be approximately equal to λ_{exc}. Spherical CdTe QDs with $\lambda = 677$ nm were used because they exhibited a strong overlap between the QD and RC spectra. Using a set of equations, the authors calculated the rates of generation of exited electrons for the QD–RC hybrid system (RHS) and for the RC alone (RRC) and found a very strong increase in the RHS compared to RRC (by a factor of about 100). This was explained by the fact that the absorption cross section of a semiconductor QD is much larger than that of an RC (Fig. 5.11b), the interaction between the QD and RC via the Förster transfer being the main mechanism resulting in enhanced optoelectronic properties. The QD in the hybrid structure supplies excitons to the RC, where they become separated within a very short time.

The hybrid structure considered here has a strong advantage for light-harvesting applications where one monolayer or a thin film is used. The amount of light energy absorbed by a thin film composed of hybrid complexes will be substantially enhanced. However, in the case of a thick film, a reduced quantum yield may be a limiting factor, and we should measure the total external quantum yield. It is also interesting to compare our hybrid system with the structure composed of metal NPs and chlorophylls. The metal–chlorophyll complexes have a strongly enhanced absorption owing to the plasmon resonance in metal NPs. However, compared to the semiconductor-based complex, the metal–photosystem complex has some disadvantages:

- A metal NP has a high absorption coefficient; hence, the entire structure has a low quantum yield (note that the absorption cross section of a metal NP is typically much larger than that of a semiconductor NP).
- The enhancement occurs only at wavelengths close to the plasmon frequency, whereas the enhancement effect in a semiconductor-based complex is expected to occur in a wide range of $\lambda < \lambda_{exc}$.

Research in energy transfer between QDs and RCs has been extended by Schmitt et al. [68] by using a series of QDs with different emission wavelengths and surface charges and several photosynthetic light-harvesting complexes, including LH1 of *Rhodospirillum rubrum* and LH2 of *Rhodobacter sphaeroides*, trimeric phycoerythrin isolated from *Gloeobacter violaceus*, trimeric phycocyanin isolated from *Spirulina platensis*, and trimeric allophycocyanin from *Arthrospira platensis*. The results of that study show that the surface charge of the QDs is not essential for obtaining a functionally coupled hybrid system with efficient excitation energy transfer from QDs to isolated pigment–protein complexes and those structural parameters govern the formation of hybrid systems with desired properties of excitation energy transfer [68].

5.5 Conclusions and Perspectives

Novel advanced nanomaterials containing highly functional biological molecules have promising potential applications in many fields, including electronics, biosensing, imaging, and energy harvesting. In this short review, we have summarized the recent progress in understanding the mechanisms of energy transfer between semiconductor QDs and photosensitive proteins in their hybrid heterostructures in an attempt to predict the possible ways to improve the biological function of photochromic biomolecules and develop novel functional materials with unique controlled photonic properties that could be used in optical switching, photovoltaics, and data storage. We have discussed the possible mechanisms of energy transfer from QDs to photosensitive biomolecules and analyzed experimental evidence for these mechanisms. We also described the principles of the engineering of these hybrid structures, the donor/acceptor parameters affecting energy transfer efficiency in them, and the biological function and functionality of the structures. Light harvesting and directional energy transfer with the use of semiconductor QDs are particularly promising approaches to developing new concepts of artificial photosynthetic systems and hybrid photovoltaic devices [61]. We believe that the novel nanobiohybrid materials will have numerous optoelectronic, photonic, and photovoltaic applications.

Acknowledgments

This work was supported by the Horizon 2020 MERA-NET program under the project ICENAP. The part of the work devoted to QD surface chemistry and QD surface ligand optimization to achieve efficient energy transfer from nano- to biosystems was supported by the Russian Science Foundation, under grant no. 14-13-01160.

We thank Mr. Vladimir Ushakov and Ms. Daria Solovyeva for their help in the preparation of the manuscript.

References

1. Agranovich, V. M., and Galanin, M. D. (1982). *Electronic Excitation Energy Transfer in Condensed Matter*, Amsterdam: North-Holland.
2. Al-Aribe, K. M., Knopf, G. K., and Bassi, A. S. (2011). Photoelectric monolayers based on self-assembled and oriented purple membrane patches, *J. Microelectromech. Syst.*, **20**, pp. 800–810.
3. Artemyev, M., Ustinovich, E., and Nabiev, I. (2009). Efficiency of energy transfer from organic dye molecules to CdSe-ZnS nanocrystals: nanorods versus nanodots, *J. Am. Chem Soc.*, **131**, pp. 8061–8065.
4. Bilan, R., Fleury, F., Nabiev, I., and Sukhanova, A. (2015). Quantum dot surface chemistry and functionalization for cell targeting and imaging, *Bioconjugate Chem.*, **26**, pp. 609–624.
5. Birge, R. R. (1990). Photophysics and molecular electronic applications of the rhodopsins, *Annu. Rev. Phys. Chem.*, **41**, pp. 683–733.
6. Birge, R. R., Gillespie, N. B., Izaguirre, E. W., Kusnetzow, A., Lawrence, A. F., Singh, D., Song, Q. W., Schmidt, E., Stuart, J. A., Seetharaman, S., and Wise, K. J. (1999). Biomolecular electronics: protein-based associative processors and volumetric memories, *J. Phys. Chem. B*, 103, **49**, pp. 10746–10766.
7. Blanton, S. A., Leheny, R. L., Hines, M. A., and Guyot-Sionnest, P. (1997). Dielectric dispersion measurements of CdSe nanocrystal colloids: observation of a permanent dipole moment, *Phys. Rev. Lett.*, **79**, pp. 865–868.
8. Bouchonville, N., Molinari, M., Sukhanova, A., Artemyev, M., Oleinikov, V., Troyon, M., and Nabiev, I. (2011). Charge-controlled assembling of bacteriorhodopsin and semiconductor quantum dots for fluorescence

resonance energy transfer-based nanophotonic applications, *App. Phys. Lett.*, **98**, 013703.

9. Bouchonville, N., Le Cigne, A., Sukhanova, A., Molinari, M., and Nabiev, I. (2013). Nano-biophotonic hybrid materials with controlled FRET efficiency engineered from quantum dots and bacteriorhodopsin, *Laser Phys. Lett.*, **10**, 085901.
10. Brizzolara, R. A. (1995). A method for patterning purple membrane using self-assembled monolayers, *BioSystems*, **35**, pp. 137–40.
11. Buhbut, S., Itzhakov, S., Tauber, E., Shalom, M., Hod, I., Geiger, T., Garini, Yu., Oron, D., and Zaban, A. (2010). Built-in quantum dot antennas in dye-sensitized solar cells, *ACS Nano*, **4**(3), pp. 1293–1298.
12. Chen, Z., Lewis, A., Takei, H., and Nebenzahl, I. (1991). Bacteriorhodopsin oriented in polyvinyl alcohol films as an erasable optical storage medium, *Appl. Opt.*, **30**, pp. 5188–5196.
13. Chen, D., Lu, Y., Sui, S., Xu, B., and Hu, K. (2003). Oriented assembly of purple membrane on solid support, mediated by molecular recognition. *J. Phys. Chem. B*, **107,** pp. 3598–3605.
14. Clapp, A. R., Medintz, I. L., Mauro, J. M., Fisher, B. R., Bawendi, M. G., and Mattoussi, H. (2004). Fluorescence resonance energy transfer between quantum dot donors and dye-labeled protein acceptors, *J. Am. Chem. Soc.*, **126**, pp. 301–310.
15. Clapp, A. R., Medintz, I. L., and Mattoussi, H. (2006). Forster resonance energy transfer investigations using quantum-dot fluorophores, *ChemPhysChem*, **7**, pp. 47–57.
16. Darst, S. A., Ahlers, M., Meller, P. H., Kubalek, E. W., Blankenburg, R., Ribi, H. O., Ringsdorf, H., and Kornberg, R. D. (1991). Two-dimensional crystals of streptavidin on biotinylated lipid layers and their interactions with biotinylated macromolecules, *Biophys. J.*, **59**, pp. 387–396.
17. Dave, B. C., Soyez, H., Miller, J. M., Dunn, B., Valentine, J. S., and Zink, J. I. (1995). Synthesis of protein-doped sol-gel SiO_2 thin films: evidence for rotational mobility of encapsulated cytochrome C, *Chem. Mater.*, **7**, pp. 1431–1434.
18. Dayal, S., Lou, Y., Samia, A. C., Berlin, J. C., Kenney, M. E., and Burda, C. (2006). Observation of non-Forster-type energy-transfer behavior in quantum dot-phthalocyanine conjugates, *J. Am. Chem. Soc.*, **128**, pp. 13974–13975.
19. Förster, T. (1948). Intermolecular energy transference and fluorescence, *Ann. Physik*, **2**, pp. 55–75.

20. Deisenhofer, J., and Michel, H. (1989). Nobel lecture. The photosynthetic reaction centre from the purple bacterium *Rhodopseudomonas viridis*, *EMBO J.*, **8**, pp. 2149–2170.
21. den Hollander, M. J., Magis, J. G., Fuchsenberger, P., Aartsma, T. J., Jones, M. R., and Frese, R. N. (2011). Enhanced photocurrent generation by photosynthetic bacterial reaction centers through molecular relays, light-harvesting complcxcs, and direct protein-gold interactions, *Langmuir*, **27**, pp. 10282–10294.
22. Förster, T. (1959). Transfer mechanism of electronic excitation, *Discuss. Faraday Soc.*, **27**, pp. 7–17.
23. Frolov, L., Rosenwaks, Y., Carmeli, C., Carmeli, I. (2005). Fabrication of a photoelectronic device by direct chemical binding of the photosynthetic reaction center protein to metal surfaces, *Adv. Mater.*, **17**, pp. 2434–2437.
24. Griep, M. H., Walczak, K., Winder, E., Lueking, D. R., and Friedrich, C. R. (2007). An integrated bionanosensing method for airborne toxin detection. *Proc. SPIE*, 6646:66460F.
25. Griep, M. H. (2008). Quantum dot/optical protein bio-nano hybrid system biosensing, in *Proceedings of the Mechanical Engineering and Engineering Mechanics*, Houghton, MI: Michigan Technological University.
26. Griep, M., Walczak, K., Winder, E., Lueking, D., and Friedrich, C. (2010). Quantum dot enhancement of bacteriorhodopsin-based electrodes, *Biosens. Bioelectron.*, **25**, pp. 1493–1497.
27. Griep, M., Martin, J., Rodriguez, V., Winder, E., Lueking, D., Mackay, R., Friedrich, C., and Karna S. P. (2011). Multi-functional protein-QD hybrid substrates for photovoltaics and real-time biosensing. *11th IEEE International Conference on Nanotechnology, Conference Publications*, Oregon, USA, pp. 233–236.
28. Govorov, A. O. (2008). Enhanced optical properties of a photosynthetic system conjugated with semiconductor nanoparticles: the role of Forster transfer, *Adv. Mater.*, **20**, pp. 4330–4335.
29. Groma, G. I., Kelemen, L., Kulcsár, A., Lakatos, M., and Váró, G. (2001). Photocycle of dried acid purple form of bacteriorhodopsin, *Biophys. J.*, **81**, pp. 3432–3441.
30. Guldi, D. M. (2002). Fullerene–porphyrin architectures; photosynthetic antenna and reaction center models, *Chem. Soc. Rev.*, **31**, pp. 22–36.
31. Gust, D., Moore, T. A., and Moore, A. L. (2001). Mimicking photosynthetic solar energy Transduction, *Acc. Chem. Res.*, **34**, pp. 40–48.

32. Hampp, N. (2000). Bacteriorhodopsin as a photochromic retinal protein for optical memories, *Chem. Rev.*, **100**, pp. 1755–1776.

33. Hampp, N. (2000). Bacteriorhodopsin: mutating a biomaterial into an optoelectronic material, *Appl. Microbiol. Biotechnol.*, **53**, pp. 633–639.

34. Haupts, U., Tittor, J., and Oesterhelt, D. (1999). Closing in on bacteriorhodopsin: progress in understanding the molecule. *Ann. Rev. Biophys. Biomol. Struct.*, **28**, pp. 367–399.

35. He, J.-A., Samuelson, L., Li, L., Kumar, J., and Tripathy, S. K. (1998). Photoelectric properties of oriented bacteriorhodopsin/polycation multilayers by electrostatic layer-by-layer assembly, *J. Phys. Chem. B*, **102**, pp. 7067–7072.

36. Heathcote, P., Fyfe, P. K., and Jones, M. R. (2002). Reaction centres: the structure and evolution of biological solar power, *Trends Biochem. Sci.*, **27**, pp. 79–87.

37. Henderson, R., Jubb, J. S., and Whytock, S. (1978). Specific labelling of the protein and lipid on the extracellular surface of purple membrane, *J. Mol. Biol.*, **123**, pp. 259–274.

38. Holder, E., Tessler, N., and Rogach, A. L. (2008). Hybrid nanocomposite materials with organic and inorganic components for opto-electronic devices, *J. Mater. Chem.*, **18**, pp. 1064–1078.

39. Hu, X. C., Damjanovic, A., Ritz, T., and Schulten, K. (1998). Architecture and mechanism of the light-harvesting apparatus of purple bacteria, *Proc. Natl Acad. Sci. U S A*, **95**, pp. 5935–5941.

40. Kalsin, A. M., Fialkowski, M., Paszewski, M., Smoukov, S. K., Bishop, K. J., and Grzybowski, B. A. (2006). Electrostatic self-assembly of binary nanoparticle crystals with a diamond-like lattice, *Science*, **312**, pp. 420–424.

41. Keszthelyi, L. (1980). Orientation of membrane fragments by electric field, *Biochim. Biophys. Acta*, **598**, pp. 429–436.

42. King, C., Barbiellin, B., Moser, D., and Renugopalakrishnan, V. (2012). Exactly soluble model of resonant energy transfer between molecules, *Phys. Rev. B*, **85**, 125106.

43. Koyama, K., Yamaguchi, N., and Miyasaka, T. (1994). Antibody-mediated bacteriorhodopsin orientation for molecular device architectures, *Science*, **265**, pp. 762–765.

44. Krivenkov, V., Samokhvalov, P., Solovyeva, D., Bilan, R., Chistyakov, A., and Nabiev, I. (2015). Two-photon-induced Förster resonance energy transfer in a hybrid material engineered from quantum dots and bacteriorhodopsin, *Opt. Lett.*, **40**, pp. 1440–1443.

45. Kühlbrandt, W. (1995). Many wheels make light work, *Nature*, **374**, pp. 497–498.
46. Li, L. S., and Alivisatos, A. P. (2003). Origin and scaling of the permanent dipole moment in CdSe nanorods, *Phys. Rev. Lett.*, **90**, pp. 097402–097405.
47. Luecke, H., Schobert, B., Richter, H. T., Cartailler, J. P., and Lanyi, J. K. (1999). Structure of bacteriorhodopsin at 1.55 A resolution, *J. Mol. Biol.*, **291**, pp. 899–911.
48. Mackowski, S. (2010). Hybrid nanostructures for efficient light harvesting, *J. Phys. Condens. Matter.*, **22**, 193102, pp. 1–17.
49. Mansour, A. N., Wong, C., and Brizzolara, R. A. (1994). Atomic structure of amorphous Al100-2xCoxCex (x = 8, 9, and 10) and Al80Fe10Ce10 alloys: an XAFS study, *Phys. Rev. B Condens. Matter.*, **50**, pp. 12401–12412.
50. Medintz, I. L., Clapp, A. R., Mattoussi, H., Goldman, E. R., Fisher, B.and Mauro, J. M. (2003). Self-assembled nanoscale biosensors based on quantum dot FRET donors, *Nat. Mater.*, **2**, pp. 630–638.
51. Miyasaka, T., Koyama, K., and Itoh, I. (1992). Quantum conversion and image detection by a bacteriorhodopsin-based artificial photoreceptor, *Science*, **255**, pp. 342–344.
52. Nabiev, I., Rakovich, A., Sukhanova, A., Lukashev, E., Zagidullin, V., Pachenko, V., Rakovich, Y. P., Donegan, J. F., Rubin, A. B., and Govorov, A. O. (2010). Fluorescent quantum dots as artificial antennas for enhanced light harvesting and energy transfer to photosynthetic reaction centers, *Angew. Chem. Int. Ed. Engl.*, **49**, pp. 7217–21.
53. Nantalaksakul, A., Reddy, D. R., Bardeen, C. J., and Thayumanavan. S. (2006). Light harvesting dendrimers, *Photosynth. Res.*, **87**, pp. 133–50.
54. Patil, A. V., Premaruban, T., Berthoumieu, O., Watts, A., and Davis, J. J. (2012). Enhanced photocurrent in engineered bacteriorhodopsin monolayer, *J. Phys. Chem. B*, **116**(1), pp. 683–689.
55. Pei, R., Cui, X., Yang, X., and Wang, E. (2001). Assembly of alternating polycation and DNA multilayer films by electrostatic layer-by-layer adsorption, *Biomacromolecules*, **2**, pp. 463–468.
56. Pons, T., Medintz, I. L., Sykora, M., and Mattoussi, H. (2006). Spectrally resolved energy transfer using quantum dot donors: ensemble and single-molecule photoluminescence studies, *Phys. Rev. B*, **73**, 245302.
57. Rakovich, A., Sukhanova, A., Bouchonville, N., Lukashev, E., Oleinikov, V., Artemyev, M., Lesnyak, V., Gaponik, N., Molinari, M., Troyon, M., Rakovich, Yu, P., Donegan, J. F., and Nabiev, I. (2010). Resonance energy transfer

improves the biological function of bacteriorhodopsin within a hybrid material built from purple membranes and semiconductor quantum dots, *Nano Lett.*, **10**, pp. 2640–2648.

58. Rakovich, A., Nabiev, I., Sukhanova, A., Lesnyak, V., Gaponik, N., Rakovich, Yu.P., and Donegan, J. F. (2013). Large enhancement of nonlinear optical response in a hybrid nanobiomaterial consisting of bacteriorhodopsin and cadmium telluride quantum dots, *ACS Nano,* **7**, pp. 2154–2160.
59. Rakovich, A., Donegan, J. F., Oleinikov, V. A., Molinari, M., Sukhanova, A., Nabiev, I., and Rakovich, Y. P. (2014). Linear and nonlinear optical effects induced by energy transfert from semiconductor nanoparticles to photosensitive biological systems, *J. Photochem. Photobiol. C: Photochem. Rev.*, **20**, pp. 17–32.
60. Renugopalakrishnan, V., Barbiellini, B., King, C., Molinari, M., Mochalov, K., Sukhanova, A., Nabiev, I., Fojan, P., Tuller, H., Chin, M., Somasundaran, P., Padrós, E., and Ramakrishna, S. (2014). Engineering a robust photovoltaic device with quantum dots and bacteriorhodopsin, *J. Phys. Chem. C*, **118**, pp. 16710–16717.
61. Rogach, A. L. (2011). Fluorescence energy transfer in hybrid structures of semiconductor nanocrystals, *Nano Today,* **6**, pp. 355–365.
62. Saab, M. B., Estephan, E., Cloitre, T., Legros, R., Cuisinier, F. J., Zimanyi, L., and Gergely, C. (2009). Assembly of purple membranes on polyelectrolyte films, *Langmuir,* **25**, pp. 5159–5167.
63. Sadhu, S., and Patra, A. (2008). Donor-acceptor systems: energy transfer from CdS quantum dots/rods to Nile Red dye, *ChemPhysChem,* **9**, pp. 2052–2058.
64. Saga, Y., and Watanabe, T. (1999). Mechanism of photocurrent generation from bacteriorhodopsin on gold electrodes, *J. Phys. Chem. B,* **103**, pp. 234–238.
65. Samia, A. C., Dayal, S., and Burda, C. (2006). Quantum dot-based energy transfer: perspectives and potential for applications in photodynamic therapy, *Photochem. Photobiol.*, **8**2, pp. 617–625.
66. Samokhvalov, P., Artemyev, M., and Nabiev, I. (2013). Current trends in colloidal synthesis of highly luminescent semiconductor nanocrystals, *Chemistry*, **19**, pp. 1534–1546.
67. Schleifenbaum, F., Blum, C., Elgass, K., Subramaniam, V., and Meixner, A. J. (2008). New insights into the photophysics of DsRed by multiparameter spectroscopy on single proteins, *J. Phys. Chem. B*, **112**, pp. 7669–7674.
68. Schmitt, F. J., Maksimov, E. G., Hatti, P., Weissenborn, J., Jeyasangar, V., Razjivin, A. P., Paschenko, V. Z., Friedrich, T., and Renger, G. (2012).

Coupling of different isolated photosynthetic light harvesting complexes and CdSe/ZnS nanocrystals via Forster resonance energy transfer, *Biochim. Biophys. Acta*, **1817**, pp. 1461–1470.

69. Schranz, M., Noll, F., and Hampp, N. (2007). Oriented purple membrane monolayers covalently attached to gold by multiple thiole linkages analyzed by single molecule force spectroscopy, *Langmuir*, **23**, pp. 11134–1138.
70. Sukhanova, A., Devy, J., Venteo, L., Kaplan, H., Artemyev, A., Oleinikov, V., Klinov, D., Pluot, M., Cohen, J. H. M., and Nabiev, I. (2004). Biocompatible fluorescent nanocrystals for immunolabeling of membrane proteins and cells, *Anal. Biochem.,* **324**, pp. 60–67.
71. Sukhanova, A., Baranov, A. V., Perova T., Cohen, J. H. M., and Nabiev, I. (2006). Controlled self-assembly of nanocrystals in polycrystalline fluorescent dendrites with energy-transfer properties, *Angew. Chem. Intl. Ed.,* **45**, pp. 2048–2052.
72. Sukhanova, A., Even-Desrumeaux, K., Kisserli, A., Tabary, T., Reveil, B., Millot, J. M., Chames, P., Baty, D., Artemyev, M., Poly, S., Oleinikov, V. A., Pluot, M., Cohen, J. H. M., and Nabiev, I. (2012). Oriented conjugates single-domain antibodies and quantum dots: toward new generation of ultra-small diagnostic nanoprobes, *NanoMedicine: NBM,* **8**, pp. 516–525.
73. Vsevolodov, N. N. (1998). *Biomolecular Electronics: An Introduction via Photosensitive Proteins*, Boston, MA: Birkhauser, pp. 71–217.
74. Weetall, H. H., and Samuelson, L. A. (1998). Optical and electrical properties of bacteriorhodopsin Langmuir-Blodgett films, *Thin Solid Films,* **312**, pp. 306–312.

Chapter 6

Ultrafast Functional Dynamics in Proteins: Local Molecular Reporters and Femtosecond 2D Spectroscopy

Andrea Cannizzo,[a] Jérémie Léonard,[b] and Stefan Haacke[b]
[a]*Institute of Applied Physics, University of Bern, Sidlerstrasse 5, CH-3012 Bern, Switzerland*
[b]*Institut de Physique et Chimie des Matériaux de Strasbourg Université de Strasbourg, CNRS UMR 7504, 23 rue du Loess, 67034 Strasbourg Cedex, France*
andrea.cannizzo@iap.unibe.ch, jeremie.leonard@ipcms.unistra.fr, haacke@unistra.fr

Uncovering local structural dynamics and fluctuations in biomolecules is one of the main challenges of modern science, with the aim of understanding how they are ultimately linked to the biological activity, which often involves global large-scale and thus slower motion of their structure. Strictly related to local and global fluctuations, the mechanisms of redistribution and dissipation of excess of vibrational and electronic energy are at the center of intensive studies aimed to understand, among all, the role they play in, for example, intraprotein or protein–DNA signaling. This chapter deals with the recent advances in ultrafast nonlinear spectroscopy of proteins with a focus on the near-UV spectral region. Here, the selective excitation and probing of natural amino acids such as tryptophans and tyrosines provides great insight

Ultrafast Dynamics at the Nanoscale: Biomolecules and Supramolecular Assemblies
Edited by Irene Burghardt and Stefan Haacke

ISBN 978-981-4745-33-8 (Hardcover), 978-981-4745-34-5 (eBook)
www.panstanford.com

into the molecular processes including energy and charge transfer that trigger the biological processes. Examples from retinal and heme proteins illustrate the wealth and precision of information provided by these spectroscopic approaches. The most recent advances in two-dimensional femtosecond near-UV spectroscopy are discussed underscoring the great potential of this technique for multichromophore systems such as proteins and DNA.

6.1 Introduction

In the last 20 years, ultrafast spectroscopy has dramatically changed our way to investigate matter. Now we can indeed follow with femtosecond resolution all the photoinducible transformations occurring between out-of-equilibrium states, and our comprehension of excited-state physics and chemistry has undergone a breathless leap [1–3]. A less evident quality is that now we can directly access times where thermal fluctuations occur, mainly time resolving spectral diffusion processes. This has given us, on the one hand, the unique capability to disentangle homogenous and inhomogeneous line-shape broadening and, on the other hand, novel methodologies to track down collective motions and long-range interactions within a molecule, which are quickly covered or even destroyed by incoherent fluctuations [3–6].

The last decade has seen an impressive improvement of ultrafast spectroscopic techniques and nowadays we have at our disposal the femtosecond time-resolved versions of virtually all the steady-state techniques (optical absorption, fluorescence emission, Raman spectroscopy, optical Kerr effect, THz spectroscopy, photoelectron spectroscopy, and even election and X-ray diffraction) and completely new spectroscopic methodologies as coherent nonlinear optical spectroscopies and microscopy, which include multidimensional photon echo (PE), coherent anti-Stokes Raman, but also chemically sensitive techniques for all-optical recognition, etc. [7–9]. These methodology developments were largely enabled and fueled by the availability and increasing versatility of femtosecond optical parametric amplifiers providing relatively powerful and stable femtosecond-tunable visible (Vis) and ultraviolet (UV) pulses, thus allowing the extension of these techniques to the whole Vis and UV regions. The latter is the spectral domain this chapter focuses on.

Fluorescent labels that chemically and structurally mimic amino acids or nucleotides are very commonly used for (time-resolved) spectroscopic investigations of the structure and dynamics of isolated biomolecules [10–12] or their complexes [13, 14], as long as they are "noninvasive" in terms of perturbing the structure and thermodynamics of the biomolecules and of hampering the biological function [15]. In the context of DNA studies, 2-aminopurine (2AP) is since long a synthetic label of choice as its absorption spectrum is red-shifted from the large extinction background due to the natural nucleotides, thus allowing for selective and label-specific excitation [16]. It is worth mentioning that important new developments beyond 2AP are under way for DNA labeling [17]. In the case of protein spectroscopy, tryptophan (Trp) and tyrosine (Tyr) residues play a similar role since they stand out among natural amino acids as presenting long-wavelength absorption bands of fairly high extinction coefficients. Trp in particular has been extensively used in ultrafast fluorescence experiments, as a local molecular probe, and to a lesser extent in transient absorption. Also fluorescent, synthetic amino acids are being actively synthesized for protein structure sensing [18].

The signal encoded in the nonlinear optical response of the biomolecules is exploited in terms of time-dependent excited-state signal amplitude, spectral shifts and dynamic line narrowing and polarization anisotropy decays. However, the conversion from these experimental features to the information on the structural dynamics of the biomolecular (protein/DNA) environment and its constituents is not straightforward since the conformational phase space is extremely large, and the exact excited-state quenching mechanism in many cases still poorly understood [11, 19–21]. The interpretation commonly resorts to a basic set of fundamental hypothesis proven on simple systems [12]. For systems with increasing complexity, intensive modeling through, for example, molecular dynamics simulations, and in particular mixed quantum mechanics/molecular mechanics (QM/MM) computational methods [22, 23], are required for an improved, possibly quantitative interpretation of UV-Vis spectroscopic data in terms of biomolecular structures [24]. This part of research will benefit in the future of advanced computational methods and increasing machine power.

In the following chapters, we review recent progress in experimental approaches and studies on ultrafast functional dynamics of proteins with emphasis on a few examples related to our own research. In the context of ultrafast photoinduced dynamics in photosensory proteins, the advantage of using UV chromophores as local reporters is emphasized when phenomena triggered by visible light are investigated (Section 6.2). Indeed, ultrafast Vis pump/UV probe spectroscopies can provide information not only on the directly photoexcited moiety but also on the hosting molecular edifice. They allow us moving our focus from the "epicenter" to the "periphery," with the aim to track down concerted motions, long distance interactions, and modulation of local dynamics by changes in distant sites, and in general on the host edifice reaction. This is highlighted in Section 6.3 for processes such as dielectric solvation, vibrational cooling and energy redistribution in prototypical heme proteins. The capability of 2D spectra to provide a "global" spectral signature of the molecule under investigation paves the way to all-optical chemical recognition methods and chemical sensitive microscopy (Section 6.4).

6.2 Photoinduced Charge Translocation and Dynamic Dielectric Response

In this first example, the capability of Trp residues to report on the dynamics around a specific site is exemplified via the retinal protein bacteriorhodopsin (bR). The experiments show that Trp's can serve there, and in many other photoreceptor proteins, as a local probe for the charge translocations occurring on the photoexcited chromophore and/or in its vicinity.

6.2.1 Ultrafast Response of Amino Acids: Example of Retinal Proteins

One classical example where Trp residues were used as local probes for the photoinduced structural dynamics is the retinal protein bR

[25, 26]. This 248 kDa protein is the photoactivated proton pump of the archebacterium *Halobacterium salinarum,* and as such is an essential element of its photosynthetic machinery in anaerobic conditions. The photocycle of bR has been studied intensively with a large panel of spectroscopic techniques [27–30] and its ground-state and intermediate structures are known with atomic precision [31–33].

Upon absorption of visible light, bR undergoes a reversible photocycle that is characterized through the absorption changes of the protonated Schiff base of retinal (PSBR) by a sequence of different intermediates:

$$bR_{568} \xrightarrow{0.5ps} J_{625} \xrightarrow{3-5ps} K_{610} \xrightarrow{2\mu s} L_{550} \xrightarrow{60\mu s} M_{412} \xrightarrow{ms} O_{640} \rightarrow bR_{568}$$

The PSBR is deprotonated during the $L_{550} \rightarrow M_{412}$ transformation, leading to the proton pumping activity that increases the proton concentration on the outer cellular side. The latter is utilized to transform ADP into ATP as the central photosynthetic activity of bR.

bR has 8 Trp and 11 Tyr residues that can be used as optical probes to investigate the protein structure, the interactions of these residues with the PSBR, and its dynamical changes during the photocycle [34]. In the 1980s the time-resolved fluorescence quenching and anisotropy of the Trp's was monitored with microsecond resolution, allowing to draw conclusions on the protein conformation changes. Excitation of Trp's at 280 nm produces an identical photocycle as under selective excitation of the PSBR in the visible, suggesting efficient energy transfer taking place from the Trp's to the PSBR. During the photocycle variations of the Trp quenching efficiency are observed, suggesting significant structural changes modulating the energy transfer efficiency. This has been clearly substantiated by the observation of the two quenching processes, one occurring on the timescale of the L_{550} formation and a slower one associated with the formation of M_{412} [35].

An important study performed with picosecond time resolution [36] provided evidence for ultrafast energy transfer from Trp's that are in close vicinity to PSBR, namely W182 and W86 that flank the PSBR linear backbone (Fig. 6.1). Upon excitation with 4 ps pulses at 263.5 nm, the global emission of the eight Trp's was time-resolved

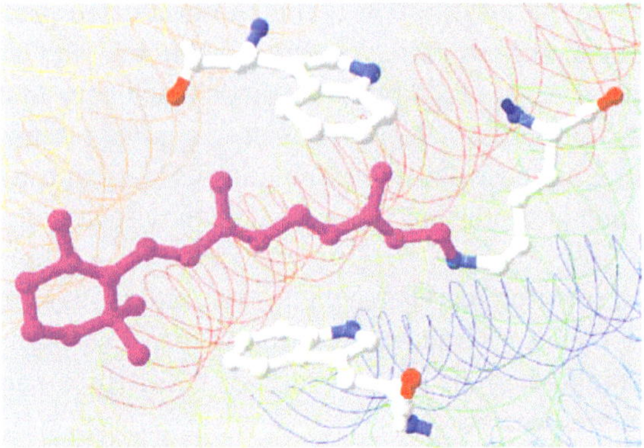

Figure 6.1 Close-up view or the retinal binding pocket in bR, according to the PDB entry 1C3W [33]. Due to a covalent bond with the N of Lys216, the protonated Schiff base of retinal (PSBR) is formed. Trp182 (above) and Trp86 (below) are at less than 5 Å distance from the PSBR backbone (purple).

providing a lifetime of 0.5 ns, which is significantly shorter than the Trp's lifetime in PSBR-lacking bacterio-opsin (1.25 ns). Clearly, and since PSBR has some high-energy transitions in the range of Trp fluorescence (300–350 nm) resonance energy transfer is at work. However, when two-photon excitation is performed with 527 nm pulses, exciting the PSBR selectively, Trp emission is observed that shows a 10 nm blue shift and most importantly, a decay time of 10 ps close to the experimental time resolution. This observation was interpreted as selective excitation of nearest-neighbor Trp's due to impulsive energy transfer from PSBR via dipole–dipole coupling, a short-range process. The Trp emission is then rapidly quenched due to reverse energy transfer, but with a sub-10 ps lifetime, as opposed to the 0.5 ns timescale when the emission is dominated by Trp residues located at larger distances from the PSBR [36].

While there is general consensus that light absorption leads to the trans-cis isomerization of the PSBR, the structural dynamics of the protein environment was addressed very little until recently. Some of the fundamentally important questions are:

- At which point in time after light absorption and by which mechanisms is the protein environment set in motion?

- Are there activation mechanisms other than PSBR isomerization?
- How does the protein environment act on the PSBR isomerization reaction to make it faster and more efficient than for PSBR in solution?

In more recent years, ultrafast mid-IR spectroscopy on bR [37–39] and other bacterial retinal proteins [40–42] has provided a few pieces of evidence that the protein is set in motion on a subpicosecond timescale. Using isotopically labeled proteins, the Diller group identified the time-resolved signatures in the amide I and amide II regions of bR and sensory rhodopsin II. While a photoinduced absorbance change is seen in amide I within the experimental time resolution (0.3 ps), the amide II region shows a slower 3 ps signal. Interestingly, partial decay possibly related to vibrational energy redistribution within the protein is observed on a 12 ps timescale [38]. In contrast to local probing via near-UV-sensitive aromatic amino acids, these experiments hint at the observed perturbations involving several amino acids and the water cluster [31, 43, 44] in close proximity to the PSBR.

The central question as to the mechanism that activates these structural changes has been revisited, as a variety of experiments performed on retinal proteins reconstituted with nonisomerizing PSBRs [45–48] indicated that the isomerization process is only half of the story. Indeed, the PSBR forms a large photoinduced dipole moment change $\Delta\mu_{\mathrm{ret}}$. Electroabsorption [49–51], hyper-Rayleigh scattering experiments [52, 53], and optical rectification [54, 55] determined values in the range of 15–30 D. This is related to the fact that the electronic state character changes from covalent to ionic [56], which implies a substantial translocation of the positive charge from the Schiff base towards the ionone head group [57–60]. Moving a portion of the positive charge from one end to the other is equivalent to adding a differential dipole moment equal to $\Delta\mu_{\mathrm{ret}}$ to the ground-state charge distribution. The general picture emerging from the femtosecond mid-IR and near-UV [61, 62] experiments on bR with isomerizing PSBR and nonisomerizing analogs, is that electronic polarization as well as the structural changes associated with isomerization are two photoactivation channels ultimately leading to free-energy storage in the protein.

6.2.2 Photoinduced Charge Transfer and Dielectric Dynamics Probed by Transient Trp Absorption

With the aim of detecting the possible time dependence of $\Delta\mu_{ret}$ with 90 fs resolution, a Vis pump, near-UV probe experiment was devised so as to monitor the differential absorption of bR's Trp residues upon excitation of the PSBR [62]. Indeed, the 15–30 D $\Delta\mu_{ret}$ generates an electric field change >10 MV/cm at the location of Trp182 and Trp86 (at less than 5 Å distance, Fig. 6.1), raising the expectation of a significant intraprotein Stark shift of the Trp S_0-L_A absorption band normally centered around 280 nm.

The dynamic Stark shift relies on the interaction between the electric dipole moments of the PSBR and of the Trp's units and the change of the latter upon excitation of the S_0–L_A transition ($\Delta\mu_{Trp}$). Since dipole–dipole interactions decrease with the sixth power of the interchromophore distance, it is obvious that only the two nearest-neighbor Trp86 and 182 will show a high enough signal. In addition, evaluating the dipole–dipole interactions readily predicts Trp86 to be much more sensitive than Trp182. Figure 6.1 and more explicitly Fig. 1 in Ref. [62] shows that the $\Delta\mu_{Trp}$ of Trp182 is almost orthogonal to $\Delta\mu_{ret}$, but the difference dipole moment of Trp86 is nearly colinear and antiparallel with $\Delta\mu_{ret}$. This leads to a decrease of the electrostatic interaction energy upon increase of $\Delta\mu_{ret}$ and thus a red shift of Trp86's absorption as observed experimentally [61–63]. A detailed study of the mutants W86F and W182F gave direct evidence for the calculated intraprotein Stark shift underpinning the role of Trp86 as a local probe for photoinduced electrostatic field changes. Indeed, both experiments and simulations show that the induced absorption peaking at 305 nm is due to the shifted absorption of Trp86 (Fig. 6.2). The Trp86-lacking mutant W86F displays a signal of much smaller amplitude at these wavelengths.

The dominant bleach signal at 270–280 nm is predominantly due to PSBR ground-state bleach at ~280 nm [36, 64], with only a minor contribution from Trp's due to excitonic coupling [61, 62]. The

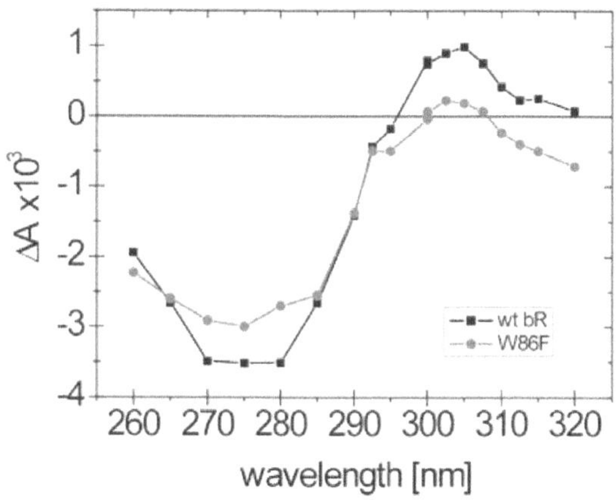

Figure 6.2 Differential near-UV absorption spectra of wt-bR and the W86F mutant for 100 fs pump–probe delay. The positive band observed at 300–310 nm in the wt protein immediately after PSBR excitation is much weaker in W86F. Spectra are reconstituted from kinetic traces at selected wavelengths [61].

smaller bleach amplitude in W86F directly demonstrates that W86 accounts only for a less than 20% contribution in this bleach band.

The time-dependent Stark effect response of Trp86 is singled out in the 305–315 nm interval, where excitonic coupling is negligible so that W86 and PSBR signals are additive [61]. It is fitted by an instrument response function (IRF)-limited rise time [61], a 0.45 ps decay, and a long-lived decay component. The absence of a rise time in this signal suggests that the torsion-induced further enhancement of $\Delta\mu_{\text{ret}}$ [65] is an event much faster than the experimental time resolution ($<$50 fs).

The 0.45 ps decay observed for the Trp86 Stark shift directly mirrors the PSBR excited-state decay (isomerization). The relatively large amplitude of the long-lived ($\gg$20 ps) component is unexpected by considering PSBR's dipole moment only, as the 13-cis (K state) and all-trans PSBR do not differ much [51]. Thus, this remaining long-lived Stark shift is assigned to a modified electrostatic interaction with the nearby environment: photoinduced structural changes located in the vicinity of W86 (e.g., hydrogen bonds, PSBR's counterion complex) or multiple smaller-scale electron density re-organizations in the protein scaffold. This is in qualitative agreement with the femtosecond mid-IR spectroscopy that provided evidence for such localized protein motion activated on a subpicosecond timescale [37, 38]. Finally, and not of least importance, the long-lived Stark shift may be an indication that a significant amount of energy is stored in the K state in form of electrostatic energy, as suggested in [31, 66, 67]. It may trigger structural rearrangements on longer timescales leading to subsequent proton transfer processes.

An alternative approach providing similar indirect insight into electronic and structural changes of the protein environment is afforded by time-resolved resonance Raman spectroscopy. When the probe wavelength is resonant with the Trp or Tyr absorption, the time-resolved Raman spectra select one specific type of amino acids [68, 69]. For bR, it was found that at least two Trp's and two Tyr's are involved in the structural response of the protein. By studying point mutations of bR with Phe's replacing Trp's, the dominant role of the three Trp's surrounding the PSBR in less than 5 Å distance was convincingly demonstrated. The differential Raman signals rise within the 3 ps time resolution of these experiments. The signals recover on a 30–35 ps timescale, which is attributed to the structural relaxation of the binding pocket, accommodating the isomerized PSBR, including H bonds that the Trp's are directly or indirectly engaged in. The Trp mutants allow then deriving some information about the possible local structural changes involved [68, 69].

6.3 Local Fluctuation and Energy Redistribution in Biomolecules

6.3.1 Solvation and Local Fluctuation

One of the first examples where ultrafast time-resolved fluorescence of Trp's was exploited to map local fluctuations in and around a protein, is the series of works by Zewail and collaborators on hydration dynamics at the protein surfaces [2, 70–73] and on intraprotein local solvation [74, 75] in several biological systems. They probed in real time the hydration dynamics of bulk water near the protein and at the protein–water interface (the so-called biological water) [2, 70, 71, 73, 76], revealing that the biological water has a completely different behavior with respect to the bulk water. They found that strong (electrostatic) interactions with the charged residues at the protein surface decrease the disorder of water at the interface, defining a rigid water structure with a much longer solvation time (10–100 ps, typically biphasic) than the bulk one ($\sim$1 ps). The former is a direct measure of surface hydration assisted by fluctuations of the protein: indeed on these timescale proteins are not rigid and their fluctuations modulate water solvation by allowing for restructuring in the new nonequilibrium state [72]. These experiments stimulated a rich computational activity, which shifted original interpretation based on water exchange towards an interplay between protein and solvent dynamics [77].

The same authors reported innovative studies on the hydration dynamics in the DNA grooves and the influence of drug binding on DNA hydration [2, 78, 79]. As observed in proteins, water molecules at the surface of DNA are dynamically ordered and authors reckon a crucial role of such water in interfacial recognition, not only of drugs but also between macromolecules. This study was carried out by following the time evolution of the Stokes shift and polarization anisotropy of the visible fluorescence of the drug, but not of the natural UV emission from nucleotide bases. Indeed their fluorescence lifetimes are too short to reveal the full temporal evolution of water solvation: regardless the specific nucleotide, their excited-state lifetimes fall in the subpicosecond timescale [80, 81], while the excess of electronic energy is disposed in water in less than 1 ps [78]. This confirms the occurrence of ultrafast internal conversion channels, which were rationalized invoking the presence of conical intersections [78, 81, 82]. See also the chapter by B. Kohler in this book. These articles also report that the excited-state deactivation and energy transfer mechanisms are more efficient in water than in other solvents. Since such a fast deactivation is supposed to play a key role in the stability of DNA and protection against UV photodamage, these studies confirm a specificity of water, as far as dynamical processes in biomolecules are concerned.

These studies have also the merit to clarify for the first time the role of biological water in site-selected recognition with real-time techniques, not only in the DNA grooves but also in binding sites of several proteins [2, 74, 83].

6.3.2 Dissipation and Redistribution of Energy in Hemoproteins

Mapping local fluctuations and identifying the functionally relevant ones is one side of the same coin, where the other is understanding how a localized excess of energy, generated during a biological function, is redistributed amongst the protein modes [84–86]. An excess of mechanical and thermal energy is generated not only upon light absorption, but also during a chemical reaction at the active site of an enzyme. Energy dissipation and redistribution are also relevant to understand protein stability and may act as possible long-range interaction mechanisms [87, 88]. When looking for "real time" studies of such processes, there are very few works. They appeared very recently and are based on UV spectroscopy of Trp and Tyr residues in hemoproteins [89–95].

Hemoproteins indeed are ideal systems to study the vibrational energy flow in proteins. The heme group is relatively isolated from the rest of the protein and intraheme intramolecular vibrational energy redistribution (IVR) is faster than the dissipation channels toward the protein edifice [93, 96, 97]. Furthermore it is characterized by extremely short-lived excited states ($<$150 fs) because of very efficient nonradiative channels [93, 94, 98]. This enables depositing optically up to 40,000 cm^{-1} (or 100 kcal/mol) of excess of vibrational energy (EVE) in a well-localized site and in few hundreds of femtoseconds (corresponding to an almost instantaneous rise of the heme temperature of 500 K [93, 94, 97]), allowing time-resolved studies of the energy redistribution processes. In addition to this thermal jump, a photoexcitation of the heme cofactor can induce an ultrafast axial ligand dissociation from the central iron atom, depending on the protein and the oxidation state of the heme group. This triggers an out-of-plane dislocation of the metal and a doming of the heme that, right after the dissociation, leaves the system in an out-of-equilibrium state. A long-lasting debate concerns the question, whether the redistribution of this impulsively released strain energy and heat occurs as a mechanical (deformation) coherent wave propagating from the center to the protein periphery or in an incoherent fashion, where the protein edifice behaves as an anharmonic, inelastic, reservoir [88, 90, 99–101]. It is worth mentioning that these structural changes at the heme site are supposed to trigger in heme- and myoglobins a global conformational change from the so-called oxy to deoxy structure [88, 102].

UV spectroscopy of Trp and Tyr is ideal to uncover these mechanisms, because these amino acids can act as intraprotein, site-specific probes to report the arrival of the perturbation.

One of the first studies was carried out by Mizutani and coworkers on myoglobin with UV resonance Raman (RR) spectroscopy (Fig. 6.3) [92, 103]. In these studies, changes of Trp and Tyr UV-RR bands allowed the authors to track down picosecond conformational changes of the side chains and environmental changes around those residues in a site-specific fashion. Based on these spectral changes, the authors proposed a detailed microscopic description of the concerted movement of protein helices, impulsively driven by changes in the heme, in the first picoseconds, and then by rearrangements toward a deoxy structure, in hundreds of picoseconds to nanoseconds. Noteworthy, they did not observe any vibrational energy propagation, excluding that the observed dynamics may be due to a transient thermal effect.

However, in the case of myoglobin, the heme cofactor is almost in contact with the solvent and the lack of intraprotein energy dissipation could be in part due to an efficient dissipation toward the external environment. To avoid such channels, several research groups moved to horse heart cytochrome c (Cyt c), a relatively small membrane-bound heme protein which plays an important role in many electron transfer processes (Fig. 6.4a) [90, 91, 93, 94]. Indeed, in contrast to myoglobin, the six-coordinated heme of Cyt c is buried well inside the protein and, according to molecular dynamics simulations [97], a very weak influence of the solvent on the energy relaxation is expected. The thermal bath of the cofactor is therefore the protein itself.

Recently, several studies appeared where this class of proteins has been investigated with different UV spectroscopies (tunable single-wavelength UV and Vis transient absorption [TA] [90];

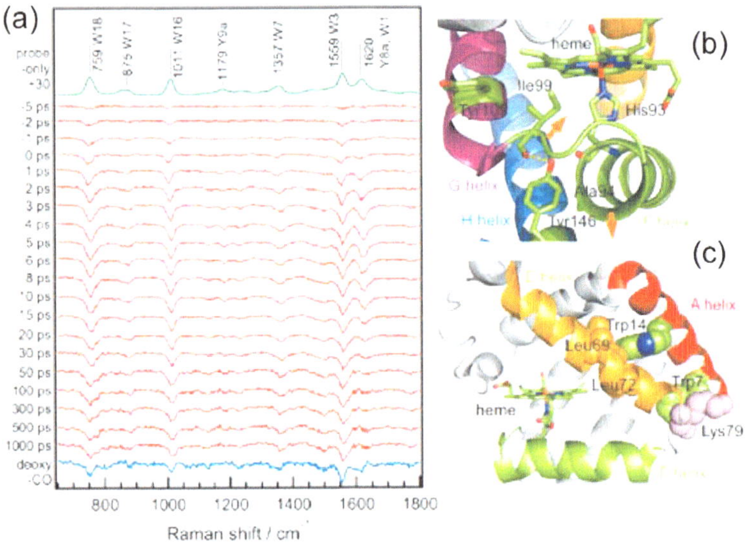

Figure 6.3 Picosecond conformational changes after ligand photodissociation in carbonmonoxy myoglobin (MbCO). Pump and Raman probes are at 408 nm and 232 nm, respectively. (a) Picosecond UV-RR difference spectra of the photodissociated horse skeletal MbCO in the range of 650–1800 cm^{-1}. The top trace is a probe-only spectrum corresponding to the UV-RR spectrum of MbCO divided by a factor of 30. The bottom trace is a deoxyMb-minus-MbCO difference spectrum. Spectra have been offset for clarity. (b and c) Crystallographic structure of horse Mb (PDB ID code 1DWR). (b) Close-up of the region near the two Tyr residues. The orange arrows indicate the directions of motions of the F helix and FG corner after CO dissociation, expected from crystallographic studies. (c) Close-up of the region near the two Trp residues. Reprinted with permission from Ref. [92]. Copyright (2007) National Academy of Sciences, U.S.A.

broadband UV- and Vis-TA and time-gated emission (TGE) [93, 94]; and anti-Stokes UV-RR on Trp residue [91]). The heme center was excited in order to study the interaction between the heme moiety of ferric and ferrous Cyt c and its single Trp residue. The latter is located at 9 Å from the heme and its fluorescence is strongly quenched by FRET to the heme group (Fig. 6.4b and Fig. 6.4c) [94]. In agreement with Vis studies they found that for the ferrous state the heme undergoes ultrafast photolysis of one of the axial ligand (a methionine labeled Met80, and highlighted in Fig. 6.4a) in less than

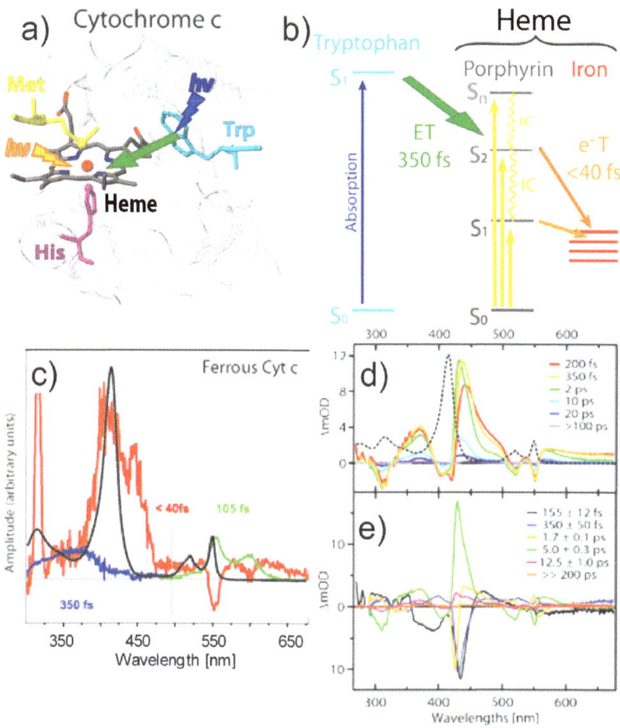

Figure 6.4 Cytochrome c (Cyt c) photophysics. (a) Crystallographic structure of ferrous Cyt c (PDB ID code 1HRC). The heme cofactor, the single tryptophan (Trp), and the two axial ligands, methionine (Met) and hisdine (His), are depicted. The two possible photoexcitation mechanisms and the Trp–heme FRET path are also shown. (b) Complete photocycle upon the possible excitation schemes. (c) Time-resolved emission upon 286 nm excitation. Only the spectral components and the relative time constants are shown: (i) the 350 fs component (blue line) is the Trp emission, quenched by FRET to the heme; (ii) the 105 fs spectrum (green line) is the Q band emission (S_1–S_0 transition); (iii) the 40 fs component (red line) is the emission stemming from the Soret band (S_2–S_0 transition) excited at 400 nm or shorter wavelength, and negative features in the Q band region describe a rise of the Q emission. The black trace depicts the stationary absorption spectrum. (d) A representative selection of transient absorption spectra at different delay times from ferrous Cyt c under 288 nm excitation. The dashed line represents the ground-state absorption spectrum in arbitrary units. (e) Decomposition of data in (d) in spectral components, each decaying with the respective time constant. Reprinted with permission from Ref. [94]. Copyright (2011) American Chemical Society.

100 fs and, after vibrational cooling within several picoseconds, it recombines back to the original sixfold coordination with a biphasic decay of 5 and 16 ps (Fig. 6.4d,e) [93, 96]. Conversely, ferric heme does not undergo any photodissociation, maintaining its sixfold coordination. Figure 6.5 illustrates the energy flow through Cyt c, measured with time-resolved Raman spectroscopy on Trp residues: relaxation and cooling show a biphasic decay with a fast, 1–3 ps, and a slow, ~10 ps, component, slightly dependent on the oxidation state (compare with Fig. 6.4) [93]. The faster component describes the heat transfer from the heme to the Trp [91, 93], while the slower is due to the statistical redistribution of the EVE among the low-frequency modes of the protein and the heat transfer to the solvent.

The pure geminate rebinding process on a picosecond timescale, the unequivocal assignment of the Trp response and the weak influence of the solvent make this protein a promising model system to investigate energy redistribution.

To explore the effect of the amount of energy deposited in the heme on the energy redistribution process, Consani et al. varied the excitation from 530 nm to 400 nm and to 288 nm, corresponding to an impulsively deposed EVE up to 100 kcal/mol (or a heme temperature increase of up to 500 K) [93]. In agreement with the results reported by Fujii et al. [91], they found that the kinetics of the energy flow is not affected by the amount of excess energy deposited, being an intrinsic feature of the protein. Accordingly, the authors asserted that the protein behaves as a linear, harmonic system that, as a cage, protects the heme and its functionality. They also followed the Trp–heme resonance energy transfer process by directly exciting the Trp with an UV pump. They observed that this process induces the Met80 detachment with almost unity quantum yield, just as under direct laser excitation. However, despite practically identical photocycles, the two excitation routes are significantly different in the way how the EVE is redistributed: with the excitation via Trp, and since IVR on the Trp is faster than energy transfer, part of the total EVE (~20 kcal/mol) is left on the amino acid generating an excited heme colder than upon direct excitation. This points out that in energy transfer processes the number of amino acids involved in the energy transfer path could play a role in facilitating the dissipation of the excess of energy.

Zang et al. also reported a distinctive 40 ps component in the decay of the Trp signal, present only in the ferrous version [90]. They argued that this was an evidence of a deformation coherently propagating through the molecule (a so-called protein quake [100, 104]), triggered by the impulsive bond breaking. As observed by the authors, a propagation time on the timescale of hundreds of picoseconds was expected [92, 105]. However neither Consani et al. nor Fujii et al. observed in the ferric and ferrous forms, respectively, any similar long component (Fig. 6.4d and Fig. 6.5, respectively). This supports instead the absence both of a thermal protein quake (i.e., a shock wave due to the heme thermal jump) and of a mechanical one (a deformation wave propagating from the heme group due to the sudden iron–ligand bond breaking). Accordingly the EVE is dissipated in a few picoseconds in an incoherent fashion from the heme to the bath, constituted of the low-frequency modes of the protein first and by the solvent later.

The localization of the vibrational energy during the flow across the protein has been related to the presence of anharmonic couplings amongst the protein vibrational modes [105]. It is tempting to relate the lack of localization to the aforementioned harmonic behavior. Moving to the lack of signature of a "mechanical protein quake," since recombination occurs in <10 ps, while, as also pointed out by Zang et al., conformational changes likely propagate in subnanosecond, the heme and its pocket could get back to the stable, ligated configuration prior to any important structural change. This would prevent the released strain energy from leaving the heme pocket. It also agrees with the lack of any long-lasting change in the protein configuration. Alternative explanations exist, including the fact that the position of the Trp residue would be unfavorable to report on a global structural change [93]. Consani et al. also reported a nonlinear dependence of the Trp response on the pump power, not present in the Vis interval, which could also be responsible for the 40 ps component reported by Zang et al. These aspects deserve still more experimental and theoretical investigation.

In conclusion of this section a few methodological notes deserve to be added.

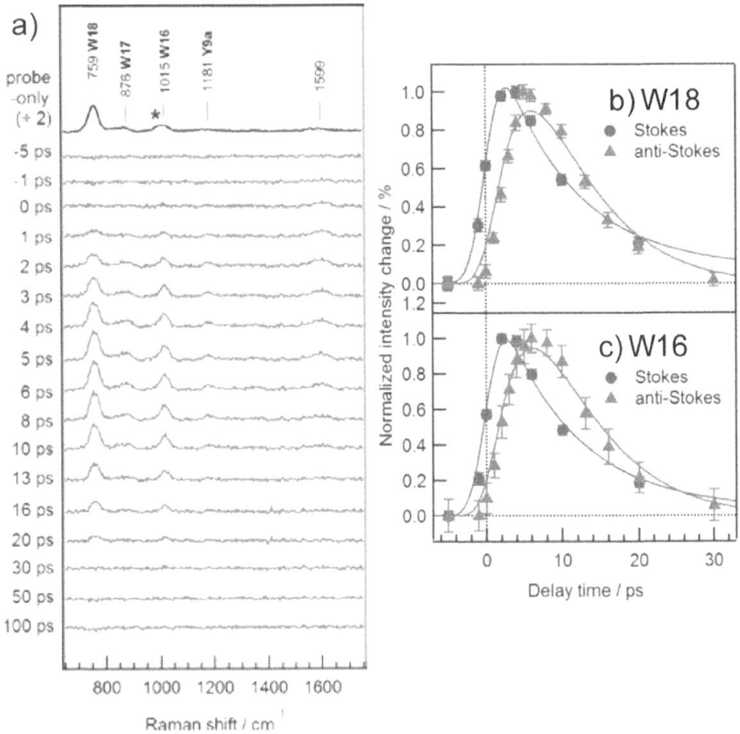

Figure 6.5 Energy flow through Cyt c. (a) Temporal intensity changes of the anti-Stokes Raman bands of Trp59 in the range of 5 to 100 ps upon Soret band excitation (405 nm) and with the Raman probe at 230 nm. Intensity kinetics of the anti-Stokes W18 and W16 lines (759 and 1015 cm^{-1} modes of Trp, respectively) are shown in (b) and (c) and compared with the respective Stokes traces (inverted in sign). Stokes intensity changes instantaneously upon photoexcitation, while anti-Stokes intensity exhibits a 1–3 ps delayed response. Increase and decrease of the anti-Stokes band intensities is assigned to the flow of excess energy from the heme to the Trp residues and the release from the Trp residues to the surroundings, respectively. Changes in the Stokes UV-RR intensities are explained as a change in the hydrogen bonding between the heme and Trp59. Reprinted with permission from Ref. [91]. Copyright (2011) American Chemical Society.

The subpicosecond energy transfer from the Trp to the heme has been found strongly dependent on the protein oxidation state, revealing a different geometry, at least of the heme pocket, in the ferrous and ferric forms. This capability of FRET Trp–heme pair to play the role of a molecular ruler has been recently used to study local structure flexibility and conformation fluctuations in myoglobin [89].

Concerning the advantage of probing heme response with UV time-resolved spectroscopy, TA signals below 360 nm have shown a spectral band characteristic of metallic ligand photolysis [93]: contrary to the usually monitored Vis bands (Soret and Q bands), it is scarcely influenced by cooling-related spectral signatures and it appears to be an ideal fingerprint to disentangle recombination processes from other relaxation mechanisms.

Last, a very promising methodology to follow vibrational energy flows was demonstrated by Fujii et al. [91], which showed for the first time anti-Stokes UV-RR spectra of aromatic amino acids in a heme protein with a time resolution of picoseconds (Fig. 6.5). As already pointed out in Section 6.2, this is a very promising tool to get information on the time evolution of the vibrational energy deposited on these residues because the intensity of anti-Stokes lines is directly related to the population of hot vibrational states. In this respect, they are a unique signature of the temperature of the probe and, in case of high-frequency modes, they are background free.

Following the flow and the dissipation of excesses of electronic and vibrational energy through a complex molecule remains a challenging task. Only the synergic use of different spectroscopies can report on the various processes involved (electron–phonon coupling, vibrational energy localization, structural changes, etc.) and, as discussed above, UV techniques have the peculiarity to report on the response of the molecular edifice on the propagation through it, using the near-UV chromophores as network nodes of known spatial distribution within the edifice.

6.4 Tracking Down Concerted Motions in Proteins and Molecular Devices: An Outlook on Coherent UV Spectroscopies

6.4.1 UV Multidimensional Spectroscopies

In the previous sections we saw the capabilities of ultrafast UV spectroscopies to provide information not only on the directly photoexcited moiety but on the hosting molecular edifice. Exploiting new UV probes (not only Trp but also other aromatic mono- or oligocyclic organic groups), we can finally move our focus from the "epicenter" to the "periphery," with the aim of tracking down concerted motions, long-distance interactions, and allosteric regulation of local dynamics.

Coherent and multidimensional spectroscopies are a novel family of methodologies, which have the potential to address these issues with even higher insightfulness.

Optical coherent spectroscopy exploits the capability of coherent sources, namely lasers, to induce in-phase optical oscillations of microscopic electric dipoles in the sample, to generate a macroscopic optical signal (PE), whose features (direction, polarization, spectrum, phase, time evolution) reveal a great variety of information on the quantum behavior of the material under study. When molecular systems are investigated, structural, chemical, dynamical properties can be unraveled [5, 7, 106–108]. Compared with other time-resolved spectroscopies, coherent spectroscopies have the unique capability to access directly fluctuation and relaxation dynamics of a molecule in its electronic ground state because, similarly to hole-burning techniques, they perturb the inhomogeneous distribution of ground-state transition frequencies and measure the recovery time of the statistically equilibrated distribution. This is a major advantage of such techniques because, except photochemistry, most of the chemistry occurs in electronic ground states rather than in the excited ones.

This approach has revolutionized modern spectroscopy and, when coupled with a heterodyne detection scheme, they give rise to time-resolved multidimensional nonlinear optical spectroscopies (2D-PE) [109–112]. They are also called pulsed Fourier transform

(FT) or heterodyne detected PE, or four-wave mixing (FWM or 4WM) spectroscopies. Despite they will not be discussed in this chapter, it deserves noting that also homodyne detected PE or 4WM techniques exist, including, among others, the classical two-pulse echo [108, 112], time-integrated and time-resolved three-pulse-stimulated PE [113], or photon echo peak shift (PEPS) [114, 115].

These techniques vary in the wavevector geometry, pulse energies, pulse timing, and detection schemes (Fig. 6.6) [108, 112]. For the sake of completeness Fig. 6.6 shows also higher orders of wave mixing techniques.

Over the past few years, there has been a growing effort aimed at developing multidimensional optical spectroscopies, which are the optical analogues of multidimensional nuclear magnetic resonance (mD-NMR) [112] ones (NOESY, COSY, etc.) [116, 117]. Nowadays, a variety of 2D vibrational (2DIR) spectroscopies have been proposed theoretically and implemented experimentally [118] in the infrared [119–122] and few years later in the near-IR and visible ranges to probe electronic transitions [4].

Their power lies in the fact that (i) they can reveal without any a priori specific knowledge couplings among different sites within the molecular system, (ii) they can separate inhomogeneous and homogeneous contributions to spectral line shapes, and (iii) they can disentangle overcrowded standard (ID) spectra (Fig. 6.7a). These techniques have been demonstrated to provide structural information and to enable tracking down processes of biological interest [4, 8, 123–129], as amyloid fibril formation [125], drug–enzyme interaction specificity [128], resonance energy transfer in photosynthetic systems [4, 130], but also electron–hole dynamics in quantum structures [131, 132] and semiconductors [133]. The superior structural information provided by 2D spectra was successfully used to perform all-optical protein recognition [9].

The extension of these techniques into the UV offers an invaluable tool to look into dynamical phenomena in macromolecules and supramolecular systems. It would indeed make possible to exploit amino acid or nucleobase residues naturally present in all proteins and DNAs or other UV chromophores constituting synthetic molecular devices, disentangling their typically overcrowded electronic spectra, and delivering a global picture of their dynamics

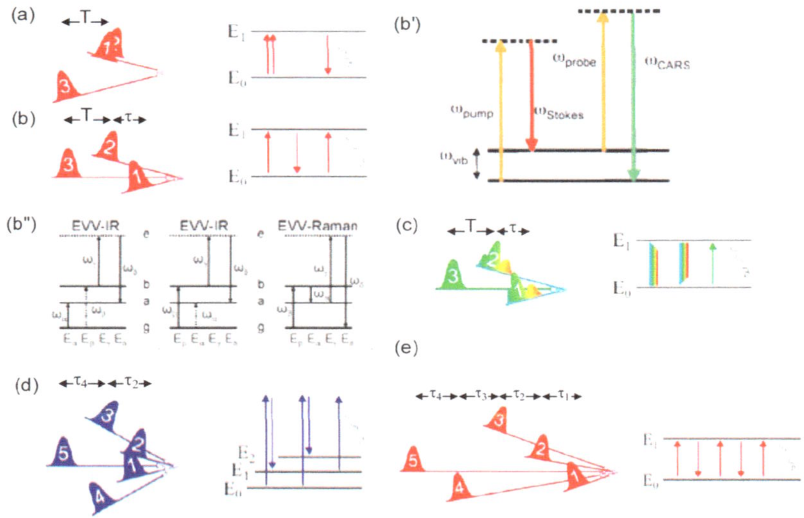

Figure 6.6 Schematic representation of the different possible nonlinear spectroscopies realized through different beam combinations with respect to phase matching and time/wavelength multiplexing. (a) Pump–probe spectroscopy in the four-wave mixing (4WM) representation. (b and c) Noncollinear 4WM for coherent spectroscopies, and general pulse shaping or coherent control 2D spectroscopy to select initial state preparations, respectively (the local oscillator field in the case of heterodyne detection is not shown. See, for instance, Fig. 6.8a). Energy-level diagrams (b') and (b") depict the 4WM processes for coherent anti-Stokes Raman spectroscopy (CARS) and electron-vibrational-vibrational (EVV) spectroscopy, respectively. The shown six-wave-mixing (6WM) experiments are 2D Raman (d) and resonant 6WM (e) spectroscopies. Adapted with permission from Ref. [112]. Copyright (2009) American Chemical Society.

and structure, via the time-dependent couplings between residues [62].

This research field is attracting more and more attention from theoretical and experimental groups. The first computational studies appeared in 2011 [134–137]. The authors predicted that UV-2D spectra contain characteristic signatures of different aromatic ammo acid transitions and of the interactions between them [136]. Transition couplings were found very sensitive to the change of residue–residue interactions and induced by residue mutations. Simulating UV-2D signals of amyloid fibrils, they explored this selective response to keep track of the formation kinetics of amyloid fibrils associated with Alzheimer disease, since UV-2D signals carry characteristic signatures of fibril size and geometry [134, 135]. They also showed that it could be possible to identify fibrillation propensity of a given protein [137].

Moving to implementing UV-2D spectrographs, there are two main technical obstacles that have prevented a full development of such techniques so far: the necessity of phase-stable pulses and of ultrabroadband UV sources (namely with $\Delta\lambda/\lambda > 0.1$). These are actually the two essential ingredients for a proper implementation of a FT-2D setup in the UV [138, 139].

These techniques require phase stability of the pulses. They were indeed initially developed in the IR region where, due to longer wavelengths, phase stability can be achieved more easily and the first experiment in the Vis range was performed in 2005 by Fleming and coworkers [4]. They exploited for the first time electronic dipoles to measure energy transfer processes and couplings in the Fenna–Matthews–Olson (FMO) light-harvesting photosynthetic protein complex (Fig. 6.7b). They also demonstrated the great potential of this technique to disentangle overcrowded ID electronic spectra. This seminal experiment was made possible thanks to the use of passive phase stabilization by diffractive optics as good as $\lambda/100$ at 600 nm (Fig. 6.8a) [109]: beams are pair-wise manipulated in such a way that any phase shift introduced by wobbling optics elements generates a compensating phase shift on the conjugated field [140]. Nowadays, inherently phase-stabilized schemes based on conventional optics exist with comparable performances [141]. Since a minimal phase stability of $\lambda/50$–$\lambda/30$ is required for a

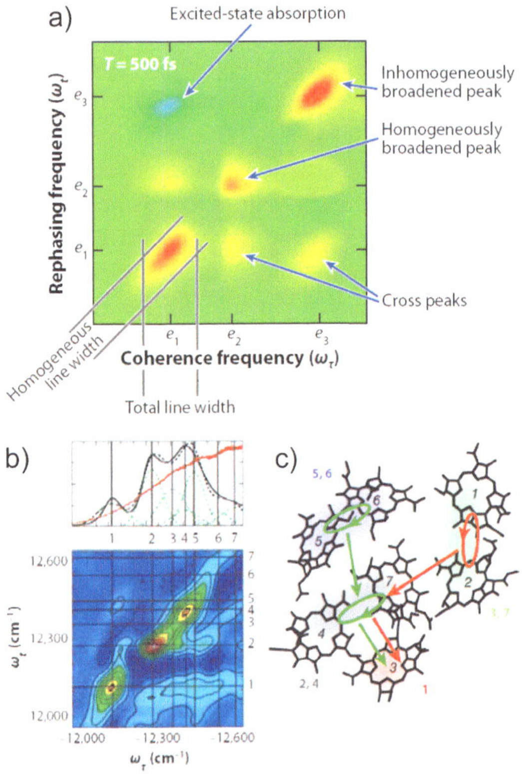

Figure 6.7 2D electronic spectroscopy: information content in a 2D spectrum. (a) Model 2D spectrum for a three-component system (el, e2, and e3). A 2D spectrum can be viewed as a map that correlates the absorption (ω_τ) and emission (ω_t) frequencies of the system. (b and c) Exciton delocalization and energy flow in the Fenna–Matthews–Olson (FMO) complex. (b) The linear absorption and $T = 0$ fs 2D spectrum of the *Chlorobium tepidum* FMO complex. Individual exciton bands are shown in green, and the red curve indicates the laser spectrum in the experiment. (c) The FMO structural arrangement of the seven bacteriochlorophyll molecules (black italic numbers) qualitatively overlaid with the delocalization patterns of the excitons (colored shading and bold numbers). Two main excitation energy transfer pathways are indicated by the red and green arrows. (a) Reprinted from Ref. [3] with permission of *Annual Review of Physical Chemistry*, **60** © by Annual Reviews, http://www.annualreviews.org. (b and c) Reprinted by permission from Macmillan Publishers Ltd: [*Nature*] Ref. [4], copyright (2005).

proper FT analysis, these novel designs have paved the way to extending coherent nonlinear techniques into the UV domain. The first fully noncollinear coherent UV 2D spectroscopy was introduced in 2010 [142]. The main credit is that they successfully proved the feasibility of multidimensional optical spectroscopy in the UV domain, between 250 nm and 375 nm. Since they used only reflective optics the concept can be transferred to all wavelength regimes. Moran and coworkers carried out UV-2D experiments to study photoinduced relaxation processes in single adenines [143] and elementary chemical reaction dynamics as electrocyclic ring opening in cycloalkenes [138].

However, these setups employed spectrally rather narrow UV pulses (~8 nm width at 287 nm), while much broader pulses are necessary to fully benefit from such a methodology. Indeed in a FT experiment, the sample has to interact simultaneously with all the spectral components of interest. The spectral information will then be retrieved by Fourier-transforming the heterodyned signal with respect to the time delays of the relative pulses. Thus, to have a 2D spectrum mapping of couplings among spectrally different chromophores, it is necessary to generate pulses broad enough to cover all the spectral region of interest. Different generation schemes for powerful (0.1–100 μJ/pulse) ultrashort (sub-10 fs) near and deep UV pulses have been proposed, based on different generation processes as achromatic phase-matching frequency doubling [144] and sum-frequency mixing [145], filamentation in rare gas-filled cells [146–148], or FWM in hollow fibers [149]. Among these schemes, achromatic phase-matching frequency doubling developed by Riedle and coworkers [144] is the most suitable for FT UV spectra's copies. Indeed with this approach UV pulses with a 270–370 nm spectrum (Fig. 6.8b) and few microjoules energy per pulse can be generated. These wavelengths match most of the chemically relevant UV region and the pulse energy is strong enough both for noncoherent spectroscopic applications (a typical pump pulse energy in an ultrafast experiment is ~0.1 μJ in a 100 um diameter spot [93]) and FWM experiments, where a few tens of nanojoules per pulse are required [4] (in this respect, they can be considered "protein friendly" techniques).

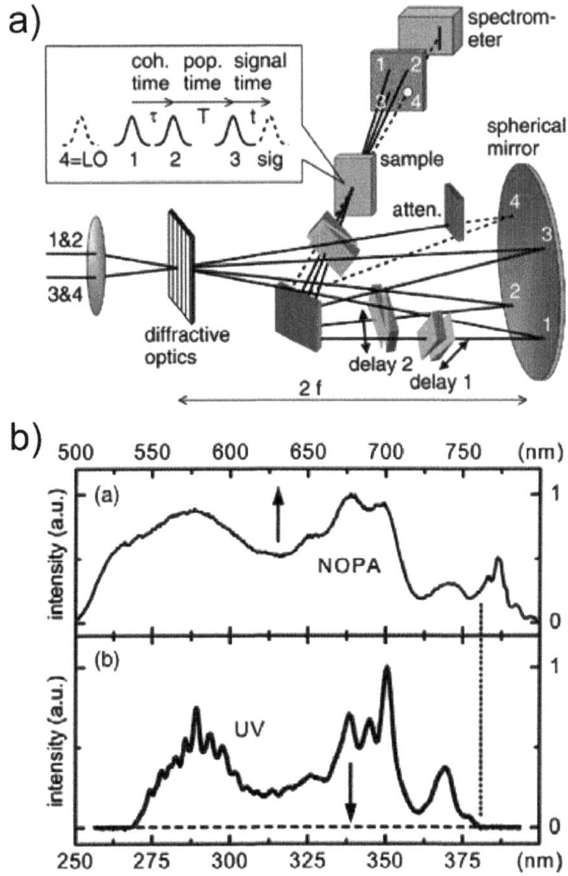

Figure 6.8 2D electronic spectroscopy solving phase stability and spectral broadness issues. (a) Inherently phase-stabilized heterodyne-detected 2D photon echo experiment (inset shows pulse timing). Phase stability between pulses 1 and 2 and between pulses 3 and 4 is achieved by generating the pairs in a diffractive beam splitter. Reprinted from Ref. [140], with the permission of AIP Publishing. (b) Visible spectrum (top) and UV spectrum generated by achromatic frequency doubling (bottom). Reprinted from Ref. [144] with permission from The Optical Society.

In conclusion of this section it is worth mentioning a broadband 2D transient absorption setup for the UV, presented in 2012 by Chergui and coworkers [150, 151]. It is essentially a pump–probe setup where a narrow band pump is varied with continuity over the spectral range 270–370 nm, while the probe is generated according to the achromatic doubling scheme by Riedle and coworkers [144]. The spectrally resolved transient absorption spectra as a function of the pump frequency constitute indeed a 2D experiment, which in principle provides the same information as 2D-PE [122]. It is a noncoherent technique with a lower time resolution (150 fs), which limits its capability of reporting on direct dipole–dipole interactions and on faster relaxation dynamics. However it has the potential to track down conformational changes and transport phenomena in macromolecules. For instance it allowed the authors to disentangle the excited-state decay pathways of the tryptophan amino acid residues in ferric myoglobins, revealing the presence of an unexpected electron transfer from the exited closest Trp to the heme in tens of picoseconds [95].

6.4.2 Pulse Shaping and Automatized UV 2D Spectrographs

Strictly connected to UV broadband pulse sources, a new generation of pulse shapers for UV application has been implemented [152–154]. This research field has been always developed in parallel with coherent spectroscopy for two main reasons. First, to keep under control temporal properties of an ultrabroadband pulse, neither grating- nor prism-based compressors are any more suitable, because of the significant contributions from higher group velocity dispersion (GVD) orders. This is particularly critical with UV pulses, where after an interaction with even a few millimeters of UV-grade silica, the time structure is dramatically affected [155]. Only the greater flexibility of a pulse shaper can, in principle, compensate for any order of GVD. Second, already in 2003 Warren and collaborators demonstrated that it is possible to implement a robust and phase-stable collinear 2D-PE setup, creating with an acousto-optic pulse shaper a collinear three-pulse sequence with well-controlled and variable interpulse delays and phases [106].

In 2009, a scheme based on a pump–probe-like geometry was proposed for Vis [156] and UV range [157], where the first pulse pair on the sample is collinear, and the generated signal is then collinear with the third pulse, which, at the same time, acts as local oscillator for the heterodyne detection (Fig. 6.9a). This greatly simplifies the phasing problem and permits rapid acquisition of 2D absorption spectra [157]. In 2013 a 2D-FT setup based on such a geometry was presented, which used fairly broad UV shaped pulses (FWHM of 10 nm at 310 nm) and a compressed supercontinuum spanning 250–720 nm as a probe pulse [158]. The potential of the setup was demonstrated on two representative molecules: pyrene and 2,2-diphenyl-5,6-benzo(2H)chromene. An extension to (ultra-) broadband UV pulses should be possible in the near future.

When compared to noncollinear 2D-PE setups based on conventional components, the collinear generation of pulse-shaped pulse pairs makes the implementation of a robust phase-stable setup definitely easier, but at the cost of renouncing the capability of the independent manipulation of the beam features (in particular polarization [159]), of a background free detection and of an optimal signal heterodyning. To fully benefit from a non collinear geometry and coherent generation and manipulation of the interaction fields, K. Nelson's group implemented a 2D-PE experiment at 800 nm based on a 2D liquid crystal (LC) spatial light modulator (SLM), allowing an automated 2D phase and amplitude pulse shaping of noncollinear, fully phase-coherent fields (Fig. 6.9b). It provides an independent control of wavevectors, pulse timings and phases, as well as spectral content of each field. This approach can generate waveforms of high complexity as multiple pulses or chirped pulses, with a superior flexibility, even when compared with conventional 2D-PE setups. Unfortunately the LC SLM technology cannot operate in the UV. Micromirror SLM based on microelectromechanical systems (MEMS) has been proposed for femtosecond pulse shaping in the UV [160], showing a potential to push the 2D SLM scheme to the UV. However, the current MEMS SLM are still performing too low [161] but recent developments on MEMS linear mirror arrays are promising [152].

At the time this chapter is written, there are no UV-2D spectrographs yet employing ultrabroadband sources either with

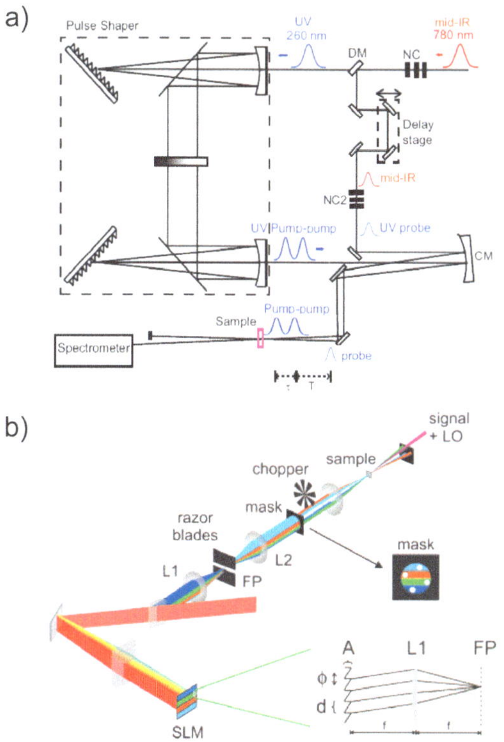

Figure 6.9 Pulse-shaping and 2D spectroscopy. (a) An automated UV2D-PE setup based on a pump–probe setup, with pulse-shaped pump beams. The UV pump at ~260 nm is generated from IR pulses by nonlinear crystals (NC) and afterward separated by a dichroic mirror (DM). It goes into a pulse shaper (the area enclosed with a dashed line) and becomes a phase-locked pump–pump pulse pair. The IR passes through the second set of nonlinear crystals (NC2) to generate the UV probe. The UV pump–pump and probe pulses are focused into the sample by the curved mirror (CM). The probe spectrum is detected with a spectrometer as a function of the pump–pump delay τ for each pump–probe delay T. (b) Noncollinear 2D-PE setup based on 2D pulse shaping. Multiple phase-stable output beams are created and overlapped at the sample in a phase-matched boxcar geometry with a 2D pulse shaper based on a 2D liquid crystal spatial light modulator (SLM in figure, while LI and L2 stand for lenses, FP for focal plan, and LO for local oscillator). With this approach, pulse timing can be modified without mechanical delay lines, and shape, phase, and spectral content of each beam can be independently specified. Pulse shaping easily provides excellent phase stability among all output beams. (a) Reprinted from Ref. [157] with permission from The Optical Society. (b) Reprinted with permission from Ref. [132]. Copyright (2007) American Chemical Society.

non collinear geometry or with collinear pulse-shaped pulses, but now both technical obstacles are overcome and the theoretical studies clarified the effective potentials of these methodologies. In addition, recent success in obtaining UV homodyne PE signals from UV dyes in different solvents [162–165] definitively proves the feasibly of UV-PE experiments.

6.4.3 All-Optical Chemically Sensitive Coherent Spectroscopies

Another context where coherent spectroscopies have proved to have important implications is in identifying chemical species with an all-optical approach. Standard vibrational spectra can, in principle, unequivocally recognize a (organic) molecule by detecting in the so-called fingerprint region of the spectrum (500–2000 cm^{-1}) [7, 166]. Coming to biological samples, congestion of the vibrational spectra grows exponentially with the molecule size and sample heterogeneity and the chemical sensitivity is dramatically restricted, as in the case of nonresonant coherent techniques (as coherent anti-Stokes Raman spectroscopy [CARS]; see Fig. 6.6b'), to classes of molecules, as proteins, lipids, and DNA.

The capability to provide both homogeneous lines and mutual anharmonic coupling makes 2D spectra a unique and easy-to-read fingerprint of a protein. Indeed, while vibrational lines can just reveal the presence of a given amino acid in the sample, the presence of anharmonic coupling reports on the nature of its neighboring amino acids. The capability to provide such rich structural information has been already proved, for example, by 2DIR techniques enabling all-optical protein recognition [8, 9]. Remarkably, since they are nonlinear spectroscopies, the signal, as for multiphoton excitation microscopy, is depth selective for 3D imaging, when coupled to a microscope [7].

As nicely shown by UV-RR experiments, the resonance effect with UV pulses gives the impressive selectivity to pick the signal from a given amino acid out of hundreds other present amino acids. The development of multidimensional experiments where UV pulses are exploited to retrieve vibrational couplings is one of the next directions where coherent UV spectroscopy could have an unforeseeable impact. In this respect a promising scheme for extension into UV is the so-called electron-vibrational-vibrational (EVV) spectroscopy scheme (Fig. 6.6b) [8, 9]. Here the first two pump pulses are in the IR region and are used to excite resonantly two different vibrational modes. The third pulse probes the polarization generated by the two IR beams. It is usually a visible, nonresonant beam such that the coupling strength of the two vibrations can be "read out" by the detection of visible photons. Relevant to UV extension, when resonant with an electronic transition, the detected signal is multiplicatively enhanced, becoming a triply resonant interaction. This methodology was shown to be able to discriminate among ten different proteins with a very high confidence (>90%), and to quantify their content, with an impressive sensitivity limit of few pmol [8, 9]. It also has been found that the intensity of combination bands is very sensitive to chemical modification of the protein as, for example, phosphorylation, paving the way to characterize posttranslation modifications of a protein. These chemical modifications occur after the protein translation and are strongly dependent on the cell environment. In this methodology, the chemical sensitivity mainly relies on the resonance selectivity of the first two pulses, and in case of overcrowed IR spectra, degradation in chemical recognition would be unavoidable. To what extent the chemical sensitivity suffers from the sample heterogeneity and molecular complexity would be spectacularly pushed further when the probe pulse is tuned to the UV. In this mixed scheme, where the first two pulses are still in the IR, signals from a specific subgroup of amino acids would be selectively enhanced. The recent advances in UV pulse shaping make possible even more sophisticated detection schemes, where the first two direct IR interactions could be replaced by pairs of UV interactions, which would coherently populate vibrational modes via coherent anti-Stokes Raman excitation (Fig. 6.6d,e). This would enhance the chemical selectivity even more, since the populated modes would be the only ones coupled to the resonant electronic transition. Thanks to the novel pulse-shaping techniques, a so complicated interaction scheme, with five pulses interacting and coherently controlled, can be achieved, as was nicely shown by Silberberg and collaborators in the case of single-pulse coherently controlled nonlinear Raman spectroscopy [7, 166].

6.5 Conclusions/Final Remarks

This chapter reviewed the recent advances in the application of UV-Vis spectroscopy with down to 10 fs resolution to a few model proteins. We discussed more particularly the usage of their intrinsic Trp and Tyr amino acids as reporter molecules for the photoinduced process upon excitation of the biologically active Vis chromophores (retinal, heme). But more generally, innumerable systems with optical transitions in the near- and deep-UV regions can be investigated with real-time techniques or manipulated with optical coherent methods. This is by itself an extraordinary breakthrough that heralds an increasing interest in these techniques. Nevertheless, as demonstrated with examples from the recent literature, their unique added value stems from the possibility to use as probes a variety of UV chromophores, most of them containing aromatic mono- or oligocyclic organic groups, for example, aromatic amino acids, nucleotide bases, imides and (di-)imines, or synthetic fluorophores, for local probing of biomolecular structures. These species are constituents of biomolecules, many molecular devices, and supramolecular structures. Exploiting them as a probe opens a new window on the (thermo)dynamical properties of the macromolecule edifice in a site-specific fashion. They provide dynamical information on intramolecular electrostatics, fluctuations and local rigidity, structural changes, anharmonic coupling, and long-range interactions, as well as vibrational and electronic coherences.

Currently most of the research activity is bio-oriented, probably because this is the community where concepts as collective motions, functional dynamics, and long-range interactions were born and are at the very base of the comprehension process. However both these concepts and the techniques herein presented can naturally be extended to any macromolecule and supramolecular systems. The capability to exploit probes transparent to the Vis radiation and sensitive to changes in the electronic state and charge localization gives these novel methodologies a great potential. Engineering of molecular devices inspired by nature (the so-called bionanotechnology [167]) will greatly benefit from this global approach to intermolecular dynamics in what is a general multicomponent system with intermolecular interactions of different strengths and natural disorder. The spectroscopic tools presented here are on the way to becoming new insightful tools for the development and optimization of photopowered nanodevices, molecular electronics, and efficient solar light harvesters or to understanding elementary steps in self-assembling processes [168–171].

Acknowledgments

AC gratefully acknowledges the support by the European Union (EEC Starting Grant 279599-FunctionalDyna), the Swiss NSF through the NCCR MUST, and the European programme COST (Action CM1202). JL and SH acknowledge funding from the French National Research Agency (contracts ANR-14-CE35-0015-01 and ANR-15-CE29-0010-01), the Strasbourg Excellence Center in Chemistry (ciFRC), and the "Investissement d'Avenir" program: Labex NIE ANR-11-LABX-0058_NIE, Labex CSC ANR-10-LABX-0026_CSC.

References

1. Zewail, A. H. (2000). Femtochemistry: atomic-scale dynamics of the chemical bond, *J. Phys. Chem. A.*, **104**, pp. 5660–5694.
2. Pal, S. K., and Zewail, A. H. (2004). Dynamics of water in biological recognition, *Chem. Rev.*, **104**, pp. 2099–2123.
3. Cheng, Y. C., and Fleming, G. R. (2009). Dynamics of light harvesting in photosynthesis, *Annu. Rev. Phys. Chem.*, **60**, pp. 241–262.
4. Brixner, T., Stenger, J., Vaswani, H. M., Cho, M., Blankenship, R. E., et al. (2005). Two-dimensional spectroscopy of electronic couplings in photosynthesis, *Nature*, **434**, pp. 625–628.
5. Mukamel, S. (2000). Multidimensional femtosecond correlation spectroscopies of electronic and vibrational excitations, *Annu. Rev. Phys. Chem.*, **51**, pp. 691–729.
6. Elsaesser, T. (2009). Two-dimensional infrared spectroscopy of intermolecular hydrogen bonds in the condensed phase, *Acc. Chem. Res.*, **42**, pp. 1220–1228.
7. Silberberg, Y. (2009). Quantum coherent control for nonlinear spectroscopy and microscopy, *Annu. Rev. Phys. Chem.*, **60**, pp. 277–292.

8. Fournier, F., Gardner, E. M., Kedra, D. A., Donaldson, P. M., Guo, R., et al. (2008). Protein identification and quantification by two-dimensional infrared spectroscopy: implications for an all-optical proteomic platform, *Proc. Natl. Acad. Sci. U S A,* **105**, pp. 15352–15357.
9. Fournier, F., Gardner, E. M., Guo, R., Donaldson, P. M., Barter, L. M. C., et al. (2008). Optical fingerprinting of peptides using two-dimensional infrared spectroscopy: proof of principle, *Anal. Biochem.*, **374**, pp. 358–365.
10. Larsen, O. F., van Stokkum, I. H., Gobets, B., van Grondelle, R., and van Amerongen, H. (2001). Probing the structure and dynamics of a DNA hairpin by ultrafast quenching and fluorescence depolarization, *Biophys. J.*, **81**, pp. 1115– 1126.
11. Larsen, O. F. A., van Stokkum, I. H. M., de Weerd, F. L., Vengris, M., Aravindakumar, C. T., et al. (2004). Ultrafast transient-absorption and steady-state fluorescence measurements on 2-aminopurine substituted dinucleotides and 2-aminopurine substituted DNA duplexes, *Phys. Chem. Chem. Phys.*, **6**, pp. 154–160.
12. Fiebig, T., Wan, C., and Zewail, A. H. (2002). Femtosecond charge transfer dynamics of a modified DNA base: 2-aminopurine in complexes with nucleotides, *ChemPhysChem*, **3**, pp. 781–788.
13. Xia, T., Wan, C., Roberts, R. W., and Zewail, A. H. (2005). RNA-protein recognition: single-residue ultrafast dynamical control of structural specificity and function, *Proc. Natl. Acad. Sci. U S A,* **102**, pp. 13013–13018.
14. Nguyen, H. N., Zhao, L., Gray, C. W., Gray, D. M., and Xia, T. (2011). Ultrafast fluorescence decay profiles reveal differential unstacking of 2-aminopurine from neighboring bases in single-stranded DNA-binding protein subsites, *Biochemistry,* **50**, pp. 8989–9001.
15. Dallmann, A., Dehmel, L., Peters, T., Mugge, C., Griesinger, C., et al. (2010). 2-Aminopurine incorporation perturbs the dynamics and structure of DNA, *Angew. Chem. Int. Ed. Engl.*, **49**, pp. 5989–5992.
16. Jones, A. C., and Neely, R. K. (2015). 2-Aminopurine as a fluorescent probe of DNA conformation and the DNA-enzyme interface, *Q. Rev. Biophys.,* **48**, pp. 244– 279.
17. Shin, D., Sinkeldam, R. W., and Tor, Y. (2011). Emissive RNA alphabet, *J. Am. Chem. Soc.,* **133**, pp. 14912–14915.
18. Sholokh, M., Zamotaiev, O. M., Das, R., Postupalenko, V. Y., Richert, L. et al. (2015). Fluorescent amino acid undergoing excited state intramolecular proton transfer for site-specific probing and imaging of peptide interactions, *J. Phys. Chem. B,* **119**, pp. 2585–2595.

19. Jean, J. M. and Hall, K. B. (2001). 2-Aminopurine fluorescence quenching and lifetimes: role of base stacking, *Proc. Natl. Acad. Sci. U S A*, **98**, pp. 37–41.

20. Liang, J. and Matsika, S. (2011). Pathways for fluorescence quenching in 2-aminopurine π-stacked with pyrimidine nucleobases, *J. Am. Chem. Soc.*, **133**, pp. 6799–6808.

21. Lobsiger, S., Blaser, S., Sinha, R. K., Frey, H.-M., and Leutwyler, S. (2014). Switching on the fluorescence of 2-aminopurine by site-selective microhydration, *Nat. Chem.*, **6**, pp. 989–993.

22. Brunk, E., and Rothlisberger, U. (2015). Mixed quantum mechanical/molecular mechanical molecular dynamics simulations of biological systems in ground and electronically excited states, *Chem. Rev.*, **115**, pp. 6217–6263.

23. Monari, A., Rivail, J. L., and Assfeld, X. (2013). Theoretical modeling of large molecular systems. advances in the local self consistent field method for mixed quantum mechanics/molecular mechanics calculations, *Acc. Chem. Res.*, **46**, pp. 596–603.

24. Voltz, K., Léonard, J., Tourón Touceda, P., Conyard, J., Chaker, Z., et al. (2016). Quantitative sampling of conformational heterogeneity of a DNA hairpin using molecular dynamics simulations and ultrafast fluorescence spectroscopy, *Nucleic Acids Res.*, **44**, pp. 3408–3419.

25. Oesterhelt, D., and Stoeckenius, W. (1971). Rhodopsin-like protein from the purple membrane of Halobacterium halobium, *Nat. New Biol.*, **233**, pp. 149–152.

26. Briand, J., Léonard, J., and Haacke, S. (2010). Ultrafast photo-induced reaction dynamics in bacteriorhodopsin and its Trp mutants, *J. Opt.*, **12**, pp. 1–14.

27. Kochendoerfer, G. G., and Mathies, R. A. (1995). Ultrafast spectroscopy of rhodopsins - photochemistry at its best!, *Isr. J. Chem.*, **35**, pp. 211–226.

28. Mathies, R. A., Lin, S. W., Ames, J. B., and Pollard, W. T. (1991). From femtoseconds to biology: mechanism of bacteriorhodopsin's light-driven proton pumps, *Annu. Rev. Biophys. Chem.*, **20**, pp. 591–518.

29. Maeda, A. (1995). Application of FTIR spectroscopy to the structural study on the function of bacteriorhodopsin, *Isr. J. Chem.*, **35**, pp. 387–400.

30. Diller, R. (2008). Primary reactions in retinal proteins, in *Biological and Medical Physics, Biomedical Engineering*, Braun, M., Gilch, P., and Zinth, W. eds., Berlin; Heidelberg: Springer, pp. 243–273.

31. Edman, K., Nollert, P., Royant, A., Belrhali, H., Pebay-Peyroula, E., et al. (1999). High-resolution X-ray structure of an early intermediate in the bacteriorhodopsin photocycle, *Nature*, **401**, pp. 822–826.
32. Pebay-Peyroula, E., Rummel, G., Rosenbusch, J. P., and Landau, E. M. (1997). X-ray structure of bacteriorhodopsin at 2.5 Angstroms from microcrystals grown in lipidic cubic phases, *Science*, **277**, pp. 1676–1681.
33. Luecke, H., Schobert, B., Lanyi, J. K., Spudich, E. N., and Spudich, J. L. (2001). Crystal structure of sensory rhodopsin II at 2.4 Angstroms: insights into color tuning and transducer interaction, *Science*, **293**, pp. 1499–1503.
34. Fukumoto, J., Hopewell, W., Karvaly, B., and El-Sayed, M. A. (1981). Time-resolved protein fluorescence studies of intermediates in the photochemical cycle of bacteriorhodopsin, *Proc. Natl. Acad. Sci. U S A*, **78**, pp. 252–255.
35. Jang, D.-J. and El-Sayed, M. A. (1989). Tryptophan fluorescence quenching as a monitor for the protein conformation changes occurring during the photocycle of bacteriorhodopsin under different perturbations, *Proc. Natl. Acad. Sci. U S A*, **86**, pp. 5815–5819.
36. Polland, H. J., Franz, M. A., Zinth, W., Kaiser, W., and Oesterhelt, D. (1986). Energy transfer from retinal to amino acids — a time-resolved study of the ultraviolet emission of bacteriorhodopsin, *Biochim. Biophys. Acta*, **851**, pp. 407–415.
37. Gross, R., Schumann, C., Wolf, M. M., Herbst, J., Diller, R., et al. (2009). Ultrafast protein conformational alterations in bacteriorhodopsin and its locked analogue BR5.12, *J. Phys. Chem. B*, **113**, pp. 7851–7860.
38. Gross, R., Wolf, M. M., Schumann, C., Friedman, N., Sheves, M., et al. (2009). Primary photoinduced protein response in bacteriorhodopsin and sensory rhodopsin II, *J. Am. Chem. Soc.*, **131**, pp. 14868–14878.
39. Herbst, J., Heyne, K., and Diller, R. (2002). Femtosecond infrared spectroscopy of bacteriorhodopsin chromophore isomerization, *Science*, **297**, pp. 822–825.
40. Amsden, J. J., Kralj, J. M., Chieffo, L. R., Wang, X., Erramilli, S., et al. (2007). Subpicosecond protein backbone changes detected during the green-absorbing proteorhodopsin primary photoreaction, *J. Phys. Chem. B*, **111**, pp. 11824–11831.
41. Neumann, K., Verhoefen, M.-K., Weber, I., Glaubitz, C., and Wachtveitl, J. (2008). Initial reaction dynamics of proteorhodopsin observed by femtosecond infrared and visible spectroscopy, *Biophys. J.*, **94**, pp. 4796–4807.

42. Rupenyan, A., van Stokkum, I. H. M., Arents, J. C., van Grondelle, R., Hellingwerf, K., et al. (2008). Characterization of the primary photochemistry of proteorhodopsin with femtosecond spectroscopy, *Biophys. J.*, **94**, pp. 4020–4030.

43. Royant, A., Edman, K., Ursby, T., Pebay-Peyroula, E., Landau, E. M., et al. (2000). Helix deformation is coupled to vectorial proton transport in the photocycle of bacteriorhodopsin, *Nature*, **406**, pp. 645–648.

44. Schobert, B., Cupp-Vickery, J., Hornak, V., Smith, S. O., and Lanyi, J. K. (2002). Crystallographic structure of the K intermediate of bacteriorhodopsin: conservation of free energy after photoisomerization of the retinal, *J. Mol. Biol.*, **321**, pp. 715–726.

45. Aharoni, A., Hou, B., Friedman, N., Ottolenghi, M., Rousso, I., et al. (2001). Non-isomerizable artificial pigments: implications for the primary light-induced events in bacteriorhodopsin, *Biochemistry (Mosc)*, **66**, pp. 1210–1219.

46. Aharoni, A., Weiner, L., Ottolenghi, M., and Sheves, M. (2000). Bacteriorhodopsin experiences light-induced conformational alterations in non-isomerizable C13=C14 pigments, *J. Biol. Chem.*, **275**, pp. 21010–21016.

47. Zhong, Q., Ruhman, S., Ottolenghi, M., Sheves, M., Friedman, N., et al. (1996). Reexamining the primary light-induced events in bacteriorhodopsin using a synthetic C13=C14- locked chromophore, *J. Am. Chem. Soc.*, **118**, pp. 12828–12829.

48. Rousso, I., Khachatryan, E., Gat, Y., Brodsky, I., Ottolenghi, M., et al. (1997). Microsecond atomic force sensing of protein conformational dynamics: implications for the primary light-induced events in bacteriorhodopsin, *Proc. Natl. Acad. Sci. U S A*, **94**, pp. 7937–7941.

49. Mathies, R. A., and Stryer, L. (1976). Retinal has a highly dipolar vertically excited singlet state: implications for vision, *Proc. Natl. Acad. Sci. U S A*, **73**, pp. 2169–2178.

50. Locknar, S. A., Chowdhury, A., and Peteanu, L. A. (2000). Matrix and temperature effects on the electronic properties of conjugated molecules: an electroabsorption study *of all-trans* retinal, *J. Phys. Chem. B.*, **104**, pp. 5816–5824.

51. Locknar, S. A., and Peteanu, L. A. (1998). Investigation of the relationship between dipolar properties and cis-trans configuration in retinal polyenes: a comparative study using stark spectroscopy and semiempirical calculations, *J. Phys. Chem. B*, **102**, pp. 4240–4246.

52. Bouevitch, O., Lewis, A., and Sheves, M. (1995). Probing bacteriorhodopsin photochemistry with nonlinear optics: comparing the

second harmonic generation of bR and the photochemically induced intermediate K, *J. Phys. Chem.*, **99**, pp. 10648–10657.

53. Huang, J. Y., Chen, Z. P. and Lewis, A. (1989). 2nd-Harmonic generation in purple membrane-poly(vinyl alcohol) films: probing the dipolar characteristics of the bacteriorhodopsin chromphore in Br570 and M412, *J. Phys. Chem.*, **93**, pp. 3314–3320.
54. Groma, G. I., Colonna, A., Lambry, J. C., Petrich, J. W., Varo, G., et al. (2004). Resonant optical rectification in bacteriorhodopsin, *Proc. Natl. Acad. Sci. U S A*, **101**, pp. 7971–7975.
55. Colonna, A., Groma, G. I., Martin, J. L., Joffre, M., and Vos, M. H. (2007). Quantification of sudden light-induced polarization in bacteriorhodopsin by optical rectification, *J. Phys. Chem. B*, **111**, pp. 2707–2710.
56. Bonacic-Koutecky, V., Koutecky, J., and Michl, J. (1987). Neutral and charged biradicals, zwitterions, funnels in S1, and proton translocation: their role in photochemistry, photophysics, and vision, *Angew Chem. Int. Ed.*, **26**, pp. 170–189.
57. Cembran, A., Bernardi, F., Olivucci, M., and Garavelli, M. (2005). The retinal chromophore/chloride ion pair: structure of the photo isomerization path and interplay of charge transfer and covalent states, *Proc. Natl. Acad. Sci. U S A*, **102**, pp. 6255–6260.
58. Frutos, L. M., Andruniów, T., Santoro, F., Ferré, N., and Olivucci, M. (2007). Tracking the excited-state time evolution of the visual pigment with multiconfigurational quantum chemistry, *Proc. Natl. Acad. Sci. U S A*, **104**, pp. 7764–7769.
59. Zadok, U., Khatchatouriants, A., Lewis, A., Ottolenghi, M., and Sheves, M. (2002). Light-induced charge redistribution in the retinal chromophore is required for initiating the bacteriorhodopsin photocycle, *J. Am. Chem. Soc.*, **124**, pp. 11844–11845.
60. Aharoni, A., Khatchatouriants, A., Manevitch, A., Lewis, A., and Sheves, M. (2003). Protein-beta-ionone ring interactions enhance the light-induced dipole of the chromophore in bacteriorhodopsin, *J. Phys. Chem. B* **107**, pp. 6221–6225.
61. Léonard, J., Portuondo-Campa, E., Cannizzo, A., van Mourik, F., van der Zwan, G., et al. (2009). Functional electric field changes in photoactivated proteins revealed by ultrafast Stark spectroscopy of the Trp residues, *Proc. Natl. Acad. Sci. U S A*, **106**, pp. 7718–7723.
62. Schenkl, S., van Mourik, F., van der Zwan, G., Haacke, S., and Chergui, M. (2005). Probing the ultrafast charge translocation of photoexcited retinal in bacteriorhodopsin, *Science*, **309**, pp. 917–920.

63. Schenkl, S., van Mourik, F., Friedman, N., Sheves, M., Schlesinger, R., et al. (2006). Insights into excited-state and isomerization dynamics of bacteriorhodopsin from ultrafast transient UV absorption, *Proc. Natl. Acad. Sci. U S A*, **103**, pp. 4101–4106.
64. Becher, B., and Ebrey, T. G. (1976). Evidence for chromophore-chromophore (exciton) interaction in the purple membrane of *Halobacterium halobium, Biochem. Biophys. Res. Commun.*, **69**, pp. 1–6.
65. Altoe, P., Cembran, A., Olivucci, M. and Garavelli, M. (2010). Aborted double bicycle-pedal isomerization with hydrogen bond breaking is the primary event of bacteriorhodopsin proton pumping, *Proc. Natl. Acad. Sci. U S A*, **107**, pp. 20172–20177.
66. Braun-Sand, S., Sharma, P. K., Chu, Z. T., Pisliakov, A. V., and Warshel, A. (2008). The energetics of the primary proton transfer in bacteriorhodopsin revisited: it is a sequential light-induced charge separation after all, *Biochim. Biophys. Acta*, **1777**, pp. 441–452.
67. Bondar, A.-N., Fischer, S., Smith, J. C., Elstner, M., and Suhai, S. (2004). Key role of electrostatic interactions in bacteriorhodopsin proton transfer, *J. Am. Chem. Soc.*, **126**, pp. 14668–14677.
68. Mizuno, M., Shibata, M., Yamada, J., Kandori, H., and Mizutani, Y. (2009). Picosecond time-resolved ultraviolet resonance Raman spectroscopy of bacteriorhodopsin: primary protein response to the photoisomerization of retinal, *J. Phys. Chem. B*, **113**, pp. 12121–12128.
69. Inada, S., Mizuno, M., Kato, Y., Kawanabe, A., Kandori, H., et al. (2013). Primary structural response in tryptophan residues of Anabaena sensory rhodopsin to photochromic reactions of the retinal chromophore, *Chem. Phys.*, **419**, pp. 65–73.
70. Peon, J., Pal, S. K., and Zewail, A. H. (2002). Hydration at the surface of the protein monellin: dynamics with femtosecond resolution, *Proc. Natl. Acad. Sci. U S A*, **99**, pp. 10964–10969.
71. Pal, S. K., Peon, J., and Zewail, A. H. (2002). Biological water at the protein surface: dynamical solvation probed directly with femtosecond resolution, *Proc. Natl. Acad. Sci. U S A*, **99**, pp. 1763–1768.
72. Qiu, W. H., Kao, Y. T., Zhang, L. Y., Yang, Y., Wang, L. J., et al. (2006). Protein surface hydration mapped by site-specific mutations, *Proc. Natl. Acad. Sci. U S A*, **103**, pp. 13979–13984.
73. Pal, S. K., Peon, J., Bagchi, B., and Zewail, A. H. (2002). Biological water: femtosecond dynamics of macromolecular hydration, *J. Phys. Chem. B*, **106**, pp. 12376–12395.
74. Qiu, W. H., Zhang, L. Y., Okobiah, O., Yang, Y., Wang, L. J., et al. (2006). Ultrafast solvation dynamics of human serum albumin: correlations

with conformational transitions and site-selected recognition, *J. Phys. Chem. B*, **110**, pp. 10540–10549.

75. Zhong, D. P., Pal, S. K., Zhang, D. Q., Chan, S. I., and Zewail, A. H. (2002). Femtosecond dynamics of rubredoxin: tryptophan solvation and resonance energy transfer in the protein, *Proc. Natl. Acad. Sci. U S A*, **99**, pp. 13–18.
76. Zhong, D. P., Pal, S. K., and Zewail, A. H. (2011). Biological water: a critique, *Chem. Phys. Lett.*, **503**, pp. 1–11.
77. Nilsson, L., and Halle, B. (2005). Molecular origin of time-dependent fluorescence shifts in proteins, *Proc. Natl. Acad. Sci. U S A*, **102**, pp. 13867–13872.
78. Pal, S. K., Peon, J., and Zewail, A. H. (2002). Ultrafast decay and hydration dynamics of DNA bases and mimics, *Chem. Phys. Lett.*, **363**, pp. 57–63.
79. Pal, S. K., Zhao, L. A., and Zewail, A. H. (2003). Water at DNA surfaces: ultrafast dynamics in minor groove recognition, *Proc. Natl. Acad. Sci. U S A*, **100**, pp. 8113–8118.
80. Peon, J., and Zewail, A. H. (2001). DNA/RNA nucleotides and nucleosides: direct measurement of excited-state lifetimes by femtosecond fluorescence up-conversion, *Chem. Phys. Lett.*, **348**, pp. 255–262.
81. Miannay, F. A., Banyasz, A., Gustavsson, T., and Markovitsi, D. (2007). Ultrafast excited-state deactivation and energy transfer in guanine-cytosine DNA double helices, *J. Am. Chem. Soc.*, **129**, pp. 14574–14575.
82. Gustavsson, T., Sarkar, N., Lazzarotto, E., Markovitsi, D., and Improta, R. (2006). Singlet excited state dynamics of uracil and thymine derivatives: a femtosecond fluorescence upconversion study in acetonitrile, *Chem. Phys. Lett.*, **429**, pp. 551–557.
83. Zhong, D. P., Douhal, A. and Zewail, A. H. (2000). Femtosecond studies of protein-ligand hydrophobic binding and dynamics: human serum albumin, *Proc. Natl. Acad. Sci. U S A*, **97**, pp. 14056–14061.
84. Elsaesser, T. and Kaiser, W. (1991). Vibrational and vibronic relaxation of large polyatomic-molecules in liquids, *Annu. Rev. Phys. Chem.*, **42**, pp. 83–107.
85. Owrutsky, J. C., Raftery, D., and Hochstrasser, R. M. (1994). Vibrational-relaxation dynamics in solutions, *Annu. Rev. Phys. Chem.*, **45**, pp. 519–555.
86. Frauenfelder, H., and Wolynes, P. O. (1985). Rate theories and puzzles of hemeprotein kinetics, *Science*, **229**, pp. 337–345.

87. Gunasekaran, K., Ma, B. Y., and Nussinov, R. (2004). Is allostery an intrinsic property of all dynamic proteins?, *Proteins*, **57**, pp. 433–443.

88. Frauenfelder, H., McMahon, B. H., Austin, R. H., Chu, K., and Groves, J. T. (2001). The role of structure, energy landscape, dynamics, and allostery in the enzymatic function of myoglobin, *Proc. Natl. Acad. Sci. U S A*, **98**, pp. 2370–2374.

89. Stevens, J. A., Link, J. J., Kao, Y. T., Zang, C., Wang, L. J., et al. (2010). Ultrafast dynamics of resonance energy transfer in myoglobin: probing local conformation fluctuations, *J. Phys. Chem. B*, **114**, pp. 1498–1505.

90. Zang, C., Stevens, J. A., Link, J. J., Guo, L., Wang, L., et al. (2009). Ultrafast proteinquake dynamics in cytochrome c, *J. Am. Chem. Soc.*, **131**, pp. 2846–2852.

91. Fujii, N., Mizuno, M., and Mizutani, Y. (2011). Direct observation of vibrational energy flow in cytochrome c, *J. Phys. Chem. B*, **115**, pp. 13057–13064.

92. Sato, A., Gao, Y., Kitagawa, T., and Mizutani, Y. (2007). Primary protein response after ligand photodissociation in carbonmonoxy myoglobin, *Proc. Natl. Acad. Sci. U S A*, **104**, pp. 9627–9632.

93. Consani, C., Bram, O., Van Mourik, F., Cannizzo, A., and Chergui, M. (2012). Energy transfer and relaxation mechanisms in cytochrome c, *Chem. Phys.*, **396**, pp. 108–115.

94. Bram, O., Consani, C., Cannizzo, A., and Chergui, M. (2011). Femtosecond UV studies of the electronic relaxation processes in cytochrome c, *J. Phys. Chem. B*, **115**, pp. 13723–13730.

95. Consani, C., Aubock, G., van Mourik, F., and Chergui, M. (2013). Ultrafast tryptophan-to-heme electron transfer in myoglobins revealed by UV 2D spectroscopy, *Science*, **339**, pp. 1586–1589.

96. Negrerie, M., Cianetti, S., Vos, M. H., Martin, J. L., and Kruglik, S. G. (2006). Ultrafast heme dynamics in ferrous versus ferric cytochrome c studied by time-resolved resonance Raman and transient absorption spectroscopy, *J. Phys. Chem. B*, **110**, pp. 12766–12781.

97. Henry, E. R., Eaton, W. A., and Hochstrasser, R. M. (1986). Molecular-dynamics simulations of cooling in laser-excited heme-proteins, *Proc. Natl. Acad. Sci. U S A*, **83**, pp. 8982–8986.

98. Demidov, A. A., Ye, X., Rosca, F., Kumar, A. T. N., Wnag, W., et al. (2001). Investigations of ultrafast kinetics and dynamic absorption in heme proteins, *Biophys. J.*, **80**, pp. 283A–283A.

99. Dantsker, D., Samuni, U., Friedman, J. M., and Agmon, N. (2005). A hierarchy of functionally important relaxations within myoglobin

based on solvent effects, mutations and kinetic model, *Biochim. Biophys. Acta*, **1749**, pp. 234–251.

100. Miyashita, O., Onuchic, J. N., and Wolynes, P. O. (2003). Nonlinear elasticity, proteinquakes, and the energy landscapes of functional transitions in proteins, *Proc. Natl. Acad. Sci. U S A*, **100**, pp. 12570–12575.
101. Metzler, R., Klafter, J., and Jortner, J. (1999). Hierarchies and logarithmic oscillations in the temporal relaxation patterns of proteins and other complex systems, *Proc. Natl. Acad. Sci. U S A*, **96**, pp. 11085–11089.
102. Frauenfelder, H., Sligar, S. G., and Wolynes, P. O. (1991). The energy landscapes and motions of proteins, *Science*, **254**, pp. 1598–1603.
103. Sato, A., and Mizutani, Y. (2005). Picosecond structural dynamics of myoglobin following photodissociation of carbon monoxide as revealed by ultraviolet time-resolved resonance Raman spectroscopy, *Biochemistry*, **44**, 14709–14714.
104. Ansari, A., Berendzen, J., Bowne, S. F., Frauenfelder, H., Iben, I. E., et al. (1985). Protein states and protein quakes, *Proc. Natl. Acad. Sci. U S A*, **82**, pp. 5000–5004.
105. Piazza, F., and Sanejouand, Y. H. (2009). Long–range energy transfer in proteins, *Phys. Biol.*, **6**, 046014.
106. Tian, P. P., Keusters, D., Suzaki, Y., and Warren, W. S. (2003). Femtosecond phase-coherent two-dimensional spectroscopy, *Science*, **300**, pp. 1553–1555.
107. Nagy, A., Prokhorenko, V., and Miller, R. J. D. (2006). Do we live in a quantum world? Advances in multidimensional coherent spectroscopies refine our understanding of quantum coherences and structural dynamics of biological systems, *Curr. Opin. Struc. Biol.*, **16**, pp. 654–663.
108. Mukamel, S. (1995). *Principles of Nonlinear Optical Spectroscopy*. Oxford Series in Optical and Imaging Sciences, New York: Oxford University Press.
109. Cowan, M. L., Ogilvie, J. P., and Miller, R. J. D. (2004). Two- dimensional spectroscopy using diffractive optics based phased-locked photon echoes, *Chem. Phys. Lett.*, **386**, pp. 184–189.
110. Jonas, D. M. (2003). Two-dimensional femtosecond spectroscopy, *Annu. Rev. Phys. Chem.*, **54**, pp. 425–463.
111. Brixner, T., Stiopkin, I. V. and Fleming, G. R. (2004). Tunable two-dimensional femtosecond spectroscopy, *Opt. Lett.*, **29**, pp. 884–886.

112. Miller, R. J. D., Paarmann, A., and Prokhorenko, V. I. (2009). Diffractive optics based four-wave, six-wave, . . . , v-wave nonlinear spectroscopy, *Acc. Chem. Res.*, **42**, pp. 1442–1451.

113. Cho, M. H., Scherer, N. F., Fleming, G. R., and Mukamel, S. (1992). Photon-echoes and related 4-wave-mixing spectroscopies using phase-locked pulses, *J. Chem. Phys.*, **96**, pp. 5618–5629.

114. deBoeij, W. P., Pshenichnikov, M. S., and Wiersma, D. A. (1996). On the relation between the echo-peak shift and Brownian-oscillator correlation function, *Chem. Phys. Lett.*, **253**, pp. 53–60.

115. Stenger, J., Madsen, D., Hamm, P., Nibbering, E. T. J., and Elsaesser, T. (2002). A photon echo peak shift study of liquid water, *J. Phys. Chem. A*, **106**, pp. 2341–2350.

116. Ernst, R. R., Bodenhausen, G., and Wokaun, A. (1987). *Principles of Nuclear Magnetic Resonance in One and Two Dimensions,* Oxfordshire, Oxford: Oxford Science; New York: Clarendon Press, Oxford University Press, xxiv, 610 p.

117. Payer, M. D. (2009). Dynamics of liquids, molecules, and proteins measured with ultrafast 2D IR vibrational echo chemical exchange spectroscopy, *Annu. Rev. Phys. Chem.*, **60**, pp. 21–38.

118. Woutersen, S., and Hamm, P. (2002). Nonlinear two-dimensional vibrational spectroscopy of peptides, *J. Phys.-Condens. Matter*, **14**, R1035–R1062.

119. Hochstrasser, R. M., Hamm, P., Lim, M., Ge, N. H., and DeGrado, W. F. (2000). Two-dimensional infrared spectra of peptide amide and N-H vibrations, *Biophys. J.*, **78**, pp. 13A–13A.

120. Hamm, P., Lim, M., DeGrado, W. F., and Hochstrasser, R. M. (2000). Pump/probe self heterodyned 2D spectroscopy of vibrational transitions of a small globular peptide, *J. Chem. Phys.*, **112**, pp. 1907–1916.

121. Londergan, C. H., Kim, Y. S., and Hochstrasser, R. M. (2005). Two-dimensional infrared spectroscopy of dipeptides in trehaloseglass, *Mol. Phys.*, **103**, pp. 1547–1553.

122. Hamm, P., Helbing, J., and Bredenbeck, J. (2008). Two-dimensional infrared spectroscopy of photoswitchable peptides, *Annu. Rev. Phys. Chem.*, **59**, pp. 291–317.

123. Manor, J., Mukherjee, P., Lin, Y. S., Leonov, H., Skinner, J. L., et al. (2009). Gating mechanism of the influenza A M2 channel revealed by 1D and 2D IR spectroscopies, *Structure*, **17**, pp. 247–254.

124. Ling, Y. L., Strasfeld, D. B., Shim, S. H., Raleigh, D. P., and Zanni, M. T. (2009). Two-dimensional infrared spectroscopy provides evidence

of an intermediate in the membrane-catalyzed assembly of diabetic amyloid, *J. Phys. Chem. B,* **113**, pp. 2498–2505.

125. Strasfeld, D. B., Ling, Y. L., Shim, S. H., and Zanni, M. T. (2008). Tracking fiber formation in human islet amyloid polypeptide with automated 2D-IR spectroscopy, *J. Am. Chem. Soc.,* **130**, pp. 6698–6699.

126. Naraharisetty, S. R. G., Kasyanenko, V. M., and Rubtsov, I. V. (2008). Bond connectivity measured via relaxation-assisted two-dimensional infrared spectroscopy, *J. Chem. Phys.,* **128**, 104502.

127. Kim, Y. S., Liu, L., Axelsen, P. H., and Hochstrasser, R. M. (2008). Two-dimensional infrared spectra of isotopically diluted amyloid fibrils from Aβ40, *Proc. Natl. Acad. Sci. U S A,* **105**, pp. 7720–7725.

128. Fang, C., Bauman, J. D., Das, K., Remorino, A., Arnold, E., et al. (2008). Two-dimensional infrared spectra reveal relaxation of the nonnucleoside inhibitor TMC278 complexed with HIV-1 reverse transcriptase, *Proc. Natl. Acad. Sci. U S A,* **105**, pp. 1472–1477.

129. Zigmantas, D., Read, E. L., Mancal, T., Brixner, T., Gardiner, A. T., et al. (2006). Two-dimensional electronic spectroscopy of the B800-B820 light-harvesting complex, *Proc. Natl. Acad. Sci. U S A,* **103**, pp. 12672–12677.

130. Engel, G. S., Calhoun, T. R., Read, E. L., Ann, T. K., Mancal, T., et al. (2007). Evidence for wavelike energy transfer through quantum coherence in photosynthetic systems, *Nature,* **446**, pp. 782–786.

131. Stone, K. W., Gundogdu, K., Turner, D. B., Li, X. Q., Cundiff, S. T., et al. (2009). Two-quantum 2D FT electronic spectroscopy of biexcitons in GaAs quantum wells, *Science,* **324**, pp. 1169–1173.

132. Vaughan, J. C., Hormrng, T., Stone, K. W., and Nelson, K. A. (2007). Coherently controlled ultrafast four-wave mixing spectroscopy, *J. Phys. Chem. A,* **111**, pp. 4873–4883.

133. Zhang, T. H., Kuznetsova, I., Meier, T., Li, X. C., Mirin, R. P., et al. (2007). Polarization-dependent optical 2D Fourier transform spectroscopy of semiconductors, *Proc. Natl. Acad. Sci. U S A,* **104**, pp. 14227–14232.

134. Jiang, J., Abramavicius, D., Falvo, C., Bulheller, B. M., Hirst, J. D., et al. (2011). Simulation of two-dimensional ultraviolet spectroscopy of amyloid fibrils, *J. Phys. Chem. B,* **114**, pp. 12150–12156.

135. Jiang, J. and Mukamel, S. (2011). Probing amyloid fibril growth by two-dimensional near-ultra violet spectroscopy, *J. Phys. Chem. B,* **115**, pp. 6321–6328.

136. Jiang, J. and Mukamel, S. (2011). Two-dimensional near-ultraviolet spectroscopy of aromatic residues in amyloid fibrils: a first principles study, *Phys. Chem. Chem. Phys.*, **13**, pp. 2394–2400.

137. Jiang, J. and Mukamel, S. (2011). Two-dimensional ultraviolet (2DUV) spectroscopic tools for identifying fibrillation propensity of protein residue sequences, *Angew. Chem. Int. Ed.*, **49**, pp. 9666–9669.

138. West, B. A., and Moran, A. M. (2012). Two-dimensional electronic spectroscopy in the ultraviolet wavelength range, *J. Phys. Chem. Lett.*, **3**, pp. 2575–2581.

139. Cannizzo, A. (2012). Ultrafast UV spectroscopy: from a local to a global view of dynamical processes in macromolecules, *Phys. Chem. Chem. Phys.*, **14**, pp. 11205–11223.

140. Brixner, T., Mancal, T., Stiopkin, I. V., and Fleming, G. R. (2004). Phase-stabilized two-dimensional electronic spectroscopy, *J. Chem. Phys.*, **121**, pp. 4221–4236.

141. Selig, U, Langhojer, F., Dimler, F., Lohrig, T., Schwarz, C., et al. (2008). Inherently phase-stable coherent two-dimensional spectroscopy using only conventional optics, *Opt. Lett.*, **33**, pp. 2851–2853.

142. Selig, U., Schleussner, C. F., Foerster, M., Langhojer, F., Nuernberger, P., et al. (2010). Coherent two-dimensional ultraviolet spectroscopy in fully noncollinear geometry, *Opt. Lett.*, **35**, pp. 4178–4180.

143. West, B. A., Womick, J. M. and Moran, A. M. (2011). Probing ultrafast dynamics in adenine with mid-UV four-wave mixing spectroscopies, *J. Phys. Chem. A*, **115**, pp. 8630– 8637.

144. Baum, P., Lochbrunner, S., and Riedle, E. (2004). Tunable sub-10-fs ultraviolet pulses generated by achromatic frequency doubling, *Opt. Lett.*, **29**, pp. 1686–1688.

145. Zhao, B. Z., Jiang, Y. L., Sueda, K., Miyanaga, N., and Kobayashi, T. (2009). Sub-15fs ultraviolet pulses generated by achromatic phase-matching sum-frequency mixing, *Opt. Express*, **17**, pp. 17711–17714.

146. Ghotbi, M., Trabs, P., and Beutler, M. (2011). Generation of high-energy, sub-20-fs pulses in the deep ultraviolet by using spectral broadening during filamentation in argon, *Opt. Lett.*, **36**, pp. 463–465.

147. Fuji, T., Horio, T., and Suzuki, T. (2007). Generation of 12 fs deep-ultraviolet pulses by four-wave mixing through filamentation in neon gas, *Opt. Lett.*, **32**, pp. 2481–2483.

148. Maioli, P., Salame, R., Lascoux, N., Salmon, E., Bejot, P., et al. (2009). Ultraviolet-visible conical emission by multiple laser filaments, *Opt. Express*, **17**, pp. 4726–4731.

149. Kida, Y., Liu, J., Teramoto, T., and Kobayashi, T. (2010). Sub-10 fs deep-ultraviolet pulses generated by chirped-pulse four-wave mixing, *Opt. Lett.*, **35**, pp. 1807–1809.

150. Aubock, G., Consani, C., Monni, R., Cannizzo, A., Van Mourik, F., et al. (2012). Femtosecond pump/supercontinuum-probe setup with 20 kHz repetition rate, optimized for broadband UV detection, *Rev. Sci. Instrum.*, **83**, 093105.

151. Aubock, G., Consani, C., van Mourik, F., and Chergui, M. (2012). Ultrabroadband femtosecond two-dimensional ultraviolet transient absorption, *Opt. Lett.*, **37**, pp. 2337–2339.

152. Extermann, J., Weber, S. M., Kiselev, D., Bonacina, L., Lani, S., et al. (2011). Spectral phase, amplitude, and spatial modulation from ultraviolet to infrared with a reflective MEMS pulse shaper, *Opt. Express*, **19**, pp. 7580–7586.

153. Krebs, N., Probst, R. A., and Riedle, E. (2010). Sub-20 fs pulses shaped directly in the UV by an acousto-optic programmable dispersive filter, *Opt. Express*, **18**, pp. 6164–6171.

154. Weber, S., Barthelemy, M., and Chatel, B. (2010). Direct shaping of tunable UV ultra-short pulses, *Appl. Phys. B*, **98**, pp. 323–326.

155. Diels, J.–C., and Rudolph, W. (1996). *Ultrashort Laser Pulse Phenomena: Fundamentals, Techniques, and Applications on a Femtosecond Time Scale.* Optics and Photonics Series, San Diego: Academic Press, xviii, 581

156. Shim, S. H., and Zanni, M. T. (2009). How to turn your pump-probe instrument into a multidimensional spectrometer: 2D IR and Vis spectroscopies via pulse shaping, *Phys. Chem. Chem. Phys.*, **11**, pp. 748–761.

157. Tseng, C. H., Matsika, S., and Weinacht, T. C. (2009). Two-dimensional ultrafast Fourier transform spectroscopy in the deep ultraviolet, *Opt. Express*, **17**, pp. 18788–18793.

158. Krebs, N., Pugliesi, I., Hauer, J., and Riedle, E. (2013). Two- dimensional Fourier transform spectroscopy in the ultraviolet with sub-20 fs pump pulses and 250-720 nm supercontinuum probe, *New J. Phys.*, **15**, 085016.

159. Read, E. L., Engel, G. S., Calhoun, T. R., Mancal, T., Ahn, T. K., et al. (2007). Cross-peak-specific two-dimensional electronic spectroscopy, *Proc. Natl. Acad. Sci. U S A*, **104**, pp. 14203–14208.

160. Hacker, M., Stobrawa, G., Sauerbrey, R., Buckup, T., Motzkus, M., et al. (2003). Micromirror SLM for femtosecond pulse shaping in the ultraviolet, *Appl. Phys. B*, **76**, pp. 711–714.

161. Rondi, A., Extermann, J., Bonacina, L., Weber, S. M., and Wolf, J. P. (2009). Characterization of a MEMS-based pulse-shaping device in the deep ultraviolet, *Appl. Phys. B*, **96**, pp. 757–761.

162. Ajdarzadeh Oskouei, A., Tortschanoff, A., Bram, O., van Mourik, F., Cannizzo, A., et al. (2010). Three pulse UV photon echo studies of molecules in solution: effect of the chirp, *J. Chem. Phys.*, **133**, 064506.

163. Ajdarzadeh Oskouei, A., Braem, O., Cannizzo, A., van Mourik, F., Tortschanoff, A., et al. (2008). Ultrafast UV photon echo peak shift and fluorescence up conversion studies of non-polar solvation dynamics, *Chem. Phys.*, **350**, pp. 104–110.

164. Ajdarzadeh Oskouei, A., Bram, O., Cannizzo, A., van Mourik, F., Tortschanoff, A., et al. (2008). Photon echo peak shift experiments in the UV: p-terphenyl in different solvents, *J. Mol. Liq.*, **141**, pp. 118–123.

165. Ajdarzadeh Oskouei, A. (2011). *Development of Ultraviolet Photon Echo Experiments for the Study of Molecular Dynamics in Solution*, PhD thesis, Ecole Polytechnique Federate de Lausanne.

166. Dudovich, N., Oron, D., and Silberberg, Y. (2002). Single-pulse coherently controlled nonlinear Raman spectroscopy and microscopy, *Nature*, **418**, pp. 512–514.

167. Goodsell, D. S. (2004). *Bionanotechnology: Lessons from Nature*, Hoboken NJ: Wiley-Liss. xii, 337 p.

168. Scholes, G. D., Fleming, G. R., Olaya-Castro, A., and van Grondelle, R. (2011). Lessons from nature about solar light harvesting, *Nat. Chem.*, **3**, pp. 763–774.

169. Ruetzel, S., Diekmann, M., Nuernberger, P., Walter, C., Engels, B., et al. (2014). Multidimensional spectroscopy of photoreactivity, *Proc. Natl. Acad. Sci. U S A*, **111**, pp. 4764–4769.

170. Nuernberger, P., Ruetzel, S. and Brixner, T. (2015). Multidimensional electronic spectroscopy of photochemical reactions, *Angew. Chem. Int. Ed.*, **54**, pp. 11368–11386.

171. Brown, K. E., Veldkamp, B. S., Co, D. T., and Wasielewski, M. R. (2012). Vibrational dynamics of a perylene-perylenediimide donor-acceptor dyad probed with femtosecond stimulated Raman spectroscopy, *J. Phys. Chem. Lett.*, **3**, pp. 2362–2366.

Section II

Theory

Chapter 7

Ultrafast Exciton Dynamics in Correlated Environments

Peter Nalbach
Fachbereich Wirtschaft & Informationstechnik, Westfälische Hochschule, Münsterstrasse 265, 46397 Bocholt, Germany
peter.nalbach@w-hs.de

7.1 Photosynthesis and Energy Transfer

Essentially all living things on earth use the sun as their energy source which is harvested by photosynthesis. Photosynthetic conversion of physical energy of sunlight into a chemical form useful for cellular processes involves many steps [1, 2]. Photosynthesis starts with the absorption of a photon by a light-harvesting pigment or chromophore forming an exciton, followed by the transfer of the exciton to the reaction center, where charge separation is initiated leading subsequently to chemical storage of the energy. In this chapter, our focus lies on the transfer dynamics which surprisingly exhibits a quantum efficiency near unity under low light conditions. Not only plants but also several bacteria make use of photosynthesis. In green sulfur bacteria such as *Chlorobaculum*

Ultrafast Dynamics at the Nanoscale: Biomolecules and Supramolecular Assemblies
Edited by Irene Burghardt and Stefan Haacke

ISBN 978-981-4745-33-8 (Hardcover), 978-981-4745-34-5 (eBook)
www.panstanford.com

tepidum light harvesting is done by an antennae complex, the chlorosome, which is a large conglomerate. The connection to the reaction center is exclusively through the small Fenna–Matthews–Olson (FMO) complex, a trimer whose monomers each contain seven chromophores [3–6]. Recently, an eighth pigment was resolved [7] which is, however, very weakly coupled to the other seven pigments and thus is of little relevance for our studies. The purpose of the FMO complex is purely to conduct the excitonic energy from the antennae to the reaction center. Typical transfer times through the FMO are on the order of picoseconds, whereas the decay of an excited chromophore (with subsequent reemission of a photon) takes on the order of nanoseconds. In addition, typically, the absorption cross section for photons is small, meaning that in general only a single exciton is present in an FMO complex. This allows to restrict later the theoretical consideration to the single excitation subspace. Furthermore, these timescales reflect the extraordinary quantum efficiency of the excitonic energy transfer in photosynthesis. However, it is still unclear why transport is so much faster than direct decay of excitons. An understanding of this question, which must lie in the design principle of photosynthetic complexes, would allow to exploit the near-unity efficiency of energy transfer within artificial light-harvesting complexes. Organic solar cells suffer from very high exciton losses during transport and thus could greatly benefit from biomimetic design principles.

Energy transfer between chromophores is mediated by the dipole–dipole coupling between the optical transition dipoles of the chromophores. The theoretical formulation for energy transfer via the dipole–dipole coupling was provided by Theodor Förster in the 1940s [8]. Förster treatment results in incoherent hopping of the excitons from chromophore to chromophore which fits expectations that quantum coherence should not be functionally relevant for a living organism at room temperature. Thermal fluctuations, however, are not necessarily enough to cause the microscopically quantum mechanical dynamics to become overdamped, since electronic dipole–dipole couplings in FMO are as high as 100 cm^{-1} $\simeq 150\ \mathrm{K}/k_{\mathrm{B}}$ and thus on the scale of room temperature. This is generally not enough to overdamp dynamics [9, 10]. Absorption data point towards large reorganization energies which result in overdamped dynamics within a Markov assumption. Nevertheless, it is a longstanding question [11, 12] whether the transfer of excitonic energy is incoherent or coherent. A few years ago, Engel et al. [13, 14] have reported long-lasting beating signals in time-resolved optical 2D spectra [15] of the FMO complex. These have been interpreted as evidence for quantum coherent energy transfer via delocalized exciton states. Quantum coherence times of more than 660 fs at 77 K [13] and more than 300 fs at physiological temperatures [14] have been reported. Similar experimental findings were observed in marine cryptophyte algae [16]. These observations sparked tremendous research activities to clarify the role of quantum coherence in living organism [17–21]. Full quantum theoretical modeling of the real-time dynamics of a FMO complex is beyond current computer power due to the large cluster size and strong non-Markovian fluctuations. Studies of the excitation dynamics, therefore, rely on simplified, that is, system–environment, models [10–12] of few chromophore sites which interact by dipolar couplings and which are exposed to environmental fluctuations of the bacteriochlorophyll molecule, the surrounding protein and the solvent [1, 12]. Weak coupling approaches have been employed to investigate the question whether quantum coherence might be relevant for the energy transfer dynamics [22–24].

We focus on the question how correlations in the environmental fluctuations promote long-lived quantum coherence and whether such a scenario is adequate for the biomolecular environments with strong vibronic modes and a surrounding strongly coupled polar solvent. In the following we review our recent works towards these goals. Following standard approaches the next section introduces briefly the general framework of system–environment models [9, 10] and two methods to treat relevant problems, namely the weak coupling approach by re-summed perturbation series (RESPET) [25–28] and the numerically exact path integral propagator approach (QUAPI) [29–34]. In the following section we study the energy transfer dynamics. At first, we show within a simplified donor–acceptor model by employing QUAPI that long-lived temporal [35–37] but also spatial [37, 38] correlations within the environment allow for very long-lived coherence and even entanglement. Simple weak coupling Markovian methods fail to describe the energy transfer properly [28, 39, 40], since they cannot capture the multiphonon processes involved [28]. Thus, in principle the biomolecular fluctuations might allow for long-lived coherence. We show then that phenomenological, highly simplified model parameters for the FMO give coherence times as observed and even longer [41, 42]. However, employing realistic model parameters obtained by spectroscopic data [43, 44], coherence times are roughly a factor of two shorter [45] than experimentally observed. Recently, an alternative interpretation was put forward where the observed coherent oscillations are caused by discrete molecular vibrations within the environment [46–48]. These vibrations must be close to resonance with excitonic excitation since high-energy discrete modes cannot generate coherent behavior [49]. Overall, a complete understanding is still elusive. We end with a short summary and outlook.

7.2 Quantum Dynamics in Fluctuating Environments

Due to the vast complexity of biomolecular systems like the FMO complex, the question whether quantum coherence is essential in a functional way for energy transfer must be answered within simplified *system–environment* models. Typically, there are some degrees of freedom of the total biomolecular system which provide the relevant functions, that is, the single excitation states of the chromophores. All other degrees of freedom are, in some sense, peripheral. The former part we refer to as *system* whereas the latter is termed *environment*. Naturally both are coupled, which results in dissipation in the system dynamics. Typically, relaxation in a quantum system is the result of energy exchange with an environment of which we have only incomplete knowledge. At the same time, decoherence or dephasing are a result of forces acting on the quantum system originating from the same environment without explicit energy exchange. The first to model dissipation in a quantum spin-$\frac{1}{2}$ system was Bloch in the 1940s introducing relaxation and decoherence rates determined by Fermi's Golden Rule from the coupling to environmental degrees of freedom. In the 1960s, Zwanzig [50] and Mori [51] formalized the separation of the system from the environment by a projection operator approach. For the treatment of the environmental degrees of freedom within such an approach, approximations are necessary. While the system evolves in time, the environment is fixed to its stable equilibrium, and only acts on the system through fluctuations. This allows a description of the environment in terms of a sum of harmonic oscillators which are bilinearly coupled to the system [52] resulting in the standard system–environment models [9, 10]. In the following, we briefly discuss the standard approach to system–environment problems. Then we describe RESPET, a method which allows to treat weak coupling between the system and environmental Markovian fluctuations. Fluctuations in biomolecular complexes, influencing energy transfer between chromophores, are, however, strongly non-Markovian with long-lived temporal and possibly spatial correlations. To treat such fluctuations, we use the QUAPI method.

7.2.1 System–Environment Models

The simplest approach to discuss the dynamics of a quantum system of interest subjected to environmental fluctuations is given by a Hamiltonian of the form

$$H(t) = H_{\mathrm{S}}(t) + H_{\mathrm{SB}} + H_{\mathrm{B}} \tag{7.1}$$

where the system of interest is bilinearly coupled via H_{SB} to an ensemble of noninteracting harmonic oscillators H_{B} describing the normal fluctuation modes of the environment. The Hamiltonian of the central system

$$H_{\mathrm{S}}(t) = \frac{\mathbf{P}^2}{2M} + V(\mathbf{Q}, t) \tag{7.2}$$

consists of a quantum particle with mass M, coordinate $\mathbf{Q}$, and momentum $\mathbf{P}$ which moves in the possibly time-dependent potential $V(\mathbf{Q}, t)$. The Hamiltonians for the environment H_{B} and the system–

environment coupling H_{SB}, that is,

$$H_{\mathrm{B}} = \sum_{j=1}^{N} \left(\frac{\mathbf{p}_j}{2m_j} + \frac{m_j \omega_j^2}{2} \mathbf{q}_j^2 \right) \tag{7.3}$$

and

$$H_{\mathrm{SB}} = \mathbf{Q} \cdot \sum_{j=1}^{N} c_j \mathbf{q}_j, \tag{7.4}$$

represent an ensemble of N harmonic oscillators with mass m_j, frequency ω_j, momenta $\mathbf{p}_j$, and position $\mathbf{q}_j$. System and environment are bilinearly coupled in the positions with coupling constant c_j. We omitted counterterms [10] which might be necessary at strong coupling to ensure stability in the harmonic environment. Since our interest is in the system's dynamics, the environment is typically assumed to be initially in thermal equilibrium at temperature T and its dynamics not to be influenced by the system. With the quantum system of interest initially in a state described by the statistical operator $\rho_{\mathrm{S}}(t_0)$, we assume a factorized total initial state

$$W(t_0) = \rho_{\mathrm{S}}(t_0) \otimes \rho_{\mathrm{B}}^{\mathrm{eq}} \tag{7.5}$$

with the canonical equilibrium statistical operator $\rho_{\mathrm{B}}^{\mathrm{eq}} = Z_{\mathrm{B}}^{-1} e^{-\beta H_{\mathrm{B}}}$ with $\beta^{-1} = k_{\mathrm{B}} T$ and $Z_{\mathrm{B}} = Tr_{\mathrm{B}}\{e^{-\beta H_{\mathrm{B}}}\}$. Other initial states result typically in deviations in the dynamics at very short times only [10, 53] except for very slow environmental fluctuations [54, 55]. It is furthermore assumed that the small quantum system of interest cannot change the state of the huge environment which accordingly has constant temperature.

7.2.1.1 Quantum Langevin equation

In order to see how the coupling to an environment of harmonic oscillators causes fluctuating forces on the quantum system of interest, one considers the equation of motion for its position $\mathbf{Q}$. The Heisenberg equations for the oscillator degrees of freedom can easily be solved leading to a quantum Langevin equation

$$M\ddot{\mathbf{Q}}(t) + M \int_{t_0}^{t} ds\, \gamma(t-s) \dot{\mathbf{Q}}(s) + V'(\mathbf{Q}, t) = \mathbf{\Gamma}(t - t_0) \tag{7.6}$$

with damping kernel

$$\gamma(t) = \frac{1}{M}\sum_{j=1}^{N} \frac{c_j^2}{m_j\omega_j^2}\cos\omega_j t$$

and external force

$$\mathbf{\Gamma}(t) = -\sum_{j=1}^{N} c_j \left[\mathbf{q}_j(t_0)\cos\omega_j t + \frac{\mathbf{p}_j(t_0)}{m_j\omega_j}\sin\omega_j t\right].$$

Notice that for simplicity, we included initial slip terms which solely depend on the system dynamics and the initial condition, into the potential for the system, such that $V \rightarrow V'(\mathbf{Q}, t)$ [56]. $\Gamma(t)$ is an operator-valued stochastic force since the initial values $\mathbf{q}_j(t_0)$ and $\mathbf{p}_j(t_0)$ reflect the thermal equilibrium of the environment. Since the environmental oscillators are harmonic, one finds for the mean force $\langle\Gamma(t)\rangle_B = 0$ and for the force-force (or environment) correlation function (for $t > s$) $\hbar L(t-s) = \langle\mathbf{\Gamma}(t)\mathbf{\Gamma}(s)\rangle_B$ and

$$L(\tau) = \sum_{j=1}^{N} \frac{c_j^2}{2m_j\omega_j}\left[\coth\left(\beta\hbar\omega_j/2\right)\cos\omega_j(\tau) - i\sin\omega_j(\tau)\right]. \quad (7.7)$$

Since the environment is assumed to be too large to be disturbed by the system, which also ensures that the Poincaré recurrence time exceeds all relevant times scales of the system, we can switch to a continuum description. Introducing furthermore the spectral density

$$J(\omega) = \pi\sum_{j=1}^{N} \frac{c_j^2}{2m_j\omega_j}\delta(\omega-\omega_j), \quad (7.8)$$

one obtains

$$\hbar L(\tau) = \frac{1}{\pi}\int_0^\infty d\omega\, J(\omega)\left[\coth\left(\frac{\beta\hbar\omega}{2}\right)\cos\omega(\tau) - i\sin\omega(\tau)\right]$$

and

$$\gamma(t) = \frac{2}{\pi M}\int_0^\infty d\omega\,\frac{J(\omega)}{\omega}\cos\omega t.$$

In the classical limit $\hbar\omega \ll k_B T$ the autocorrelation function (Eq. 7.7) becomes $\hbar L(t-s) = \langle\mathbf{\Gamma}(t)\mathbf{\Gamma}(s)\rangle_B = Mk_B T\gamma(t-s)$ and the quantum Langevin equation is formally identical to the classical one. The quantum Langevin equation description nicely shows that our

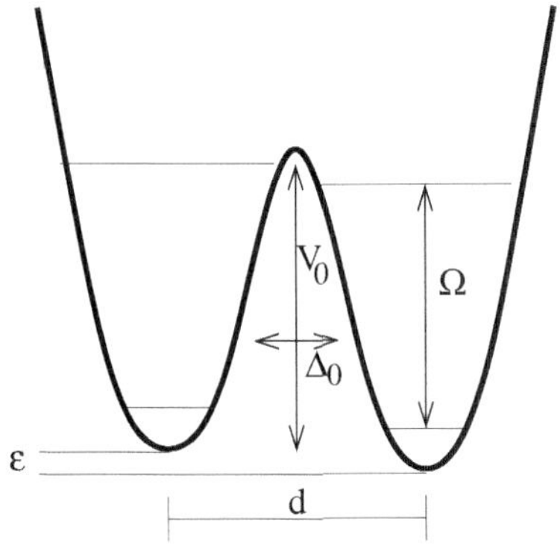

Figure 7.1 Double-well potential.

basic idea to model environmental fluctuations acting on a quantum system of interest can be described within a system–environment model as introduced above. For actual studies of the dynamics of the system, we refer in practice to other methods introduced in the following.

7.2.1.2 Spin-boson model

So far, we described the system degree of freedom as a particle within a potential, defined by continuous momentum and position variables. Very often, the system of interest can be further truncated to a finite set of states, in the simplest case even to a quantum two-level system. A natural example is a particle moving in a double-well potential (see Fig. 7.1). If both wells only differ in their energy minimum by an asymmetry ϵ and this asymmetry and temperature $k_{\mathrm{B}}T$ are both much smaller than the excitation energy Ω in each

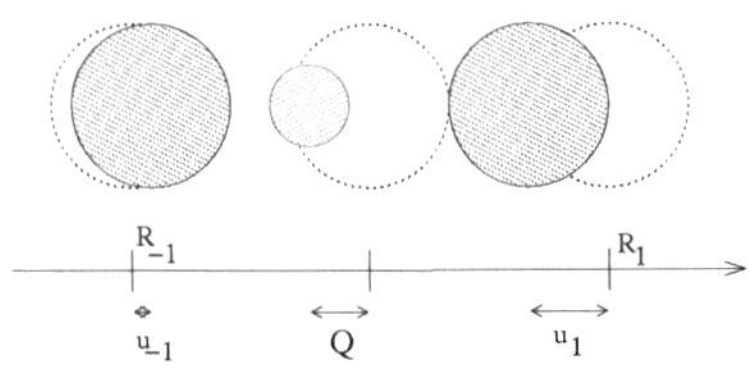

Figure 7.2 Substitutional defect in between two host ions.

separate well, then a description including only the lowest state of each potential well is sufficient [9, 10].

If the potential barrier V_0 between both wells is infinite, the eigenstates in each well, particle left $|l\rangle$ and particle right $|r\rangle$, would be the eigenstates of the full problem. For a finite barrier $V_0 < \infty$ the particle can tunnel between the two wells and the eigenstates are superpositions of the states $|l\rangle$ and $|r\rangle$ with a Hamiltonian

$$H_S = \frac{\Delta_0}{2}\sigma_x + \frac{\epsilon}{2}\sigma_z$$

with position $\sigma_z = |l\rangle\langle l| - |r\rangle\langle r|$, tunneling operator $\sigma_x = |l\rangle\langle r| + |r\rangle\langle l|$ and tunneling element Δ_0. Employing a path integral approach including a WKB approximation one finds $\Delta_0 \approx e^{-S}$ with $S \simeq d\sqrt{2mV_0}$.

Some caution is advised when reducing a physical problem of a particle in a double-well potential embedded in an environment to a two-level system coupled to environmental fluctuations. We briefly illustrate the procedure for a substitutional defect in a 1D chain of atoms as shown in Fig. 7.2. The complete potential $V = V_{\text{lattice}} + W(Q, \{u_j\})$ for the defect (red circle in Fig. 7.2) is given by a sum of the potential of the lattice V_{lattice} without the defect and the interaction of the defect with the lattice $W(Q, \{u_j\})$. Assuming small displacements u_j of the atoms around their stable equilibrium positions R_j allows to expand the potential in a series in u_j. Restricting consideration to quadratic order and introducing normal modes results in the standard system–environment Hamiltonian, that is, Eqs. 7.2–7.4, with coupling constants $\lambda^{(k)}(Q)$ and wavevector k of the respective normal mode. Assuming that the resulting potential $V(Q)$ for the particle represents a double well with potential minima at $\pm d/2$ and that Δ_0, ϵ, $k_B T \ll \Omega$, allows then to describe the defect coupled to lattice vibrations by the spin-boson Hamiltonian (q_k and p_k rescaled to absorb the mass m)

$$H = \frac{\Delta_0}{2}\sigma_x + \frac{\epsilon}{2}\sigma_z + \sigma_z \sum_k \lambda_k q_k + \frac{1}{2}\sum_k \left(p_k^2 + \omega_k^2 q_k^2\right) \qquad (7.9)$$

with coupling constants $\lambda_k = 2\lambda^{(k)}(\pm d/2) = -2i(\Lambda/\sqrt{N})\sin(ka)$ for lattice constant a. Similar results are obtained for a substitutional defect in a 3D crystal whose phonon modes can be described within a Debye model with a density of modes $\rho(\omega) \propto \omega^2$.

Using linear dispersion we obtain explicitly the spectral function (Eq. 7.8) as $J(\omega) = \eta\omega^3 e^{-\omega/\omega_D}$ with coupling constant η, Debye frequency ω_D and assuming an exponential suppression of modes with frequency beyond ω_D. Strictly speaking the mapping to the spin-boson Hamiltonian is valid for vibrations with $\omega \lesssim \Omega$ only. Higher-frequency modes modify the tunneling element Δ_0 and cannot be treated within a spin-boson formulation but must be treated before the mapping [9]. Here, we therefore assumed for simplicity that $\Omega \simeq \omega_D$.

The spin-boson Hamiltonian (Eq. 7.9) is a powerful model which allows to discuss many physical problems like energy transfer in donor–acceptor pairs or light-harvesting complexes, charge dynamics in a double quantum dot, chemical reactions, tunneling systems in crystals and glasses, to name but a few. Many tools have been developed to treat spin-boson dynamics, two of which I introduce in the following. As can already be seen in the quantum Langevin equation the influence of the environment can be fully described knowing just the spectral function (Eq. 7.8). Generally, the spectral function (Eq. 7.8) is assumed to be a continuous function in the limit of $N \to \infty$

$$J(\omega) = 2\pi\alpha\omega_s^{1-s}\omega^s e^{-\omega/\omega_c} \tag{7.10}$$

with a coupling constant α, a cutoff frequency ω_c beyond which fluctuations are strongly suppressed and a relevant energy scale ω_s to render α unitless. The spectral functions are named in relation to the spectral exponent s, that is, ohmic (subohmic, super Ohmic) for $s = 1$ ($s < 1$, $s > 1$). Additionally, molecular systems often exhibit additional narrow peaks in the fluctuations spectrum resulting from intramolecular vibrations [57].

7.2.1.3 Bloch equations

Before I discuss the methods that will be employed later, I briefly present a phenomenological approach to discuss the dissipative dynamics of a quantum two-level system which goes back to Bloch in the 1940s. To this end, we switch to a diagonal basis, in which the system Hamiltonian is diagonal, that is, $H_d = R^\dagger H R$ with

$$H_d = \frac{E}{2}\tau_x + (u\tau_z + v\tau_x)\sum_k \lambda_k q_k + \frac{1}{2}\sum_k \left(p_k^2 + \omega_k^2 q_k^2\right) \tag{7.11}$$

with $E = \sqrt{\Delta^2 + \epsilon^2}$ and $R = \exp(i\phi\sigma_y/2)$, $\phi = \arctan(\epsilon/\Delta_0)$, $u = \cos\phi$, $v = \sin\phi$. I use the Pauli matrices τ_i here in order to highlight the basis change in regard to Eq. 7.9 where σ_z corresponds to a real position.

Bloch proposed that the quantum dynamics of the statistical operator of the two-level system $\rho(t) = \frac{1}{2}\left(\mathbb{1} + \sum_j r_j(t)\tau_j\right)$ follows the quantum von Neumann equation with two additional features. The coupling $\propto u\tau_z$ in Eq. 7.11 can induce transitions in the two-level system, and therefore relaxation. Thermodynamics dictates that the system will relax towards thermal equilibrium, that is, $\rho^{\mathrm{eq}} = \frac{1}{2}(\mathbb{1} + r_x^{\mathrm{eq}}\tau_x)$ with $r_x^{\mathrm{eq}} = -\tanh(\beta E/2)$. This relaxation as well as the energy fluctuations due to the coupling $\propto v\tau_x$ will also result in a loss of coherence. The Bloch equations incorporate such a behavior on a purely phenomenological basis into the dynamical equations

$$\begin{aligned} \partial_t r_x(t) &= -\gamma_1[r_x(t) - r_x^{\mathrm{eq}}] \\ \partial_t r_y(t) &= -\gamma_2 r_y(t) - E r_z(t) \\ \partial_t r_z(t) &= +E r_y(t) - \gamma_2 r_z(t) \end{aligned} \quad (7.12)$$

thereby defining the relaxation rate γ_1 and the decoherence rate γ_2 which are determined using Fermi's Golden Rule.

The main goal of the research field of open or dissipative quantum dynamics is on one hand to determine the rates given specific fluctuation spectra and on the other hand to obtain and understand deviations from Bloch equations as a result of the coupling to the environment. In case of weak system–environment coupling the dynamics is typically Markovian and the Bloch equations are a good description. Far less understood is the dissipative dynamics for driven quantum systems [34, 59–61] relevant for quantum control schemes and for strongly correlated fluctuations as typically encountered in biomolecular systems.

7.2.2 Methods: RESPET and QUAPI

7.2.2.1 Weak coupling approximation

For weak system–environment coupling, that is, $\alpha \ll 1$ (see 7.8), the dynamics can be addressed in a re-summed perturbation series approach (RESPET) [25–28].

The spin-boson dynamics, governed by the Hamiltonian Eq. 7.9 or Eq. 7.11, is determined by the von Neumann equation

$$\partial_t W(t) = -i[H, W(t)] =: \mathcal{L}W(t) \qquad (7.13)$$

with the full statistical operator $W(t)$, the initial condition W_0 (Eq. 7.5) and Liouville operator $\mathcal{L}$. A formal solution is achieved by the time evolution superoperator $\mathcal{U}(t)$ with

$$\mathcal{U}(t)W_0 = W(t) \quad \text{and} \quad \mathcal{U}(t) = \exp(\mathcal{L}t). \qquad (7.14)$$

For a perturbative approach in the coupling between system and environment we split the Hamiltonian $H_0 = H - H_{\text{SB}}$ with $H_{\text{SB}} = \sigma_z \sum_k \lambda_k q_k$ and likewise the Liouvillian $\mathcal{L} = \mathcal{L}_0 + \mathcal{L}_{\text{SB}}$. The time evolution operator obeys a Dyson equation

$$\mathcal{U}(t, t_0) = \mathcal{U}_0(t, t_0) + \int_{t_0}^{t} ds \mathcal{U}_0(t, s)\mathcal{L}_{\text{SB}}\mathcal{U}_0(s, t_0)$$
$$+ \int_{t_0}^{t} ds \int_{t_0}^{s} ds' \mathcal{U}_0(t, s)\mathcal{L}_{\text{SB}}\mathcal{U}_0(s, s')\mathcal{L}_{\text{SB}}\mathcal{U}(s', t_0)$$

with $\mathcal{U}_0(t, t_0) = e^{\mathcal{L}_0(t-t_0)}$. Integrating over the environmental degrees of freedom in the Dyson equation results in the effective time evolution superoperator

$$\mathcal{U}_{\text{eff}}(t, t_0) = \mathcal{U}_{\text{S}}(t, t_0) + \int_{t_0}^{t} ds \int_{t_0}^{s} ds' \mathcal{U}_{\text{S}}(t, s)\mathcal{M}(s, s')\mathcal{U}_{\text{eff}}(s', t_0)$$

with the memory kernel $M(s, s') = \langle \mathcal{L}_{\text{SB}}\mathcal{U}_0(s, s')\mathcal{L}_{\text{SB}} \rangle_{\text{B}}$ where $\langle \cdot \rangle_{\text{B}} = \text{Tr}\left\{ \cdot \rho_{\text{B}}^{\text{eq}} \right\}$. Notice that $\langle \mathcal{L}_{\text{SB}} \rangle_{\text{B}} = 0$ and that we considered the memory kernel in lowest order (one-loop approximation) in the coupling between system and environment only. The effective time evolution superoperator $\mathcal{U}_{\text{eff}}$ describes the reduced dynamics of the system. With the Laplace transformation defined as $f(z) = i \int_0^\infty dt e^{izt} f(t)$ the Dyson equation is solved by

$$\mathcal{U}_{\text{eff}}(z) = \left(\mathcal{U}_{\text{S}}^{-1}(z) + \mathcal{M}(z) \right)^{-1}. \qquad (7.15)$$

Both $\mathcal{U}_{\text{S}}^{-1}(z)$ and $\mathcal{M}(z)$ are readily determined [27, 28]. Employing a secular approximation and a pole approximation within the Laplace back transformation, one obtains the standard Markovian results, that is, dynamics as predicted by the Bloch equations with rates $\gamma_2 = \gamma_1/2 + \Gamma_2$ and

$$\gamma_1 = \frac{1}{2} u^2 J(E) \coth\left(\frac{1}{2}\beta E\right) \quad \text{and} \quad \Gamma_2 = \left[v^2 J(\omega) \coth\left(\frac{1}{2}\beta\omega\right) \right]_{\omega \to 0}. \qquad (7.16)$$

Herein, Γ_2 is the dephasing rate due to low-frequency fluctuations. Our treatment results additionally in a renormalization of the tunneling element frequency, that is, $\Delta_0 \to \Delta$ with

$$\Delta - \Delta_0 = 2\,\Re\left[\int_0^\infty d\omega J(\omega)\coth(\beta\omega/2)\frac{-z}{z^2-\omega^2}\right].$$

These results are in line with results from Bloch–Redfield, Lindblad, and other weak coupling approaches. Deviations between the various weak coupling treatments emerge in higher order in the system–environment coupling only. When such corrections become sizable, more sophisticated methods are needed.

7.2.2.2 Correlated environmental fluctuations

The real-time dynamics of a dissipative quantum system described by the Hamiltonian Eq. 7.1 can be determined numerically in an exact way by using the quasi-adiabatic propagator path integral (QUAPI) [29–34]. Specifically, QUAPI allows to study non-Markovian dynamics resulting from either strong system–environment coupling, nontrivial spectral functions or strongly correlated environmental fluctuations. Furthermore, QUAPI is ideally suited to treat driven situations with a time-dependent system Hamiltonian.

The key point of QUAPI is that it determines the reduced density operator $\rho(t)$ of the system iteratively starting from an initially prepared state $\rho(t_0)$ at time t_0 and then evolving it in N discrete time steps of length $\delta t = (t - t_0)/N$, that is, $t_k = k \cdot \delta t + t_0$. The short-time propagator $U(t_{k+1}, t_k)$ for the full Hamiltonian is approximated by a symmetric Trotter splitting into a part depending on the system Hamiltonian

$$U_S(t_{k+1}, t_k) = \mathcal{T}\exp\left\{-\frac{i}{\hbar}\int_{t_k}^{t_{k+1}} ds\, H_S(s)\right\}$$

and a part involving the environment plus coupling leading to

$$U(t_{k+1}, t_k) \simeq e^{-\frac{i}{\hbar}(H_{SB}+H_B)\frac{\delta t}{2}} \cdot U_S(t_{k+1}, t_k) \cdot e^{-\frac{i}{\hbar}(H_{SB}+H_B)\frac{\delta t}{2}}.$$

The full short-time propagator describes time evolution over a Trotter time slice δt by $W(t_{k+1}) = U(t_{k+1}, t_k)W(t_k)U^{-1}(t_{k+1}, t_k)$. The symmetric Trotter slicing is exact only in the limit $\delta t \to 0$ and thus introduces a finite Trotter error of order δt^3 in a single time step. Thus over the full time evolution a Trotter error of order δt^2 accumulates [62] which must be eliminated by choosing δt small enough to achieve convergence.

The system short-time propagator can be determined either analytically for finite-dimensional time-independent systems or numerically employing a standard fourth order Runge–Kutta scheme with adaptive stepsize. More problematic is the calculation of the environment short-time propagator since it depends explicitly on the system coordinate due to H_{SB}. We will discuss the case of a system with an M-dimensional Hilbert space with a generic system–environment coupling $H_{\mathrm{SB}} = \hat{\sigma} \cdot \sum_\nu \lambda_\nu q_\nu$ with generalized system coordinate $\hat{\sigma}$. For the spin-boson case $\hat{\sigma} = \sigma_z$. With the eigenvectors $|\hat{\sigma}_j\rangle$ of the generalized coordinate with eigenvalue $\hat{\sigma}_j$ $(1 \leq j \leq M)$ we construct $\mathbb{1} = \sum_j |\hat{\sigma}_j\rangle\langle\hat{\sigma}_j| \otimes \mathbb{1}_{\mathrm{B}}$. Inserting this in the full sliced time evolution operator $U(t = t_N, t_0)$ in between system and environment short-time propagators turns the system coordinate into a numerical parameter for the calculation of the environment short-time propagators at the cost of N sums over $1 \leq j \leq M$ leading to the reduced density operator

$$\begin{aligned}
\langle\hat{\sigma}_{l_1}|\rho(t)|\hat{\sigma}_{l_2}\rangle &= \langle\hat{\sigma}_{l_1}|\langle U(t = t_N, t_0)W(t_0)U^{-1}(t = t_N, t_0)\rangle_{\mathrm{B}}|\hat{\sigma}_{l_2}\rangle \\
&= \sum_{j_0} \cdots \sum_{j_N} \sum_{j'_0} \cdots \sum_{j'_N} \delta_{j_N, l_1}\delta_{j'_N, l_2} \cdot I^{(N)}(\hat{\sigma}_{j_0}, \hat{\sigma}_{j'_0}, \ldots, \hat{\sigma}_{j_N}, \hat{\sigma}_{j'_N}) \cdot \\
&\cdot \langle\hat{\sigma}_{j_N}|U_{\mathrm{S}}(t, t-\delta t)|\hat{\sigma}_{j_{N-1}}\rangle \cdots \langle\hat{\sigma}_{j_1}|U_{\mathrm{S}}(t_0+\delta t, t_0)|\hat{\sigma}_{j_0}\rangle\langle\hat{\sigma}_{j_0}|\rho(t_0)|\hat{\sigma}_{j'_0}\rangle \cdot \\
&\quad \cdot \langle\hat{\sigma}_{j'_0}|U_{\mathrm{S}}^{-1}(t_0+\delta t, t_0)|\hat{\sigma}_{j'_1}\rangle \cdots \langle\hat{\sigma}_{j'_{N-1}}|U_{\mathrm{S}}(t, t-\delta t)|\hat{\sigma}_{j'_N}\rangle \qquad (7.17)
\end{aligned}$$

with the Feynman–Vernon influence functional $I^{(N)}(\hat{\sigma}_{j_0}, \ldots, \hat{\sigma}_{j'_N})$ in its discrete form [63]. Equation 7.17 represents a discrete path integral where we sum over all possible paths thereby keeping the system coordinate fixed during the time intervals δt but allowing transitions at the times t_k. The influence functional can be represented by

$$I^{(N)}(\ldots) = \exp\left\{-\frac{1}{\hbar}\sum_{k=0}^{N}\sum_{k'=0}^{k}[\hat{\sigma}_{j_k} - \hat{\sigma}_{j'_k}]\left[\eta_{kk'}\hat{\sigma}_{j_{k'}} - \eta^*_{kk'}\hat{\sigma}_{j'_{k'}}\right]\right\}$$

where the coefficients $\eta_{kk'}$ are environment memory kernels closely related to the environment correlation function Eq. 7.7. Explicit expression are found in Ref. [29, 30].

To explicitly determine the influence functional one must realize that the environment correlation function Eq. 7.7 decays with time. For an Ohmic spectrum and an environment at finite temperature the decay is exponential at long times. Thus, $\eta_{kk'}$ is exponentially small for $k - k' > K$ and accordingly we can neglect them and approximate the influence functional by

$$I^{(N)}(\ldots) \simeq \exp\left\{ -\frac{1}{\hbar} \sum_{k=0}^{N} \sum_{\substack{k'= \\ \max\{0,k-K\}}}^{k} [\hat{\sigma}_{j_k} - \hat{\sigma}_{j'_k}] \left[\eta_{kk'} \hat{\sigma}_{j_{k'}} - \eta^*_{kk'} \hat{\sigma}_{j'_{k'}} \right] \right\}.$$

Inserting this into Eq. 7.17 for the reduced density operator and reordering the sums we obtain

$$\begin{aligned} \langle \hat{\sigma}_{l_1} | \rho(t) | \hat{\sigma}_{l_2} \rangle &= A_N(\hat{\sigma}_{l_1}, \hat{\sigma}_{l_2}; \hat{\sigma}, \ldots, \hat{\sigma}) \cdot \\ &\quad \cdot \exp\left\{ -\frac{1}{\hbar} [\hat{\sigma}_{j_N} - \hat{\sigma}_{j'_N}] \left[\eta_{NN} \hat{\sigma}_{j_N} - \eta^*_{NN} \hat{\sigma}_{j'_N} \right] \right\} \end{aligned}$$

with arbitrary but fixed $\hat{\sigma}$. Here A_k is the reduced density tensor which depends on $2K$ arguments $\hat{\sigma}_{j_k}, \hat{\sigma}_{j'_k}, \ldots, \hat{\sigma}_{j_{k+K}}, \hat{\sigma}_{j'_{k+K}}$ and obeys the recursion relation

$$A_{k+1}(\hat{\sigma}_{j_{k+1}}, .., \hat{\sigma}_{j'_{k+1+K}}) = \sum_{j_k, j'_k} A_k(\hat{\sigma}_{j_k}, .., \hat{\sigma}_{j'_{k+K}}) \cdot \Lambda_k(\hat{\sigma}_{j_k}, .., \hat{\sigma}_{j'_{k+1+K}})$$

with

$$\begin{aligned} \Lambda_k(\hat{\sigma}_{j_k}, .., \hat{\sigma}_{j'_{k+1+K}}) &= \langle \hat{\sigma}_{j_{k+1}} | U_{\mathrm{S}}(t_{k+1}, t_k) | \hat{\sigma}_{j_k} \rangle \langle \hat{\sigma}_{j'_k} | U_{\mathrm{S}}^{-1}(t_{k+1}, t_k) | \hat{\sigma}_{j'_{k+1}} \rangle \cdot \\ &\quad \cdot \prod_{k'=k}^{k+K} \exp\left\{ -\frac{1}{\hbar} [\hat{\sigma}_{j_{k'}} - \hat{\sigma}_{j'_{k'}}] \left[\eta_{k'k} \hat{\sigma}_{j_k} - \eta^*_{k'k} \hat{\sigma}_{j'_k} \right] \right\} \end{aligned}$$

and initial condition

$$A_0(\hat{\sigma}_{j_0}, \ldots, \hat{\sigma}_{j'_{0-K+1}}) = \langle \hat{\sigma}_{j_0} | \rho(t_0) | \hat{\sigma}_{j'_0} \rangle. \tag{7.18}$$

Notice that the notion tensor describes the multidimensional character of the array only and does not refer to transformation properties of tensors.

The scheme introduced above can be implemented numerically very efficiently. Two approximation steps were employed, namely the Trotter time slicing and neglecting environment correlations $\eta_{kk'}$ with $k - k' > K$. The latter approximation typically fails for $T = 0$ or environmental fluctuations retaining very long memory times.

Nevertheless in many physical situations it proves a very successful scheme. The recipe to achieve exact results is the following: (i) Determine the specific behavior of the environment correlations and fix a memory time τ_{mem} up to which the correlations are taken fully into account. A practical criterion can be $\eta_{K0}/\eta_{00} \leq 10^{-3}$. (ii) Select an initial value for the Trotter slicing δt and a value for the maximal considered time steps K which fulfill $\tau_{mem} = K\delta t$. Since the overall Trotter error scales like δt^2 for small enough δt, convergence can be obtained by extrapolation to $\delta t \to 0$. (iii) Ensure convergence of the results with respect to K.

In energy transfer in photosynthetic light-harvesting complexes, each eigenstate of the generalized coordinate couples to separate but correlated environments. Denoting the environments by indices α, β, cross correlations cause nonvanishing cross memory $\eta_{kk'}^{(\alpha\beta)} \neq 0$ which can be included in the influence functional [38] by

$$I^{(N)}(\hat{\sigma}_{j_0}, \ldots, \hat{\sigma}_{j'_N}) = \exp\left\{ -\frac{1}{\hbar} \sum_{\alpha,\beta} \sum_{k=0}^{N} \sum_{k'=0}^{k} [\hat{\sigma}_{j_k} - \hat{\sigma}_{j'_k}] \left[\eta_{kk'}^{(\alpha\beta)} \hat{\sigma}_{j_{k'}} - \eta_{kk'}^{*(\alpha\beta)} \hat{\sigma}_{j'_{k'}} \right] \right\}. \tag{7.19}$$

Such correlated fluctuations acting on separate parts of the system mediate interactions between these system parts. Prominent examples of such interactions are the RKKY interaction between nuclear spins mediated by a coupling to conduction electrons or the weak interactions between tunneling systems in glasses mediated by their coupling to phonons [64].

7.3 Excitation Energy Transfer Dynamics

Energy transfer between two pigments in photosynthesis [1, 2] as well as in artificial donor–acceptor pairs is typically mediated by radiationless Förster type processes [8] due to dipole–dipole couplings. Surprisingly, natural photosynthesis shows an almost perfect quantum yield. It is a longstanding question whether energy transfer is best described by an incoherent hopping model or whether quantum coherent dynamics is present. In terms of system–environment models this relates to the question whether the dynamics is in the strong coupling overdamped or the weak coupling coherent regime. One of the experimentally best studied systems is the FMO complex [3, 4] which occurs in green sulfur bacteria such as *Prosthecochloris aestuarii* and *Chlorobaculum tepidum* and serves as exciton conductor between the antenna complex and the reaction center. The FMO protein is a trimer where the three identical subunits consist of seven chromophoric sites which are fully characterized [5, 6]. Typical transport times for excitons through FMO are of the order of picoseconds. Recent 2D femtosecond spectroscopy [13, 14, 65] on the FMO complex revealed oscillatory behavior on the same timescale which was interpreted as electronic quantum coherent oscillations. This raises the important question why the strong environmental fluctuations are insufficient to fully decohere the quantum transport. Alternative interpretations of the experiments in terms of vibronic effects [46–48] were also put forward. At the heart of both cases is the accuracy of our knowledge of the environmental fluctuations.

We investigate the excitonic energy transfer dynamics in a reduced model within a system–environment approach. Since environmental fluctuations due to the proteins and the solvent are strongly non-Markovian, we employ the numerically exact quasi adiabatic path integral method (QUAPI). Estimates from solvation models and experimental data [66–68] suggest that the spectra exhibit rather long temporal correlations. Within simple donor–acceptor models we show in the next section that long quantum coherence times (beyond 200 fs) result from such temporally correlated environmental fluctuations [28, 35–37]. Furthermore, such non-Markovian environments generate long-lived entanglement [35]. Experimentally, little is known about spatial correlations in the environment but they can also cause long quantum coherence times [38]. Using more realistic models for a typical light-harvesting complex, as the FMO complex, however, we show in the following section that coherence times are not longer than 200 fs at room temperature [42, 45, 49] favoring an incoherent transfer model. We furthermore rule out a scenario of quantum coherence in the FMO exciton dynamics protected by the discrete high-energy molecular modes [49]. The influence of discrete low-energy modes is currently heavily discussed as likely explanation for the observed longtime oscillatory dynamics.

7.3.1 Energy Transfer in Donor–Acceptor Systems

To model excitation energy transfer we employ the simplest model of Frenkel excitons which transfer their excitation energy between a donor and an acceptor site, leading to the standard Hamiltonian [12]

$$H_{\mathrm{da}} = \frac{\Delta_0}{2}\{|d\rangle\langle a| + |a\rangle\langle d|\} + \epsilon_d|d\rangle\langle d| + \epsilon_a|a\rangle\langle a| \\ + \sum_{j=a/d}|j\rangle\langle j|\sum_{\kappa}\nu_{\kappa}^{(j)}q_{j,\kappa} + \sum_{j=a/d}H_{\mathrm{B}j}\,. \qquad (7.20)$$

The state $|d\rangle$ ($|a\rangle$) denotes the exciton localized at the donor (acceptor) site, Δ_0 is the respective electronic coupling between donor and acceptor and ϵ_d (ϵ_a) is the excited electronic energy of the donor (acceptor), respectively. We neglect fluctuations in the electronic coupling [69–71] and only consider fluctuations in the site energies which typically dominate. Each chromophoric site is coupled to its local environment $H_{\mathrm{B}j}$, (Fig. 7.3), as can be derived from molecular Hamiltonians of pigment–protein complexes [72, 73]. In such a system, electronic de-excitation of a donor site and excitation of an acceptor site take place via nonequilibrium environmental states in accordance with the Franck–Condon principle and the environmental modes associated with each site then relax to the respective equilibrium states resulting in site-dependent reorganization dynamics.

Within a chemical solvation model for the dielectric fluctuations of the protein–solvent environment of chromophores, its spectral distribution has an Ohmic form [66–68], that is, $J(\omega) = 2\pi\alpha\omega\exp(-\omega/\omega_c)$ with coupling constant α, which is related to the characteristic quantities of the solvent, such as the dipole moment, the frequency-dependent dielectric constant and the Debye relaxation time. Typical energy scales in biomolecules are $\Delta_0 \simeq 0.2$–2 meV for resonant energy transfer between red and green chromophores or $\Delta_0 \simeq 46$–100 meV for the light-harvesting complex LH-II in the bacteria chlorophyll molecule in purple bacteria [67], while the environmental energy scale is $\hbar\omega_c \simeq 2-8$ meV for water at room temperature. Moreover, for typical free

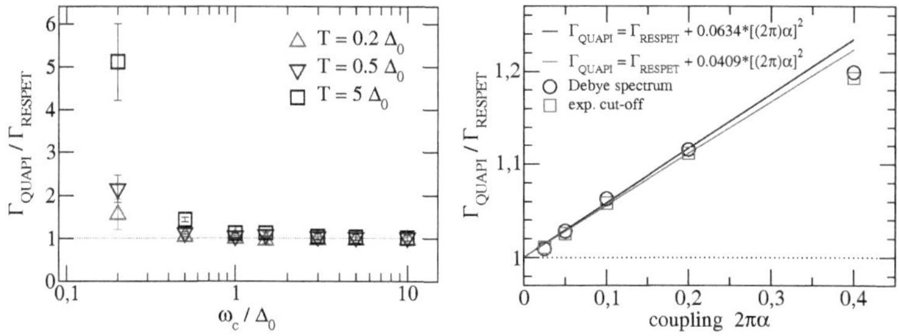

Figure 7.3 The ratio $\Gamma_{\text{QUAPI}}/\Gamma_{\text{RESPET}}$ (left) versus cutoff frequency ω_c for three different temperatures for a fixed weak coupling $\alpha = 0.1/(2\pi)$ and (right) versus coupling constant α at $\frac{1}{2}T = \frac{1}{2}\Delta_0 = \omega_c$. The lines are linear fits to the data to the four lowest coupling strengths.

chromophores in water at room temperature, $\alpha \simeq 0.1-1$ [67]. Thus, typically $\Delta_0 \gtrsim \omega_c$ implying environmental temporal correlations or memory on timescales longer than Δ_0^{-1}.

7.3.1.1 Multiphonon transitions

To test the accuracy of weak coupling approaches for this parameter regime we simulate the population dynamics of a donor–acceptor system with Hamiltonian (Eq. 7.20) and Ohmic spectral function employing the numerically exact QUAPI and the weak coupling RESPET approach. Using RESPET we analytically derive the time evolution of the population difference in our donor–acceptor model to lowest order in the system–environment coupling. Thereby all non-Markovian corrections are determined to the lowest order, but they turn out to be negligible at weak coupling. Employing QUAPI we observe qualitatively similar but quantitatively very different behavior. In both cases we determine the decoherence rates. For $\alpha = 0.1/(2\pi)$ Fig. 7.3 (left panel) plots the ratio $\Gamma_{\text{QUAPI}}/\Gamma_{\text{RESPET}}$ of decoherence rates determined via QUAPI and via RESPET versus ratio ω_c/Δ_0 for three temperatures T. At $\Delta_0 \sim T \gtrsim \omega_c$ the numerically observed damping rate differs from the one-phonon rate of the RESPET calculation by up to a factor of 5. This difference increases with decreasing cutoff frequency ω_c and increasing temperature. Figure 7.3 (right panel) plots $\Gamma_{\text{QUAPI}}/\Gamma_{\text{RESPET}}$ versus the system–environment coupling α for fixed $\frac{1}{2}T = \frac{1}{2}\Delta_0 = \omega_c$.

This clearly shows that this discrepancy depends quadratically on the system–environment coupling showing the onset of two-phonon processes. In these calculations, we investigate an Ohmic spectrum with exponential cutoff function (red squares) and alternatively with Debye cutoff function $J(\omega) = 2\pi\alpha\omega\omega_c^2/(\omega_c^2+\omega^2)$. Both spectra show identical qualitative behavior but small quantitative differences. Thus nonperturbative multiphonon transitions induced by the slow environment yield significant contributions to the decoherence rate of the exciton dynamics. We conclude accordingly that a weak coupling approximation fails resulting from the suppression of the one-phonon process due to the suppressed spectral density for fluctuations resonant with the exciton dynamics. The quantitative results thus strongly depend on the high-frequency tail of the solvent fluctuation spectra, that is, at $J(\omega)$ for $\omega > \omega_c$, emphasizing the importance of a detailed knowledge of the spectrum also in the high-frequency regime. This renders any perturbative approach to energy transfer in these structures highly questionable [28, 39, 40].

There are several numerically exact methods to treat system–environment problems. We have employed QUAPI and the cumulant time-nonlocal master equation approach [36], which also is a numerically exact treatment, in order to compare numerical performance when calculating the real-time quantum dynamics of a biomolecular donor–acceptor system (Eq. 7.20) in the presence of slow environmental fluctuations. Both approaches yield coinciding results in the relevant crossover regime from weak to strong electronic couplings for a Debye fluctuation spectrum, displaying coherent as well as incoherent transitions. Importantly both approaches perform equally well. Going to low temperatures and studying environments with other spectra like exponentially cutoff spectra (as in 7.10), the cumulant time-nonlocal master equation approach suffers considerably in performance whereas QUAPI is indifferent to the chosen spectral function and environmental temperature barring lowest temperatures. However, QUAPI performance scales exponentially with system size whereas the cumulant time-nonlocal master equation approach scales more favorably here. Thus we expect QUAPI to be the ideal method to study the FMO complex with seven chromophores and a well-known but superohmic fluctuation spectrum [43, 44].

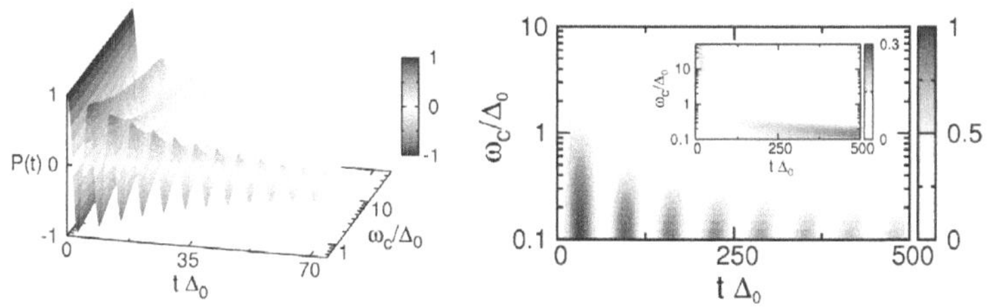

Figure 7.4 (Left) Population difference $P(t)$ for a single donor–acceptor pair and full cross-over from a Markovian environment, that is, $\omega_c \gg \Delta_0$, to a non-Markovian environment, that is, $\omega_c \ll \Delta_0$. Parameters are $k_B T = 0.1\hbar\Delta_0$, and $\alpha = 0.1$. (Right) Negativity $N(t)$ as a function of ω_c for $J = 0.1\Delta$ (main) and $J = 0$ (inset), for $\alpha = 0.01$ and $k_B T = 0.1\hbar\Delta_0$.

7.3.1.2 Coherence due to temporal correlations

Next, we investigate explicitly coherence times in the biomolecular energy transfer in order to study whether the experimentally observed oscillation times of up to a picosecond might reasonably explained as electronically coherent oscillations [35]. Figure 7.4 (left) plots the population difference $P(t) = \langle |d\rangle\langle d| - |a\rangle\langle a| \rangle$ versus time and ratio ω_c/Δ_0 for $k_B T = 0.1\hbar\Delta_0$, and $\alpha = 0.1$. For large cutoff frequencies the coherent oscillations decay very rapidly whereas for $\omega_c/\Delta_0 \lesssim 1$ many coherent oscillations are observable despite the strong coupling $\alpha = 0.1$. Small cutoff frequencies, that is, $\omega_c/\Delta_0 \lesssim 1$, imply environmental temporal correlations, that is, in Eq. 7.7, and accordingly memory on timescales longer than Δ_0^{-1}. Thus long-time environmental temporal correlations in the protein–solvent environment allow for long coherence times in the biomolecular energy transfer.

We furthermore studied two coupled donor–acceptor systems, where we allowed the excitons from each donor to reach each acceptor, with Hamiltonian $H_{2\text{da}} = H_{2\text{da,S}} + H_{\text{SB}} + H_{\text{B}}$ and

$$H_{2\text{da,S}} = \sum_{i=1,2} H_{\text{da},i} + J\ \{|d_1\rangle\langle a_2| + |a_2\rangle\langle d_1| + |d_2\rangle\langle a_1| + |a_1\rangle\langle d_2|\}. \tag{7.21}$$

Figure 7.4 (right) plots the negativity $N(t)$ (color coded) [74, 75], a measure of quantum entanglement, as a function of ω_c for $J = 0.1\Delta$ (main) and $J = 0$ (inset), for $\alpha = 0.01$ and $k_B T = 0.1\hbar\Delta_0$. Starting from a two-exciton state with vanishing negativity, we observe for small cutoff frequency oscillatory behavior with (re-)emergence and vanishing of entanglement. The inset shows the emergence of entanglement with subsequent very slow decay even for vanishing coupling J. Here, however, we assumed the fluctuations between donor and acceptor to be also spatially correlated. Thus, long-time environmental temporal correlations generate entanglement in multichromophoric systems and thus could give rise even to nontrivial quantum effects [22–24].

7.3.1.3 Influence of spatial correlations

Quantum coherent transfer is also strongly supported by spatial correlations in the environmental fluctuations [38] which are due either to propagating modes or to local fluctuations with a finite localization length. We consider two types of spatial correlations. One the one hand we coupled both donor and acceptor, to a single common environment of propagating modes (similar to phonons in condensed matter) with according spectral function for the intersite fluctuations between donor and acceptor

$$J_{da}(\omega) = 2\alpha\omega e^{-\omega/\omega_c}\frac{\sin(\omega t_0)}{\omega t_0} \quad \text{with} \quad t_0 = \frac{r_{da}}{v} \tag{7.22}$$

and the sound velocity v and donor–acceptor distance r_{da}. The sound velocity of the solvent determines the wavelength of the environmental modes, which, in turn, has to be compared to the spatial distance of the chromophore sites. When the wavelength exceeds the distance between donor and acceptor sites, we find a long-lived coherence since the energies of donor and acceptor fluctuate in phase and their energy difference is constant as sketched in the inset of Fig. 7.5 (left). The main Fig. 7.5 (left) plots the occupation of the donor for a single donor–acceptor system versus time and speed of sound. With increasing speed of sound and thus wavelength, the coherence time increases.

On the other hand we considered donor and acceptor coupled to local fluctuations with a finite localization length ξ with according spectral function for the intersite fluctuations between donor and acceptor

$$J_{da}^{\text{loc}}(\omega) = 2\alpha\omega e^{-\omega/\omega_c} e^{-r_{da}/\xi} \tag{7.23}$$

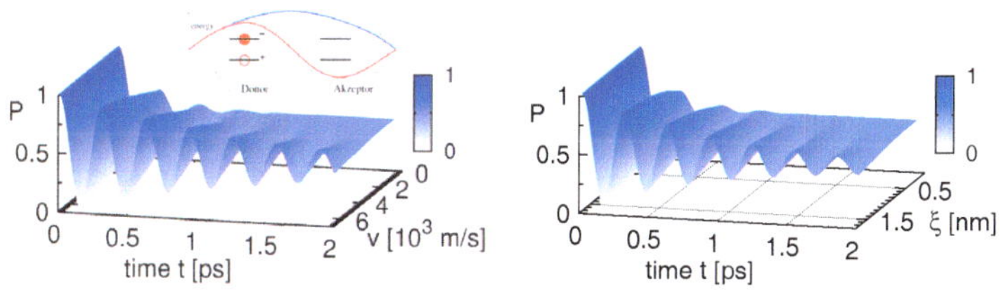

Figure 7.5 Occupation difference of donor and acceptor for $\omega_c = \Delta = 106\ \text{cm}^{-1}$, $T = 152$ K, $\alpha = 0.08$ and $r_{da} = 3.8$ Å versus time and speed of sound assuming propagating environmental modes (left) or versus time and localization length ξ assuming localized environmental modes (right).

Figure 7.5 (right) plots the occupation of the donor for a single donor–acceptor system versus time and localization length ξ. Again, we observe that coherence times increase substantially once the localization lengths of the local fluctuations become longer than the distance between donor and acceptor.

In addition, we consider two spatially separated donor–acceptor pairs (Eq. 7.21) with vanishing J under the influence of propagating environmental modes. As discussed in Section 7.2.2.2 correlated environmental fluctuations acting on separate parts of a system, here the two donor–acceptor systems, mediate interactions. As long as the donor–acceptor distance r_{da} in a single pair is less than the distance r between the two pairs, we observe qualitatively similar results as for single donor–acceptor systems as visible in Fig. 7.6 (left) where we plot the probability $P_{d_1,d_2}(t)$ of both excitons being at the respective donor sites versus time and sound velocity, assuming propagating environmental modes with donor–acceptor distance $r_{da} = 3.8$ Å and distance between both donors (acceptors) $r = 38$ Å. Parameters are chosen to be $\omega_c = \Delta = 106\ \text{cm}^{-1}$, $T = 15.2$ K, and $\alpha = 0.04$. Environmental fluctuations with increasing spatial correlations (increasing v) have a decreasing decohering effect on the quantum energy transfer. The situation changes once the two donors and/or two acceptors are closer to each other than the respective partner in the donor–acceptor pair. Increasing speed of sound and thus wavelength of the environmental modes, we increase the spatial range of correlations which now increase the decohering influence. This can be observed in Fig. 7.6 (right) which

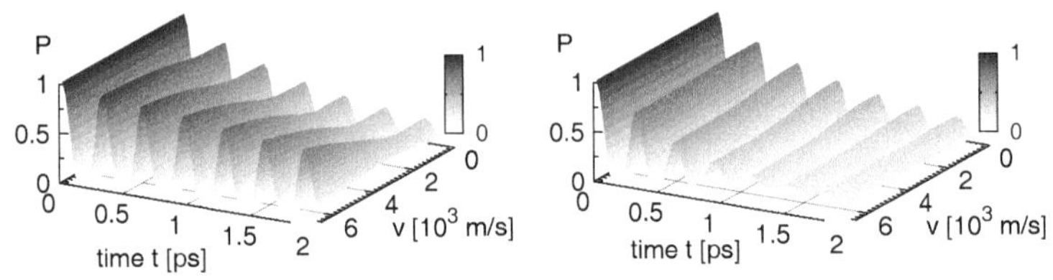

Figure 7.6 Probability $P_{d_1,d_2}(t)$ of both exciton being at the respective donor sites versus time and sound velocity, assuming propagating environmental modes. Parameters are chosen to be $\omega_c = \Delta = 106\ \text{cm}^{-1}$, $T = 15.2$ K, and $\alpha = 0.04$. The distances are $r_{da} = 3.8$ Å and $r = 38$ Å (left) and $r_{da} = 38$ Å and $r = 3.8$ Å (right).

plots again $P_{d_1,d_2}(t)$ for the same parameters as before but distances $r_{da} = 38$ Å and $r = 3.8$ Å.

7.3.2 Exciton Dynamics in the FMO Complex

The FMO complex is a trimer consisting of identical, weakly interacting monomers [44], each containing seven bacteriochlorophyll*a* (BChl*a*) molecular sites which transfer excitons. The recently resolved eigth site [7] is only very weakly coupled to the first seven and thus irrelevant for our consideration. The pigments are embedded in a large protein complex. Each pigment can be reduced to its two lowest electronic levels and their excited states are electronically coupled along the complex. Recombination is negligibly slow (~ns) compared to exciton transfer times (~ps). Thus, the excitation dynamics is reliably described within the one-exciton subspace. The coupling of the seven excited levels leads to the Hamiltonian

$$H_{\text{FMO}} = \begin{pmatrix} 240 & -87.7 & 5.5 & -5.9 & 6.7 & -13.7 & -9.9 \\ & 315 & 30.8 & 8.2 & 0.7 & 11.8 & 4.3 \\ & & 0 & -53.5 & -2.2 & -9.6 & 6.0 \\ & & & 130 & -70.7 & -17.0 & -63.3 \\ & & & & 285 & 81.1 & -1.3 \\ & & & & & 435 & 39.7 \\ & & & & & & 245 \end{pmatrix} \qquad (7.24)$$

in units of cm^{-1} in site representation [44] for an FMO monomer of *C. tepidum*. We define the lowest site energy of pigment 3 as reference. Each chromophore is coupled to a separate environment leading to

$$H = H_{\mathrm{FMO}} + \sum_{j=1}^{7} |j\rangle\langle j| \sum_{\kappa} \nu_{\kappa}^{(j)} q_{j,\kappa} + \sum_{j=1}^{7} \frac{1}{2} \sum_{\kappa} \left(p_{j,\kappa}^2 + \omega_{j,\kappa}^2 q_{j,\kappa}^2 \right). \tag{7.25}$$

We assume that fluctuations at different sites are identical but spatially uncorrelated as suggested by extensive molecular dynamics studies [76]. Employing QUAPI we study the quantum dynamics in the FMO complex and determine the coherence times after an initial excitation of site 1 or 6 which are believed to be the entrance sites for excitonic energy since they lie closest to the chlorosome, its antenna complex. We study the dynamics for several available spectral functions for the environmental fluctuations [42, 45, 49] which is, as seen above, the key quantity determining the coherence properties of the energy transfer.

Ishizaki and Fleming [41] (IF model) use a single Ohmic spectrum with a Debye cutoff function for each site independently

$$J_{IF}(\omega) = 2\lambda \frac{\omega \omega_{c,IF}}{\omega_{c,IF}^2 + \omega^2} \tag{7.26}$$

with reorganization energy $\lambda = 35\ \mathrm{cm}^{-1}$ and Debye relaxation time (or inverse cutoff frequency) $\omega_{c,IF}^{-1} = 50$ fs. This spectrum is depicted in Fig. 7.7 (left) by the dark green dot-dashed line (lower left corner). It includes only a continuous part describing the solvent fluctuations and omits all discrete vibronic modes from the BChl*a* or the embedding protein. Cho et al. [5] (CF model) use an exponentially cutoff Ohmic spectrum

$$J_{CF}(\omega) = \pi \frac{\lambda}{\omega_{c,CF}} \omega e^{-\omega/\omega_{c,CF}} \tag{7.27}$$

with $\lambda = 35\ \mathrm{cm}^{-1}$ but $\omega_{c,CF} = 50\ \mathrm{cm}^{-1}$ ($\omega_{c,CF}^{-1} = 106$ fs) which is depicted in Fig. 7.7 (left) by the red dashed line (lower left corner). Wendling et al. [43] have provided fluorescence line-narrowing measurements at low temperature and have obtained the so far still most detailed experimentally determined fluctuation spectrum for FMO. Adolphs and Renger [44] (AR model) give an explicit form of this spectrum, thereby focusing only on the discrete vibronic state with lowest frequency besides the continuous contributions, leading to

$$J_{AR}(\omega) = \pi \omega^2 S_0 g_0(\omega) + \pi \omega^2 S_H \delta(\omega - \omega_H) \tag{7.28}$$

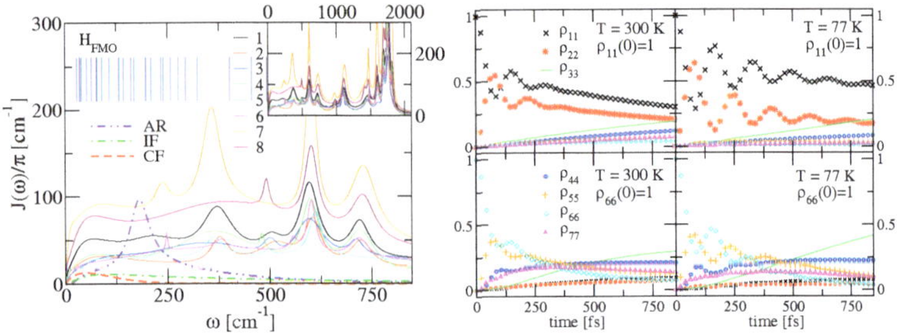

Figure 7.7 (Left) Environmental spectral functions for FMO trimer in the OK model are plotted by a full line for each site j. Additionally, the spectra of CF, IF, and AR models are shown. The vertical blue lines in the upper-left corner depict the energy differences in the Hamiltonian (Eq. 7.24). (Right) Time-dependent occupation probabilities for $T = 300$ K and 77 K with $\rho_{11}(0) = 1$ and $\rho_6(0) = 1$ in the IF model.

with $S_0 = 0.5$, $S_H = 0.22$, $\omega_H = 180$ cm^{-1} and

$$g_0(\omega) = 6.105 \cdot 10^{-5} \cdot \frac{\omega^3}{\omega_1^4} e^{-\sqrt{\omega/\omega_1}} + 3.8156 \cdot 10^{-5} \cdot \frac{\omega^3}{\omega_2^4} e^{-\sqrt{\omega/\omega_2}}$$

with $\omega_1 = 0.575$ cm^{-1} and $\omega_2 = 2$ cm^{-1}. This spectrum is plotted in Fig. 7.7 (left) as the purple dot-dot-dashed line. There is a continuous underlying contribution plus a single overlying vibronic peak. Again independent but identical fluctuations at each site are assumed. We furthermore broaden the unphysical δ-peak and use instead a Lorentzian peak

$$\pi \omega^2 S_H \delta(\omega - \omega_H) \rightarrow \pi S_H \omega_H \frac{\omega \gamma_p}{(\omega - \omega_H)^2 + \gamma_p^2}$$

with width $\gamma_p = 29$cm^{-1} taken as a typical value from the Olbrich et al. [58] spectra (see below).

Recently, Olbrich et al. [58] (OK model) have determined by extensive molecular dynamical simulations the fluctuations of the site energies in the FMO. This allowed them to extract detailed fluctuation spectral functions $J_j(\omega)$ for each site j separately. They observe different spectra for each site. The spectra consist of two continuous contributions resulting from solvent fluctuations and up to 13 discrete vibrational modes. Unfortunately, the continuous solvent contributions are overestimated [77]. We plot the spectra in Fig. 7.7 (left). The spectra show discrete vibrational modes up to energies around 1800 cm^{-1} (see inset in Fig. 7.7, left). In the upper left corner (vertical blue) of Fig. 7.7 (left) the energy differences between the eigenstates in the Hamiltonian (Eq. 7.24) are indicated by vertical lines. These energy differences are (at least within a one-phonon picture) the relevant energy quanta which the FMO electronic excitations want to exchange with the environment. These energy differences do not exceed about 500 cm^{-1} and accordingly, we expect little influence of the higher energetic modes for the excitation dynamics. Note that despite the strong temporal correlations present in all these fluctuation spectra, the dynamics is nevertheless to a good approximation Markovian [78]. However, weak coupling approaches still fail since they neglect the multiphonon character of the exciton–environment interaction.

7.3.2.1 Coherence times in the FMO complex

To understand whether the vibrational fluctuations of the protein and the pigments in the FMO complex allow for the experimentally observed electronic coherence times, we employ QUAPI to determine the time-dependent populations of the FMO pigment sites, that is, the time-dependent reduced density operator of the FMO (Eq. 7.24). We study two temperatures, that is, $T = 300$ K (physiological temperature) and $T = 77$ K (typical experimental temperature). Both the pigments BChl 1 and BChl 6 are oriented towards the baseplate protein and are thus believed to be initially excited (entrance sites) [79]. Thus, we consider two cases, that is, $\rho_{11}(0) = 1$ and $\rho_{66}(0) = 1$. We focus on the transient coherence effects and thus do not include an additional sink at the exit site BChl 3.

At first, we consider the IF model, that is, environmental fluctuations as used by Ishizaki and Fleming. In Fig. 7.7 (right), we show the time-dependent pigment occupation probabilities $\rho_{jj}(t)$. For $\rho_{66}(0) = 1$ at room temperature, coherent oscillations last up to 300 fs and at 77 K to about 600 fs. For $\rho_{11}(0) = 1$, coherence is supported longer due to the strong electronic coupling between sites 1 and 2. At room temperature, it survives for up to about 400 fs and for 77 K at least up to 800 fs. Our results coincide with results by Ishizaki and Fleming [41] and are compatible with experimentally observed oscillation times. Secondly, we investigate the time-dependent populations of the FMO pigment sites in the AR model with environmental fluctuation spectra (Eq. 7.28). In Fig. 7.8 (left), we show the time-dependent pigment occupation probabilities $\rho_{jj}(t)$. For $\rho_{66}(0) = 1$ at room temperature, coherent oscillations are completely suppressed and at 77 K they last up to about 250 fs. For $\rho_{11}(0) = 1$, coherence is supported longer. At room temperature, it survives for up to about 200 fs and for 77 K up to 500 fs at most. Thus, coherence times are shorter than experimentally observed. We emphasize that coherence features are a transient property and, hence, not only the low-frequency bath modes, but also the discrete vibrational modes are relevant (100 fs corresponds to $\sim$333 cm^{-1}). Since the OK model overestimates the continuous solvent contribution we do not consider it here. At last, we treat the CF model with an exponentially cutoff spectrum. This model results in the longest temporal correlations within the environment and converged results could only be obtained by vast numerical effort. We, therefore, focused on the dynamics of $\rho_{11}(t)$ for $T = 300$ K and $\rho_{11}(0) = 1$ plotted in Fig. 7.8(upper right)

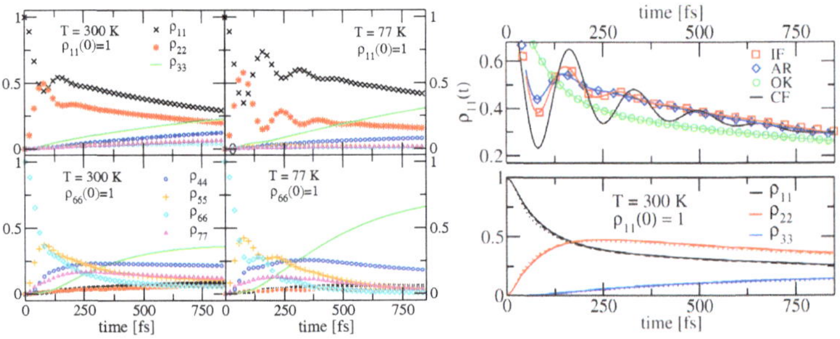

Figure 7.8 (Left) Time-dependent occupation probabilities of all seven FMO sites for $T = 300$ K, 77 K with $\rho_{11}(0) = 1$ and $\rho_6(0) = 1$ in the AR model (left). (Upper right) Time-dependent occupation probability $\rho_{11}(t)$ at $T = 300$ K with $\rho_{11}(0) = 1$ in the CF, IF, OK, and AR models in comparison. (Lower right) Time-dependent occupation probabilities of sites 1, 2 and 3 versus time for $T = 300$ K with $\rho_{11}(0) = 1$ in the OK model. Full/dashed/dotted lines show results with the full/only vibrational modes up to energies 750 cm^{-1}/only the continuous solvent contribution of the fluctuation spectrum.

by the black full line. Here, we observe the longest coherence time exceeding 800 fs at room temperature. Figure 7.8 (upper right) also plots the results from the previous models in comparison.

We thus find at room temperatures coherences times of 800, 400 and maximally 200 fs for the CF, IF, and AR model, respectively. Experimentally quantum coherence beatings in 2D optical femtosecond spectroscopy have been observed on timescales of more than 660 fs at 77 K [13] and 300 fs at physiological temperatures [14]. The two phenomenological spectra of the CF and the IF model yield substantially different coherence times. The AR model spectra are determined directly from experiments and thus currently are the most accurate form available for the spectral density realized in nature. It includes vibronic effects via a strongly localized Huang–Rhys mode. The resulting coherence time is clearly shorter than the experimentally observed beating times. In total, we must conclude that theoretical models cannot support the interpretation of the quantum beatings in the 2D electronic spectra as purely electronic coherence. However, due to the rather large uncertainties in the environmental fluctuation spectra, theoretical models can also not refute that interpretation.

7.3.2.2 High-energy vibrations

Although the continuous background induced by the solvent is overestimated in the OK model the molecular discrete modes in the FMO complex are revealed in great detail and we can clarify the question whether the discrete vibrational molecular modes in the FMO complex can in principle reintroduce a quantum mechanically coherent time evolution by including the discrete high-energy modes. In more detail, we investigate whether by including the discrete modes, quantum coherent dynamics can prevail while it would be entirely overdamped without these modes.

For this, we study the influence of the discrete molecular modes in detail by successively adding them one by one to the environmental fluctuation spectrum and calculating their effects on the population dynamics which we again determine employing QUAPI. In Fig. 7.8(lower right) we plot the time-dependent pigment occupation probabilities $\rho_{jj}(t)$ for site 1, 2 and 3 at $T = 300$ K with initially $\rho_{11}(0) = 1$. The dotted lines show the population dynamics including only the continuous solvent fluctuations. As expected due to the large spectral weight the dynamics is completely overdamped. Next, we consecutively add the discrete molecular modes to the spectra and investigate their effects on the population dynamics. We show, for clarity, only two cases. The dashed lines show the dynamics for spectra including all modes with peak frequencies up to 750 cm^{-1} and the full lines show the dynamics for the full fluctuation spectra including all molecular modes. The effects on the dynamics due to the discrete vibrational modes are small for times up to a picosecond. Thus our major observation is that their influence on the actual population dynamics is negligible. We observe only overdamped dynamics. Thus, in the FMO complex the discrete molecular high-energy modes which couple to the electronic single exciton subspace cannot *reintroduce* quantum coherence again beyond times which are determined by the strong solvent fluctuations. By this, we can rule out a scenario of quantum coherence in the FMO exciton dynamics protected by discrete high-energy molecular mode. Only when the dynamics is coherent already without the discrete modes, its coherence might be additionally enhanced. Low-energy discrete modes with energies not exceeding typical excitonic transition energies cannot be ruled out to be responsible for the experimentally observed coherence features.

7.4 Summary and Outlook

It is a longstanding question whether quantum coherence is functionally relevant for excitonic energy transfer in biomolecular complexes, that is, in natural photosynthesis. Recent experimentally observed beatings on the picosecond time scale in time-resolved optical 2D spectra have been interpreted as evidence for quantum coherent energy transfer via delocalized exciton states. We have investigated the question how quantum coherence can survive over such long times in these biomolecular environments with strong vibronic modes and a surrounding strongly coupled polar solvent. Subsequently, we have questioned whether the interpretation of the observed oscillations as electronic quantum coherence is adequate.

Fluctuations in biomolecular complexes influencing energy transfer, for example, in natural photosynthesis, are strongly non-Markovian exhibiting temporal correlations on times scales comparable to the transfer times. Employing a numerically exact path integral propagator approach we have shown that such temporally correlated fluctuations in principle allow for very long lived quantum coherence and even the generation of entanglement. A comparative study with a weak coupling approach revealed that multiphonon processes dominate in the biomolecular environmental fluctuations which cannot be treated by any weak coupling method. Alternatively to temporal correlations within the environmental fluctuations spatial correlations also weaken the decohering influence of the fluctuations, equally leading to long coherence times.

Using oversimplified phenomenological fluctuation spectra for the FMO complex we accordingly observe coherent dynamics on timescales, as experimentally observed. However, employing more realistic model parameters obtained by spectroscopic data, we find coherence times on the order of hundred femtoseconds in contrast to picoseconds experimentally observed. Thus, the oscillations observed in ultrafast nonlinear spectroscopy in the FMO complex are too long lived to originate from the temporal correlations present, and spatially correlated fluctuations are believed to be irrelevant in the FMO complex. Alternative interpretations have been proposed where the observed coherent oscillations are caused by discrete molecular vibrations employing long coherence times. These vibrations must be close to resonance with excitonic excitations since we could show that high-energy discrete modes cannot generate coherent behavior. To finally clarify the origin of the long lived beating dynamics in the ultrafast nonlinear spectroscopy in the FMO complex, we need detailed environmental fluctuation spectra, that is, of the continuous solvent contributions but also the discrete vibrational modes close to resonance with excitonic excitations. This would allow to include the vibronic effects into the picture to check whether it allows a consistent explanation for the data.

Acknowledgments

I wish to thank M. Thorwart for continuous intellectual support.

References

1. H. van Amerongen, L. Valkunas, and R. van Grondelle, *Photosynthetic Excitons* (World Scientific, Singapore, 2000).
2. R.E. Blankenship, *Molecular Mechanisms of Photosynthesis* (Blackwell, Oxford, 2002).
3. R.E. Fenna and B.W. Matthews, *Nature*, **258**, 573 (1975).
4. Y.-F. Li, W. Zhou, R.E. Blankenship, and J.P. Allen, *J. Mol. Biol.*, **271**, 456 (1997).
5. M. Cho, H.M. Vaswani, T. Brixner, J. Stenger, and G.R. Fleming, *J. Phys. Chem. B*, **109**, 10542 (2005).
6. M.T.W. Milder, B. Brüggemann, R. van Grondelle, and J.L. Herek, *Photosynth. Res.*, **104**, 257 (2010).
7. D.E. Tronrud, J. Wen, L. Gay, and R.E. Blankenship, *Photosynth. Res.*, **100**, 79-87 (2009).
8. T. Förster, *Ann. Phys.*, **2**, 55 (1948).
9. A.J. Leggett, S. Chakravarty, A.T. Dorsey, M.P.A. Fisher, A. Garg, W. Zwerger, *Rev. Mod. Phys.*, **59**, 1 (1987).
10. U. Weiss, *Quantum Dissipative Systems* (World Scientific, Singapore, 1999).
11. P. Reineker, *Exciton Dynamics in Molecular Crystals and Aggregates*, Springer Tracts in Modern Physics, Vol. 94 (Springer, Berlin, 1982).
12. V. May and O. Kühn, *Charge and Energy Transfer Dynamics in Molecular Systems* (Weinheim, Wiley-VCH, 1999).
13. G.S. Engel, T.R. Calhoun, E.L. Read, T.K. Ahn, T. Mancal, Y.C. Cheng, R.E. Blankenship, and G.R. Fleming, *Nature*, **446**, 782 (2007).
14. G. Panitchayangkoon, D. Hayes, K.A. Fransted, J.R. Caram, E. Harel, J. Wen, R.E. Blankenship, and G.S. Engel, *Proc. Natl. Acad. Sci. U S A*, **107**, 12766 (2010).
15. T. Brixner, J. Stenger, H.M. Vaswani, M. Cho, R.E. Blankenship, G.R. Fleming, *Nature*, **434**, (2005) 625.
16. E. Collini, C.Y. Wong, K.E. Wilk, P.M.G. Curmi, P. Brumer, and G.D. Scholes, *Nature*, **463**, 644 (2010).

17. G.D. Scholes, *J. Phys. Chem. Lett.*, **1**, 2–8 (2010).
18. G.D. Scholes, G.R. Fleming, and A. Olaya-Castro, *Nat. Chem.*, **3**, 763 (2011).
19. A. Ishizaki and G.R. Fleming, *Ann. Rev. Condens. Matter Phys.*, **3**, 333 (2012).
20. K.E. Dorfman, D.V. Voronine, S. Mukamel, and M.O. Scully, *Proc. Natl. Acad. Sci. U S A*, **110**, 2746–2751 (2013).
21. P. Nalbach and M. Thorwart, *Proc. Natl. Acad. Sci. U S A*, **110**, 2693 (2013) (invited commentary).
22. M. Mohseni, P. Rebentrost, S. Lloyd, and A. Aspuru-Guzik, *J. Chem. Phys.*, **129**, 174106 (2008).
23. M.B. Plenio and S.F. Huelga, *New J. Phys.*, **10**, 113019 (2008).
24. A. Olaya-Castro, C.F. Lee, F. Fassioli, and N.F. Johnson, *Phys. Rev. B*, **78**, 085115 (2008).
25. H. Horner, *Eur. Phys. J. B* **18**, 453 (2000).
26. P. Nalbach, O. Terzidis, K.A. Topp, and A. Würger, *J. Phys.: Condens. Matter*, **13**, 1467–1483 (2001).
27. P. Nalbach, *Phys. Rev. B*, **66**, 134107 (2002).
28. P. Nalbach and M. Thorwart, *J. Chem. Phys.*, **132**, 194111 (2010).
29. N. Makri, D.E. Makarov, *J. Chem. Phys.*, **102**, 4600 (1995).
30. N. Makri, D.E. Makarov, *J. Chem. Phys.*, **102**, 4611 (1995).
31. N. Makri, E. Sim, D. Makarov, M. Topaler, *Proc. Natl. Acad. Sci. U S A*, **93** (1996) 3926.
32. M. Thorwart, P. Reimann, P. Jung, and R.F. Fox, *Chem. Phys.*, **235**, 61 (1981).
33. M. Thorwart, P. Reimann, and P. Hänggi, *Phys. Rev. E*, **62**, 5808 (2000).
34. P. Nalbach and M. Thorwart, *Phys. Rev. Lett.*, **103**, 220401 (2009).
35. M. Thorwart, J. Eckel, J.H. Reina, P. Nalbach, and S. Weiss, *Chem. Phys. Lett.*, **478**, 234–237 (2009).
36. P. Nalbach, A. Ishizaki, G.R. Fleming and M. Thorwart, *New J. Phys.*, **13**, 063040 (2011).
37. P. Nalbach and M. Thorwart, Quantum coherence and entanglement in photosynthetic light-harvesting complexes, in *Quantum Efficiency in Complex Systems, Part I: Biomolecular Systems*, Semiconductors and Semimetals, Vol. 83, E. Weber, M. Thorwart, and U. Würfel (eds.) (Academic Press, 2010).
38. P. Nalbach, J. Eckel, and M. Thorwart, *New J. Phys.*, **12**, 065043 (2010).

39. A. Ishizaki, and G.R. Fleming, *J. Chem. Phys.*, **130**, 234110 (2009).
40. A. Ishizaki, and G.R. Fleming, *J. Chem. Phys.*, **130**, 234111 (2009).
41. A. Ishizaki, and G.R. Fleming, *Proc. Natl. Acad. Sci. U S A*, **106**, 17255 (2009).
42. P. Nalbach and M. Thorwart, *J. Phys.: Conf. Ser.*, **376**, 012025 (2012).
43. M. Wendling, T. Pullerits, M.A. Przyjalgowski, S.I.E. Vulto, T.J. Aartsma, R. van Grondelle, and H. van Amerongen, *J. Phys. Chem. B*, **104**, 5825 (2000).
44. J. Adolphs and T. Renger, *Biophys. J.*, **91**, 2778 (2006).
45. P. Nalbach, D. Braun and M. Thorwart, *Phys. Rev. E*, **84**, 041926 (2011).
46. T. Mancal, N. Christensson, V. Lukes, F. Milota, O. Bixner, H.F. Kauffmann, and J. Hauer, *J. Phys. Chem. Lett.*, **3**, 1497 (2012).
47. S. Polyutov, O. Kühn, and T. Pullerits, *Chem. Phys.*, **394**, 21 (2012).
48. N. Christensson, H.F. Kauffmann, T. Pullerits, and T. Mancal, *J. Phys. Chem. B*, **116**, 7449 (2012).
49. P. Nalbach and M. Thorwart, *J. Phys. B: At. Mol. Opt. Phys.*, **45**, 154009 (2012).
50. R. Zwanzig, *J. Chem. Phys.*, **33**, 1338 (1960).
51. H. Mori, *Prog. Theor. Phys.*, **33**, 127 (1965).
52. A.O. Caldeira and A.J. Leggett, *Ann. Phys.*, **149**, 374 (1983).
53. A. Lucke, C.H. Mak, R. Egger, J. Ankerhold, J. Stockburger, and H. Grabert, *J. Chem. Phys.*, **107**, 8397 (1997).
54. P. Nalbach and M. Thorwart, *Phys. Rev. B*, **81**, 054308 (2010).
55. P. Nalbach and M. Thorwart, *Phys. Rev. B*, **87**, 014116 (2013).
56. P. Hänggi, Generalized Langevin equations: a useful tool for the perplexed modeller of nonequilibrium fluctuations? In *Stochasitc Dynamics*, Vol. 484 of *Lecture Notes in Physics*, L. Schimansky-Geyer and Th. Pöschel (eds.) (Springer, Heidelber, 1997).
57. C.A. Mujica-Martinez, P. Nalbach, and M. Thorwart, *Phys. Rev. Lett.*, **111**, 016802 (2013).
58. C. Olbrich, J. Strümpfer, K. Schulten, and U. Kleinekathöfer, *J. Chem. Phys. Lett.*, **2**, 1771 (2011).
59. B. Thimmel, P. Nalbach, and O. Terzidis, *Eur. Phys. J. B*, **9**, 207–214 (1999).
60. P. Nalbach and M. Thorwart, *Chem. Phys.*, **375**, 234 (2010).
61. P. Nalbach, J. Knörzer and S. Ludwig, *Phys. Rev. B*, **87**, 165425 (2013).
62. R.M. Fye, *Phys. Rev. B*, **33**, 6271 (1986).
63. R.P. Feynman and F.L. Vernon, Jr., *Ann. Phys. (N.Y.)*, **24**, 118 (1963).

64. P. Nalbach, D.D. Osheroff, and S. Ludwig, *J. Low Temp. Phys.*, **137**, 395 (2004).
65. V.I. Prokhorenko, A.R. Holzwarth, F.R. Nowak, and T.J. Aartsma, *J. Phys. Chem. B*, **106**, 9923 (2002).
66. J. Gilmore and R.H. McKenzie, *J. Phys.: Condens. Matter*, **17**, 1735 (2005).
67. J. Gilmore and R. McKenzie, *Chem. Phys. Lett.*, **421**, 266 (2006).
68. J. Gilmore and R. McKenzie, *J. Phys. Chem. A*, **112**, 2162 (2008).
69. S. Jang, *J. Chem. Phys.*, **127**, 174710 (2007).
70. D. Beljonne, C. Curutchet, G.D. Scholes, and R.J. Silbey, *J. Phys. Chem. B*, **113**, 6583 (2009).
71. P. Nalbach, I. Pugliesi, H. Langhals, and M. Thorwart, *Phys. Rev. Lett.*, **108**, 218302 (2012).
72. T. Renger, V. May, and O. Kühn, *Phys. Rep.*, **3**, 137 (2001).
73. A. Ishizaki, T.R. Calhoun, G.S. Schlau-Cohen, and G.R. Fleming, *Phys. Chem. Chem. Phys.*, **12**, 7319 (2010).
74. A. Peres, *Phys. Rev. Lett.*, **77** (1996) 1413.
75. M. Horodecki, P. Horodecki, R. Horodecki, *Phys. Lett. A*, **223** (1996) 1.
76. C. Olbrich, J. Strumpfer, K. Schulten, and U. Kleinekathöfer, *J. Chem. Phys. B*, **115**, 758 (2011).
77. Private communication with U. Kleinekathöfer.
78. C. Mujica-Martinez, P. Nalbach, and M. Thorwart, *Phys. Rev. E*, **88**, 062719 (2013).
79. J. Wen, H. Zhang, M.L. Gross, and R.E. Blankenship, *Proc. Natl. Acad. Sci. U S A*, **106**, 6134 (2009).

Chapter 8

Excitation Energy Transfer in Light-Harvesting Systems: Theory, Models, and Application

Pengfei Huo[a,b] and David F. Coker[b]

[a]*Department of Chemistry, University of Rochester, Rochester, NY 14627-0216, USA*
[b]*Department of Chemistry, Boston University, 590 Commonwealth Avenue, Boston, MA 02215, USA*
huo@chem.rochester.edu, coker@bu.edu

In this chapter, we use the recently developed partial linearized density matrix (PLDM) propagation approach to explore the excitation energy transfer dynamics in various light-harvesting systems. We demonstrate that PLDM propagation can capture both short-time nonequilibrium coherent dynamics, and recover long-time thermal populations that result from relaxation of initial excitation in the model dissipative environment of protein–pigment photosynthetic light-harvesting complexes. We find that the energy transfer rate can be optimized as a function of the strength of the system–environment interactions. Despite the fact that experiments show signatures of coherent dynamics at short times we find that such features play little role in controlling the energy transfer rate. We

Ultrafast Dynamics at the Nanoscale: Biomolecules and Supramolecular Assemblies
Edited by Irene Burghardt and Stefan Haacke

ISBN 978-981-4745-33-8 (Hardcover), 978-981-4745-34-5 (eBook)
www.panstanford.com

show that the presence of coherent beating in chromophore network populations, which matches closely with features observed in nonlinear 2D spectroscopy signals, is strongly dependent on the way the excitation is injected into the multichromophore network. The artificial initial conditions employed in the experiments enhance the coherence and with more realistic initialization the coherent features are significantly diminished. We show that a simple kinetic model can be used to understand the population dynamics and relaxation to thermal equilibrium in these systems. Our calculations employing various realistically parameterized model Hamiltonians for photosynthetic light-harvesting systems explore the effects initialization, the influence of strongly coupled dimers that result in quantum coherent beating, and the effect of correlated bath fluctuations in diagonal and off-diagonal elements of the system–bath Hamiltonian over realistic parameter ranges for these systems.

8.1 Introduction

Photosynthetic systems like the various well studied bacterial models [1–7] involve assemblies of many chromophore, such as bacteriochlorophyll molecules, arrayed in structured complexes embedded in protein scaffoldings. These ubiquitous structures have evolved to function as extraordinarily efficient solar energy capture, and transfer systems that funnel electronic excitation into reaction centers (RCs) where long term energy storage is initiated. Remarkably, recent multidimensional nonlinear spectroscopic experiments reveal that the early stages of the excitation energy transfer process in these nanoscale energy "transmission grids" involve quantum coherent superpositions of vibrational and electronic states that give rise, at least at early times, to observed quantum beating between the excitonic states involved in the energy transfer processes [3]. This signature of coherent excitation energy transfer has even been found at ambient temperature [1, 2]. Even more surprising is the suggestion, from the interpretation of these signals, that the protein scaffolding that supports the chromophores seems to move in a collective fashion producing correlated fluctuation in the energy levels of the different chromophores that effectively protects the electronic coherence [4].

Exactly what role, if any, this fundamentally quantum behavior plays in the functioning of these structures is as yet unclear, but many imaginative suggestions have been made [1, 3, 4, 8–15]. However, whether this coherence appears only due to the special initial conditions of the spectroscopic experimental set up, or does the coherence actually play a significant role to assist excitation transport is still under debate [16–19, 21]. For example, the ultrafast spectroscopic experiments directly promote the ground-state wave packet to the excited state in a Frank–Condon excitation that give rises to the strong coherent beating signals. If the initial wave packet is adapted to the excited-state potential by relaxation, the coherent beating will be dramatically suppressed [20]. Also, the set up for the experiments is often, by necessity, highly artificial so making statements about the in vivo function of the much more complex biological system can be problematic. For example, in the pioneering early studies of the Fenna–Matthews–Olsen (FMO) excitonic wire complex [4], the trimer pigment–protein complex is removed from its biological context, in which it acts as a conduit between the baseplate of the light-harvesting chlorosome and the RC where charge separation is initiated, and placed in aqueous solution. The details of how the connections are made between these different units, and how the excitation energy is initially injected into the excitonic wire turn out to have a profound influence on the response and it has been shown in realistic model studies [18, 26] that the coherent beating observed in the experiments is a property of the artificial preparation and initial excitation used in the experiments. When energy is allowed to flow into the wire with a more realistic initialization involving energy transfer from other chromophores at the interfaces between the various units (e.g., like the "eighth chromophore" that bridges between the baseplate and the FMO complex) a steady incoherent flow of population is observed rather than the artificial coherent population beating found in the experiments [18, 26, 27]. While the experiments indeed point to intriguing quantum coherent properties of ultrafast response of the individual excitonic circuit elements their interconnection in biological systems seems to wash out these intrinsically quantum properties and the overall behavior of the composite biological system can be largely understood in terms of incoherent energy transfer processes. Though the excitation transfer inside a small complex of chromophores may occur coherently on the few picosecond timescale, the inter complex transfer, which is most often the rate determining step, typically on the tens of picosecond time scale, occurs incoherently due to the large separation and weak coupling between different complexes [7].

The experimental findings raise interesting questions about whether or not there are circumstances in systems like these in which the environments might be engineered to protect long-lived coherence, and if this might be utilized to influence energy transfer in some technologically interesting ways or perhaps such mechanisms might even enable the manipulation of long-lived quantum coherent states of these systems in ways that may be useful, like how quantum coherence transfer can be employed in various NMR spectroscopies. As mentioned above, whether long-lived coherence can actually play a significant role to enhance the rate of energy transfer through a quantum network is a debatable issue.

On the other hand, the energy transfer rate through such multiscale chromophore networks can be optimized by exploiting processes analogous to Kramers turnover effects observed in barrier crossing phenomena [30] that are also related to stochastic resonance- or noise-assisted quantum transport phenomena [11, 13]. In the limit that the dissipation is small, so too are the energy-level fluctuations of the different chromophores, and the energy transfer rate is controlled by the intersite couplings in the bare system Hamiltonian. As the dissipation is increased the site energy-level fluctuations become larger and as the energy gaps of different coupled sites occasionally fluctuate into resonance with one another, the energy transfer rate can be enhanced significantly through these stochastic resonances. Even stronger dissipation results in excitation trapping giving rise to the turnover like behavior and slower energy transfer. Theoretical studies on various model photosynthetic light-harvesting complexes [17, 18, 29] scanning local environmental coupling have demonstrated that these types of effects are paramount in controlling and optimizing the energy transfer rates in these systems. These studies further suggest that in the absence of correlation between fluctuations in the baths of different chromophores that the presence of coherent superposition states seems to have little effect on energy transfer rates [17, 18, 29, 31]. These studies, however, do indicate that the presence of different types of correlation in the fluctuations of the baths of different chromophores can lead to enhanced coherence and simultaneously increased energy transfer rates. These special correlations are generally highly artificial and they are not expected to be relevant in natural photosynthetic systems [19, 32] but it may be possible to engineer synthetic nanostructured systems that can take advantage of such behavior.

Nevertheless, it is critically important to have a reliable quantum simulation methodology that can accurately describe the transition between short-time nonequilibrium coherent quantum dynamics in dissipative systems and their passage to long-time thermal equilibrium so that we can understand how artificial nanostructures might be engineered to take advantage of quantum coherent electronic and vibrational dynamics.

Reliable theoretical methods that can treat many coupled chromophores interacting with a nanostructured collective bath are required to address these intriguing questions. However, standard perturbation theory based methods, such as Förster resonance energy transfer (FRET) theory [30, 33, 34], which assumes the electronic coupling between chromophores is smaller than the electron–phonon coupling, or the Redfield equation [30, 35, 37, 38], which assumes electron–phonon coupling is small compare to the electronic coupling between chromophores, cannot accurately describe the exciton dynamics in multichromophore–protein systems that involve moderate interchromphore–dipole coupling and chromophore–exciton-vibrational environmental coupling [35, 39]. Many approximate approaches like the Redfield equation methods assume the Markovian approximation [30] and ignore memory effects of the bath that can influence the system response [35, 36]. The secular approximation is also often made in which it is further assumed that the dynamics of the diagonal and off-diagonal elements of the density matrix evolve independently. Also many approximation schemes ignore the possibility of multiple coherent interfering pathways for energy flow entirely and treat the relaxation as an incoherent rate process between the various states.

To address the questions raised in the experiments on long-lived quantum coherent dynamics in photosynthetic systems, theoretical methods must be able to describe both short-time coherent dynamics and long-time incoherent transfer and thermalization correctly. The mean-field approaches like the Ehrenfest trajectory methods [22, 23], the linearized semiclassical initial value representation (LSC-IVR) approach [40, 41], the Poisson bracket mapping equation (PBME) implementation of mixed quantum-classical dynamics [24, 42, 43], or the stochastic Liouville Equation approach [31] generally give less accurate descriptions of long-time excitation energy transfer. Such approaches typically only give reliable dynamics for a few hundred fentoseconds for typical parameter ranges in these photosynthetic systems and give erroneous results for the 10ps timescales expected for thermalization [16, 36]. A method based on a hybrid of the Ehrenfest dynamics (that can give short-time coherent dynamics correctly) and the NIBA equation (which normally gives correct thermal long-time limits) has been proposed [39], and found to generate more accurate results for a model two level dissipative system, than either of these methods independently, but the general treatment of multilevel systems with such hybrid approaches requires special care [39].

Numerically exact methods that can be used to explore the condense phase dynamics and especially the exciton dynamics in light-harvesting systems include: the hierarchical coupled reduced master equation (HCRME) [10, 44, 45] and its rescaled version [46, 47] and the quasi-adiabatic propagator path integral (QUAPI) [48–50]. However, these methods are normally limited by particular form of model Hamiltonian and can have large computational costs. The multiconfiguration time-dependent Hartree (MCTDH) approach and its multilayer formulation [51–53] offers an exact scheme that can include arbitrary spectral densities. These approaches, however, are most easily implemented in a wave function formulation rather than a finite temperature density matrix description, which can be expensive to implement numerically. Approximate methodologies such as the cumulant time-nonlocal master equation [50], non-

Markovian quantum jump (NMQJ) [54, 55], generalized Bloch–Redfield (GBR) [18, 29, 56], polaronic quantum master equation [57–59] and non-Markovian quantum state diffusion (NMQSD) [27, 60] methods are being intensely investigated to try to obtain accurate results that are benchmarked against the exact methods mentioned above but with significantly lower computational costs. More details on this progress in methodology development for treating excitation energy transfer dynamics is presented in a recent review [61].

In this chapter, we use the recently developed partial linearized density matrix (PLDM) propagation method [62–64] to explore the excitation energy transfer dynamics in various biological photosynthetic light-harvesting systems. The scheme uses a linearized approximation in the difference between the forward and backward paths of the heavy nuclear DOF that represent the amplitudes in the density matrix propagation. For the light electronic DOF, on the other hand, the partial linearized approach retains both forward and backward amplitudes and so includes important interference effects. The approach is nonperturbative and does not require making the Markovian approximation. Moreover, it does not require any particular form for the Hamiltonian, and thus can be applied to treat general system–bath interactions as well as arbitrary forms for the bath. In addition, it has been show to give both short-time coherent dynamics and long-time thermal equilibration reliably for moderate system–bath coupling situations such as those found for many model photosynthetic light-harvesting systems [62]. This method will be derived and explained in Section 8.2.

This chapter is organized as following: First, a brief summary of the PLDM method is provided in Section 8.2. The general system–bath model Hamiltonian and the procedure to sample bath parameters according to a given arbitrary spectral density is provided in Section 8.3. Next, in Section 8.4, the seven-state model Hamiltonian for the FMO complex [10, 39, 47, 49, 60] is used to demonstrate the capability of the PLDM approach to give both short time coherent dynamics and the long-time transient energy transfer rate and thermal equilibration correctly [62]. In Section 8.5 we explore the full eight-site model of the FMO complex and compare the dynamics resulting from realistic in vivo–like initial excitation of the connecting chromophore BChl8 with the initial excitation of the BChl1–BChl2 strongly coupled dimer. In Section 8.6, we explore the stochastic resonance behavior that determines the optimal excitation energy transfer rate, here however, we adopt the full eight-site model of the FMO complex that enables us to study the more realistic in vivo initialization for the excitation energy transfer process in this system [18, 26, 27]. For this system we find that different initial conditions can have a profound influence on the dynamics. For example, some initial conditions yield strongly coherent energy transfer, while for others the transfer is essentially incoherent.

One of the characteristic structural motifs found in these quantum energy transfer networks is the electronically coupled dimer and with direct initial excitation of such a dimer structure, the long-time coherent beating signature is observed. In Section 8.7, we demonstrate that if such dimer structures are replaced by a single mean state that is coupled to the rest of the network in the same way as each state in the dimer, the coherent beating pattern can be removed from the signal but the over all rate of energy transfer from the initially occupied mean state to the rest of the network accurately captures the average behavior of the fully coherently coupled system. To further explore this behavior we adapt a realistic model of the phycocyanin 645 (PC645) from Chroomonas CCMP270 [1, 17].

One of the proposed mechanisms for prolonging the lifetime of coherent superposition states in condensed phase multichromophore systems is the idea that there are long-wavelength low-frequency motions that cause correlated fluctuations in the excitation energies of the different chromophores. In Section 8.8, we explore this proposition using generalized system–bath models that include different types of correlation in both diagonal and off-diagonal intersite excitonic couplings, and their influence on decoherence and excitation energy transfer rates. Finally we summarize our main conclusions in Section 8.9.

We should point out that, various other interesting topics, such as how discrete, special frequency, bath modes can enhance the excitation energy transfer rate [58, 60, 70], the calculation of the 2D spectroscopy [105, 108–110], atomistic molecular dynamics (MD) simulation to obtain spectral densities [103, 104, 107] and exciton dynamics [24, 55, 106] are not discussed in this chapter. Also, there are interesting investigations of excitation energy transfer dynamics in various other biological light-harvesting systems, such as B850 in LH II [34, 65, 66], LHCII [5, 67, 68], PE545 [1, 6, 69, 70], double-exciton dynamics in FMO [71], and the interaction of exciton transfer and charge separation dynamics in extended harvesting network RC systems [72], as well as energy transfer in DNA [73], and vibrational excitation transfer in peptides [25] that all involve similar physical processes in different contexts and readers are referred to these references for further explanation. A particularly relevant discussion of how coupling to vibrational modes of the chromophores and their surrounding environments can influence energy transfer in light harvesting is presented in Refs. [127, 128].

8.2 Partial Linearized Density Matrix Propagation

The mixed quantum-classical strategy for simulating systems in which quantum coherent dynamics and electronically nonadiabatic transitions play important roles is to describe the nuclear degrees of freedom (DOF) classically or semiclassically, while still treating the electronic DOF fully quantum mechanically with an evolution operator that is parameterized by trajectories of the nuclear DOF [74]. Though many schemes for implementing this idea have been developed [75], fundamental questions arise about the accuracy of dynamics methods that treat electronic and nuclear DOF on different dynamical footing [78], and which can give an incorrect description of the back-reaction from the classical system to quantum system [76], often leading to an incorrect treatment of the long-time thermal equilibrium of the quantum state populations [77]. To overcome this difficulty the mapping Hamiltonian idea, which exactly maps the discrete quantum states onto continuous coordinates, has been proposed [79, 80] and enables consistent treatment for all DOF. This idea replaces the evolution of the electronic subsystem by the dynamics of a system of fictitious mapping harmonic oscillators. With this approach, for example, the quantum amplitude transfer operator transforms as $|\alpha\rangle\langle\gamma| \rightarrow \hat{a}_\alpha^\dagger \hat{a}_\gamma$, where $\hat{a}_\gamma = (1/\sqrt{2\hbar})\left(\hat{q}_\gamma + i\,\hat{p}_\gamma\right)$ and $\hat{a}_\gamma^\dagger = (1/\sqrt{2\hbar})\left(\hat{q}_\gamma - i\,\hat{p}_\gamma\right)$. A general electronic Hamiltonian in the diabatic representation,

$\hat{h}_{el} = \sum_{\alpha,\gamma} |\alpha\rangle\langle\alpha|\hat{h}_{el}(\hat{R})|\gamma\rangle\langle\gamma|$, can thus be rewritten in this mapping formulation as:

$$\hat{h}_m(\hat{R}) = \frac{1}{2\hbar}\sum_{\alpha} h_{\alpha,\alpha}(\hat{R})(\hat{q}_\alpha^2 + \hat{p}_\alpha^2 - \hbar)$$
$$+\frac{1}{2\hbar}\sum_{\gamma\neq\alpha} h_{\alpha\gamma}(\hat{R})(\hat{q}_\alpha\hat{q}_\gamma + \hat{p}_\alpha\hat{p}_\gamma) \quad (8.1)$$

Here, $\hat{P}$,$\hat{R}$ and $\hat{p}$,$\hat{q}$ represent the nuclear and mapping oscillator DOF. Direct implementation of this mapping Hamiltonian by stationary phase approximation and classical trajectories encounters a fatal problem when $q_\alpha^2 + p_\alpha^2 < \hbar$, since some of the classical DOF can evolve on an inverted potential surface $-h_{\alpha,\alpha}(R)$ [81, 82]. Moreover, the population $\rho_{\alpha\alpha} = \hat{a}_\alpha^\dagger\hat{a}_\alpha = 1/(2\hbar)(\hat{q}_\alpha^2 + \hat{p}_\alpha^2 - \hbar)$ is not guaranteed to have a positive expected value when the approach is implemented approximately, for example, with the LSC-IVR applied to multilevel systems [40, 82]. To overcome these problems here we use the coherent state representation of the mapping DOF [84].

The essential quantity of interest here is the evolution of the density matrix:

$$\langle R_t n_t|\hat{\rho}(t)|R_t' n_t'\rangle = \sum_{n_0,n_0'}\int dR_0 dR_0' \langle R_t n_t|e^{-\frac{i}{\hbar}\hat{H}t}|R_0 n_0\rangle$$
$$\times\left\langle R_0 n_0|\hat{\rho}(0)|R_0' n_0'\rangle\langle R_0' n_0'|e^{\frac{i}{\hbar}\hat{H}t}|R_t' n_t'\right\rangle \quad (8.2)$$

where the total Hamiltonian is $\hat{H} = \hat{P}^2/2M + h_m(\hat{R}, \hat{p}, \hat{q})$. The propagator matrix elements can be written in discrete path integral form:

$$\langle R_N m_{n_t}|e^{-\frac{i}{\hbar}\hat{H}t}|R_0 m_{n_0}\rangle = \int\prod_{k=1}^{N-1} dR_k \frac{dP_k}{2\pi\hbar}\frac{dP_N}{2\pi\hbar} e^{\frac{i}{\hbar}S_0}$$
$$\times\left\langle n_t|e^{-\frac{i}{\hbar}\epsilon\hat{h}_m(R_{N-1})}\ldots e^{-\frac{i}{\hbar}\epsilon\hat{h}_m(R_0)}|n_0\right\rangle \quad (8.3)$$

where

$$S_0 = \epsilon\sum_{k=1}^{N}\left[P_k\frac{(R_k - R_{k-1})}{\epsilon} - \frac{P_k^2}{2M}\right]$$

and

$$T_{[n_t,n_0]} = \left\langle n_t|e^{-\frac{i}{\hbar}\epsilon\hat{h}_m(R_{N-1})}....e^{-\frac{i}{\hbar}\epsilon\hat{h}_m(R_0)}|n_0\right\rangle$$

is the nuclear path-dependent quantum transition amplitude.

In the coherent state representation [83] (using the arbitrary coherent state width parameter as $\gamma = 1/2$ and setting units so that $\hbar$=1), the transition amplitude can be expressed as [84, 85]:

$$T_{[n_t,n_0]} = \int dq_0 dp_0 \frac{1}{4}(q_{n_t} + ip_{n_t})(q_{n_0} - ip_{n_0})c_t e^{iS_1(t)} \times e^{-\frac{i}{2}\sum_\alpha (q_{\alpha t}p_{\alpha t} - q_{\alpha 0}p_{\alpha 0})} e^{-\frac{1}{2}\sum_\alpha (q_{\alpha 0}^2 + p_{\alpha 0}^2)} \tag{8.4}$$

where $c_t = e^{-\frac{i}{2\hbar}\int_0^t d\tau \sum_\alpha \hat{h}_{\alpha\alpha}(R)}$, $S_1(t) = \int_0^t L_1(\tau)d\tau$, with $L_1 = \sum_\alpha p_\alpha \dot{q}_\alpha - h_m^{cl}(R) + \frac{1}{2}\sum_\alpha h_{\alpha\alpha}(R)$ and

$$h_m^{cl}(R, p, q) = \frac{1}{2}\sum_\alpha h_{\alpha\alpha}(R)\left(p_\alpha^2 + q_\alpha^2\right) + \frac{1}{2}\sum_{\gamma\neq\alpha} h_{\gamma\alpha}(R)\left(p_\gamma p_\alpha + q_\gamma q_\alpha\right) \tag{8.5}$$

The term $(\hbar/2)\sum_\alpha h_{\alpha\alpha}(R)$ in the action that gives rise to the problem of inverting the potential when deriving classical trajectories from a stationary phase approximation to the propagator, can be eliminated as this term is cancelled exactly by the prefactor, c_t [84, 85].

According to Eq. 8.2, the forward and backward propagators must be combined to evolve the density matrix and this can lead to phase cancellation. Accurate representation of such cancellation in numerical implementation often presents problems. Approximate schemes such as forward-backward SC-IVR [41, 86] can alleviate these difficulties. Alternatively, to circumvent these problems, we follow the idea of partial "linearization" [85, 87–89] in the nuclear DOF that begins by transforming the forward and backward nuclear path variables, R and R', to mean and difference path variables: $\bar{R} = (R + R')/2$ and $Z = (R - R')$ respectively (with a similar transformation for the nuclear environment momenta, $\bar{P} = (P + P')/2$ and $Y = (P - P')$). The nuclear kinetic action difference can then be rewritten exactly as:

$$(S_0 - S_0') = \bar{P}_N Z_N - \bar{P}_1 Z_0 - \sum_{k=1}^{N-1}(\bar{P}_{k+1} - \bar{P}_k)Z_k - \sum_{k=1}^{N}\left[\frac{\epsilon}{m}\bar{P}_k - (\bar{R}_k - \bar{R}_{k-1})\right]Y_k \tag{8.6}$$

The central approximation with this approach involves truncating the phase difference in the combined transition amplitude terms to linear order in Z, based on the assumption that for short times, forward and backward paths will remain close to each other. Expanding the action difference to first order in Z, and using the approximation that $\sum_\alpha p_\alpha \dot{q}_\alpha - h_m^{cl}(\bar{R}, p, q) \sim \sum_\alpha p_\alpha \dot{q}_\alpha - h_m^{cl}(R, p, q) = \frac{1}{2}\frac{d}{d\tau}\left(\sum_\alpha p_\alpha q_\alpha\right)$, we find that the transition amplitude phase can be rewritten as:

$$\begin{aligned}(S_1 - S_1') = \int_0^t \Bigg[& \frac{1}{2}\frac{d}{d\tau}\sum_\alpha (p_{\alpha\tau}q_{\alpha\tau} - p'_{\alpha\tau}q'_{\alpha\tau}) \\ & + \frac{1}{2}\left(\nabla_{\bar{R}} h_m^{cl}(\bar{R}_\tau, p_\tau, q_\tau) + \nabla_{\bar{R}} h_m^{cl}(\bar{R}_\tau, p'_\tau, q'_\tau)\right) Z_\tau \\ & + \mathcal{O}(Z_\tau^2)\Bigg] d\tau \end{aligned} \tag{8.7}$$

With this expression the first term in Eq. 8.7 cancels the boundary terms in $T_{[n_t,n_0]}$ given in Eq. 8.4 with similar cancellation in the backward path transition amplitude, $T'_{[n'_0,n'_t]}$.

Combining these approximate forward and backward phase factors ($e^{i/\hbar(S_0-S'_0)}T_{[n_t,n_0]}T'_{[n'_0,n'_t]}$) and performing the analytic integrals over $Z_0 \ldots . Z_{N-1}$ in the resulting linearized expression for the evolving density matrix, we arrive at the final approximate expression for $\rho(t)$:

$$\begin{aligned} & \left\langle \bar{R}_N + \frac{Z_N}{2} n_t | \hat{\rho}(t) | \bar{R}_N - \frac{Z_N}{2} n'_t \right\rangle \qquad (8.8) \\ & = \sum_{n_0, n'_0} \int d\bar{R}_0 dq_0 dp_0 dq'_0 dp'_0 G_0 G'_0 \\ & \times \frac{1}{4}(q_{n_t} + ip_{n_t})(q_{n_0} - ip_{n_0})\frac{1}{4}(q'_{n'_t} - ip'_{n'_t})(q'_{n'_0} + ip'_{n'_0}) \\ & \times \int \prod_{k=1}^{N-1} d\bar{R}_k \frac{d\bar{P}_k}{2\pi\hbar}\frac{d\bar{P}_N}{2\pi\hbar}(\hat{\rho})_W^{n_0,n'_0}(\bar{R}_0, \bar{P}_1) e^{\frac{i}{\hbar}\bar{P}_N Z_N} \\ & \times \prod_{k=1}^{N-1} \delta\left(\frac{\bar{P}_{k+1} - \bar{P}_k}{\epsilon} - F_k\right) \prod_{k=1}^{N} \delta\left(\frac{\bar{P}_k}{M} - \frac{\bar{R}_k - \bar{R}_{k-1}}{\epsilon}\right) \end{aligned}$$

Here, $G_0 = e^{-\frac{1}{2}\sum_\alpha (q_{\alpha 0}^2 + p_{\alpha 0}^2)}$ and $G'_0 = e^{-\frac{1}{2}\sum_{\alpha'} (q'^2_{\alpha' 0} + p'^2_{\alpha' 0})}$ provide the initial distributions for the forward and backward mapping variables that satisfy equations of motion determined by the mapping Hamiltonian parameterized by the time dependence of the mean nuclear path, for example, $\dot{q}_{n_t} = \partial h_m^{cl}(\bar{R}_t)/\partial p_{n_t}$ and $\dot{p}_{n_t} = -\partial h_m^{cl}(\bar{R}_t)/\partial q_{n_t}$, and the nuclear trajectories are determined by a "mean field" like force:

$$F_k = -\frac{1}{2}\left[\nabla_{\bar{R}_k} h_m^{cl}(\bar{R}_k, p_k, q_k) + \nabla_{\bar{R}_k} h_m^{cl}(\bar{R}_k, p'_k, q'_k)\right] \tag{8.9}$$

Finally, the initial distribution of the mean nuclear DOF is determined by the partial Wigner transform with respect to the bath variables of the initial density:

$$(\hat{\rho})_W^{n_0,n'_0}(\bar{R}_0, \bar{P}_1) = \int dZ_0 \left\langle \bar{R}_0 + \frac{Z_0}{2} n_0 |\hat{\rho}| \bar{R}_0 - \frac{Z_0}{2} n'_0 \right\rangle e^{-\frac{i}{\hbar}\bar{P}_1 Z_0} \tag{8.10}$$

For simplicity, here we consider factorized initial conditions, where $\rho_0 = \rho_{bath}^{eq}(R)\rho_{sys}$, however generalization to real correlated equilibrium initial conditions within the mapping description has been addressed [90].

Numerical implementation of the above expressions is straightforward, and includes the sampling of classical DOF from $(\hat{\rho})_W^{n_0,n'_0}$ $(\bar{R}_0, \bar{P}_1)$, and mapping variables are sampled from the Gaussian functions. The product of δ-functions in Eq. 8.8 is equivalent to a time-stepping prescription for the evolution of the bath variables. The force expression in Eq. 8.9 means the classical trajectory is governed by a mean classical mapping Hamiltonian $1/2[\hat{h}_m^{cl}(\bar{R}, p, q) + \hat{h}_m^{cl}(\bar{R}, p', q')]$.

We should point out that this "mean field" approach is quite different from "Ehrenfest dynamics," or equivalently LSC-IVR where the force acting on the classical DOF is

$$F^{LSC-IVR} = F^{Eh} = -\sum_{\alpha} \frac{1}{2\hbar}\left[(q_\alpha^2 + p_\alpha^2 - 1)\nabla_R h_{\alpha,\alpha}\right.$$
$$\left. + \sum_{\gamma \neq \alpha}(q_\alpha q_\gamma + p_\alpha p_\gamma)\nabla_R h_{\alpha,\gamma}\right]$$

(with initial conditions $q_\alpha = p_\alpha = \delta_{\alpha\gamma}$ if $|\psi(0)\rangle = |\gamma\rangle$) [86]. Also, by following a different sequence of canceling and linearizing in the phase difference, a "surface hopping" (SH)-like version of the present theory has been derived, known as iterative linearized density matrix (ILDM) propagation [91] and it has been tested in a variety of challenging high dimensional general quantum dynamical problems [16, 64, 91, 92]. While ILDM is highly accurate, its convergence with large numbers of iterations is problematic for longer times with extended coupling regions [16]. The behavior of the current approach, outlined in the examples below seems to be a significant improvement. Moreover, one can also use the current expression as a short-time approximation and develop an iterative propagation scheme, and such an approach has been recently explored [64].

8.3 Model Hamiltonian

The model Hamiltonian used to describe the single exciton energy transfer dynamics in dissipative environments like protein, lipids and aqueous solutions has the coupled system–bath model form. In this system–bath Hamiltonian model, the excitation of the chromophores is represented as the zeroth-order system Hamiltonian, which involves terms representing localized molecular excitation together with interchromophore coupling between these localized single excitation states:

$$\hat{H}_0 = \sum_{\alpha=1}^{N_{\text{state}}} \epsilon_\alpha |\alpha\rangle\langle\alpha| + \sum_{\alpha\neq\gamma}^{N_{\text{state}}} \Delta_{\alpha,\gamma} |\alpha\rangle\langle\gamma| \tag{8.11}$$

Here the ket $|\alpha\rangle$ represents the state of excitation of the multichromophore system in which chromophore α has been photoexcited to its first excited state, while all other chromophores remain in their ground states, ϵ_α is the excitation energy of chromophore α, and the coupling between the different localized singly excited states, $\Delta_{\alpha\gamma}$, can be represented in terms of the interaction of the transition moments of the different chromophores, so that within the transition dipole approximation (a useful starting point for systems with localized excitations on molecules that are separated by distances on the order of a nanometer), the interchromophore electronic coupling has the form:

$$\Delta_{\alpha\gamma} = [\mu_\alpha \cdot \mu_\gamma - 3(\mu_\alpha \cdot \mathbf{n}_{\alpha\gamma})(\mu_\gamma \cdot \mathbf{n}_{\alpha\gamma})]/\epsilon R_{\alpha\gamma}^3 \tag{8.12}$$

where, μ_α, say, is the $1 \leftarrow 0$ transition dipole moment vector of chromophore, α, and $\mathbf{n}_{\alpha\gamma} = \mathbf{R}_{\alpha\gamma}/R_{\alpha\gamma}$, with $R_{\alpha\gamma} = |\mathbf{R}_{\alpha\gamma}|$, the interchromophore distance.

The chromophores are embedded in a thermally fluctuating environment like a protein or a solvent. Such environments may be represented by a bath of independent harmonic oscillators with Hamiltonian $\hat{H}_b$, in which each oscillator with a different frequency is linearly coupled to the electronic excitation of a given chromophore described by a system–bath coupling Hamiltonian, $\hat{H}_{s-b}$. In the models used here we suppose that each chromophore α is surrounded by its own local environment described by $n^{(\alpha)}$ independent harmonic oscillators that are coupled bilinearly to the excitation of chromophore α, to represent the thermal fluctuation of this chromophore's excitation energy. We further assume, in this initial model, that the individual chromophore excitation energy fluctuations are uncorrelated (though this assumption will be relaxed in the model studies of Section 8.8) so the bath and system–bath Hamiltonian terms have the following forms:

$$\hat{H}_b = \sum_{\alpha=1}^{N_{\text{state}}} \sum_{i=1}^{n^{(\alpha)}} \frac{1}{2} \left[P_i^{(\alpha)2} + \omega_i^{(\alpha)2} R_i^{(\alpha)2} \right] \tag{8.13}$$

$$\hat{H}_{s-b} = \sum_{\alpha=1}^{N_{\text{state}}} \sum_{i=1}^{n^{(\alpha)}} c_i^{(\alpha)} R_i^{(\alpha)} |\alpha\rangle\langle\alpha| \tag{8.14}$$

Here, $c_i^{(\alpha)}$ controls the strength of the interaction of excited chromophore α with the ith mode of its independent bath, and the mass for the bath oscillator is set to unity.

Completing the squares for each system–bath coupling contribution results in a counterterm, H_{counter}, [21, 30, 93] that can be included as an energy offset and has the form:

$$\hat{H}_{\text{counter}} = \sum_{\alpha=1}^{N_{\text{state}}} \sum_{i=1}^{n^{(\alpha)}} \frac{c_i^{(\alpha)2}}{2\omega_i^{(\alpha)2}} |\alpha\rangle\langle\alpha| \tag{8.15}$$

The system part of the Hamiltonian can thus be represented as $\hat{H}_s = \hat{H}_0 + \hat{H}_{\text{counter}}$, and the total Hamiltonian is written as the sum of the system, the bath, and their interaction:

$$\hat{H}_{\text{tot}} = \hat{H}_s + \hat{H}_b + \hat{H}_{s-b} \tag{8.16}$$

In addition to the dynamical fluctuations described by the above system–bath model Hamiltonian, the chromophore site energies can vary from one region to another due to different local static disorder giving rise to inhomogenous broadening. This static disorder can be modeled by adding Gaussian random noise to the site energies and averaging over this static distribution, as in Refs. [34, 65, 93]. Processes such as excitation loss, for example, through radiative relaxation, excitation energy loss through states that behave as "sinks" for leakage of population through charge separation and recombination processes etc., can be modeled by adding imaginary parts to the diagonal terms of the Hamiltonian [11, 13]. Alternatively, rather than considering these processes through such effective model terms, they can be explicitly incorporated through a set of charge transfer (CT) states in the model Hamiltonian so that the single exciton manifold can couple directly to the CT state manifold within such representations [94]. Further, in the final result sections of this review the possible effects of correlation between energy fluctuations at different chromophore sites will be explored using generalized system–bath models [29, 95–100]. Also we study the possibility that the bath couples to the system in such a way as to produce fluctuations in the off-diagonal electronic coupling matrix elements [19, 32, 101, 102] and we explore the influence of such terms on the population relaxation and decoherence dynamics later in Section 8.8.

With this harmonic bath bilinearly coupled to the exciton state model, we can express the thermally averaged site α excitation energy fluctuation correlation function as $C^{\beta}_{\alpha\alpha}(t) = \langle\langle\alpha|\hat{H}_{s-b}(0)|\alpha\rangle\langle\alpha|\hat{H}_{s-b}(t)|\alpha\rangle\rangle_{\beta}$ with the following detailed expression [30],

$$\begin{aligned}C^{\beta}_{\alpha\alpha}(t) &= \int dR \sum_{\gamma} \langle R|\langle\gamma|\hat{\rho}_{\beta} \sum_{i=1}^{n^{(\alpha)}} c_i^{(\alpha)} \hat{R}_i^{(\alpha)}(0) \\ &\quad \times \sum_{i=1}^{n^{(\alpha)}} c_i^{(\alpha)} \hat{R}_i^{(\alpha)}(t)|\gamma\rangle|R\rangle \\ &= \sum_{i=1}^{n^{(\alpha)}} c_i^{(\alpha)2} \langle \hat{R}_i^{(\alpha)}(0) \hat{R}_i^{(\alpha)}(t)\rangle_{\beta} \end{aligned} \tag{8.17}$$

In the above $\langle\ldots\rangle_{\beta}$ indicates the equilibrium thermal average obtained by tracing the product of the quantity of interest and the

Boltzmann density operator $\hat{\rho}_\beta = \exp[-\beta\hat{H}]$ over all DOF, and here $\beta = 1/k_B T$.

Depending on if we choose to treat the bath harmonic oscillators quantum mechanically or classically we obtain different results for the thermal average bath oscillator coordinate correlation functions contained in the above expression, in particular we find [30]:

$$\langle \hat{R}_i^{(\alpha)}(0)\hat{R}_i^{(\alpha)}(t)\rangle_\beta^{qu} = [\coth(\beta\hbar\omega_i^{(\alpha)}/2)\cos\omega_i^{(\alpha)}t \\ -i\sin\omega_i^{(\alpha)}t]\hbar/2\omega_i^{(\alpha)} \tag{8.18}$$

and

$$\langle R_i^{(\alpha)}(0)R_i^{(\alpha)}(t)\rangle_\beta^{cl} = \cos\omega_i^{(\alpha)}t/\beta\omega_i^{(\alpha)2} \tag{8.19}$$

The principal quantity that enables us to relate these thermal average correlation functions to the linear system–bath coupling coefficients, c_i^α, in the model Hamiltonian is the spectral density, defined as follows[a]

$$J^{(\alpha)}(\omega) = \frac{\pi}{2}\sum_i \frac{c_i^{(\alpha)2}}{\omega_i^{(\alpha)}}\delta(\omega - \omega_i^{(\alpha)}) \tag{8.20}$$

With this definition it is straightforward to show that the quantum correlation function can be transformed to give the spectral density according to the result:

$$J^{(\alpha)}(\omega) = \frac{2}{\hbar}\tanh(\hbar\beta\omega/2)\int_0^\infty dt\mathrm{Re}[C_{\alpha\alpha}(t)]\cos(\omega t) \tag{8.21}$$

From the definition in Eq. 8.20 it is clear that since the spectral density contains only Hamiltonian parameters defining the interactions between the system and its environment it cannot depend on temperature. The correlation functions, as thermal averages, must

[a]In many references, the system–bath Hamiltonian is written in terms of phonon bath creation and annihilation operators, for example, $H_{s-b} = \sum_{\alpha=1}^{N_{\text{state}}}\sum_{i=1}^{n^{(\alpha)}}\hbar\omega_i^{(\alpha)}g_i^{(\alpha)}(b_i^{(\alpha)\dagger} + b_i^{(\alpha)})|\alpha\rangle\langle\alpha|$, where $(b_i^{(\alpha)\dagger} + b_i^{(\alpha)}) = \sqrt{2m\omega^{(\alpha)})_i/\hbar}q_i^{(\alpha)}$, so $H_{s-b} = \sum_{\alpha=1}^{N_{\text{state}}}\sum_{i=1}^{n^{(\alpha)}}\sqrt{2m\omega_i^{(\alpha)}\hbar}\omega_i^{(\alpha)}g_i^{(\alpha)}q_i^{(\alpha)}|\alpha\rangle\langle\alpha|$. Comparing with the expression in Eq. 8.13, we find $c_i^{(\alpha)} = \sqrt{2\hbar m\omega_i^{(\alpha)}}\omega_i^{(\alpha)}g_i^{(\alpha)}$. Note with this definition the spectral density is usually defined as $J(\omega) = \frac{1}{2}\sum_j \frac{c_j^2}{\hbar m\omega_j}\delta(\omega - \omega_j)$, so there is a π factor difference. With the relation between c_i^α and g_i^α, $J(\omega) = \sum_j g_j^2\omega_j^2\delta(\omega - \omega_j)$.

of course depend on temperature so the $\tanh(\hbar\beta\omega/2)$ factor in the above expression must cancel the temperature dependence in the correlation function and make the true harmonic bath spectral density independent of temperature.

In many calculations of the spectral density the site energy fluctuation correlation functions are determined by QM/MM simulations [24, 55, 103–105] in which classical mechanics determines the dynamics of the bath degrees of freedom. It has recently been shown [107, 111, 112] that using the quantum result in Eq. 8.21 to process QM/MM correlation functions leads to spurious temperature-dependent spectral density results and that classical simulation data is actually more consistently treated using the following classical result to extract temperature independent spectral densities

$$J_{cl}^{(\alpha)}(\omega) = \beta\omega \int_0^{\infty} dt C_{\alpha\alpha}^{cl}(t)\cos(\omega t) \tag{8.22}$$

In classical limit, where $\hbar \to 0$ or $\beta \to 0$, Eq. 8.20 thus leads to a quasi-classical relation between the spectral density and correlation function:

$$\begin{aligned} C_{\alpha\alpha}^{cl} &= k_B T \sum_{l=1}^{n^{(\alpha)}} \frac{c_l^{(\alpha)2}}{\omega_l^{(\alpha)2}} \cos(\omega_l^{(\alpha)} t) \\ &= \frac{k_B T}{\pi} \int_0^{\infty} d\omega \frac{J_{cl}^{(\alpha)}(\omega)}{\omega} \cos(\omega t) \end{aligned} \tag{8.23}$$

The linear relationship between the spectral density and correlation function thus exists in both the classical and quantum case.[b]

If we express the density of state (DOS) of the bath for chromophore α as $\rho^{(\alpha)}(\omega) = \sum_i \delta(\omega - \omega_i^{(\alpha)})$, then from Eq. 8.20 we obtain

$$c_i^{(\alpha)^2} = \omega_i^{(\alpha)} 2J^{(\alpha)}(\omega_i^{(\alpha)})/\pi\rho^{(\alpha)}(\omega_i^{(\alpha)}) \tag{8.24}$$

Also, we have the following expressions from which we can sample the frequencies of our bath modes:

$$\int_0^{\omega_j^{(\alpha)}} \rho^{(\alpha)}(\omega)d\omega = j; \int_0^{\omega_m^{(\alpha)}} \rho^{(\alpha)}(\omega)d\omega = n^{(\alpha)} \tag{8.25}$$

[b] For simplicity, throughout we use this classical relationship between the correlation functions and the corresponding spectral densities. Switching to the quantum expression will not influence the results presented here since we use the spectral density directly, not the correlation function, and both quantum and classical expression maintain the linear relationship between correlation function function and spectral density.

where $n^{(\alpha)}$ is total number of the sampled modes and $\omega_m^{(\alpha)}$ is the maximum frequency of the modes in the bath of chromophore α. Once the frequency of the modes are determined by inverting Eq. 8.25, the coupling constants of the different modes can be obtained from Eq. 8.24. In the real numerical calculation, the detail form of $\rho^{(\alpha)}(\omega)$ will not influence the final result as long as enough modes is sampled. Normally, we can choose $\rho(\omega) = aJ(\omega)/\omega$ [113, 114], or $\rho(\omega) = n^{(\alpha)}/\sqrt{\omega\omega_m}$ [115]. As an example of the implementation of this approach in the theoretical study of the FMO complex outlined in the sections below an ohmic spectral density with a lorentzian truncation, or Debye spectral density form was employed to fit the experimental data on this system by Ishizaki and Fleming [10]. The precise model form is thus: $J(\omega) = 2\lambda\omega\tau_c/(1+\omega^2\tau_c^2)$, with $\tau_c = 50$ fs and $\lambda = 35$ cm^{-1}. In this model the same spectral density is assumed for all chromophores [10]. With the choice of the first $\rho(\omega)$ and the Eq. 8.25, we have $a = n/(2\lambda\tan^{-1}(\omega_m\tau_c))$, $c_j = 2\sqrt{\lambda\tan^{-1}(\omega_m\tau_c)/\pi n}\,\omega_j$ and $\omega_j = \tan(j\tan^{-1}(\omega_m\tau_c)/n)/\tau_c$[c] Actually, this sampling procedure can be generalized for arbitrary shape of spectral densities numerically and applied, for example, to those obtained directly from the MD simulations [24, 55, 103, 106]. In this way the effects of site-dependent spectral densities can be explored [112].

8.4 Short-Time Coherent Dynamics and Long-Time Thermal Equilibrium from PLDM Propagation

In this section, we demonstrate that the PLDM propagation approach can treat short-time coherent dynamics as well as recovering the long-time relaxation to thermal equilibrium. To show this we employ a realistically parameterized system–bath model of the FMO light-harvesting complex that was fitted to reproduce the experimental 2D nonlinear coherent optical response. The model

[c]Another commonly used spectral density, known as debye spectral density, has the form $J(\omega) = \frac{\pi}{2}\zeta\omega e^{-\omega/\omega_c}$ [85, 92, 123], then $a = N_b/[\zeta\omega_c(1-e^{-\omega_m/\omega_c})]$, $\omega_0 = \frac{1}{\zeta a} = \frac{\omega_c}{N_b}(1-e^{-\omega_m/\omega_c})$, and $\omega_j = -\omega_c\ln(1-j\omega_0/\omega_c)$ and $c_j = \omega_j\sqrt{\zeta\omega_0 M_j}$.

includes seven coupled states in which the electronic excitation is localized on one of the seven different chromophores in the FMO monomer system. As mentioned above the biologically active FMO complex is a trimer structure that bridges between the baseplate of the chlorosome and the RC and each monomer in the in vivo structure actually contains eight chlorophyll chromophores. In the ultrafast two dimensional nonlinear spectroscopy experiments, however, the FMO trimer complex is removed from its biological context and incorporated in aqueous solution so that the peripheral BChl8 chromophores that are responsible for forming the excitonic connection between the baseplate and each seven chromophore FMO monomer complex, become destabilized and can be lost from the solvated structure. These experiments thus probe the coherent excitation energy transfer dynamics of the seven chromophore system, while the in vivo biological system involves a fundamentally different system that is fed excitation via the bridging eighth chromophore. These different structures and different initial conditions result in fundamentally different relaxation dynamics for the in vivo system and the artificial aqueous solvated FMO trimer system explored in the experiments.

The basic structure of the monomer showing the seven internal chromophores and the eighth bridging BChl, as well as a diagrammatic representation of the energy levels of the principal excitation energy transfer pathway of the complex from the BChl8 source chromophore to the BChl3 sink chromophore, which makes its way around the edge of the pigment–protein complex, involving the strongly coupled BChl1–BChl2 dimer as intermediate states, is depicted in Fig. 8.1a. The parameters for the seven-state model system Hamiltonian used here are detailed in Ref. [10]. We should emphasize that the master equation or influence functional approaches, which treat the bath implicitly, are also capable of reproducing the short-time coherent dynamics and relaxation to thermal equilibrium for this model system [10, 47, 49, 60]. However, most trajectory-based approaches, which explicitly propagate the bath coordinates (and potentially could be used in an all atom description of such problems), normally cannot give a satisfactory results for the long-time relaxation behavior in such model systems [20, 22, 23, 40, 43]. To the best of our

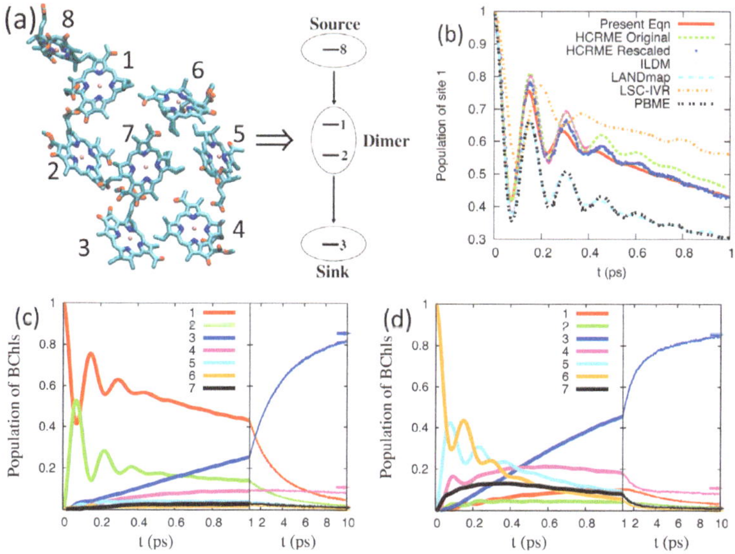

Figure 8.1 (a) Atomistic view of the chromophores in FMO. (b) Population obtained by different methods for the seven-state FMO model [10] with initial excitation at site 1 at $T = 77$ K, comparing results obtained with HCRME [10], rescaled-HCRME [47], LSC-IVR [40], PBME [43], and ILDM [16] approaches. (c and d) Population of all seven sites with initially excitation at 1 (c) or 6 (d). Arrows at long times show expected equilibrium BChl populations.

knowledge, the PLDM propagation approach is currently one of the few trajectory-based methods that can reliably capture both the short-time coherent behavior and long-time thermalization for the realistic parameterization of the seven-state FMO model, and as such should give accurate results when embedded in MD-like simulations as compared to other available trajectory based methods [24].

The other panels of Fig. 8.1 show population dynamics results computed with the seven-state FMO model Hamiltonian at $T = 77$ K using different initial conditions [62]. Panel (b) compares the evolution of initially occupied site 1 population computed using various approximate methods. PLDM propagation gives results in excellent agreement with results from the HCRME [10, 47] and ILDM approaches [16] that, in principle, are numerically exact. The PLDM approach accurately captures both the short-time coherent dynamics and, as demonstrated in Fig. 8.1c,d, the method is capable of generating stable converged results at very long times, as the system and bath exchange energy and approach thermal equilibrium. The Boltzmann equilibrium populations of the site basis states are also displayed in the figure as arrows at long times. We see that by 10 ps the state populations are accurately approaching their equilibrium values. The other approximate implementations of the mapping Hamiltonian description of this system, such as those provided, for example, by the LSC-IVR [40] and PBME [43] approaches, on the other hand, fail to give the correct thermal equilibrium limit at long times (e.g. LSC-IVR [40] gives negative populations). Similar equilibration problems have been reported with the stochastic Liouville equation approach [31], the original Ehrenfest dynamics scheme [22, 23] and the Haken–Strobl–Reineker approximation [32] for such complex quantum dissipative systems.

While the long-time thermal equilibrium state achieved with these different initial conditions should be independent of the sampled path to equilibrium if our dynamics algorithm is reliable, as we indeed find with these studies, the actual nonequilibrium dynamics explored from the different initial conditions is observed to be profoundly different and provides a significant way to influence the overall energy transfer rate as reported, for example, by the rate of increase in population of the BChl3 state, which is the sink to the RC. With the seven-state Hamiltonian model we see that initializing in the BChl1 state results in relatively slower initial energy transfer to the BChl3 state compared with the situation when state BChl6 is initially excited where we see the rate almost double by comparison. The dynamics ensuing after BChl1 initial excitation is characterized by a single strongly coupled dimer transferring excitation to state BChl3. According to the seven-state model Hamiltonian in which $\epsilon_1 = 12410$ cm^{-1}, $\epsilon_2 = 12530$ cm^{-1}, $\epsilon_3 = 12210$ cm^{-1}, $\Delta_{12} = -87.7$ cm^{-1}, $\Delta_{23} = 30.8$ cm^{-1} [10], the coupling from the initially excited dimer to the sink state, BChl3, occurs primarily through state BChl2, which, according to the parameters listed above, is higher in energy than state BChl1 in this model so the dynamics must take the excitation up over a barrier before depositing it into the BChl3 sink state. Since the BChl2 state is higher in energy its population remains relatively smaller throughout the dynamics (see Fig. 8.1c) and if we assume a simple first-order kinetic model for the relaxation so that the rate at which BChl3 is populated is proportional to population of the state that directly feeds it, that is, $d[\text{BChl3}]/dt \sim k_{23}\,[\text{BChl2}]$ we expect smaller energy transfer rates when the feeder state population is smaller, and here the feeder rate constant k_{23} is proportional to the square of the magnitude of the electronic coupling, $\Delta_{23} = 30.8\ \text{cm}^{-1}$ [10], between the feeder and sink states according to Fermi golden rule–like arguments.

The dynamics resulting from initially exciting BChl6 displayed in Fig. 8.1d is significantly different and involves energy transfer between two coherently coupled dimers: from the BChl6–BChl5 dimer to the BChl4–BChl7 dimer, and then subsequent transfer from the BChl4 feeder state to BChl3 sink state. The electronic coupling between the feeder state, BChl4, and the BChl3 sink state for this pathway is significantly larger $\Delta_{43} = -53.5\ \text{cm}^{-1}$, and comparing the square magnitudes gives a larger rate constant by almost a factor of 3 over that for the 2–3 feed of the pathway explored with BChl1 initial excitation. Since the populations of the primary feed states (BChl2 or BChl4 in the BChl1 and BChl6 initialized pathways, respectively) reach similar values ($\sim$0.2) we see that the initial rates of energy transfer to BChl3 along these pathways differ by about a factor of 2–3 as expected from the differences in Fermi golden rule rate constants and this simple kinetic analysis. The fact that the BChl1 state initialized dynamics also produces a small population in the more strongly coupled feeder state BChl4 enhances the rate of energy transfer to BChl3 along this pathway making the difference in rates between the two pathways slightly less than the factor of three predicted by this simple model.

The quantum coherent beating of the populations of the strongly coupled states involved in the various dimers in the system are clearly observed in the dynamics computed with the different photo excited initial conditions explored here. According to the simple kinetic model that we have used above to understand the observed differences in the energy transfer rates for the different model pathways, however, this higher-frequency beating should have little influence on the slower energy transfer dynamics that essentially sees only the average effect of the feeder state populations rather than any influence of the detailed transient coherent oscillations.

8.5 The Eight-Site FMO Complex: A Model for in vivo Initial Excitation

The studies of the seven-BChl model of the FMO complex outlined in the previous section suggest that there exists two nearly independent energy transfer pathways [10]. With this model, sites BChl1 and BChl6 are assumed to be possible locations for accepting the excitation from the chlorosome. From there, the energy is subsequently funneled either from site BChl1 to BChl2 (pathway 1) or from site BChl6 to sites BChl5, BChl7, and BChl4 (pathway 2). The terminal state for either route is site BChl3, from which the excitation is subsequently transferred to the RC (see Fig. 8.1a). The couplings within each of the pathways are much larger than the couplings between the two, which implies that the two routes are nearly independent. However, a more careful crystallographic analysis of the FMO complex and its connection to the chlorosome demonstrates that the individual monomer subunits each contain eight BChls [26]. The eighth BChl resides on the surface of the protein complex, and it has been suggested that this additional chromophore is often lost during sample preparation for the spectroscopic experiments [2, 3]. Additionally, Renger and coworkers argued [26] that the eighth BChl provides the most efficient pathway for exciton injection into the FMO complex as a result of its position and orientation with respect to the chlorosome. If this is correct and site BChl8 serves as the primary acceptor of excitation energy from the chlorosome, then a preferential energy transfer route emerges through pathway BChl8–BChl1–BChl2–BChl3, instead of the alternative pathway 2, due to the fact that site BChl8 is only moderately strongly coupled to site BChl1 ($\Delta_{81} = 38\ \text{cm}^{-1}$ [18]) and there is no significant coupling to BChl6.

In Fig. 8.2 we show PLDM propagation results for excitation energy transfer dynamics with the eight-state model for FMO [18], using various initial conditions, including (a) starting from site BChl8, (b) starting from site BChl1, and (c) starting from site BChl6, all with $T = 77$ K. In addition to including the extra BChl8 state, the eight-state FMO model has another important feature that distinguishes it from the published seven-state model that was

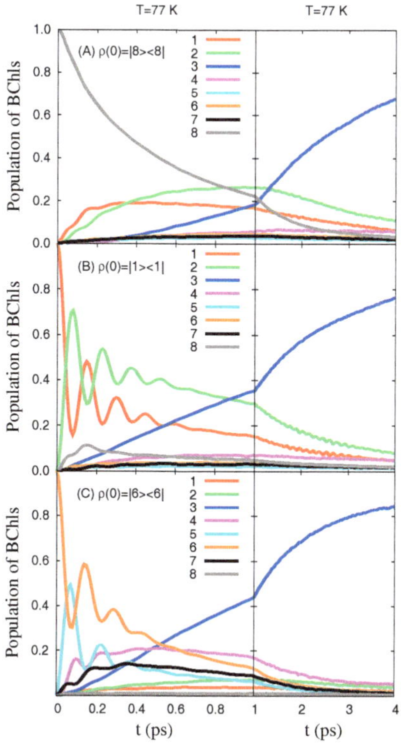

Figure 8.2 Population dynamics for the eight-chromophore model of FMO at $T = 77$ K with chromophores at different sites, initial excitation. (A) Site 8, (B) site 1, and (C) site 6. Site energy, electronic coupling, and the parameters for spectral density are adopted from Refs. [18, 26, 27]

used in the studies of the previous section. Unlike the seven-state model in which state BChl2 is higher in energy than state BChl1 and forms a barrier to energy transfer, in the eight-state model the chromophores are organized energetically in a strictly downhill fashion with the adjacent coupled chromophores stacked so that their energies are in the order BChl8 > BChl1 > BChl2 > BChl3.

In the top panel, Fig. 8.2a, we see that the initial excitation of BChl8 undergoes an exponential decay that slowly feeds the population to state BChl1, which passes it on to BChl2 with little evidence of any coherent population oscillation between the states of the BChl1–BChl2 dimer. The transient population of state BChl2 levels off at ~0.2 on the 1ps timescale and this results in essentially the same slower energy transfer dynamics to state BChl3 as we observed with the seven-state model in the previous section. Initially exciting state BChl1, as displayed in the middle panel of Fig. 8.2b, on the other hand, results in a transient that shows strong coherent oscillation between the component chromophores of the BChl1–BChl2 dimer. Now, with the "downhill" eight-state FMO model, because state BChl2 is lower in energy it can build up a higher initial transient population ~0.4, which, as expected from our simple kinetic model, results in an energy transfer rate to BChl3 that is about twice that of the situation we find with BChl8 initial excitation described above. A similarly rapid rate is observed with the BChl6 initial excitation in Fig. 8.2c, and results for essentially the same reason as noted for the seven-state model, that is, stronger coupling between the BChl4 feed state and the BChl3 sink state. These off-diagonal parts of the model Hamiltonian are slightly different between the seven- and eight-state models [18].

The results presented here for the more realistic in vivo initial conditions of exciton transport in FMO involving BChl8 initial excitation, as suggested in Ref. [26], indicate that coherence effects should be strongly suppressed compared to the artificial situation explored in the experiments that initially excite BChl1. Thus the observed population beating in the experiments might be strongly dependent on this particular, and perhaps artificial, initial excitation.

A final point about the environmental initial conditions and their influence on the calculated coherent population beating is also worth mentioning [20]. As the initial condition of the nuclear environmental coordinates are sampled according to the thermal distribution of the harmonic bath that is completely uncoupled from the system the magnitude of the population oscillations can be enhanced by this unphysical initial bath preparation. However, if one shifts the initial nuclear coordinates to approximately represent the effects of the system–bath coupling on the initial bath distribution thus approximating the true equilibrium initial condition for both system and bath, which may be a more reasonable mimic to the initial conditions in natural process, the magnitude of the population oscillations will be suppressed [20].

8.6 Strength of System–Environment Coupling Optimizes Energy Transfer Dynamic Turnover Behavior

It has been suggested that the maximum excitation energy transfer rate is achieved by a delicate balance between coherence and dissipation [10]. In this section, however, we will demonstrate that the optimized transfer rate is achieved due to a Kramers-like dynamical turnover effect [30] and not influenced significantly by the presence of coherent beating. The system that is ideally suited to show this is the eight-state FMO model outlined in the previous section. By initially exciting either BChl1 or BChl8 we can switch between coherent population beating accompanying the energy transfer dynamics and simple incoherent exponential growth or decay like energy transfer as we have already demonstrated in Section 8.5. By scanning the system–bath coupling strength, that is, the effective dissipative frictional interaction between the bath and the system, we can explore the variation of energy transfer rate from the under damped to over damped regime when quantum coherent beating is present as with BChl1 initial excitation and similarly when it is not present, for example, with BChl8 initial excitation.

The results are shown in Fig. 8.3 with the initial state population (BChl1 left panel, and BChl8 right panel) shown as solid curves, and the population sink state BClh3 displayed using dashed curves. The different colors correspond to varying the solvent reorganization energy $\lambda(\alpha)$ that controls the system–bath coupling strength of chromophore α from 0.2 to 5.0 times it's original value, λ_0, while leaving the environmental coupling of all other chromophores at λ_0. The temperature for this study is $T = 300$ K. We see that as the local chromophore–bath coupling is initially increased, the rate of Bchl3 population growth increases for 0.2, 0.5, and optimizes for $1.0\lambda_0$. Further increasing the feeder state environmental coupling beyond this point (2.0 and 5.0 λ_0) causes the energy transfer rate to state BChl3 to decrease below its optimal value. We see that the same type of turnover behavior is observed independent of whether the pathway shows coherent population oscillations, as with BChl1 initialization, or an incoherent pathway is used as with

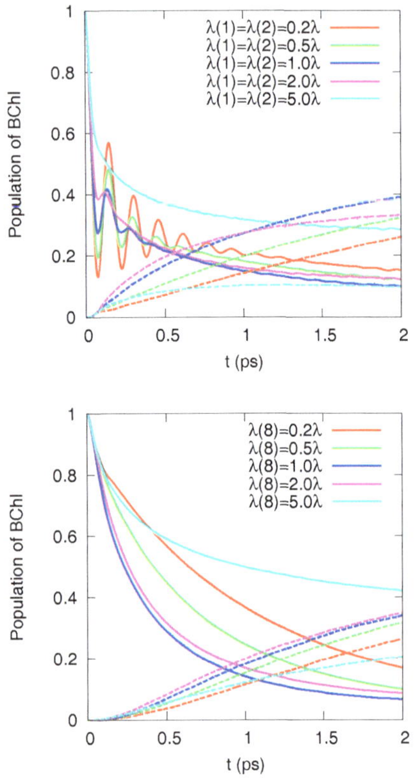

Figure 8.3 Stochastic resonance behavior of exciton transfer in eight-chromophore FMO model. $T = 300$ K, and the parameters for the Hamiltonian are taken as from Ref. [18]. Left: Initial excitation from site 1, the solvent reorganization energy for states 1 and 2 are scanned from 0.2 to 5 times of original λ, and for other sites the solvent reorganization energy is kept the same as the original value λ_0. Right: Initial excitation from site 8. Only site 8's solvent reorganization energy are scanned.

BChl8 initialization. This observation is contrary to the argument that it is a balance between coherence and friction that controls the maximum energy transfer rate [10].

The reason for this turnover is similar to the behavior underlying the Kramers turnover [30] in energy barrier crossing phenomena [17]. This type of optimization behavior with increasing system–bath coupling strength (also known as environment or noise assisted quantum transport) has also been observed in simple models of light-harvesting networks treated with approximate theories [11, 13, 18, 29, 56]. This behavior can be understood in terms of the ideas underlying stochastic resonance processes: For a pair of states whose static excitation energies are separated by $\epsilon_{\alpha,\beta}$ with small electronic coupling $\Delta_{\alpha,\beta}$ say, if the solvent reorganization energy is weak, the fluctuations in the site energies driven by the bath dynamics will be small compared to the electronic excitation energy gap and there will be little population transfer. As the system–bath coupling is increased and the fluctuations in site energies become on the order of the static excitation energy gap there will be many times when the fluctuations drive the site energies into resonance and the finite electronic coupling will lead to appreciable population transfer and maximization of the energy transfer rate between the sites for some value of system–bath coupling. At stronger system–bath coupling still, the larger fluctuations in site energies drive the system out of resonance more often and we see a decrease in population transfer in the network. The fact that the results presented in Fig. 8.3 reveal precisely these signatures of stochastic resonance behavior in this realistic, non-Markovian model with an accurate treatment of the quantum dynamics is suggestive that the simple ideas underlying the phenomenon of stochastic resonance are robust in determining the optimization behavior of these nanostructured quantum dynamical light-harvesting network systems and that the quantum coherent beating does little to optimize the energy transfer rates in these systems.

8.7 Excitation Dynamics in Phycocyanin 645

Another interesting system for which detailed nonlinear multidimensional spectroscopy studies have revealed long-lived quantum coherent beating in the excitation transfer dynamics even at room temperature is phycocyanin PC645 system from the algae Chroomonas CCMP270 [1]. This system involves bilin chromophores that are bound to the local protein environment, rather than chlorophyll molecules that intercalate into the protein as with FMO, for example. With this PC645 system, the bilin chromophores are bound to the protein scaffolding so the spectral density that controls the dissipation has very different characteristics to the chlorophyll systems and the bilins also have typical inter chromophore distances that are more than twice those found in chlorophyll networks. The geometry of these bilin pigment–protein complexes is displayed in Fig. 8.4a. The corresponding energy-level connectivity diagram is presented in Fig. 8.4b.

In Fig. 8.4c we present the long-time excitation energy transfer dynamics in PC645 calculated using PLDM propagation. Here we initially excite one of the high energy, strongly electronically coupled bilin molecules that forms the DBV dimer core in the center of the complex, and the excitation follows a downhill transfer process, which builds up population in intermediate MBV states before it finally reaches to the peripheral PCB states at the edge of the complex that act as sinks to RCs. The coherent beating in this system lasts around 500 fs at elevated temperature, and subsequently the site populations show transient relaxation as the energy moves downhill to the PCB bilins, accurately reproducing the general characteristics observed in the experimental studies. At long times (around 15 ps), the populations reach their thermal distribution. Stochastic resonance behavior has also been investigated in detail for this system [17], and scans of λ values of all sites reveals that the maximum energy transfer rate for different sites occur at different values of local bath λ parameters due to the particular position and role of each chromophore in the PC645 network. Similar site-dependent stochastic resonance optimization behavior has also been found in FMO [13].

To directly explore the effect of the initially prepared coherent superposition of DBV dimer states on the rate of energy transfer to the peripheral chromophore sink states, we ran exploratory calculations with a Hamiltonian in which the DBV dimer states were replaced by a single state with excitation energy that was the mean of the DBV site energies and which had the same coupling strengths to the rest of the chromophore network and to the environment. In Fig. 8.4d we compare the populations of the peripheral sink chromophores as functions of time computed in the presence of the fully coherently coupled DBV dimer states (solid curves), with the

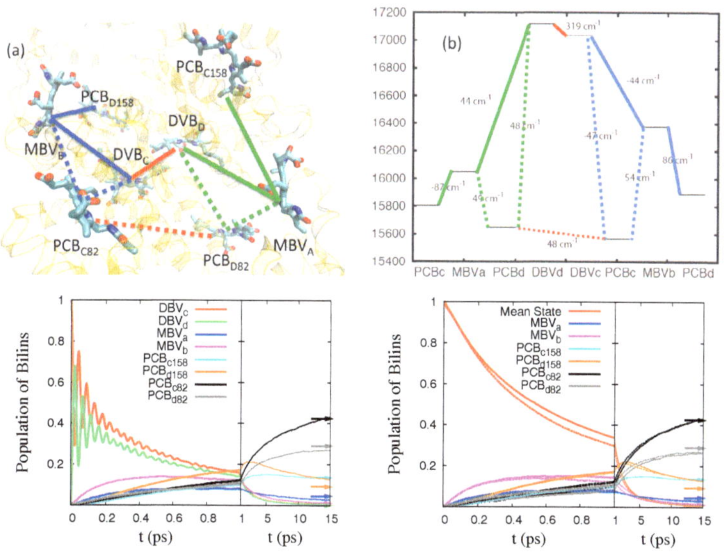

Figure 8.4 (a) Structure of phycocyanin PC645, showing predominately α-helical protein scaffolding with embedded bilin chromophores. Lines connect chromophores that participate in various coherences observed in the dissipative quantum dynamics of the system. (b) Energy-level "connectivity diagram" for PC645. Horizontal line segments give site energies in cm^{-1} of chromophore sites listed along the x axis. Lines connecting the ends of these site energy levels are labeled with the major off-diagonal electronic couplings between sites. The color coding is consistent with structural connections in (a). (c) Long-time PLDM calculation for PC645 up to 15 ps. Thermal populations computed from independent equilibrium calculations are indicated as colored arrows and correspond well to long-time limit populations. (d) Comparison of populations of peripheral acceptor chromophores as functions of time when the initially excited, coherently coupled DBV dimer states are replaced with a single effective mean state, removing initial coherence. Temperature T = 294 K. Solid curves give site populations observed with coherently coupled DBV dimer initial states; dashed curves show site populations obtained when the initial mean DBV state is used.

populations obtained when the DBV dimer is replaced by a single "mean state" where there can be no initially excited coherence. From the figure it is clear that there is no difference between the original model and the mean-state model after 5 ps, where the final equilibrium is established at around 15 ps. Thus the influence of the initial coherence on the excitation transfer dynamics is only limited to short times and the long-time excitation transfer dynamics is essentially unaffected by the presence of the coherent population beating in this system.

8.8 Correlations between Site Energy Fluctuations

The 2D photon echo experiments of Lee et al. [4] show a long-lived beating signal between bacteriochlorophyll (BChl or B) and bacteriopheophytin (BPhy, or H) states that is associated with $|H\rangle\langle B|$ coherent dynamics in the RC of *Rhodobacter sphaeroides*. These workers found that they could fit this spectral dynamics reasonably by assuming the fluctuations in the B and H site energies were strongly correlated, and suggested that in-phase energy fluctuations could be responsible for preserving coherence [4]. Many recent publications construct site energy correlation models and explore the role of such correlation in preserving the coherence and its influence on excitation energy transfer dynamics [29, 95–100]. Further, Chen and Silbey realized that correlations could also exist between site energy and interchromophore transition dipole coupling that can enrich the behavior of coherent features in the dynamics and influence transfer rates in interesting ways in these systems [19, 32, 101, 102]. In this final results section, we construct models that include these different types of correlation and explore their influence on both the magnitude and life time of coherent population beating in these systems and the effects on excitation energy transfer rate. Our results reported here use the PLDM propagation approach and thus go beyond many of the previous studies that use much more approximate propagation methods. This should enable us to easily discern the differences between real effects of correlated fluctuations and the influence of approximations in treating the dissipative quantum dynamics.

To describe the site energy correlations that have not been captured in the independent bath model in Section 8.3, we use the common bath model [57, 73] outlined below, which assumes that chromophores H and B have their own harmonic baths, and we also introduce a set of modes that are "common" to both chromophores so the common bath model Hamiltonian has the form:

$$\hat{H}_b^{com} = \sum_{\alpha=H,B} \sum_{l=1}^{n^{(\alpha)}} \frac{1}{2} \left[p_l^{(\alpha)2} + \omega_l^{(\alpha)2} q_l^{(\alpha)2} \right] + \sum_{m=1}^{n_{com}^{(HB)}} \frac{1}{2} \left[P_m^{(HB)2} + \Omega_m^{(H,B)2} Q_m^{(HB)2} \right]$$

$$\hat{H}_{s-b}^{com} = \left[\sum_{l=1}^{n^{(H)}} \tilde{c}_l^{(H)} q_l^{(H)} + \sum_{m=1}^{n_{com}^{(HB)}} \tilde{C}_m^{(H)} Q_m^{(HB)} \right] |H\rangle\langle H| + \left[\sum_{l=1}^{n^{(B)}} \tilde{c}_l^{(B)} q_l^{(B)} + \sum_{m=1}^{n_{com}^{(HB)}} \tilde{C}_m^{(B)} Q_m^{(HB)} \right] |B\rangle\langle B| \tag{8.26}$$

In the common bath model $q_l^{(H)}$, $q_l^{(B)}$ represent the lth independent bath oscillator that couple to the $|H\rangle$ or $|B\rangle$ states respectively, with coupling strengths $\tilde{c}_l^{(H)}$ and $\tilde{c}_l^{(B)}$. The mth common bath oscillator mode $Q_m^{(HB)}$ couples to both the $|H\rangle$ and $|B\rangle$ states simultaneously. With this model, these common bath modes give rise to correlated energy-level fluctuations when the common bath coupling constants $\tilde{C}_m^{(H)}$ and $\tilde{C}_m^{(B)}$ are not zero.

The approach employed in fitting the experimental data [4] assumes a model for the correlated fluctuations in site excitation energies of the B and H chromophores based on the strength of correlation, a, specified by the following relationship between the different spectral densities $j^{(HB)}(\omega) = |a|\sqrt{j_{\text{tot}}^{(H)} j_{\text{tot}}^{(B)}}(\omega)$ where the spectral densities, for example, $j_{\text{tot}}^{(H)}(\omega)$ and $j^{(HB)}(\omega)$ determine the coupling constants in Eq. 8.26 according to analogous results in Eq. 8.20. The value of the correlation strength defined in this way obtained from the fit to experimental data [4] is $a = 0.9$.

Next, we introduce another situation in which the intersite off-diagonal coupling bath and the bath that causes fluctuations in site energies are correlated. For simplicity, we assume that all the modes that couple to $|H\rangle\langle H|$ and $|B\rangle\langle B|$ also couple to $|H\rangle\langle B|$ and $|B\rangle\langle H|$. However, the modes coupled to $|H\rangle\langle H|$ and the modes coupled to

$|B\rangle\langle B|$ are independent. With the above assumption, the bilinear system–bath and harmonic bath Hamiltonians can be written as:

$$\hat{H}_{s-b} = \sum_{l=1}^{n^{(H)}} c^{(H)}{}_l \hat{q}_l^{(H)} |H\rangle\langle H| + \sum_{l=1}^{n^{(B)}} c^{(B)}{}_l \hat{q}_l^{(B)} |B\rangle\langle B|$$
$$+ \left\{ k^{(H)} \sum_{l=1}^{n^{(H)}} c_l^{(H)} \hat{q}_l^{(H)} + k^{(B)} \sum_{l=1}^{n^{(B)}} c_l^{(B)} \hat{q}_l^{(B)} \right\}$$
$$\times [|H\rangle\langle B| + |B\rangle\langle H|] \tag{8.27}$$

and

$$\hat{H}_b = \hat{H}_b^H + \hat{H}_b^B$$
$$= \sum_{l=1}^{n^{(H)}} \frac{1}{2} \left[\hat{p}_l^{(H)2} + \omega_l^{(H)2} \hat{q}_l^{(H)2} \right] \tag{8.28}$$
$$+ \sum_{l=1}^{n^{(B)}} \frac{1}{2} \left[\hat{p}_l^{(B)2} + \omega_l^{(B)2} \hat{q}_l^{(B)2} \right]$$

respectively, and $\left[\hat{H}_b^H, \hat{H}_b^B\right] = 0$, due to the independence of different chromophore baths. With this model, $k^{(B)}$ and $k^{(H)}$ simply modulate the strength with which the independent $\{\hat{q}_l^{(B)}\}$ and $\{\hat{q}_l^{(H)}\}$ bath degrees of freedom influence the intersite off-diagonal electronic coupling. The assumed form of the bath coordinate–dependent electronic coupling with this model is thus: $\Delta_{HB}(\{\hat{q}_l^{(H)}\}, \{\hat{q}_l^{(B)}\}) = \Delta_{HB}^{(0)} + \{k^{(H)} \sum_{l=1}^{n^{(H)}} c_l^{(H)} \hat{q}_l^{(H)} + k^{(B)} \sum_{l=1}^{n^{(B)}} c_l^{(B)} \hat{q}_l^{(B)}\}$. The physical idea underlying this form is that $\Delta_{HB}^{(0)}$ arises from the interaction between transition dipoles in the reference geometry, as described in Eq. 8.12. The second term in the above expression represents, for example, the process by which modes in the bath of chromophore B, that is, $\{\hat{q}_l^{(B)}\}$, modify the transition dipole moment, μ_B, and effect the electronic coupling with chromophore H as the bath causes fluctuations about the reference geometry.

In Fig. 8.5a, we demonstrate that the relative sign of the $k^{(H)}$ and $k^{(B)}$ can either enhance or suppress the excitation energy transfer rate, and similarly influence the magnitude and relaxation time scale of the oscillatory behavior of populations (that result from oscillations in coherence density matrix elements). In the left panel of Fig. 8.5, for example, we see that when $k^{(H)} = -0.5$ and $k^{(B)} = 0.5$

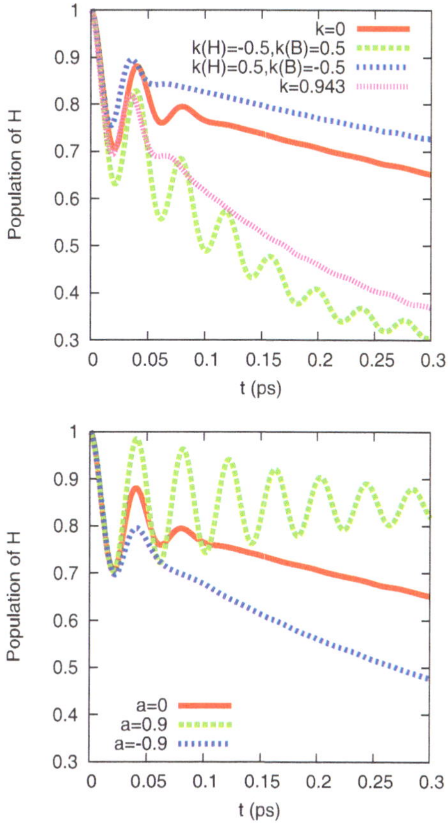

Figure 8.5 Population of donor site H for model Hamiltonian, including the site energy–off-diagonal coupling correlations (left panel, obtained using model Hamiltonian in Eq. 8.27), and site energy–site energy correlations (right panel, obtained using model Hamiltonian in Eq. 8.26). Various curves show results of full quantum dynamics calculations with different signs of the fluctuation terms in the different model Hamiltonians. The magnitude of the population oscillations can be either enhanced (left: for $k^{(H)} = -0.5$ and $k^{(B)} = 0.5$; or right: $a > 0$) or suppressed (if these quantities have opposite signs). The same trends in oscillatory behavior are observed for the coherence density matrix elements ρ_{HB} (not shown). The left panel also includes results for the independent bath model that incorporates additional independent modes that are bilinearly coupled to off-diagonal electronic Hamiltonian matrix elements that describe excitation energy transfer between sites (magenta curve, see text).

the population of the initially excited H chromophore decays rapidly compared with the $k = 0$ case (i.e., the situation of independent baths coupled only to the different sites). At the same time, the oscillations of the population are also of larger amplitude compared to the $k = 0$ situation suggesting stronger long-lived coherent dynamics in this case. These observations are consistent with the expectations from the Markovian model theory provided by Chen and Silbey [32]. On the other hand, when ($k^{(H)} = 0.5$, $k^{(B)} = -0.5$) the excitation energy transfer rate decreases and the coherent oscillatory features damp out very quickly compared to the case of independent site energy baths (i.e., the $k = 0$ case).

As another point of comparison, in the left panel of Fig. 8.5, we also explore the situation involving a separate independent off-diagonal coupling bath for which the Hamiltonian takes the form:

$$\hat{H}_{s-b}^{ind(J)} = \sum_{l=1}^{n^{(H)}} c_l^{(H)} q_l^{(H)} |H\rangle\langle H| + \sum_{l=1}^{n^{(B)}} c_l^{(B)} q_l^{(B)} |B\rangle\langle B| + \sum_{l=1}^{n^{(J)}} kc^{(J)}{}_l q_l^{(J)} \left[|H\rangle\langle B| + |B\rangle\langle H|\right]$$

where $q_l^{(J)}$ represents another set of independent harmonic bath modes that are bilinearly coupled to the off-diagonal electronic coupling elements. By setting the off-diagonal correlation function in this independent bath model equal to that of the correlated bath model (details see Ref. [19]), we obtain the relationship between the parameter k and k_H and k_B.

As one can see, the independent off-diagonal bath model gives excitation energy transfer dynamics that results in site populations that essentially decay at the same rate as observed for the correlated bath case when $k^{(H)} = -0.5$ and $k^{(B)} = 0.5$. No coherent population oscillations, however, are observed in the absence of correlation. This comparison suggests that the optimal rate of energy transfer for the fully correlated off-diagonal bath model Hamiltonian of Eq. 8.27 can be controlled by presence of solvent driven fluctuations in intersite coupling and that, while correlation in site energy fluctuations and off-diagonal coupling are not critical in determining the magnitude of the overall relaxation rate, such correlations can lead to long-lived oscillations in population that are the signature of coherent quantum dynamics.

The right panel of Fig. 8.5 presents results obtained using the generalized correlated site energy fluctuation model presented in Eq. 8.26. From these results we see that increasing the positive correlation can enhance the magnitude of the population oscillations signaling long-lived coherent dynamics, while decreasing the rate of energy transfer. Increasing the negative correlation, on the other hand, gives the opposite trend, that is, an increased rate of incoherent energy transfer.

The important observation from this demonstration is that depending on the nature of correlations in the model Hamiltonian the presence of coherent beating features does not necessarily signal enhanced or diminished rate of energy transfer. Comparing results of the left and right panels in Fig. 8.5, we see that the population beating can be enhanced by appropriate choice of model parameters and the energy transfer can either be suppressed (right) or enhanced (left). Thus we draw the conclusion that coherence can be viewed as a consequence of the structure of the Hamiltonian and the initial conditions of the dynamics. The results presented here demonstrate that, typically, Hamiltonian terms that correlate site energy fluctuations will enhance coherent beating but result in slower energy transfer. However, when the model Hamiltonian incorporates correlation between environmental modes that influence fluctuations in diagonal and off-diagonal terms a variety of different scenarios are found including the possibility of simultaneously enhancing coherent effects and energy transfer.

8.9 Concluding Remarks

In this chapter we outline a new methodology, PLDM propagation [62] that can accurately describe the short-time coherent dynamics and long-time thermal equilibration in a consistent framework. With this powerful methodology for treating dissipative quantum dynamics, we show that models of photosynthetic light-harvesting pigment–protein chromophore networks, with realistic parameterizations can show coherent dynamics that depend on the strength of the electronic and environmental couplings and on the initial conditions of excitation. Moreover, by scanning the strength of the system–bath interaction the excitation energy transfer rate can be optimized as a result of the very general phenomenon of stochastic resonance, or environmental or noise-assisted quantum transport. Finally we have shown that various types of correlation between the modes that affect fluctuations of different chromophore excitation energies and environment-dependent electronic couplings can be useful for tuning the lifetime of coherence and adjusting the rate of energy transfer in these systems. Understanding how material properties might be engineered to take advantage of these findings could be extremely important for new nanotechnological applications in energy and quantum information science.

Acknowledgements

We gratefully acknowledge support for this research from the National Science Foundation under grants CHE-0911635 and CHE-1301157, as well as support from Science Foundation Ireland under grant no. 10/IN.1/ I3033. DFC acknowledges the support of his Stokes Professorship in Nanobiophysics from Science Foundation Ireland. We also acknowledge a grant of supercomputer time from Boston University's Office of Information Technology and Scientific Computing and Visualization.

References

1. E. Collini, C.Y. Wong, K.E. Wilk, P.M.G. Curmi, P. Brumer, and G.D. Scholes, *Nature*, **463**, 644 (2010).
2. G. Panitchayangkoona, D. Hayes, K.A. Fransted, J.R. Caram, E. Harel, J. Wen, R.E. Blankenship, and G.S. Engel, *Proc. Natl. Acad. Sci.*, **107**, 12766 (2010).
3. G.S. Engel, T.R. Calhoun, E.L. Read, T.-K. Ahn, T. Mancal, Y.-C. Cheng, R.E. Blankenship, and G.R. Fleming, *Nature*, **446**, 782 (2007).
4. H. Lee, Y.-C. Cheng, and G.R. Fleming, *Science*, **316**, 1462 (2007).

5. G.S. Schlau-Cohen, A. Ishizaki, T.R. Calhoun, N.S. Ginsberg, M. Ballottari, R. Bassi, and G.R. Fleming, *Nat. Chem.*, **4**, 389 (2012).
6. C.Y. Wong, R.M. Alvey, D.B. Turner, K.E. Wilk, D.A. Bryant, P.M.G. Curmi, R.J. Silbey and G.D. Scholes, *Nat. Chem.*, **4**, 396 (2012).
7. G. Raszewski, W. Saenger, and T. Renger, *J. Am. Chem. Soc.*, **130**, 4431 (2008).
8. G.D. Scholes, G.R. Fleming, A. Olaya-Castro, and R. van Grondelle, *Nat. Chem.*, **3**, 763 (2012).
9. G.D. Scholes, *J. Phys. Chem. Lett.* **1**, 2 (2010).
10. A. Ishizaki and G.R. Fleming, *Proc. Natl. Acad. Sci.*, **106**, 17255 (2009).
11. P. Rebentrost, M. Mohseni, and A. Aspuru-Guzik, *J. Phys. Chem. B*, **113**,9942 (2009); P. Rebentrost, M. Mohseni, I. Kassal, S. Lloyd, and A. Aspuru-Guzik, *New. J. Phys.*, **11**, 033003 (2009).
12. P. Rebentrost and A. Aspuru-Guzik, *J. Chem. Phys.*, **134**, 101103 (2011).
13. F. Caruso, A.W. Chin, A. Datta, S.F. Huelga, and M.B. Plenio, *J. Chem. Phys* **131**, 105106 (2009); M.B. Plenio and S.F. Huelga, *New J. Phys.*, **10**, 113019 (2008); F. Caruso, A.W. Chin, A. Datta, S.F. Huelga, and M.B. Plenio, *Phys. Rev. A.*, **81**, 062346 (2010); A.W. Chin, A. Datta, F. Caruso, S.F. Huelga, and M.B. Plenio, *New J. Phys* **12**, 065002 (2010).
14. M. Sarovar, Y-C. Cheng, and K.B. Whaley, *Phys. Rev. E*, **83**, 011906 (2011).
15. L.A. Pachon and P. Brume, *J. Phys. Chem. Lett.*, **2**, 2728 (2011).
16. P. Huo and D.F. Coker, *J. Chem. Phys.*, **133**, 184108, (2010).
17. P. Huo and D.F. Coker, *J. Phys. Chem. Lett.*, **2**, 825 (2011).
18. J. Moix, J. Wu, P. Huo, D.F. Coker, and J. Cao, *J. Phys. Chem. Lett.*, **2**, 3045 (2011).
19. P. Huo and D.F. Coker, *J. Chem. Phys.*, **136**, 115102, (2012).
20. T.C. Berkelbach, T.E. Markland, and D.R. Reichman, *J. Chem. Phys.*, **136**, 084104 (2012).
21. L. Mühlbacher and U. Lleinekathöfer, *J. Phys. Chem. B*, **116**, 3900 (2012).
22. A. Ishizaki and G.R. Fleming, *J. Phys. Chem. B*, **115**, 6227 (2011).
23. M. Aghtar, J. Liebers, J. Strmpfer, K. Schulten, and U. Kleinekathöfer, *J. Chem. Phys.*, **136**, 214101 (2012).
24. H.W. Kim, A. Kelly, J.W. Park, and Y.M. Rhee, *J. Am. Chem. Soc.*, **134**, 11640 (2012).
25. M. Kobus, P.H. Nguyen, G. Stock, *J. Chem. Phys.*, **134**, 124518 (2011).

26. M.S. am Busch, F. Müh, M. El-Amine Madjet, and T. Renger, *J. Phys. Chem. Lett.*, **2**, 93 (2011).
27. G. Ritschel, J. Roden, W.T. Strunz, A. Aspuru-Guzik, and A. Eisfeld, *J. Phys. Chem. Lett.*, **2**, 2912 (2011).
28. J. Adolphs and T. Renger, *Biophys. J.*, **91**, 2778 (2006).
29. J. Wu, F. Liu, Y. Shen, J. Cao, and R.J. Silbey, *New J. Phys.*, **12**, 105012 (2010).
30. A. Nitzan, *Chemical Dynamics in Condensed Phases* (Oxford University Press, New York, 2006).
31. X. Chen and R.J. Silbey, *J. Phys. Chem. B*, **115**, 5499 (2011).
32. X. Chen and R.J. Silbey, *J. Chem. Phys.*, **132**, 204503 (2010).
33. S. Jang, M.D. Newton, and R.J. Silbey, *Phys. Rev. Lett.*, **92**, 218301 (2004).
34. S. Jang, M.D. Netwton, and R.J. Silbey, *J. Phys. Chem. B*, **111**, 6807 (2007).
35. A. Ishizaki and G.R. Fleming, *J. Chem. Phys.*, **130**, 234110 (2009).
36. A. Ishizaki and G.R. Fleming, *J. Chem. Phys.*, **130**, 234111 (2009).
37. D.J. Heijs, V.A. Malyshev, and J. Knoester, *Phys. Rev. Lett.*, **95**, 177402 (2005).
38. M. Schroder, U. Kleinekathofer, and M. Schreiber, *J. Chem. Phys.*, **124**, 084903 (2006).
39. T.C. Berkelbach, T.E. Markland, and D.R. Reichman, *J. Chem. Phys.*, **136**, 034113 (2012).
40. G. Tao and W.H. Miller, *J. Phys. Chem. Lett.*, **1**, 891 (2010).
41. W.H. Miller, *J. Chem. Phys.*, **136**, 210901 (2012).
42. H. Kim, A. Nassimi, and R. Kapral, *J. Chem. Phys.*, **129**, 084102 (2008); A. Nassimi, S. Bonella, and R. Kapral, *J. Chem. Phys.*, **133**, 134115 (2010); A. Kelly, R. van Zon, J. Schofield, and R. Kapral, *J. Chem. Phys.*, **136**, 084101 (2012).
43. A. Kelly and Y.M. Rhee, *J. Phys. Chem. Lett.*, **2**, 808 (2011); W.-G. Lee, A. Kelly, and Y.M. Rhee, *Bull. Korean Chem. Soc.*, **33**, 933 (2012).
44. A. Ishizaki and Y. Tanimura, *J. Phys. Soc. Jpn.* **74**, 3131–3135 (2005).
45. J. Strümpfer and K. Schulten, *J. Chem. Theory Comput.*, **8**, 2808 (2012).
46. Q. Shi, L.P. Chen, G.J. Nan, R.X. Xu, and Y.J. Yan, *J. Chem. Phys.*, **130**, 084105 (2009).
47. J. Zhu, Sabre Kais, P. Rebentrost, and A. Aspuru-Guzik, *J. Phys. Chem. B*, **115**, 1531 (2011); J. Zhu, S. Kais, A. Aspuru-Guzik, S. Rodriques, B. Brock, and P.J. Love, *J. Chem. Phys.*, **137**, 074112 (2012).
48. N. Makri and D.E. Makarov, *J. Chem. Phys.*, **102**, 4600 (1995); N. Makri and D.E. Makarov, *J. Chem. Phys.*, **102**, 4611 (1995).

49. P. Nalbach, D. Braun, and M. Thorwart *Phys. Rev. E*, **4**, 041926 (2011).
50. P. Nalbach, A. Ishizaki, G.R. Fleming, and M. Thorwart, *New. J. Phys.*, **13**, 063040 (2011).
51. H. Tamura, I. Burghardt and M. Tsukada, *J. Phys. Chem. C*, **115**, 10205 (2011).
52. H. Wang and M. Thoss, *J. Chem. Phys.*, **119**, 1289 (2003).
53. M. Thoss, I. Kondov, and H. Wang, *Chem. Phys.*, **304**, 169 (2004).
54. R. Chakraborty, P. Rebentrost, and A Aspuru-Guzik, *J. Chem. Phys.*, **131**, 184102 (2009).
55. S. Shim, P. Rebentrost, S. Valleau, A. Aspuru-Guzik, *Biophys. J.*, **102**, 649 (2012).
56. J. Cao and R.J. Silbey, *J. Phys. Chem. A*, **113**, 13826 (2009).
57. S. Jang, Y.-C. Cheng, D.R. Reichman, and J.D. Eaves, *J. Chem. Phys.*, **129**, 101104 (2008); S. Jang, *J. Chem. Phys.*, **131**, 164101 (2009); *ibid*, **135**, 034105 (2011); L. Yang, S. Caprasecca, B. Mennucci, and Seogjoo Jang, *J. Am. Chem. Soc.*, **132**, 16911 (2010);
58. A. Kolli, A. Nazir, and A. Olaya-Castro, arXiv:1106.2784 [quant-ph]
59. Eric N. Zimanyi and R.J. Silbey, *Phil. Trans. R. Soc. A*, **370**, 3620 (2012).
60. G. Ritschel, J. Roden, W.T. Strunz, and A. Eisfeld, *New J. Phys.*, **13**, 113034 (2011).
61. L.A. Pachón ab and P. Brumer, *Phys. Chem. Chem. Phys.*, **14**, 10094 (2012).
62. P. Huo and D.F. Coker, *J. Chem. Phys.*, **135**, 201101 (2011).
63. P. Huo and D.F. Coker, *Mol. Phys.*, **110**, 1035 (2012).
64. P. Huo and D.F. Coker, *J. Chem. Phys.*, **137**, 22A535 (2012).
65. J. Strümpfer and K. Schulten, *J. Chem. Phys.*, **131**, 225101 (2009).
66. S.-H. Yeh, J. Zhu, S. Kais, arXiv:1204.5262 [physics.chem-ph]
67. G.S. Schlau-Cohen, T.R. Calhoun, N.S. Ginsberg, E.L. Read, M. Ballottari, R. Bassi, R. van Grondelle, and G.R. Fleming, *J. Phys. Chem. B*, **113**,15352 (2009); G.S. Schlau-Cohen, T.R. Calhoun, N.S. Ginsberg, M. Ballottari, R. Bassi, and G.R. Fleming, *Proc. Natl. Acad. Sci.*, **107**, 13276 (2010).
68. V.I. Novoderezhkin, M.A. Palacios, H. van Amerongen, and R. van Grondelle, *J. Phys. Chem. B*, **109**, 10493 (2005).
69. H. Hossein-Nejad, C. Curutchet, A. Kubica, and G.D. Scholes, *J. Phys. Chem. B*, **115**, 5243 (2011).
70. A. Kolli, E.J. O'Reilly, G.D. Scholes, and A. Olaya-Castro, arXiv:1203.5056v1 [physics.bio-ph]

71. D. Abramavicius, B. Palmieri, and S. Mukamel, *Chem. Phys.*, **357**, 79 2009
72. D. Abramavicius and S. Mukamel, *J. Chem. Phys.*, **133**, 064510 (2010).
73. A.G. Dijkstra and Y. Tanimura, *New J. Phys.*, **2010**(12),055005.
74. P. Pechukas, *Phys. Rev.*, **181**, 174 (1969).
75. J.C. Tully, *J. Chem. Phys.*, **93**, 1061 (1990); J.C. Tully, *Classical and Quantum Dynamics in Condensed Phase Simulations*, G. Ciccotti, B. Berne, and D. Coker (eds.) (World Scientific, Dordrecht, 1998), p. 489.
76. J.C. Tully, *J. Chem. Phys.*, **137**, 22A301 (2012).
77. P.V. Parandekar and J.C. Tully, *J. Chem. Phys.*, **122**, 094102 (2005); P.V. Parandekar and J.C. Tully, *J. Chem. Theory Comput.*, **2**, 229 (2006).
78. J. Bader and B. Berne, *J. Chem. Phys.*, **100**, 8359 (1994); B. Berne, J. Bader, and P. Hanggi, *J. Chem. Phys.*, **104**, 1111 (1996); S. Egorov, E. Rabani, and B. Berne, *J. Phys. Chem. A.*, **103**, 10978 (1999).
79. W.H. Miller and C.W. McCurdy, *J. Chem. Phys.*, **69**, 5163 (1978); H.D. Meyer and W.H. Miller, *J. Chem. Phys.*, **70**, 3214 (1979); C.W. McCurdy, H.D. Meyer, and W.H. Miller, *J. Chem. Phys.*, **70**, 3177 (1979).
80. G. Stock and M. Thoss, *Phys. Rev. Lett.*, **78**, 578 (1997); G. Stock and M. Thoss, *Phys. Rev. A.*, **59**, 64 (1999).
81. S. Bonella, and D.F. Coker, *J. Chem. Phys.*, **114**, 7778 (2001); S. Bonella and D.F. Coker, *Chem. Phys.*, **268**, 189 (2001).
82. E.A. Coronado, J. Xing, and W.H. Miller, *Chem. Phys. Lett.*, **349**, 512 (2001).
83. M.F. Herman and E. Kluk, *Chem. Phys.*, **91**, 27 (1984); W.H. Miller, *Mol. Phys.*, **100**, 397 (2002).
84. S. Bonella and D.F. Coker, *J. Chem. Phys.*, **118**, 4370 (2003).
85. S. Bonella and D.F. Coker, *J. Chem. Phys.*, **122**, 194102 (2005); S. Bonella, D. Montemayor, and D.F. Coker, *Proc. Natl. Acad. Sci.*, **102**, 6715 (2005).
86. N. Ananth, C. Venkataraman, and W.H. Miller, *J. Chem. Phys.*, **127**, 084114 (2007); G. Tao and W.H. Miller, *J. Chem. Phys.*, **130**, 184108 (2009); H. Wang, M. Thoss, K.L. Sorge, R. Gelabert, X. Giménez, and W.H. Miller, *J. Chem. Phys.*, **114**, 2562 (2001).
87. X. Sun, H. Wang, and W.H. Miller *J. Chem. Phys.*, **109**, 7064 (1998).
88. J.A. Poulsen, G. Nyman, and P.J. Rossky, *J. Chem. Phys.*, **119**, 12179 (2003); J.A. Poulsen and G. Nyman, *J. Phys. Chem. A*, **108**, 8743 (2004); J.A. Poulsen, G. Nyman, and P.J. Rossky, *Proc. Natl. Acad. Sci.*, **102**, 6709 (2005).

89. Q. Shi and E. Geva, *J. Phys. Chem. A*, **107**, 9059 (2003); Q. Shi and E. Geva, *J. Chem. Phys.*, **118**, 8173 (2003); Q. Shi and E. Geva, *J. Chem. Phys.*, **119**, 9030 (2003). Q. Shi and E. Geva, *J. Chem. Phys.*, **118**, 7562 (2003).
90. Z. Ma and D.F. Coker, *J. Chem. Phys.*, **128**, 244108 (2008); N. Ananth and T.F. Miller III, *J. Chem. Phys.*, **133**, 234103 (2010).
91. E.R. Dunkel, S. Bonella, and D.F. Coker, *J. Chem. Phys.*, **129**, 114106 (2008).
92. P. Huo, S. Bonella, L. Chen, and D.F. Coker, *Chem. Phys.*, **370**, 87 (2010).
93. J. Strümpfer and K. Schulten, *J. Chem. Phys.*, **137**, 065101 (2011).
94. D. Abramavicius and S. Mukamel, *J. Chem. Phys.*, **2010**(133), 184501/1–184501/13.
95. A. Ishizaki and G.R. Fleming, *New J. Phys.*, **12**, 055004 (2010).
96. P. Nalbach, J. Eckel, and M. Thorwart, *New J. Phys.*, **12**, 065043 (2010); P. Nalbach and M. Thorwart, *New J. Phys.*, **11**, 085001 (2009); P. Nalbach and M. Thorwart, *New J. Phys.*, **12**, 194111 (2010).
97. F. Fassioli, A. Nazir, and A. Olaya-Castro, *J. Phys. Chem. Lett.*, **1**, 2139 (2010). F. Fassioli and A. Olaya-Castro, *New J. Phys.*, **12**, 085006 (2010).
98. D. Abramavicius and S. Mukamel, *J. Chem. Phys.*, **134**, 174504 (2011).
99. J. Strümpfer and K. Schulten, *J. Chem. Phys.*, **134**, 095102 (2011).
100. E. Hennebicq, D. Beljonne, C. Curutchet, G.D. Scholes, and R.J. Silbey, *J. Chem. Phys.*, **130**, 214505 (2009).
101. S.M. Vlaming and R.J. Silbey, *J. Chem. Phys.*, **136**, 055102 (2012).
102. Seogjoo Jang, *J. Chem. Phys.*, **127**, 174710 (2007); L. Yang, M. Devi, and S. Jang, *J. Chem. Phys.*, **137**, 024101 (2012).
103. C. Olbrich, J. Strümpfer, K. Schulten, and U. Kleinekathöfer, *J. Phys. Chem. Lett.*, **2** 1771 (2011).
104. C. Olbrich and U. Kleinekathfer, *J. Phys. Chem. B*, **114**, 12427 (2010); C. Olbrich, J. Strmpfer, K. Schulten, and U. Kleinekathöfer, *J. Phys. Chem. B*, **115**, 758 (2011).
105. C. Olbrich, T.L.C. Jansen, J. Liebers, M. Aghtar, J. Strümpfer, K. Schulten, J. Knoester, and U. Kleinekathöfer, *J. Phys. Chem. B*, **115** 8609 (2011).
106. T. Fujita, J.C. Brookes, S.K. Saikin, and A. Aspuru-Guzik, *J. Phys. Chem. Lett.*, **3**, 2357 (2012).
107. Y.Jing, R. Zheng, H. Li, and Q. Shi, *J. Phys. Chem. B*, **116**, 1164 (2011).
108. D. Abramavicius, B. Palmieri, D. Voronine, F. Sanda, and S. Mukamel, *Chem. Rev.*, **109**, 2350 (2009); S. Mukamel, D. Abramavicius, L. Yang, W. Zhuang, I.V. Schweigert, and D.V. Voronine, *Acc. Chem. Res.*, **42**, 553 (2009).

109. L. Chen, R. Zheng, Y. Jing, and Qiang Shi, *J. Chem. Phys.*, **134**, 194508 (2011).

110. J. Yuen-Zhou, J.J. Krich, and A. Aspuru-Guzik, *J. Chem. Phys.*, **136**, 234501 (2012); J. Yuen-Zhou, J.J. Krich, M. Mohseni, and A. Aspuru-Guzik, *Proc. Natl. Acad. Sci.*, **108**, 17615 (2011); J. Yuen-Zhou and A. Aspuru-Guzik, *J. Chem. Phys.*, **134**, 134505 (2011).

111. S. Valleau, A. Eisfeld, and A. Aspuru-Guzik, *J. Chem. Phys.*, **137**, 224103 (2012).

112. E. Rivera, D. Montemayor, M. Masia, and D.F. Coker, *J. Phys. Chem. B*, **117**, 5510 (2013).

113. N. Makri, *J. Phys. Chem. B*, **103**, 2823 (1999).

114. D. Mac Kernan, G. Ciccotti, and R. Kapral, *J. Chem. Phys.*, **116**, 2346 (2002); R. Kapral and G. Ciccotti, *J. Chem. Phys.*, **110**, 8919 (1999).

115. H. Wang, X. Song, D. Chandler, and W.H. Miller, *J. Chem. Phys.*, **110**, 4828 (1999).

116. E. Romero, I.H.M. van Stokkum, V.I. Novoderezhkin, J.P. Dekker, and R van Grondelle, *Biochemistry*, **49**, 4300 (2010).

117. V.I. Novoderezhkin, E.G. Andrizhiyevskaya, J.P. Dekker, and Rienk van Grondelle, *Biophys. J.*, **89**, 1464 (2005); V.I. Novoderezhkin, J.P. Dekker, and R. van Grondelle, *Biophys. J.*, **93**, 1293 (2007).

118. G. Raszewski, W. Saenger, and T. Renger, *Biophys. J.*, **88**, 986 (2005); G. Raszewski, B.A. Diner, E. Schlodder, and T. Renger, *Biophys. J.*, **95**, 105 (2008).

119. Vladimir I. Novoderezhkin, Elisabet Romero, Jan P. Dekker, and Rienk van Grondelle, *ChemPhysChem*, **12**, 681 (2011).

120. A. Damjanovic, I. Kosztin, U. Kleinekathöfer, and Klaus Schulten, *Phys. Rev. E*, **65**, 031919 (2002).

121. D.E. Makarov and N. Makri, *Chem. Phys. Lett.*, **221**, 482 (1994).

122. R. Egger and C.H. Mak, *Phys. Rev. B*, **50**, 15210 (1994).

123. K. Thompson and N. Makri, *J. Chem. Phys.*, **110**, 1343 (1999).

124. A. Golosov and D.R. Reichman, *J. Chem. Phys.*, **114**, 1065 (2001).

125. Q. Shi, L. Chen, G. Nan, R.X. Xu, and Y. Yan, *J. Chem. Phys.*, **130**, 084105 (2009).

126. A.J. Leggett, S. Chakravarty, A.T. Dorsey, M.P.A. Fisher, A. Garg, and W. Zwerger, *Rev. Mod. Phys.*, **59**, 1 (1987).

127. Mi Kyung Lee and David F. Coker, *J. Phys. Chem. Lett.*, **7**, 3171–3178 (2016).

128. Mi Kyung Lee, Pengfei Huo, and David F. Coker, *Annu. Rev. Phys. Chem.*, **67**, 639–668 (2016).

Chapter 9

Bridging the Gap between Coherent and Incoherent Resonance Energy Transfer Dynamics by Quantum Master Equations in the Polaron Picture

Seogjoo Jang
Department of Chemistry and Biochemistry, Queens College, City University of New York, USA
Seogjoo.Jang@qc.cuny.edu

9.1 Introduction

If quantum effects play significant roles in biological environments, the most likely situation is when they interact with quantum particles or are quantum-mechanically excited by external agents. The light-harvesting and energy transfer processes in the early stage of photosynthesis [1, 2] are well-established examples of such case. However, whether quantum effects in these systems can be encapsulated into force parameters/rate constants such that conventional description based on classical dynamics/rate processes are sufficient, or there are in fact intriguing quantum-dynamical

Ultrafast Dynamics at the Nanoscale: Biomolecules and Supramolecular Assemblies
Edited by Irene Burghardt and Stefan Haacke

ISBN 978-981-4745-33-8 (Hardcover), 978-981-4745-34-5 (eBook)
www.panstanford.com

processes underlying seemingly classical average behavior, is not clearly understood yet.

The fundamental mechanism underlying the energy transfer in photosynthesis is so-called resonance energy transfer (RET) or excitation energy transfer (EET) [3–5] which can be used interchangeably although they have different connotations. Given a pair of molecules, a donor in an excited state and an acceptor in the ground state, the RET reaction can swap their electronic excitations within timescales of 100 fs and 10 ps if the distance between the donor and the acceptor is less than 10 nm.

The first quantum-mechanical theory for RET was proposed in 1948 by Förster [6, 7] who derived the now famous Förster's expression. This expression sums up all of the major effects within parameters that can be determined by independent spectroscopic measurement. Similar theories for other coupling mechanisms have also been developed [3, 7, 8], which will be here called collectively as Förster's resonance energy transfer (FRET) along with the original theory. Numerous experimental and theoretical works based on those theories have been made [4, 5, 9, 10].

FRET, in essence, is nothing but Fermi's golden rule between two electronic states interacting through resonance electronic coupling. There is no question in its utility [11–13]. However, the assumptions underlying the theory in general fall short in describing RET processes of natural photosynthetic systems and large organic molecules being investigated in modern experiments, and much improvement and refinement is necessary. There have been some theoretical advances addressing these issues [2, 14–21] but mostly still within the assumption of incoherent transfer.

An essential element for quantitative description of RET in photosynthetic light-harvesting complexes is clarifying the effect of quantum coherence due to moderately or strongly coupled pigment molecules. Given that such molecules can be grouped together and intergroup electronic couplings remain weak, the RET process between different groups may be treated within assumptions comparable to those behind FRET. Multichromophoric generalization of FRET (MC-FRET) [16, 20] is appropriate for this case and has been shown to provide a reasonable explanation of experimental data for RET dynamics from the B800 unit to the

B850 unit in the light-harvesting 2 (LH2) complex of purple bacteria [2, 17, 18, 22].

In more general situation where such grouping is not obvious, it becomes necessary to account for the quantum coherence explicitly in real time dynamics calculation. Advances in two-dimensional electronic spectroscopy (2DES) [23, 24] have also realized direct real time observation of coherent quantum dynamics possible, motivating further development in the theories of RET. From a theoretical point of view, unambiguous clarification of the 2DES results is very challenging. Let alone the difficulty of quantum-dynamical simulation of large scale systems, unravelling the contribution of electronic quantum coherence out of systems going through nonequilibrium relaxation and exhibiting significant ensemble dephasing is a complex task.

There are various quantum dynamics methods that can be utilized for the simulation of molecular systems in condensed environments [25–28], among which the quantum master equation (QME) approach has its particular merit of being efficient and scalable. The QME approach has been very successful in the field of nuclear magnetic resonance (NMR). Due to the weak coupling of spins to their environments and the slowness of the spin flipping compared to other molecular motions, the QME used for the modeling of NMR signals is a rather simple one based on weak system–bath coupling (second order) and the Markovian approximation. The corresponding QME is better known as the Bloch–Redfield equation [29, 30].

Theories [11, 31–35] adapting the Bloch–Redfield equation for coherent resonance energy transfer (CRET) have been developed and applied to appropriate systems. However, when the electron–phonon coupling is significant and there are multitudes of dynamics with different timescales, it becomes necessary to go beyond the assumptions of the Markovian bath and/or the weak system–bath coupling.

While the theory and numerical implementation of the second-order non-Markovian QME are now well-established [36, 37], development of higher-order approximations [28, 38–45] continues to be a challenging theoretical task. A straightforward way to accomplish this is to expand the formal QME with respect to the (bare) system–bath interaction Hamiltonian [28], but the resulting time evolution equations become intractable for general system–bath couplings as the order increases. An alternative approach is to introduce a transformation that renormalizes the system–bath interaction, which takes care of some terms up to the infinite order, and then to employ low-order approximation. The polaron transformation introduced by Holstein [46] is an attractive choice for such purpose.

The theory of polaron was pioneered by Holstein [46, 47] to treat charge transfer in organic molecular crystals in the 1960s and later extended by Silbey and coworkers to treat RET in the 1970s [15, 48–51]. This provides a general framework to describe both strong and weak electron–phonon coupling limits within a unified formalism, but full dynamical equations have not been fully developed. It was only recently that a compact and explicit expression for polaronic quantum master equation (PQME) at the second-order level was completely derived, which was led to further development lately [52–58].

In particular, the PQME approach was extended for multi-chromophoric CRET (MC-CRET) theory [54]. This allows addressing coherent exciton dynamics, multiphonon effects, and non-Markovian effects for general multichromophore systems beyond the perturbation theory and weak system–bath interaction limit while being equivalent to the latter in its numerical overhead. This has great potential as a tool to elucidate the effects of electronic coherence and its interaction with environments on RET dynamics in photosynthetic complexes. A detailed derivation and discussion at this level of theory is provided in this chapter.

9.2 System–Bath Hamiltonian and Formally Exact Quantum Master Equation

Consider the following system Hamiltonian consisting of N quantum states.

$$H_s = \sum_{l=1}^{N} E_l |l\rangle\langle l| + \sum_{l\neq l'}^{N} J_{ll'} |l\rangle\langle l'| \,. \tag{9.1}$$

Here, in the context of the RET dynamics in single exciton space, $|l\rangle$ represents the electronic state where only the lth site or molecule is excited while all others are in the ground electronic state, and $J_{ll'}$ is the electronic resonance coupling between states $|l\rangle$ and $|l'\rangle$. The remaining degrees of freedom are defined as the bath with Hamiltonian H_b. Assuming that the interaction between system and bath can be represented by an additive term H_{sb}, the total Hamiltonian can be expressed as

$$H = H_s + H_{sb} + H_b \,. \tag{9.2}$$

Let us denote the total density operator representing the system and bath at time t as $\rho(t)$. This is governed by the following quantum Liouville equation:

$$\frac{d}{dt}\rho(t) = -i\mathcal{L}\rho(t)\,, \tag{9.3}$$

where $\mathcal{L}\rho(t) = [H, \rho(t)]/\hbar$, the commutator divided by $\hbar$. Let us define the zeroth-order Hamiltonian as

$$H_0 = H_s + H_b \,. \tag{9.4}$$

Then, in the interaction picture with respect to H_0, Eq. 9.3 becomes

$$\frac{d}{dt}\rho_I(t) = -i\mathcal{L}_{sb,I}(t)\rho_I(t)\,, \tag{9.5}$$

where $\rho_I(t) = e^{iH_0t/\hbar}\rho(t)e^{-iH_0t/\hbar}$ and $\mathcal{L}_{sb,I}(t)$ is the Liouville operator (commutator with the Hamiltonian divided by $\hbar$) for

$$H_{sb,I}(t) = e^{iH_0t/\hbar}H_{sb}e^{-iH_0t/\hbar}\,. \tag{9.6}$$

In general, exact numerical solution of Eq. 9.5 is not possible because of (infinitely) large size of the bath. If the primary interest is in the system part, we only need to focus on the following reduced system density operator:

$$\sigma_I(t) = Tr_b\{\rho_I(t)\}\,. \tag{9.7}$$

The equation governing the time evolution of this reduced density operator is in general called QME. The well-known Bloch–Redfield equation is a second-order QME in the Markovian bath limit. First, let us review the derivation of the formally exact QME employing a projection operator technique.

Let us assume that we can define a projection superoperator $\mathcal{P}$ such that

$$\mathcal{P}\mathcal{L}_{sb,I}(t)\mathcal{P} = 0 \tag{9.8}$$

and its complement $\mathcal{Q} = 1 - \mathcal{P}$. A detailed form of $\mathcal{P}$ will be provided below, which is not necessary in obtaining the formal solution. Applying $\mathcal{P}$ to Eq. 9.5 and inserting the identity, $\mathcal{P} + \mathcal{Q} = 1$, between $\mathcal{L}_{sb,I}(t)$ and $\rho_I(t)$, we obtain the following time evolution equation

$$\frac{d}{dt}\mathcal{P}\rho_I(t) = -i\mathcal{P}\mathcal{L}_{sb,I}(t)\mathcal{Q}\rho_I(t)\,, \tag{9.9}$$

where the identity of Eq. 9.8 has been used. Applying $\mathcal{Q}$ to Eq. 9.5 and inserting the identity, $\mathcal{P} + \mathcal{Q} = 1$, between $\mathcal{L}_{sb,I}(t)$ and $\rho_I(t)$, we also obtain

$$\frac{d}{dt}\mathcal{Q}\rho_I(t) = -i\,\mathcal{Q}\mathcal{L}_{sb,I}(t)(\mathcal{P} + \mathcal{Q})\rho_I(t)\,. \tag{9.10}$$

A formal solution of this equation is

$$\begin{aligned}\mathcal{Q}\rho_I(t) = {} & \exp_{(+)}\left[-i\int_0^t d\tau\,\mathcal{Q}\mathcal{L}_{sb,I}(\tau)\right]\mathcal{Q}\rho_I(0) \\ & -i\int_0^t d\tau\,\exp_{(+)}\left[-i\int_\tau^t d\tau'\mathcal{Q}\mathcal{L}_{sb,I}(\tau')\right] \\ & \times\mathcal{Q}\mathcal{L}_{sb,I}(\tau)\mathcal{P}\rho_I(\tau)\,,\end{aligned} \tag{9.11}$$

where $\exp_{(+)}$ denotes exponential expansion of the exponent with chronological time ordering (where operators at later times are placed on the lefthand side). Inserting this into Eq. 9.9, we obtain the following time evolution equation for the projected portion of the total density operator:

$$\begin{aligned}\frac{d}{dt}\mathcal{P}\rho_I(t) = {} & -i\mathcal{P}\mathcal{L}_{sb,I}(t)\exp_{(+)}\left[-i\int_0^t d\tau\,\mathcal{Q}\mathcal{L}_{sb,I}(\tau)\right]\mathcal{Q}\rho_I(0) \\ & -\int_0^t d\tau\mathcal{P}\mathcal{L}_{sb,I}(t)\exp_{(+)}\left[-i\int_\tau^t d\tau'\mathcal{Q}\mathcal{L}_{sb,I}(\tau')\right] \\ & \times\mathcal{Q}\mathcal{L}_{sb,I}(\tau)\mathcal{P}\rho_I(\tau)\,.\end{aligned} \tag{9.12}$$

At this point, let us introduce the following definition of the projection operator:

$$\mathcal{P}(\cdot) = \rho_b Tr_b\{(\cdot)\}\,, \tag{9.13}$$

and assume that

$$Tr_b\left\{\rho_b H_{sb,I}(t)\right\} = 0\,. \tag{9.14}$$

The definition Eq. 9.13 together with Eq. 9.14 ensures that the condition of Eq. 9.8 is satisfied. It is important to note that the condition of Eq. 9.14 is not as restrictive as it appears to be, and can always be satisfied by appropriate definition of H_s. In other words, in case $Tr_b\{\rho_b H_{sb}\}$ is not zero, this nonzero term can always be added to H_s and the difference from the original H_{sb} can be defined as a new H_{sb}. With this renormalized H_{sb} in mind, we can assume that Eq. 9.14 can always be satisfied.

Using the definition of Eq. 9.13 in Eq. 9.12 and taking trace of this over the bath, we obtain the following formally exact QME:

$$\begin{aligned}\frac{d}{dt}\sigma_I(t) &= -iTr_b\left\{\mathcal{L}_{sb,I}(t)\exp_{(+)}\left[-i\int_0^t d\tau\mathcal{Q}\mathcal{L}_{sb,I}(\tau)\right]\mathcal{Q}\rho_I(0)\right\}\\ &\quad-\int_0^t d\tau Tr_b\left\{\mathcal{L}_{sb,I}(t)\exp_{(+)}\left[-i\int_\tau^t d\tau'\mathcal{Q}\mathcal{L}_{sb,I}(\tau')\right]\right.\\ &\quad\left.\times\mathcal{Q}\mathcal{L}_{sb,I}(\tau)\rho_b\right\}\sigma_I(\tau)\,.\end{aligned} \tag{9.15}$$

This QME, although exact, is not useful in practice because $\exp_{(+)}[-i\int_\tau^t d\tau'\mathcal{Q}\mathcal{L}_{sb,I}(\tau')]$ cannot be determined without full information on the system and bath degrees of freedom.

9.3 QME in the Weak System–Bath Coupling Limit

9.3.1 Time-Nonlocal Equation

For weak system–bath interaction, an approximation can be made such that only terms exact up to the second order of $\mathcal{L}_{sb,I}(t)$ are kept in Eq. 9.15. The resulting equation is the well-known second-order time-nonlocal QME shown below.

$$\begin{aligned}\frac{d}{dt}\sigma_I(t) &= -iTr_b\left\{\mathcal{L}_{sb,I}(t)\mathcal{Q}\rho_I(0)\right\}\\ &\quad-\int_0^t d\tau Tr_b\left\{\mathcal{L}_{sb,I}(t)\mathcal{Q}\mathcal{L}_{sb,I}(\tau)\mathcal{Q}\rho_I(0)\right\}\\ &\quad-\int_0^t d\tau Tr_b\left\{\mathcal{L}_{sb,I}(t)\mathcal{Q}\mathcal{L}_{sb,I}(\tau)\rho_b\right\}\sigma_I(\tau)\,.\end{aligned} \tag{9.16}$$

Employing the fact that $Tr_b\{\mathcal{L}_{sb,I}(t)\rho_b\} = 0$, some of $\mathcal{Q}$ can be removed in the above equation. The resulting expression is as follows:

$$\frac{d}{dt}\sigma_I(t) = -iTr_b\left\{\mathcal{L}_{sb,I}(t)\rho_I(0)\right\}$$
$$-\int_0^t d\tau Tr_b\left\{\mathcal{L}_{sb,I}(t)\mathcal{L}_{sb,I}(\tau)\mathcal{Q}\rho_I(0)\right\}$$
$$-\int_0^t d\tau Tr_b\left\{\mathcal{L}_{sb,I}(t)\mathcal{L}_{sb,I}(\tau)\rho_b\right\}\sigma_I(\tau). \quad (9.17)$$

Further simplification of Eq. 9.17 is possible if the bath can be modeled as a set of harmonic oscillators and the system–bath interaction is linear in the displacements of the oscillators as follows:

$$H_b = \sum_n \hbar\omega_n\left(b_n^\dagger b_n + \frac{1}{2}\right), \quad (9.18)$$

$$H_{sb} = \sum_{l=1}^{N}\sum_n \hbar\omega_n g_{n,l}(b_n + b_n^\dagger)|l\rangle\langle l|, \quad (9.19)$$

where all $g_{n,l}$'s are assumed to be real here, but extension to the case of complex coupling constants can be made if necessary. In addition, for the following initial condition:

$$\rho(0) = \sigma(0)\rho_b, \quad (9.20)$$

which corresponds to the physical situation where the excited state at time zero is created by a short pulse with duration much shorter than the timescale of the bath, all the inhomogeneous terms in Eq. 9.17 disappears. As a result, Eq. 9.17 can be expressed as

$$\frac{d}{dt}\sigma_I(t) = -\int_0^t d\tau Tr_b\left\{\mathcal{L}_{sb,I}(t)\mathcal{L}_{sb,I}(\tau)\rho_b\right\}\sigma_I(\tau)$$
$$= -\sum_l\sum_{l'}\int_0^t d\tau C_{ll'}(t-\tau)[S_l(t), S_{l'}(\tau)\sigma_I(\tau)]$$
$$+\text{H.c.}, \quad (9.21)$$

where H.c. denotes Hermitian conjugates of all the previous terms, $S_l(t) = e^{iH_st/\hbar}|l\rangle\langle l|e^{-iH_st/\hbar}$, and

$$C_{ll'}(t) = \sum_n \omega_n^2 g_{n,l} g_{n,l'} Tr_b\left\{(b_n e^{-i\omega_n t} + b_n^\dagger e^{i\omega_n t})(b_n + b_n^\dagger)\rho_b\right\} \quad (9.22)$$

Introducing the following spectral density,

$$\mathcal{J}_{ll'}(\omega) = \sum_n g_{n,l} g_{n,l'} \delta(\omega - \omega_n)\omega^2 , \tag{9.23}$$

and calculating the trace over the bath explicitly, Eq. 9.22 can be expressed as

$$\mathcal{C}_{ll'}(t) = \int_0^\infty d\omega \mathcal{J}_{ll'}(\omega) \left(\coth\left(\frac{\beta\hbar\omega}{2}\right) \cos(\omega t) - i \sin(\omega t) \right) . \tag{9.24}$$

9.3.2 Time-Local Equation

The time-local equation is often preferred in a real-time solution because of its efficiency for numerical calculation. A standard approach to derive a time-local QME is to expand formal cumulant solution for $\sigma(t)$, from which a second-order time-local QME can be obtained. However, instead of this procedure, we can take a short cut and replace $\sigma_I(\tau)$ with $\sigma_I(t)$ in Eq. 9.17. This procedure leads to a correct result because $\sigma_I(t)$ is equivalent to $\sigma_I(\tau)$ up to the zeroth-order of $H_{sb,I}(t)$. The resulting second-order time-local QME has the following expression:

$$\begin{aligned}
\frac{d}{dt}\sigma_I(t) = & -i Tr_b \left\{ \mathcal{L}_{sb,I}(t)\rho_I(0) \right\} \\
& - \int_0^t d\tau Tr_b \left\{ \mathcal{L}_{sb,I}(t)\mathcal{L}_{sb,I}(\tau)\mathcal{Q}\rho_I(0) \right\} \\
& - \int_0^t d\tau Tr_b \left\{ \mathcal{L}_{sb,I}(t)\mathcal{Q}\mathcal{L}_{sb,I}(\tau)\rho_b \right\} \sigma_I(t) .
\end{aligned} \tag{9.25}$$

For the bath Hamiltonian and system–bath terms respectively given by Eqs. 9.18 and 9.19 and for the initial condition given by Eq. 9.20, inhomogeneous terms disappear and Eq. 9.25 can be simplified as

$$\frac{d}{dt}\sigma_I(t) = -\sum_l \sum_{l'} \mathcal{R}_{ll'}(t)\sigma_I(t) , \tag{9.26}$$

with

$$\mathcal{R}_{ll'}(t)\sigma_I(t) = \sum_l \sum_{l'} \int_0^t d\tau\, \mathcal{C}_{ll'}(t-\tau)[S_l(t), S_{l'}(\tau)\sigma_I(t)] + \text{H.c.} , \tag{9.27}$$

where H.c. represents Hermitian conjugates of all the previous terms.

It is often implicitly assumed that the second-order time-local QME is less accurate than the time-nonlocal QME because the former appears to involve an additional approximation, replacing $\sigma(\tau)$ with $\sigma(t)$, which is often invoked based on timescale separation argument. However, this is not true in general because how one obtains one or the other is irrelevant to how significant remaining higher-order terms are. In fact, there are counterexamples suggesting that the former performs better than the latter [59, 60]. An analysis based on the fourth-order QME [38] indicates that this may be because the second-order time-local QME contains some fourth-order contributions of the time-nonlocal QME. Thus, care should be taken and more theoretical study is needed.

9.4 QME in the Polaron Picture

The second-order QMEs described in the previous section have great practical value because of simplicity and ease of numerical implementation. However, their accuracy degrades as the system–bath coupling (or electron–phonon) coupling increases. Furthermore, they do not approach the correct limit of strong system–bath coupling where the RET dynamics becomes incoherent, and are not appropriate for intermediate regime where the transition from coherence to incoherence occurs. In addressing this issue, an attractive approach is to consider the second-order QME followed by polaron transformation [46–50].

For the convenience of formulation, the system Hamiltonian, Eq. 9.1, can be divided into population and coherence terms as follows:

$$H_s = H_s^p + H_s^c\,, \tag{9.28}$$

where $H_s^p = \sum_{l=1}^{N} E_l|l\rangle\langle l|$ and $H_s^c = \sum_{l\neq l'}^{N} J_{ll'}|l\rangle\langle l'|$. We also assume that the bath Hamiltonian and system–bath Hamiltonian are given by Eqs. 9.18 and 9.19.

Let us introduce the following generator:

$$G = \sum_{l=1}^{N}\sum_{n} g_{n,l}(b_n^\dagger - b_n)|l\rangle\langle l|\,. \tag{9.29}$$

The polaron transformation is generated by this operator, and the corresponding transformation operator e^{-G} can be shown to be unitary in the combined space of system and bath. Thus, application of e^{-G} to $H = H_s+H_{sb}+H_b$ results in the following total Hamiltonian in the polaron picture:

$$\tilde{H} = e^G H e^{-G} = H_s^p + \tilde{H}_s^c + e^G(H_{sb} + H_b)e^{-G}\,, \tag{9.30}$$

where the fact that $e^G H_s^p e^{-G} = H_s^p$ has been used and

$$\tilde{H}_s^c = e^G H_s^c e^{-G} = \sum_{l \neq l'}^{N} J_{ll'}\theta_l^\dagger \theta_{l'} |l\rangle\langle l'|\,, \tag{9.31}$$

with $\theta_l = e^{-\sum_n g_{n,l}(b_n^\dagger - b_n)}$. In Eq. 9.30,

$$e^G(H_{sb} + H_b)e^{-G} = H_b - \sum_l \sum_n \hbar\omega_n g_{n,l}^2 |l\rangle\langle l|\,. \tag{9.32}$$

Thus, introducing

$$\tilde{H}_s^p = \sum_{l=1}^{N} \tilde{E}_l |l\rangle\langle l|\,, \tag{9.33}$$

with $\tilde{E}_l = E_l - \sum_n \hbar\omega_n g_{n,l}^2$, Eq. 9.30 can be expressed as

$$\tilde{H} = \tilde{H}_s^p + \tilde{H}_s^c + H_b\,. \tag{9.34}$$

In a similar manner, we can define the total density operator in the polaron picture as $\tilde{\rho}(t) = e^G\rho(t)e^{-G}$. The quantum Liouville equation governing the time evolution of this is

$$\frac{d}{dt}\tilde{\rho}(t) = -i\tilde{\mathcal{L}}\tilde{\rho}(t) = -i(\tilde{\mathcal{L}}_s^p + \tilde{\mathcal{L}}_s^c + \mathcal{L}_b)\tilde{\rho}(t)\,, \tag{9.35}$$

where $\tilde{\mathcal{L}}_s^p$, $\tilde{\mathcal{L}}_s^c$, and $\mathcal{L}_b$ are quantum Liouville operators corresponding to $\tilde{H}_s^p$, $\tilde{H}_s^c$, and H_b.

Out of the transformed Hamiltonian $\tilde{H}$, Eq. 9.34, let us define the zeroth- and the first-order terms as follows:

$$\tilde{H}_0 = \tilde{H}_s^p + \langle \tilde{H}_s^c \rangle_b + H_b \equiv \tilde{H}_{0,s} + H_b\,, \tag{9.36}$$

$$\tilde{H}_1 = \tilde{H}_s^c - \langle \tilde{H}_s^c \rangle_b\,, \tag{9.37}$$

with $\langle \tilde{H}_s^c \rangle_b = Tr_b\{\tilde{H}_s^c \rho_b\}$. This term represents the average effect of the system-bath interaction and can be calculated explicitly. In Eq. 9.36, $\tilde{H}_{0,s} = \tilde{H}_s^p + \langle \tilde{H}_s^c \rangle_b$ represents the system part of the renormalized zeroth-order Hamiltonian. An explicit expression for this can be found as follows.

$$\tilde{H}_{0,s} = \sum_{l=1}^{N} \tilde{E}_l |l\rangle\langle l| + \sum_{l\neq l'}^{N} J_{ll'} w_{ll'} |l\rangle\langle l'| , \tag{9.38}$$

where

$$w_{ll'} = \langle \theta_l^\dagger \theta_{l'} \rangle = \langle \theta_{l'}^\dagger \theta_l \rangle = e^{-\sum_n \coth\left(\frac{\beta\hbar\omega_n}{2}\right)\delta g_{n,ll'}^2/2} , \tag{9.39}$$

with $\delta g_{n,ll'} = g_{n,l} - g_{n,l'}$. The first-order term $\tilde{H}_1$, defined by Eq. 9.37, can also be calculated and has the following expression.

$$\tilde{H}_1 = \sum_{l\neq l'}^{N} J_{ll'} \tilde{B}_{ll'} |l\rangle\langle l'| , \tag{9.40}$$

where

$$\tilde{B}_{ll'} = \theta_l^\dagger \theta_{l'} - w_{ll'} = e^{\sum_n \delta g_{n,ll'}(b_n^\dagger - b_n)} - w_{ll'} . \tag{9.41}$$

Let us consider the dynamics in the interaction picture with respect to $\tilde{H}_0$ (not H_0). First, the interaction picture representation of $\tilde{H}_1$ becomes

$$\tilde{H}_{1,I}(t) \equiv e^{i\tilde{H}_0 t/\hbar} \tilde{H}_1 e^{-i\tilde{H}_0 t/\hbar} = \sum_{l\neq l'}^{N} J_{ll'} \tilde{B}_{ll'}(t) \mathcal{T}_{ll'}(t) , \tag{9.42}$$

where

$$\tilde{B}_{ll'}(t) = e^{iH_b t/\hbar} \tilde{B}_{ll'} e^{-iH_b t/\hbar} , \tag{9.43}$$

$$\mathcal{T}_{ll'}(t) = e^{i\tilde{H}_{0,s} t/\hbar} |l\rangle\langle l'| e^{-i\tilde{H}_{0,s} t/\hbar} . \tag{9.44}$$

The interaction picture representation of the polaron transformed total density operator is defined as $\tilde{\rho}_I(t) = e^{i\tilde{\mathcal{L}}_0 t} \tilde{\rho}(t)$, which evolves according to

$$\frac{d}{dt}\tilde{\rho}_I(t) = -\tilde{\mathcal{L}}_{1,I}(t)\tilde{\rho}_I(t) , \tag{9.45}$$

where $\tilde{\mathcal{L}}_{1,I}(t)$ is the Liouville operator for $\tilde{H}_{1,I}(t)$.

As in the case of Eq. 9.7, we can define the reduced system density operator as follows:

$$\tilde{\sigma}_I(t) \equiv Tr_b \{\tilde{\rho}_I(t)\} . \tag{9.46}$$

Then, employing the same projection operator $\mathcal{P}$ defined by Eq. 9.13, we obtain the following second-order time-nonlocal equation:

$$\begin{aligned}\frac{d}{dt}\tilde{\sigma}_I(t) = &-iTr_b\left\{\tilde{\mathcal{L}}_{1,I}(t)\tilde{\rho}(0)\right\}\\ &-\int_0^t d\tau\, Tr_b\left\{\tilde{\mathcal{L}}_{1,I}(t)\tilde{\mathcal{L}}_{1,I}(\tau)\mathcal{Q}\tilde{\rho}(0)\right\}\\ &-\int_0^t d\tau\, Tr_b\left\{\tilde{\mathcal{L}}_{1,I}(t)\tilde{\mathcal{L}}_{1,I}(\tau)\rho_b\right\}\tilde{\sigma}_I(\tau), \qquad (9.47)\end{aligned}$$

where the fact that $Tr_b\left\{\tilde{\mathcal{L}}_{1,I}(t)\rho_b\right\} = 0$ and $\tilde{\rho}_I(0) = \tilde{\rho}(0)$ has been used.

The time-local equation for Eq. 9.47 can be obtained by replacing $\tilde{\sigma}_I(\tau)$ with $\tilde{\sigma}_I(t)$. The resulting QME can be expressed as follows [53]:

$$\frac{d}{dt}\tilde{\sigma}_I(t) = -\tilde{\mathcal{R}}(t)\tilde{\sigma}_I(t) + \tilde{\mathcal{I}}(t), \qquad (9.48)$$

where

$$\tilde{\mathcal{R}}(t) = \int_0^t d\tau\, Tr_b\{\tilde{\mathcal{L}}_{1,I}(t)\tilde{\mathcal{L}}_{1,I}(\tau)\rho_b\}, \qquad (9.49)$$

$$\begin{aligned}\tilde{\mathcal{I}}(t) &= \tilde{\mathcal{I}}^{(1)}(t) + \tilde{\mathcal{I}}^{(2)}(t)\\ &= -iTr_b\{\tilde{\mathcal{L}}_{1,I}(t)\mathcal{Q}\rho(0)\}\\ &\quad -\int_0^t d\tau\, Tr_b\{\tilde{\mathcal{L}}_{1,I}(t)\tilde{\mathcal{L}}_{1,I}(\tau)\mathcal{Q}\tilde{\rho}(0)\}. \qquad (9.50)\end{aligned}$$

In the above expression, $\tilde{\mathcal{I}}^{(1)}(t)$ and $\tilde{\mathcal{I}}^{(2)}(t)$ are terms first and second-order in $\tilde{\mathcal{L}}_{1,I}(t)$, respectively.

More detailed expression for $\tilde{\mathcal{R}}(t)\tilde{\sigma}_I(t)$ in Eq. 9.48 can be found by inserting expressions for $\tilde{H}_{1,I}(t)$, $\tilde{B}_{ll'}(t)$, and $\mathcal{T}_{ll'}(t)$, Eqs. 9.42–9.44, into the Liouville operator $\tilde{\mathcal{L}}_{1,I}(t)$ and taking the system operators out of the trace over the bath. This results in the following expression:

$$\begin{aligned}\tilde{\mathcal{R}}(t)\tilde{\sigma}_I(t) = &\frac{1}{\hbar^2}\sum_{l\neq l'}^{N}\sum_{m\neq m'}^{N} J_{ll'}J_{mm'}\int_0^t d\tau\langle\tilde{B}_{ll'}(t)\tilde{B}_{mm'}(\tau)\rangle\\ &\times[\mathcal{T}_{ll'}(t), \mathcal{T}_{mm'}(\tau)\tilde{\sigma}_I(t)] + \text{H.c.}, \qquad (9.51)\end{aligned}$$

where $\langle\tilde{B}_{ll'}(t)\tilde{B}_{mm'}(\tau)\rangle$ is the time correlation function of $\tilde{B}_{ll'}(t)$, which is defined by Eqs. 9.41 and 9.43. Explicit calculation of this bath correlation function leads to

$$\langle\tilde{B}_{ll'}(t)\tilde{B}_{mm'}(\tau)\rangle = w_{ll'}w_{mm'}\left(e^{-\mathcal{K}_{ll',mm'}(t-\tau)} - 1\right), \qquad (9.52)$$

where $w_{ll'}$ is defined by Eq. 9.39 and

$$\mathcal{K}_{ll',mm'}(t) = \sum_n \delta g_{n,ll'} \delta g_{n,mm'} \left(\coth\left(\frac{\beta\hbar\omega}{2}\right) \cos(\omega t) - i \sin(\omega t) \right). \tag{9.53}$$

It is obvious from the above definition that $\mathcal{K}_{ll',mm'}(t) = -\mathcal{K}_{ll',m'm}(t) = -\mathcal{K}_{l'l,mm'}(t) = \mathcal{K}_{l'l,m'm}(t)$. Employing a spectral density $\mathcal{J}_{ll',mm'}(\omega)$, which can be defined in the same way as Eq. 9.23 by replacing $g_{n,l}$ and $g_{n,l'}$ with $\delta_{gn,ll'}$ and $\delta_{gn,mm'}$, respectively, we can also express Eq. 9.53 as

$$\mathcal{K}_{ll',mm'}(t) = \int_0^\infty d\omega \frac{\mathcal{J}_{ll',mm'}(\omega)}{\omega^2} \left(\coth\left(\frac{\beta\hbar\omega}{2}\right) \cos(\omega t) - i \sin(\omega t) \right). \tag{9.54}$$

A comparison of this with the analogue of Eq. 9.24 shows that

$$\mathcal{C}_{ll',mm'}(t) = \frac{d^2}{dt^2} \mathcal{K}_{ll',mm'}(t) . \tag{9.55}$$

To obtain more detailed expression for inhomogeneous terms in Eqs. 9.50, specification of the initial exciton state needs to be made first. Let us assume the following general initial exciton state:

$$|I\rangle = \sum_{p=1}^{N} I_p |p\rangle , \tag{9.56}$$

where I_p's are complex numbers that can be arbitrary except for the normalization condition $\sum_p |I_p|^2 = 1$. On the other hand, the bath is assumed to be in the canonical equilibrium for the ground electronic state. Thus, the initial condition for the total density operator is given by

$$\rho(0) = |I\rangle\langle I|\rho_b = \sum_{p,\,p'=1}^{N} I_p I^*_{p'} |p\rangle\langle p'|\rho_b . \tag{9.57}$$

Physically, the above initial condition corresponds to sudden creation of a single exciton while all the bath degrees of freedom remain in the original equilibrium for the ground electronic state. For excitation pulses with the time duration of about 100 fs or less, this is a reasonable approximation.

For the initial total density operator of Eq. 9.57,

$$\tilde{\rho}(0) = e^{G} \rho(0) e^{-G} = \sum_{p,\,p'} I_p I^*_{p'} \theta^\dagger_p \rho_b \theta_{p'} |p\rangle\langle p'| . \tag{9.58}$$

This expression can be used for the calculation of $\mathcal{I}^{(1)}(t)$ in Eq. 9.50. For the calculation of $\mathcal{I}^{(2)}(t)$, $\mathcal{Q}\tilde{\rho}(0)$ is needed, which can be expressed as

$$\mathcal{Q}\tilde{\rho}(0) = e^{G}\rho(0)e^{-G} - \rho_b Tr_b\{e^{G}\rho(0)e^{-G}\}$$
$$= \sum_{p,\,p'=1}^{N} \tilde{\sigma}^{pp'}(0)\delta\rho_{b,\,pp'}\,, \tag{9.59}$$

where

$$\tilde{\sigma}^{pp'}(0) \equiv I_p I_{p'}^* w_{pp'}|p\rangle\langle p'|\,, \tag{9.60}$$

$$\delta\rho_{b,\,pp'} \equiv \frac{1}{w_{pp'}}\theta_p^{\dagger}\rho_b\theta_{p'} - \rho_b\,. \tag{9.61}$$

By summing $\tilde{\sigma}^{pp'}(0)$ over p and p', it is easy to show that $\sum_{p,\,p'=1}^{N}\tilde{\sigma}^{pp'}(0) = \tilde{\sigma}(0) = Tr_b\{\tilde{\rho}(0)\}$.

Using the definition of $\tilde{\mathcal{L}}_{1,I}(t)$ given by Eq. 9.45 and the expression for $Q\tilde{\rho}(0)$, Eq. 9.59, the first-order term in Eq. 9.50 can be expressed as

$$\tilde{\mathcal{I}}^{(1)}(t) = -\frac{i}{\hbar}\sum_{l\neq l'}^{N}\sum_{p,\,p'=1}^{N} J_{ll'}[\mathcal{T}_{ll'}(t),\,\tilde{\sigma}_{pp'}(0)]$$
$$\times\frac{1}{w_{pp'}}Tr_b\{\tilde{B}_{ll'}(t)\theta_p^{\dagger}\rho_b\theta_{p'}\}\,, \tag{9.62}$$

where the definition of Eq. 9.60 has been used. For the calculation of Eq. 9.62, let us first introduce the following operator:

$$Y_{p'p}^{ll'}(t) \equiv \theta_{p'}e^{\sum_n \delta g_{n,ll'}(b_n^{\dagger}e^{i\omega_n t} - b_n e^{-i\omega_n t})}\theta_p^{\dagger}\,. \tag{9.63}$$

Employing the cyclic symmetry of the trace operation, the trace over the bath in Eq. 9.62 can be expressed as

$$Tr_b\{\tilde{B}_{ll'}(t)\theta_p^{\dagger}\rho_b\theta_{p'}\} = Tr_b\{Y_{p'p}^{ll'}(t)\rho_b\} - w_{ll'}w_{pp'}\,. \tag{9.64}$$

To evaluate this, we can first express Eq. 9.63 as

$$Y_{p'p}^{ll'}(t) = e^{i\sum_n(g_{n,p}+g_{n,p'})\delta g_{n,ll'}\sin(\omega_n t)}$$
$$\times e^{\sum_n\{(\delta g_{n,ll'}e^{i\omega_n t} - \delta g_{n,p'p})b_n^{\dagger} - (\delta g_{n,ll'}e^{-i\omega_n t} - \delta g_{n,p'p})b_n\}}\,, \tag{9.65}$$

where the following identity that holds between two operators satisfying $[A,\,[A,\,B]] = [B,\,[A,\,B]] = 0$ has been used.

$$e^A e^B = e^{A+B}e^{[A,\,B]/2}\,. \tag{9.66}$$

Combining Eqs. 9.64 and 9.65 and using the following identity that holds for any complex function $\alpha(t)$:

$$Tr_b\{e^{\sum_n \alpha_n(t)b_n^\dagger - \alpha_n^*(t)b_n}\rho_b\} = e^{-\frac{1}{2}\sum_n \coth(\frac{\beta\hbar\omega_n}{2})|\alpha_n(t)|^2}, \qquad (9.67)$$

we obtain the following expression for Eq. 9.64:

$$\begin{aligned} Tr_b\{\tilde{B}_{ll'}(t)\theta_p^\dagger\rho_b\theta_{p'}\} &= w_{pp'}w_{ll'} \\ &\times \left(e^{i\sum_n(g_{n,p}+g_{n,p'})\delta g_{n,ll'}\sin(\omega_n t)}\right. \\ &\times \left. e^{\sum_n \coth(\frac{\beta\hbar\omega_n}{2})\cos(\omega_n t)\delta g_{n,ll'}\delta g_{n,p'p}} - 1\right). \end{aligned} \qquad (9.68)$$

Inserting this expression into Eq. 9.62, we can express the first-order inhomogeneous term as follows:

$$\begin{aligned} \tilde{\mathcal{I}}^{(1)}(t) = &-\frac{i}{\hbar}\sum_{l\neq l'}^{N}\sum_{p,p'=1}^{N} J_{ll'}w_{ll'}[\mathcal{T}_{ll'}(t), \tilde{\sigma}^{pp'}(0)] \\ &\times \left\{e^{-\mathcal{K}_{ll',pp'}(t)}f_{p',ll'}(t) - 1\right\}, \end{aligned} \qquad (9.69)$$

where $\mathcal{K}_{ll',pp'}(t)$ is defined by Eq. 9.53 and

$$f_{p',ll'}(t) = e^{2i\sum_n g_{n,p'}\delta g_{n,ll'}\sin(\omega_n t)}. \qquad (9.70)$$

The second-order inhomogeneous term in Eq. 9.50 can be evaluated in a similar manner. First, employing the definition of the Liouville operator given by Eq. 9.45 and using the expression for $\mathcal{Q}\rho(0)$, Eq. 9.59, we obtain

$$\begin{aligned} \tilde{\mathcal{I}}^{(2)}(t) = &-\frac{1}{\hbar^2}\sum_{l\neq l'}^{N}\sum_{m\neq m'}^{N}\sum_{p,p'=1}^{N} J_{ll'}J_{mm'} \\ &\times \int_0^t d\tau\, Tr_b\{\tilde{B}_{ll'}(t)\tilde{B}_{mm'}(\tau)\delta\rho_{b,pp'}\} \\ &\times [\mathcal{T}_{ll'}(t), \mathcal{T}_{mm'}(\tau)\tilde{\sigma}_{pp'}(0)] + \text{H.c.}, \end{aligned} \qquad (9.71)$$

where

$$\begin{aligned} &Tr_b\left\{\tilde{B}_{ll'}(t)\tilde{B}_{mm'}(\tau)\delta\rho_{b,pp'}\right\} \\ &= Tr_b\left\{\left(e^{\sum_n \delta g_{n,ll'}(b_n^\dagger e^{i\omega_n t} - b_n e^{-i\omega_n t})} - w_{ll'}\right)\right. \\ &\times \left(e^{\sum_n \delta g_{n,mm'}(b_n^\dagger e^{i\omega_n \tau} - b_n e^{-i\omega_n \tau})} - w_{mm'}\right) \\ &\times \left.\left(\frac{1}{w_{pp'}}\theta_p^\dagger\rho_b\theta_{p'} - \rho_b\right)\right\}. \end{aligned} \qquad (9.72)$$

In the above expression, the terms being averaged over ρ_b are similar to those that appear in the components of the homogeneous term, Eq. 9.52. Calculating these terms first, Eq. 9.72 can be expressed as

$$
\begin{aligned}
& Tr_b\left\{\tilde{B}_{ll'}(t)\tilde{B}_{mm'}(\tau)\delta\rho_{b,pp'}\right\} \\
&= \frac{1}{w_{pp'}} Tr_b\left\{e^{\sum_n \delta g_{n,ll'}(b_n^\dagger e^{i\omega_n t}-b_n e^{-i\omega_n t})}\right. \\
&\times \left. e^{\sum_n \delta g_{n,mm'}(b_n^\dagger e^{i\omega_n \tau}-b_n e^{-i\omega_n \tau})}\theta_p^\dagger \rho_b \theta_{p'}\right\} \\
&- Tr_b\left\{e^{\sum_n \delta g_{n,ll'}(b_n^\dagger e^{i\omega_n t}-b_n e^{-i\omega_n t})}\right. \\
&\times \left. e^{\sum_n \delta g_{n,mm'}(b_n^\dagger e^{i\omega_n \tau}-b_n e^{-i\omega_n \tau})}\rho_b\right\} \\
&- w_{mm'}w_{ll'}\left(e^{-\mathcal{K}_{ll',pp'}(t)} f_{p',ll'}(t) - 1\right) \\
&- w_{ll'}w_{mm'}\left(e^{-\mathcal{K}_{mm',pp'}(\tau)} f_{p',mm'}(\tau) - 1\right),
\end{aligned}
\tag{9.73}
$$

where the definitions for $\mathcal{K}_{ll',pp'}(t)$ and $f_{p',ll'}(t)$ are respectively given by Eqs. 9.53 and 9.70.

The terms being averaged over $\theta_p^\dagger \rho_b \theta_p$ in Eq. 9.72 can be calculated in a manner similar to Eq. 9.68. In other words, applying the identity of Eq. 9.66 repeatedly to put all the operators within one exponent and applying the identity of Eq. 9.67, the first term in Eq. 9.73 can be calculated and expressed as

$$
\begin{aligned}
& Tr_b\left\{e^{\sum_n \delta g_{n,ll'}(b_n^\dagger e^{i\omega_n t}-b_n e^{-i\omega_n t})}\right. \\
&\times \left. e^{\sum_n \delta g_{n,mm'}(b_n^\dagger e^{i\omega_n \tau}-b_n e^{-i\omega_n \tau})}\theta_p^\dagger \rho_b \theta_{p'}\right\} \\
&= w_{pp'}w_{ll'}w_{mm'}e^{-\mathcal{K}_{ll',mm'}(t-\tau)-\mathcal{K}_{ll',pp'}(t)-\mathcal{K}_{mm',pp'}(\tau)} \\
&\times f_{p',ll'}(t) f_{p',mm'}(\tau).
\end{aligned}
\tag{9.74}
$$

Combining all of the above the expressions, we obtain the following final expression for Eq. 9.72:

$$
\begin{aligned}
& Tr_b\left\{\tilde{B}_{ll'}(t)\tilde{B}_{mm'}(\tau)\delta\rho_{b,pp'}\right\} \\
&= w_{ll'}w_{mm'}\left\{e^{-\mathcal{K}^{ll'}_{mm'}(t-\tau)}\right. \\
&\times \left(e^{-\mathcal{K}_{ll',pp'}(t)-\mathcal{K}_{mm',pp'}(\tau)} f_{p',ll'}(t) f_{p',mm'}(\tau) - 1\right) \\
&\left. - e^{-\mathcal{K}_{ll',pp'}(t)} f_{p',ll'}(t) - e^{-\mathcal{K}_{mm',pp'}(\tau)} f_{p',mm'}(\tau) + 2\right\}.
\end{aligned}
\tag{9.75}
$$

Inserting this into Eq. 9.71, we find that

$$\tilde{\mathcal{I}}^{(2)}(t) = -\sum_{p,p'=1}^{N}\sum_{l\neq l'}^{N}\sum_{m\neq m'}^{N}\frac{J_{ll'}J_{mm'}}{\hbar^2}$$
$$\times\int_0^t d\tau\,\mathcal{F}^{pp'}_{ll',mm'}(t,\tau)[\mathcal{T}_{ll'}(t),\mathcal{T}_{mm'}(\tau)\tilde{\sigma}^{pp'}(0)]$$
$$+\mathrm{H.c.}, \tag{9.76}$$

where

$$\mathcal{F}^{pp'}_{ll',mm'}(t,\tau) = w_{ll'}w_{mm'}\left\{e^{-\mathcal{K}_{ll',mm'}(t-\tau)}\right.$$
$$\times\left(e^{-\mathcal{K}_{ll',pp'}(t)-\mathcal{K}_{mm',pp'}(\tau)}f_{p',ll'}(t)f_{p',mm'}(\tau)-1\right)$$
$$\left.-e^{-\mathcal{K}_{ll',pp'}(t)}f_{p',ll'}(t)-e^{-\mathcal{K}_{mm',pp'}(\tau)}f_{p',mm'}(\tau)+2\right\}. \tag{9.77}$$

This completes the derivation of the second-order time-local QME in the polaron picture.

9.5 Practical Issues

9.5.1 Physical Observables in the Polaron Picture

Numerical implementation of the PQME can be as efficient as that of the original QME, while being applicable to much wider regime that covers both incoherent and coherent RET dynamics. The price for such advantage is that the expressions for the kernels are more complicated in the former. However, more fundamental issue is that only the diagonal elements (in the site excitation basis) of the reduced system density operator $\tilde{\sigma}(t)$ remain the same as those of $\sigma(t)$. The off-diagonal elements of $\tilde{\sigma}(t)$ are no longer the same as those of $\sigma(t)$, and there is no easy way to relate them. This is because polaron transformation introduces extra phonon entanglement terms for the off-diagonal elements of the total density operator, the detailed information of which gets lost by the trace over the bath. Thus, care should be taken in proper physical interpretation of $\tilde{\sigma}(t)$. A detailed description of how to obtain physical information out of the interaction picture version of the reduced system density operator in the polaron picture, $\tilde{\sigma}_I(t)$, is provided below.

Given the following general operator A of the system:

$$A = \sum_{l,l'=1}^{N} A_{ll'} |l\rangle\langle l'| \,, \tag{9.78}$$

it transforms into the following form in the polaron picture:

$$\tilde{A} = e^{G} A e^{-G} = \sum_{l,l'=1}^{N} A_{ll'} \theta_l^{\dagger} \theta_{l'} |l\rangle\langle l'| \,. \tag{9.79}$$

In the interaction picture with respect to $\tilde{H}_0$, this again becomes

$$\begin{aligned} \tilde{A}_I(t) &= e^{i\tilde{H}_0 t/\hbar} \tilde{A} e^{-\tilde{H}_0 t/\hbar} \\ &= \sum_{l,l'=1}^{N} A_{ll'} e^{i\sum_n \delta g_{n,ll'}(b_n^{\dagger} e^{i\omega_n t} - b_n e^{-i\omega_n t})} \\ &\quad \times e^{i\tilde{H}_{0,s} t/\hbar} |l\rangle\langle l'| e^{-i\tilde{H}_{0,s} t/\hbar} \,. \end{aligned} \tag{9.80}$$

Thus, the measured value of A at time t is given by

$$\begin{aligned} a(t) &\equiv Tr\left\{\tilde{A}_I(t)\tilde{\rho}_I(t)\right\} \\ &= \sum_{l,l'=1}^{N} A_{ll'} Tr\{e^{i\sum_n \delta g_{n,ll'}(b_n^{\dagger} e^{i\omega_n t} - b_n e^{-i\omega_n t})} \\ &\quad \times e^{i\tilde{H}_{0,s} t/\hbar} |l\rangle\langle l'| e^{-i\tilde{H}_{0,s} t/\hbar} \tilde{\rho}_I(t)\} \,. \end{aligned} \tag{9.81}$$

In general, the operators within the trace in the above expression cannot be disentangled into the product of system and bath parts. The exception to this is the case where the operator is diagonal in the site basis such that $A_{ll'} = \delta_{ll'} A_l$. In this case, $a(t)$ can be calculated from the reduced system density operator $\tilde{\sigma}_I(t) = Tr_b\{\rho_I(t)\}$ as shown below.

$$\begin{aligned} a(t) &= \sum_l A_l Tr\{e^{i\tilde{H}_{0,s} t/\hbar} |l\rangle\langle l| e^{-i\tilde{H}_{0,s} t/\hbar} \tilde{\rho}_I(t)\} \\ &= \sum_l A_l Tr_s\{e^{i\tilde{H}_{0,s} t/\hbar} |l\rangle\langle l| e^{-i\tilde{H}_{0,s} t/\hbar} Tr_b\{\tilde{\rho}_I(t)\}\} \\ &= \sum_l A_l Tr_s\{e^{i\tilde{H}_{0,s} t/\hbar} |l\rangle\langle l| e^{-i\tilde{H}_{0,s} t/\hbar} \tilde{\sigma}_I(t)\} \,. \end{aligned} \tag{9.82}$$

A good example for the above quantity is the following population operator:

$$P_l \equiv |l\rangle\langle l| \,, \tag{9.83}$$

which provides primary information on the exciton flow dynamics. In the interaction and polaron picture, this transforms to

$$P_{l,I}(t) = e^{i\tilde{H}_{0,s}t/\hbar}|l\rangle\langle l|e^{-i\tilde{H}_{0,s}t/\hbar} . \tag{9.84}$$

Thus, the population of excitation at the lth chromophore at time t can be calculated from the reduced system density operator as follows:

$$p_l(t) = Tr_s\left\{P_{l,I}(t)\tilde{\sigma}_I(t)\right\} . \tag{9.85}$$

9.5.2 Assessment of Quantum Coherence

To assess the effect of quantum coherence, comparison of the population from the PQME with that based on the following Pauli master equation can be made.

$$\frac{dp_l^h(t)}{dt} = \sum_{m\neq l}\left\{k_{m\to l}^h(t)\,p_m^h(t) - k_{l\to m}^h(t)\,P_l^h(t)\right\} , \tag{9.86}$$

with the rate kernel given by the following non-Markovian version of FRET theory [19]:

$$k_{l\to m}^h(t) = 2\frac{(J_{lm}w_{lm})^2}{\hbar^2}\mathrm{Re}\int_0^t d\tau e^{i(\tilde{E}_l-\tilde{E}_m)\tau/\hbar}\left(e^{\mathcal{K}_{lm,lm}(\tau)} - 1\right) . \tag{9.87}$$

In the above rate expression, note that the divergent component that appears due to the neglect of the lifetime of the excited state has been subtracted within the integrand. This regularization removes the singularity that can exist for superohmic spectral density. For Ohmic spectral density, w_{lm}^2 and $e^{\mathcal{K}_{lm,lm}(\tau)}$ should be combined together so as to have finite value and the extra regularization term does not make contribution. However, consideration of this Ohmic case is not of interest because the PQME becomes equivalent to Eq. 9.86 in this limit unless a variational approach that transform only part of bath is employed [56, 58].

9.5.3 Numerical Implementation

Given the explicit expressions for the kernel and inhomogeneous term of the second-order PQME as derived in the previous section, it is possible to conduct numerical time evolution of the QME in any basis of system Hamiltonian by utilizing appropriate unitary transformation from the site-local basis.

While the choice of basis does not affect the numerical result, it can determine the easiness of actual numerical implementation. The basis of renormalized exciton states, the eigenstates of $\tilde{H}_{0,s}$, is advantageous in this respect because the system time evolution operator for the zeroth-order Hamiltonian becomes diagonal in this case. Full and detailed expressions for $\tilde{\mathcal{R}}(t)$ and $\tilde{\mathcal{I}}(t)$ have been derived in this basis. With these, explicit integration of all the matrix elements of $\tilde{\mathcal{R}}(t)$ and $\tilde{\mathcal{I}}(t)$ as 1D functions of time can be made during the propagation of the reduced density operator. Further numerical details can be found in Refs. [53, 54].

9.6 Model Calculations

The feature and capability of the second-order PQME is illustrated here for two- and three-state systems coupled to a harmonic oscillator bath.

9.6.1 Two-State System

Let us consider the case where the system consists of two molecules or two parts of a molecule, namely, $N = 2$ in Eq. 9.1. Instead of denoting them as 1 and 2, we will call one as the donor (D) of RET and the other as the acceptor (A). The excitation energy of D is E_D and that of A is E_A. The resonance coupling between the two is denoted as J. The quantum-mechanical state where only D is excited is represented by $|D\rangle$ and that where only A is excited is represented by $|A\rangle$. With these, the bare system Hamiltonian in the single exciton space can be expressed as

$$H_s = E_D|D\rangle\langle D| + E_A|A\rangle\langle A| + J\left(|D\rangle\langle A| + |A\rangle\langle D|\right). \qquad (9.88)$$

The bath is modeled by an infinite set of harmonic oscillators and the system–bath coupling is assumed to be linear in the bath coordinates. Thus, the bath Hamiltonian is given by Eq. 9.18 and the system–bath Hamiltonian by Eq. 9.19, which for the present case takes the following form:

$$H_{sb} = \sum_n \hbar\omega_n (b_n + b_n^\dagger)(g_{n,D}|D\rangle\langle D| + g_{n,A}|A\rangle\langle A|)\,. \qquad (9.89)$$

For simplicity, it is assumed that the spectral densities of D and A are identical and there are no common modes. However, it should be stressed that the theory can be applied to more general situation and the common modes can sometimes make significant contributions. It is also assumed that the spectral density has superohmic form with exponential cutoff. Thus,

$$\mathcal{J}_{D(A)}(\omega) = \sum_n \delta(\omega - \omega_n)\omega_n^2 g_{nD(A)}^2 = \frac{\eta}{3!}\frac{\omega^3}{\omega_c^2}e^{-\omega/\omega_c}\,. \qquad (9.90)$$

In all the calculations, it is assumed that $\omega_c = 100\ \text{cm}^{-1}$, which reflects the low-frequency phonon modes of protein environments in light-harvesting complexes.

Figure 9.1 shows the time-dependent population of the acceptor, $P_A(t)$, for $J = 50\ \text{cm}^{-1}$. Three different choices of $\eta = 0.5, 1,$ and 5, and three different energetic conditions, $E_D - E_A = 100,$ 0, and $-100\ \text{cm}^{-1}$, were tested. Panel (a) in Fig. 9.1 is for $T = 100$ K and panel (b) is for $T = 300$ K. Both results with and without the inhomogeneous term $\tilde{\mathcal{I}}(t)$ are shown. The differences between the two are not significant in the present case. Also compared are the population based on the incoherent FRET dynamics, Eq. 9.86.

It is apparent that the population exhibits significant coherence lasting much longer than 1 ps for the weakest system–bath coupling limit ($\eta = 0.5$). However, with the increase of the system–bath coupling, the coherence disappears quickly. Within an order of magnitude increase in η, almost complete agreement can be seen between CRET and FRET. It is also important to note that the presence of coherence in population dynamics does not necessarily mean more efficient population transfer in the steady-state limit.

Figure 9.1b shows similar results at a higher temperature of $T = 300$ K. Even at this room temperature, clear signature of quantum coherence can be seen for $\eta = 0.5$ and $\eta = 1$, although very weak in the latter case, which suggests that experimental observation of electronic quantum coherence that lasts up to 1 ps at room temperature is not that surprising. However, comparison of the steady-state limits between CRET and FRET shows that such

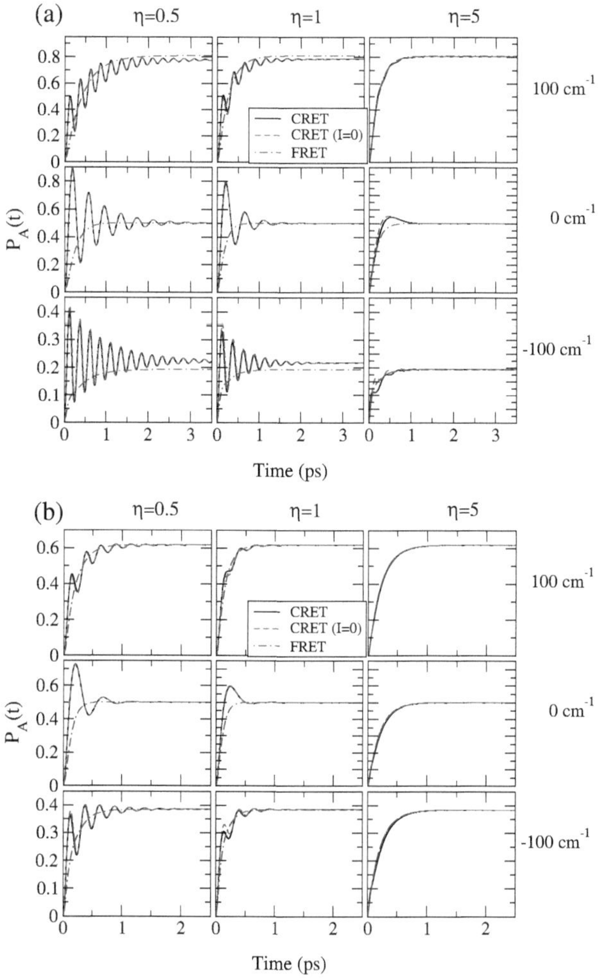

Figure 9.1 Time-dependent donor population, $P_A(t)$ for $J = 50\ \text{cm}^{-1}$. (a) $T = 100$ K and (b) $T = 300$ K. For each case, the top, middle, and bottom rows are for $E_D - E_A = 100, 0, -100\ \text{cm}^{-1}$, respectively. Each column corresponds to a different value of η, as indicated.

transient coherence in population dynamics is somewhat irrelevant to the efficiency of population transfer. At this temperature and for moderately small electronic coupling, FRET appears to work well in predicting the steady-state limit.

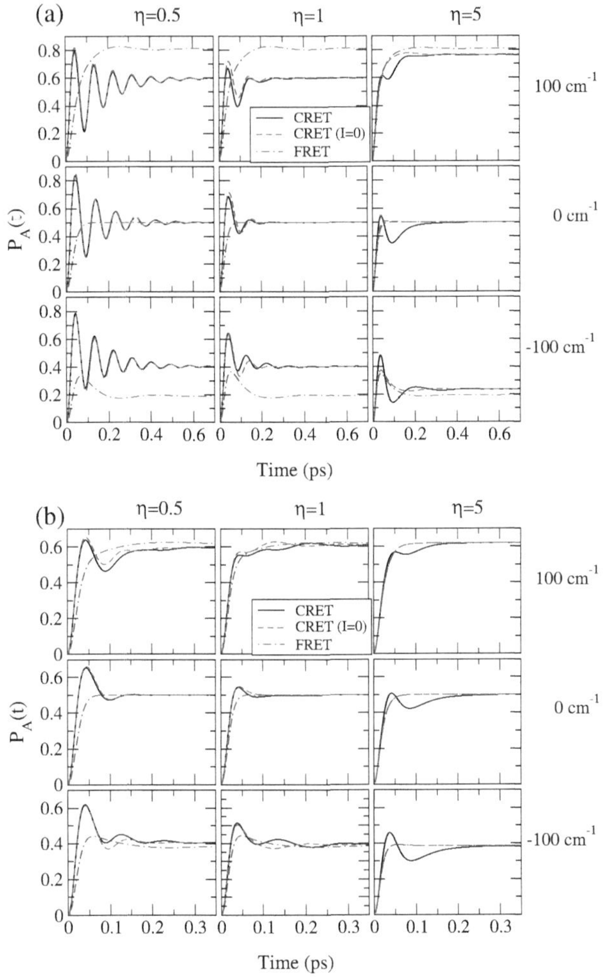

Figure 9.2 Time-dependent donor population, $P_A(t)$ for $J = 200\ \text{cm}^{-1}$. All the conventions are the same as in Fig. 9.1.

Figure 9.2 shows similar results for larger electronic coupling of J $= 200\ \text{cm}^{-1}$. The steady-state limit is reached much quicker in this case, within less than 500 fs in all the cases. Here, the discrepancy between CRET and FRET in the stationary limit is very large for $\eta = 0.5$ at $T = 100$ K, which clearly demonstrates that the coherent delocalization of the exciton enables overcoming the energetic barrier and beating the detailed balance. In the downhill situation ($E_D - E_A = 100\ \text{cm}^{-1}$), the quantum coherence reduces the transfer efficiency. In the uphill situation ($E_D - E_A = -100\ \text{cm}^{-1}$), it enhances the efficiency. However, these quantum effects are diminished significantly as the system–bath coupling and temperature increases. On the other hand, Fig. 9.2b shows that combined effects of quantum coherence and nonequilibrium result in substantial transient signals that last up to about 200 fs.

9.6.2 Three-State System

Let us consider the case where the system consists of three molecules or three parts of a molecule, namely, $N = 3$ in Eq. 9.1. As in the previous section, the donor and acceptor are denoted as D and A, which are joined through a bridge (B) molecule with excitation energy of E_B.

There is no direct coupling between D and A. The resonance coupling between D and B is assumed to be the same as that between B and A. Thus, denoting the state where B is excited as $|B\rangle$, the system Hamiltonian in the present case can be expressed as

$$\begin{aligned} H_s &= E_D|D\rangle\langle D| + E_B|B\rangle\langle B| + E_A|A\rangle\langle A| \\ &\quad + J\left(|D\rangle\langle B| + |B\rangle\langle D| + |B\rangle\langle A| + |A\rangle\langle B|\right). \end{aligned} \tag{9.91}$$

As in the case of the previous section, the bath is modeled by an infinite set of harmonic oscillators and the system–bath coupling is assumed to be linear in the bath coordinates. Thus, the bath Hamiltonian is given by Eq. 9.18 and the system–bath Hamiltonian by Eq. 9.19, which for the present case takes the following form:

$$\begin{aligned} H_{sb} &= \sum_n \hbar\omega_n(b_n + b_n^\dagger)(g_{n,D}|D\rangle\langle D| \\ &\quad + g_{n,B}|B\rangle\langle B| + g_{n,A}|A\rangle\langle A|). \end{aligned} \tag{9.92}$$

For simplicity, it is assumed that the spectral densities of D, B, and A are all identical and there are no common modes. The same superohmic spectral density is assumed. Thus,

$$\mathcal{J}_X(\omega) = \sum_n \delta(\omega - \omega_n)\omega_n^2 g_{nX}^2 = \frac{\eta}{3!}\frac{\omega^3}{\omega_c^2}e^{-\omega/\omega_c}, \tag{9.93}$$

where $X = D$, B, or A and $\omega_c = 100\ \text{cm}^{-1}$.

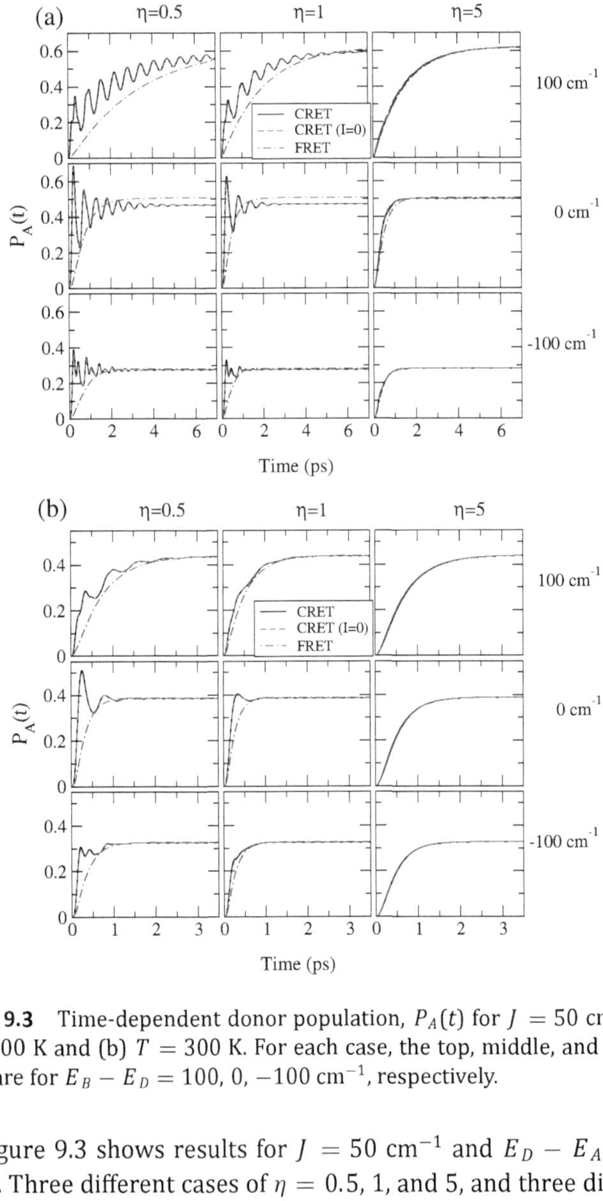

Figure 9.3 Time-dependent donor population, $P_A(t)$ for $J = 50\ \text{cm}^{-1}$. (a) $T = 100$ K and (b) $T = 300$ K. For each case, the top, middle, and bottom rows are for $E_B - E_D = 100, 0, -100\ \text{cm}^{-1}$, respectively.

Figure 9.3 shows results for $J = 50\ \text{cm}^{-1}$ and $E_D - E_A = 50\ \text{cm}^{-1}$. Three different cases of $\eta = 0.5$, 1, and 5, and three different energies of the bridge state $E_B - E_D = 100, 0, -100\ \text{cm}^{-1}$ were tested. As in the case of the two-state system, coherent population dynamics lasting much longer than 1 ps can be observed for $\eta =$

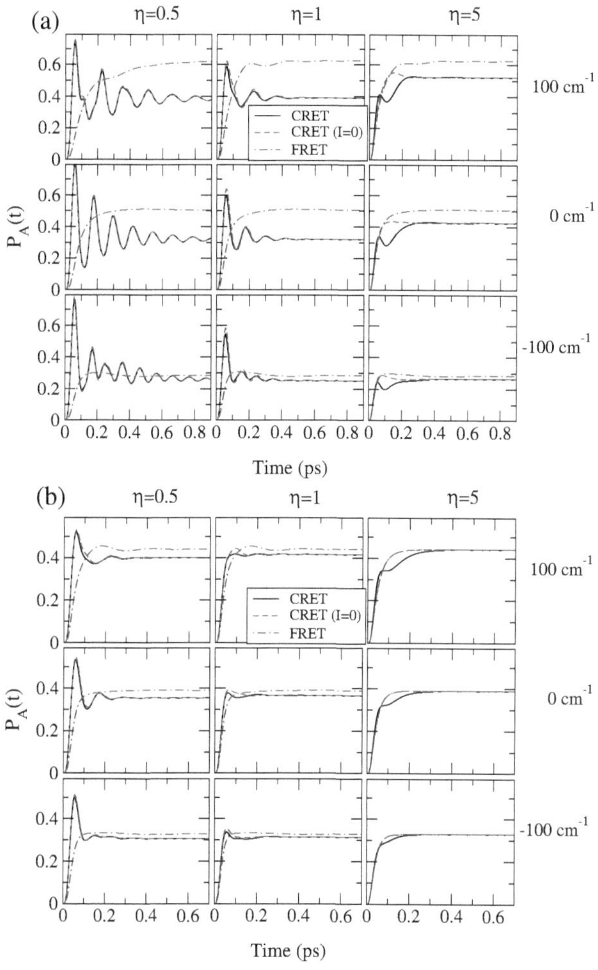

Figure 9.4 Time-dependent donor population, $P_A(t)$ for $J_{DB} = J_{BA} = 200$ cm^{-1}. Other conventions are the same as in Fig. 9.3.

0.5. The initial population dynamics is shown to be much more insensitive to the energy of the bridge for the case of CRET, which suggests that the main benefit of utilizing quantum coherence is in its robustness against disorder. However, such positive effects are shown to diminish quickly with the increase of system–bath coupling and temperature.

Figure 9.4 shows similar results for larger electronic coupling of $J = 200$ cm^{-1}. As in the case of a two-state system, the dynamics becomes accelerated significantly. The steady-state limit is reached within about 500 fs at 100 K and within about 200 fs at 300 K.

For weak ($\eta = 0.5$) and moderately weak ($\eta = 1$) system–bath couplings, the initial population transfer based on CRET is much faster than FRET especially at $T = 100$ K. However, the steady-state limit population $P_A(t)$ based on the former is always smaller than that based on the latter. This can be understood considering that E_A is lower than E_D. Thus, coherent delocalization offsets populations dictated by the classical detailed balance condition, even when the interaction between D and A is mediated by B.

9.7 Applications

The theory has been applied to actual molecular system called dithiaanthracenophane (DTA) as shown in Fig. 9.5a. Earlier theoretical work [62] on DTA suggested that the electronic coupling between the two anthracene units, based on experimental fitting of the period of oscillation, is 14.5 cm^{-1}, which appeared to be close to their [62] theoretical estimate based on transition dipole interaction, about 20 cm^{-1}. However, standard electronic structure calculation [61] showed that the actual electronic coupling should be at least 50 cm^{-1}, even including the solvation effect. The measured oscillation period and dephasing rate cannot be reconciled with this value of electronic coupling within the Redfield equation approach [30, 32], which assumes weak system–bath coupling. However, within the PQME approach, it was possible to explain the experimental observation without introducing any empirical parameter.

Figure 9.5b shows some of the calculation results. The system Hamiltonian is the same as Eq. 9.88 except that $J' = w_h J$, with $w_h = 0.449$, is used to reflect the renormalization of the electronic coupling due to the high-frequency mode [61]. It is assumed that $E_D = E_A$. For each value of J, the values of η and ω_c were chosen such that the corresponding reorganization energy is 400 cm^{-1} and the period of oscillation remains about 1.2 ps.

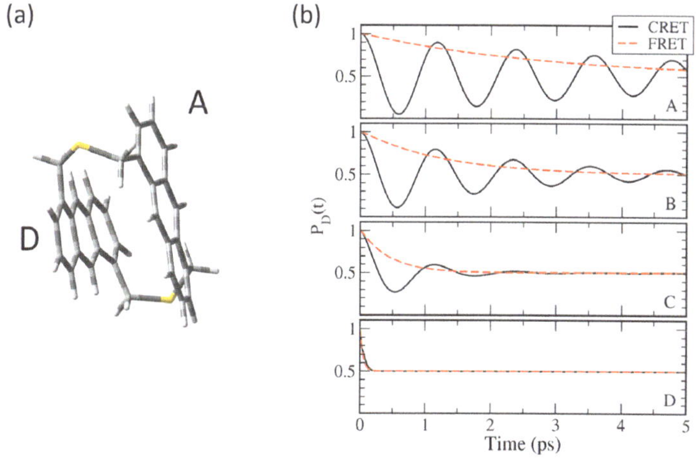

Figure 9.5 (a) The structure of dithia-anthracenophane (DTA). The donor (D) and the acceptor (A) in this molecule correspond to the anthracene units. (b) Time-dependent population dynamics of $|D\rangle$ for different choices of $J' = w_h J$, where $w_h = 0.449$ is the renormalization factor due to high-frequency mode. The values of $(J, \eta, \omega_c/2\pi c)$ for each case are as follows: (A) (53.5 cm^{-1}, 2.4, 500 cm^{-1}); (B) (66.8 cm^{-1}, 3, 400 cm^{-1}); C (100 cm^{-1}, 4, 300 cm^{-1}); D (334 cm^{-1}, 6, 200 cm^{-1}). Adapted with permission from Ref. [61]. Copyright (2010), American Chemical Society.

Among the four results, the cases (A) and (B) seem to best represent the actual RET dynamics in the DTA molecule. The implication of these results is that the bare electronic coupling can be in the range of 50–100 cm^{-1} and the apparent slow oscillation is the result of loose polaronic effects of the molecular vibration and the solvent. In addition, this also shows that the renormalization of the electronic coupling and dephasing comes from the same source and cannot be assumed to be independent of each other.

The DTA has some amount of torsional flexibility, but it has not been clearly understood whether such torsional modulation has significant effect on the dephasing and relaxation dynamics. A theory [63] of inelastic coherent resonance energy transfer (I-CRET) that can account for this effect was developed recently by extending the PQME approach. A series of model calculations based on this theory indicates that the torsional modulation can enhance the dephasing but is not at a significant level for the case of DTA.

9.8 Summary and Outlook

The second-order PQME as a quantum-dynamical method that can bridge coherent and incoherent RET dynamics was presented, and its capability and promise were demonstrated for model systems and an application to DTA.

To make the present chapter complete and self-contained, a concise derivation of formally exact QME and their conventional second-order approximations were also provided. Thus, the present work also serves as a minireview of QME formalism in general, with special emphasis on PQME.

Model calculations and applications were limited to relatively simple systems in order to demonstrate clearly important features captured by the PQME. However, it can be applied to much larger systems with tens of quantum states. Such applications will provide new insights into the role of electronic quantum coherence and electronic-vibrational couplings in the RET dynamics.

Despite its promise, the PQME is relatively new method and much effort has to be made in order to understand its capability and limitations in more detail. For example, to what extent the second-order PQME can be trusted and serves as significant improvement over conventional second-order QME is not yet well established. For this, systematic numerical studies for a broad range of model systems is are needed.

There is much room for further improvement of the method. Approximation up to the second order has been tested, but expansion up to the fourth order is possible and may result in significant improvement. In addition, variational PQME [56, 58] has been shown to solve one important issue of PQME, overrelaxation for a sluggish bath. Further refinement of variational PQME seems possible and will be an important theoretical direction to explore.

Acknowledgments

This work was supported by the US National Science Foundation CAREER award (grant no. CHE-0846899), the Office of Basic Energy

Sciences, Department of Energy (grant no. DE-SC0001393), and the Camille Dreyfus Teacher Scholar Award. SJ acknowledges Dr. Ahsan Nazir for helpful comments.

References

1. R. E. Blankenship, *Molecular Mechanisms of Photosynthesis* (Blackwell Science, Oxford, 2002).
2. S. Jang and Y.-C. Cheng, *WIREs Comput. Mol. Sci.*, **3**, 84 (2013).
3. V. M. Agranovič and M. D. Galanin, *Electronic Excitation Energy Transfer in Condensed Matter* Vol. 3 (North-Holland, 1982).
4. R. Silbey, *Annu. Rev. Phys. Chem.*, **27**, 203 (1976).
5. G. D. Scholes, *Annu. Rev. Phys. Chem.*, **54**, 57 (2003).
6. T. Förster, *Ann. Phys.*, **6**, 55 (1948).
7. T. Förster, *Modern Quantum Chemistry. Part III: Action of Light and Organic Crystals*, O. Sinanoglu, eds. (Academic Press, New York, 1965).
8. D. L. Dexter, *J. Chem. Phys.*, **21**, 836 (1953).
9. J. B. Birks, *Excited States of Biological Molecules* (John Wiley & Sons, London, 1976).
10. D. L. Andrews and A. A. Demidov, *Resonance Energy Transfer* (Wiley, New York, 1999).
11. V. M. Agranovich and R. M. Hochstrasser, *Spectroscopy and Excitation Dynamics of Condensed Molecular Systems*, Vol. 4 (North-Holland, 1983).
12. L. Stryer and R. P. Haugland, *Proc. Natl. Acad. Sci. U S A*, **58**, 719 (1967).
13. T. Heyduk, *Curr. Opin. Biotechnol.*, **13**, 292 (2002).
14. T. Soules and C. Duke, *Phys. Rev. B*, **3**, 262 (1971).
15. B. Jackson and R. Silbey, *J. Chem. Phys.*, **78**, 4193 (1983).
16. H. Sumi, *J. Phys. Chem. B*, **103**, 252 (1999).
17. G. D. Scholes, X. J. Jordanides, and G. R. Fleming, *J. Phys. Chem.*, **105**, 1640 (2001).
18. A. Kimura and T. Kakitani, *J. Phys. Chem. A*, **111**, 12042 (2007).
19. S. Jang, Y. Jung, and R. J. Silbey, *Chem. Phys.*, **275**, 319 (2002).
20. S. Jang, M. D. Newton, and R. J. Silbey, *Phys. Rev. Lett.*, **92**, 218301 (2004).
21. S. Jang, *J. Chem. Phys.*, **127**, 174710 (2007).
22. S. Jang, M. D. Newton, and R. J. Silbey, *J. Phys. Chem. B*, **111**, 6807 (2007).
23. G. S. Engel et al., *Nature*, **446**, 782 (2007).

24. E. Collini et al., *Nature*, **463**, 644 (2010).
25. W. H. Miller, *J. Phys. Chem. A*, **105**, 2942 (2001).
26. M. Ben-Nun, J. Quenneville, and T. J. Martínez, *J. Phys. Chem. A*, **104**, 5161 (2000).
27. N. Makri, *Annu. Rev. Phys. Chem.*, **50**, 167 (1999).
28. H.-P. Breuer, B. Kappler, and F. Petruccione, *Ann. Phys.*, **291**, 36 (2001).
29. R. K. Wangsness and F. Bloch, *Phys. Rev.*, **89**, 728 (1953).
30. A. G. Redfield, *IBM J. Res. Dev.*, **1**, 19 (1957).
31. J. M. Jean and G. R. Fleming, *J. Chem. Phys.*, **103**, 2092 (1995).
32. K. Wynne and R. Hochstrasser, *J. Raman Spectrosc.*, **26**, 561 (1995).
33. A. Matro and J. A. Cina, *J. Phys. Chem.*, **99**, 2568 (1995).
34. T. Renger, V. May, and O. Kühn, *Phys. Rep.*, **343**, 137 (2001).
35. M. Yang and G. R. Fleming, *Chem. Phys.*, **275**, 355 (2002).
36. U. Kleinekathöfer, *J. Chem. Phys.*, **121**, 2505 (2004).
37. A. Pomyalov and D. J. Tannor, *J. Chem. Phys.*, **123**, 204111 (2005).
38. S. Jang, J. Cao, and R. J. Silbey, *J. Chem. Phys.*, **116**, 2705 (2002).
39. B. B. Laird, J. Budimir, and J. L. Skinner, *J. Chem. Phys.*, **94**, 4391 (1991).
40. B. B. Laird and J. L. Skinner, *J. Chem. Phys.*, **94**, 4405 (1991).
41. A. A. Golosov and D. R. Reichman, *Chem. Phys.*, **296**, 129 (2004).
42. M.-L. Zhang, J. K. Being, and E. Geva, *J. Chem. Phys.*, **125**, 044106 (2006).
43. R.-X. Xu and Y. Yan, *Phys. Rev. E*, **75**, 031107 (2007).
44. M. Schröder, M. Schreiber, and U. Kleinekathöfer, *J. Chem. Phys.*, **126**, 114102 (2007).
45. R. Doll, D. Zueco, M. Wubs, S. Kohler, and P. Hänggi, *Chem. Phys.*, **347**, 243 (2008).
46. T. Holstein, *Ann. Phys.*, **8**, 325 (1959).
47. T. Holstein, *Ann. Phys.*, **8**, 343 (1959).
48. R. Silbey and R. A. Harris, *J. Chem. Phys.*, **80**, 2615 (1984).
49. R. Harris and R. Silbey, *J. Chem. Phys.*, **83**, 1069 (1985).
50. S. Rackovsky and R. Silbey, *Mol. Phys.*, **25**, 61 (1973).
51. D. Yarkony and R. Silbey, *J. Chem. Phys.*, **65**, 1042 (1976).
52. S. Jang, Y.-C. Cheng, D. R. Reichman, and J. D. Eaves, *J. Chem. Phys.*, **129**, 101104 (2008).
53. S. Jang, *J. Chem. Phys.*, **131**, 164101 (2009).
54. S. Jang, *J. Chem. Phys.*, **135**, 034105 (2011).

55. A. Nazir, *Phys. Rev. Lett.*, **103**, 146404 (2009).
56. D. P. McCutcheon and A. Nazir, *J. Chem. Phys.*, **135**, 114501 (2011).
57. A. Kolli, A. Nazir, and A. Olaya-Castro, *J. Chem. Phys.*, **135**, 154112 (2011).
58. E. N. Zimanyi and R. J. Silbey, *Phil. Trans. R. Soc. A*, **370**, 3620 (2012).
59. M. Palenberg, R. Silbey, C. Warns, and P. Reineker, *J. Chem. Phys.*, **114**, 4386 (2001).
60. L. Chen, R. Zheng, Q. Shi, and Y. Yan, *J. Chem. Phys.*, **131**, 094502 (2009).
61. L. Yang, S. Caprasecca, B. Mennucci, and S. Jang, *J. Am. Chem. Soc.*, **132**, 16911 (2010).
62. I. Yamazaki, S. Akimoto, T. Yamazaki, S.-i. Sato, and Y. Sakata, *J. Phys. Chem. A*, **106**, 2122 (2002).
63. L. Yang, M. Devi, and S. Jang, *J. Chem. Phys.*, **137**, 024101 (2012).

Chapter 10

Theory of Metal Nanoparticle–Affected Optical and Transport Properties in Supramolecular Complexes

Yuan Zhang, Yaroslav Zelinskyy, Gerold Kyas, and Volkhard May
Institut für Physik, Humboldt-Universität zu Berlin,
Newtonstraße 15, D-12489 Berlin, Germany
may@physik.hu-berlin.de

10.1 Introduction

Within the last two decades the investigation of metal nanoparticles (MNPs) has attracted enormous interest in physics, physical chemistry and even molecular biology (for a recent review see [1, 2]). Considering MNPs with an extension ranging from some nm up to some 100 nm, a multitude of experimental arrangements and effects have been studied. The central observation here is the strong response of the MNP electrons to external electromagnetic perturbations. This response takes place via the formation of surface plasmons representing collective motions of the metal conduction band electrons. Since the number of involved electrons

Ultrafast Dynamics at the Nanoscale: Biomolecules and Supramolecular Assemblies
Edited by Irene Burghardt and Stefan Haacke

ISBN 978-981-4745-33-8 (Hardcover), 978-981-4745-34-5 (eBook)
www.panstanford.com

is huge, the MNP absorption and scattering cross sections are large. The incoming field is drastically altered, subwavelength localization of electromagnetic energy appears, and high-intensity hot spots of field strength are formed near the surface of the MNP. When placing molecules, molecular complexes, biomolecules, semiconductor quantum dots, or nanocrystals in the proximity of an MNP, their optical and transport properties are essentially modified [1]. The simplest effect would be the increased optical excitation of a quantum system near an MNP due to the field enhancement of an externally applied laser pulse.

Optical properties of spherical MNPs have been the subject of Mie–Gans theory dating back to the beginning of the 20th century. The discovery of surface-enhanced Raman scattering in 1974 renewed the interest in MNPs. Nowadays the properties of differently shaped MNPs and arrangements of MNPs are of interest. They may act as nanoantennas and may localize the radiation field of the visible region on a nanometer scale. So-called metamaterials are formed by a huge and regular arrangement of MNPs. Of similar interest are arrangements of MNPs with molecules extending up to large proteins and pigment–protein complexes. A detailed understanding of MNP-enhanced absorption, emission, and scattering, as well as intermolecular transport of charge and energy, should be gained. Potential applications lie, for example, in the field of photovoltaics, molecular electronics, molecular biology, and medicine. To achieve insight into MNP-induced processes in molecular systems, a rather large number of theoretical studies have been carried out.

It is the intention of the following discussion to present a consistent theory of optical and transport properties of single molecules or supramolecular complexes (SCs) placed in the proximity of MNPs. Figure 10.1 displays an example for such an SC–MNP system where a linear SC (a molecular chain) is positioned near a single spherical MNP. Our approach will be based on an uniform quantum formulation of the whole system. Therefore, we do not offer a survey of the existing literature but refer to other treatments and theoretical attempts only when necessary. The theory presented in the following considers the electronic excitations of the MNP and the SC and their mutual coupling as well as the coupling to an external

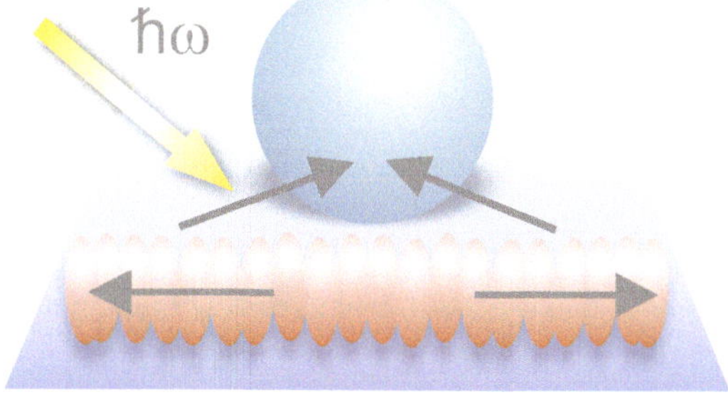

Figure 10.1 Scheme of a spherical MNP placed beside a linear arrangement of molecules. Yellow arrow: optical excitation; small gray arrows: possible excitation energy transfer pathways. Reprinted from Ref. [7], with permission from John Wiley & Sons, Inc.

laser field. We concentrate on the presence of a single spherical MNP and on a spatial extension of the system where the SC–MNP coupling can be described via instantaneous Coulomb interaction (absence of propagation effects). The theory does not require the computation of an MNP-induced local field by a solution of Maxwell's equations. Here, external field–induced dynamics in the SC–MNP system are accounted for by the solution of density matrix equations defined for the coupled system. This offers a consistent description of optical and transport properties of the system. The presentation is based on our former work of Refs. [3–11]. It will end up in an SC–MNP electronic energy–level scheme, as shown in Fig. 10.2.

To offer a somewhat more detailed insight in our general approach we start with the general Hamiltonian suitable to describe the coupled SC–MNP system in the presence of an external laser field. It reads

$$H(t) = H_{\text{sc}} + H_{\text{mnp}} + H_{\text{sc-mnp}} + H_{\text{field}}(t) \,. \tag{10.1}$$

The Hamiltonian of the SC

$$H_{\text{sc}} = \sum_{n} H_n + \frac{1}{2} \sum_{n,n'} V_{nn'} \tag{10.2}$$

covers a set of single molecules counted by n with the respective Hamiltonian H_n. They shall be characterized by an electronic ground state $\varphi_{ng}(x)$ and an excited state $\varphi_{ne}(x)$ (the set of electron coordinates is abbreviated by x). If necessary these two states

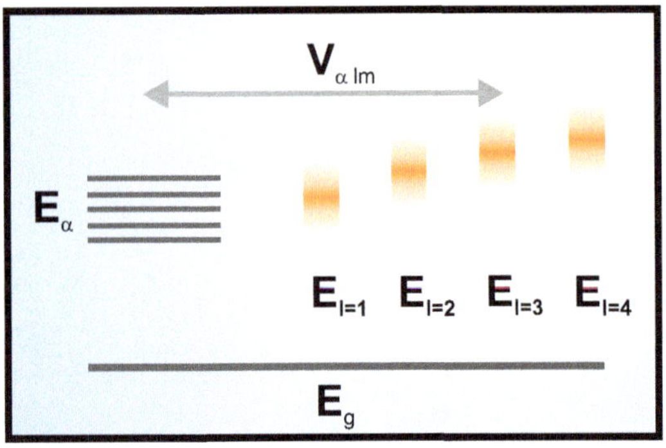

Figure 10.2 Energy-level scheme of an SC–MNP system like that shown in Fig. 10.1. The SC is represented by its exciton energies E_α (to visualize them only a restricted number has been shown; E_g denotes the SC–MNP ground-state energy). The multipole excitation energies E_l referring to the spherical MNP are drawn as broadened levels (orange) reflecting their short lifetime. They are connected with the exciton levels via the energy exchange coupling with matrix elements $V_{\alpha,lm}$. Reprinted with permission from Ref. [6]. Copyright (2012) American Chemical Society.

can be extended by related vibrational wave functions. However, in the following we will concentrate on the electronic degrees of freedom. $V_{nn'}$ is the intermolecular Coulomb coupling. Often $H_{\rm sc}$ can be reduced to a Frenkel–exciton Hamiltonian (see [12] and the subsequent discussion in Section 10.6). The MNP Hamiltonian

$$H_{\rm mnp} = \sum_j \frac{\mathbf{p}_j^2}{2m_{\rm el}} + \frac{1}{2}\sum_{j,j'} \frac{e^2}{|\mathbf{y}_j - \mathbf{y}_{j'}|} + \sum_j v_{\rm ion}(\mathbf{y}_j) \quad (10.3)$$

describes a set of $N_{\rm el}$ conduction band electrons. It includes the Coulomb coupling among them, and characterizes the effect of the metal ions by the local potential $v_{\rm ion}$. The related MNP electron wave functions are denoted as $\psi_k(y)$ with quantum number k and with the set of electron coordinates y. The MNP electron eigenenergies E_k represent a broad continuum covering collective and individual excitations, that is, surface plasmons, volume plasmons, and single electron–hole excitations. Their coupling to the SC excitations is described by

$$H_{\text{sc-mnp}} = H^{\rm (inst)}_{\text{sc-mnp}} + H^{\rm (ret)}_{\text{sc-mnp}} \,. \quad (10.4)$$

First, this part of the overall Hamiltonian includes the instantaneous Coulomb interaction among all charges of the SC and the MNP. Here it is sufficient to concentrate on the interaction among the electrons (see the related discussion in Section 10.3, $\mathbf{x}_{nj}$ are the electron coordinates of molecule n),

$$H_{\text{sc-mnp}}^{(\text{inst})} = \sum_{n} \sum_{j,j'} \frac{e^2}{|\mathbf{x}_{nj} - \mathbf{y}_{j'}|} \, . \tag{10.5}$$

Second, the retarded part of the SC–MNP coupling, Eq. 10.4, is mediated by transversal polarized photons and includes the Hamiltonian of free photons

$$H_{\text{sc-mnp}}^{(\text{ret})} = \sum_{\lambda,\mathbf{k}} \hbar(\hat{h}_{\lambda\mathbf{k}}^{(+)} a_{\lambda\mathbf{k}} + \hat{h}_{\lambda\mathbf{k}}^{(-)} a_{\lambda\mathbf{k}}^{+}) + \sum_{\lambda,\mathbf{k}} \hbar\omega_{\mathbf{k}}(a_{\lambda\mathbf{k}}^{+} a_{\lambda\mathbf{k}} + 1/2) \, . \tag{10.6}$$

The expression covers creation and annihilation operators of photons $a_{\lambda\mathbf{k}}^{+}$ and $a_{\lambda\mathbf{k}}$, respectively (with polarization λ and wave vector $\mathbf{k}$). The photon energy is denoted by $\hbar\omega_{\mathbf{k}} = \hbar c|\mathbf{k}|$. The operators $\hat{h}_{\lambda\mathbf{k}}^{(\pm)}$ describe the coupling to the SC–MNP system and are detailed in Section 10.5.4.

As already indicated, the treatment of the SC–MNP as a uniform quantum system shall take into account its small extension with respect to photon wavelengths in the optical range. Those correspond to the excitation energies addressed by the external laser pulses of the SC as well as the MNP. Therefore, retardation effects mediated by the quantized transversal photon field are of less importance. As a consequence $H_{\text{sc-mnp}}^{(\text{ret})}$ is treated in perturbation theory while the instantaneous Coulomb interaction $H_{\text{sc-mnp}}^{(\text{inst})}$ has to be considered beyond a perturbational approach. For example, this guarantees that the SC–MNP emission spectrum addresses far-field photons (in plane wave expansion) but includes the complete effect of the instantaneous Coulomb interaction. Finally, we present the coupling to external laser fields

$$H_{\text{field}}(t) = -\hat{\boldsymbol{\mu}}\mathbf{E}_{\text{ext}}(t). \tag{10.7}$$

The quantity $\hat{\boldsymbol{\mu}}$ is the dipole operator of the whole SC–MNP system. Although the MNP may display multipole excitations, the coupling to MNP dipole excitations is here sufficient (the external field slowly varies across the MNP).

Having introduced the general framework of the subsequent discussions, we next concentrate on a quantum description of the collective electronic MNP excitations. This discussion will result in strongly lifetime broadened MNP multipole excitations as presented in Fig. 10.2. Afterward, the SC–MNP coupling is derived (gray arrows in Fig. 10.2). Concerning the standard approach of introducing Frenkel exciton levels of the SC we refer to the literature (see, for example, Ref. [12]).

10.2 Collective Excitations of Metal Nanoparticle Electrons

The descriptions of MNP electrons can be divided into approaches directly based on the Hamiltonian H_{mnp}, and in quantization schemes which start from macroscopic equations for the electron polarization field. Kinetic Boltzmann equations for the MNP conducting electrons have been also used in this context (see, for example, [13, 14]). These treatments are based on the assumption that such a "second quantization" of the electron motion in an MNP gives correct results. This can be proven when coupling the MNP excitations to those of a single molecule or of other MNP. In the present description as detailed below the coupling of excited states either of the SC or of the MNP to unexcited ones is determined by so-called transition densities relating the electronic ground state to the excited one. Respective MNP transition densities are missing in the second quantization treatments making the whole approach questionable. Therefore, we will proceed in a more microscopic way.

The MNP theories which are based directly on H_{mnp} cover a theory resulting in dipole plasmons [15–17]. Higher multipole plasmons are accounted for in the theory of [18–23] which borrows ideas from nuclear physics (theory of vibrating nuclei). Higher multipole plasmons are also included in the approach of [24] which considers the MNP electron density in an RPA-like approximation (we also refer to the Green's function treatment of [25]).

10.2.1 Description of Dipole Plasmons

It has been demonstrated in Ref. [15] that the MNP Hamiltonian H_{mnp}, introduced in Eq. 10.3, can be transformed to an expression characterizing surface plasmon excitations coupled to the remaining intrinsic electron motion of the MNP (individual electron–hole pair generation). Such a separation is achieved by introducing an electron center of mass motion (collective motion) with coordinate $\mathbf{R} = \sum_j \mathbf{y}_j / N_{\rm el}$ and a relative motion (intrinsic motion) $\mathbf{r}_j = \mathbf{y}_j - \mathbf{R}$. This results in the following separation of the MNP Hamiltonian

$$H_{\rm mnp} = H_{\rm mnp}^{\rm (intr)} + \frac{\mathbf{P}^2}{2N_{\rm el}m_{\rm el}} + U_{\rm pl}(\mathbf{R}) + \Big(U\left(\mathbf{R}, \{\mathbf{r}_j\}\right) - U_{\rm pl}(\mathbf{R})\Big) . \quad (10.8)$$

The original Hamiltonian transformed to the electron center of mass is denoted as $H_{\rm mnp}^{\rm (intr)}$. The potential $U\left(\mathbf{R}, \{\mathbf{r}_j\}\right)$ accounts for the coupling between the collective and intrinsic electron motion. It is obtained from the difference of metal ion potentials

$$U\left(\mathbf{R}, \{\mathbf{r}_j\}\right) = \sum_j \left(v_{\rm ion}(\mathbf{R} + \mathbf{r}_j) - v_{\rm ion}(\mathbf{r}_j)\right) . \quad (10.9)$$

The expectation value of this potential with respect to the intrinsic electron ground-state wave function ψ_0 results in the following expression (the plasmon potential)

$$U_{\rm pl}(\mathbf{R}) = \langle\psi_0|U\left(\mathbf{R}, \{\mathbf{r}_j\}\right)|\psi_0\rangle = \int d^3\mathbf{r}\; v_{\rm ion}(\mathbf{r})\left(n_{\rm el}(\mathbf{r} - \mathbf{R}) - n_{\rm el}(\mathbf{r})\right) . \quad (10.10)$$

In the second part of this equation we introduce the single-particle density of the intrinsic electron motion $n_{\rm el}(\mathbf{r}) = N_{\rm el} \int dr\; \delta(\mathbf{r} - \mathbf{r}_1)|\psi_0(r)|^2$. The plasmon potential follows as the interaction energy of the displaced electron density $n_{\rm el}(\mathbf{r} - \mathbf{R})$ with the metal ion potential relative to the energy of the undisplaced density $n_{\rm el}(\mathbf{r})$. Accordingly, the part $\mathbf{P}^2/2N_{\rm el}m_{\rm el} + U_{\rm pl}(\mathbf{R})$ of Eq. 10.8 defines the collective surface plasmon excitations. Following Ref. [17] and introducing a quasi-particle expansion of $H_{\rm mnp}^{\rm (intr)}$ (describing intrinsic electron–hole pair excitations) the difference potential $U\left(\mathbf{R}, \{\mathbf{r}_j\}\right) - U_{\rm pl}(\mathbf{R})$ describes the coupling between plasmon excitations and individual electron–hole pair excitations. The main consequence of this coupling is the fast plasmon decay (on a 10 fs time scale).

Considering a spherical MNP and expanding $U_{\rm pl}$ with respect to powers of $\mathbf{R}$ one obtains to lowest order $U_{\rm pl}(\mathbf{R}) \approx N_{\rm el}m_{\rm el}\omega_{\rm dp}^2\mathbf{R}^2/2$. The dipole plasmon frequency reads $\omega_{\rm dp}^2 = 4\pi e^2 \int d^3\mathbf{r}\;\; n_{\rm el}(\mathbf{r})n_{\rm ion}(\mathbf{r})/3N_{\rm el}m_{\rm el}$, where $n_{\rm ion}(\mathbf{r})$ is the metal ion density deduced from $v_{\rm ion}(\mathbf{r})$. Using the standard Mie frequency of dipole plasmons $\omega_1^{(M)\,2} = 4\pi n_{\rm bulk} e^2/3m_{\rm el}$ (with bulk electron density $n_{\rm bulk}$) one may write $\omega_{\rm dp} = \omega_1^{(M)}\sqrt{1 - \delta N_{\rm el}/N_{\rm el}}$. The reduced value of $\omega_{\rm dp}$ compared to $\omega_1^{(M)}$ is related to the number $\delta N_{\rm el}$ of electrons tunneling through the MNP surface (so-called spill-out electrons). This reduction of $\omega_{\rm dp}$ roughly corresponds to the experimental observation (higher orders of the $U_{\rm pl}(\mathbf{R})$ expansion and energy shifts due to the plasmon intrinsic electron–hole coupling improve this result; more details can be found in [15–17]).

Noting the lowest order expression of $U_{\rm pl}(\mathbf{R})$, the collective MNP electron motion appears as an isotropic 3D harmonic oscillator. Its transition dipole moments are easily calculated. Considering the separation of $H_{\rm mnp}$, Eq. 10.8, as a system–bath separation of dissipative quantum dynamics the whole approach can be directly embedded in a density matrix theory with the coupling potential $U(\mathbf{R}, \{\mathbf{r}_j\}) - U_{\rm pl}(\mathbf{R})$ considered in second order of perturbation theory. Resulting plasmon decay rates then directly describe their short lifetime due to the excitation of intrinsic electron–hole pairs (for more details see also [3]).

10.2.2 Description of Multipole Plasmons

While the consideration of dipole plasmons as presented in the preceding section is rather straightforward, higher multipole excitations could not be described in this scheme so far. As is well known, however, those are essential whenever near-field effects come into play. This is the case if the MNP interacts with a molecule which is in its close proximity or if two or more MNPs interact with each other. Also MNP radiative decay may display multipole contributions to the emitted photons. Consequently, a Hamiltonian like

$$H_{\rm pl} = \sum_N E_N |N\rangle\langle N| \tag{10.11}$$

is needed which covers all collective plasmon excitations (with energy E_N and state vector $|N\rangle$). In the case of a spherical MNP one expects multipolar indices lm ($l = 0, 1, \ldots; m = -l, \ldots, l$) as quantum numbers and a similarity of the E_{lm} with the classical Mie frequencies (the bulk plasma frequency is $\omega_{\rm pl}^2 = 4\pi n_{\rm bulk} e^2/m_{\rm el}$)

$$\omega_l^{(\rm M)} = \sqrt{l/(2l+1)}\,\omega_{\rm pl}\,. \tag{10.12}$$

A respective approach has been advocated in Refs. [18–21, 23] which is based on the introduction of an effective Hamiltonian and an analysis of particular MNP energy moments. The weakness of the subsequent derivation of $H_{\rm pl}$ is, however, that the Hamiltonian does not follow directly by a sequence of approximations from $H_{\rm mnp}$, Eq. 10.3, but is only defined by its eigenvalues E_N and matrix elements of its eigenstates $|N\rangle$.

According to Ref. [20] collective excited state and energies of the MNP electron system can be determined by starting with the ansatz $\psi(y) = \exp(i\hat{Q})\psi_0(y)$ for the excited-state wave function (y again abbreviates the whole set of MNP electron coordinates). The ansatz implies that the excited state is obtained from the ground state by a deformation caused by the operator $\hat{Q}$. The latter should be of the form $\hat{Q} = \sum_j q(\mathbf{y}_j)$, where the $q(\mathbf{y}_j)$ are single-electron operators which only depend on the electron coordinates. In the case of a spherical MNP it becomes possible to mimic multipole excitations of the MNP electron system if the various $q(\mathbf{y})$ are identified with multipole operators

$$q_{lm}(\mathbf{y}) = \sqrt{4\pi/(2l+1)}\,y^l Y_{lm}(\vartheta, \varphi)\,. \tag{10.13}$$

The y, ϑ, and φ are the spherical coordinates of $\mathbf{y}$, and the $Y_{lm}(\vartheta, \varphi)$ denote spherical harmonics. Since an ansatz is used for the electronic wave function we can only expect to get that part of the energy spectrum which is compatible with the chosen type of wave functions.

A time-dependent variant of $\psi(y)$ is introduced to demonstrate that it represents a good approximation of the collective-state eigenfunctions of $H_{\rm mnp}$. It takes the form $\psi(y;\alpha(t), \beta(t)) = \exp(i\alpha(t)\hat{Q})\exp(im_{\rm el}\beta(t)[H_{\rm mnp}, \hat{Q}]_-)\psi_0(y)$ [20]. Introducing a variational principle (which is equivalent to the time-dependent Schrödinger equation) equations of motion for the time-dependent functions $\alpha(t)$ and $\beta(t)$ are obtained. Those are interpreted as a Lagrangian formulation of the dynamics with generalized coordinates and velocities. Then, one changes to canonical equations with canonical variables defining a Hamilton function. A particular choice of the canonical variables [20] results in a Hamilton function which is understood to reproduce in a good approximation those eigenvalues of $H_{\rm mnp}$ which correspond to collective excitations generated by the operator $\hat{Q}$. Introducing a harmonic approximation of the respective potential energy an oscillator frequency ω can be directly deduced and is identified with the approximate eigenvalues of $H_{\rm mnp}$, Eq. 10.3. The square of the oscillator frequency ω is obtained as the ratio of energy moments m_3/m_1, which are specific examples of the general type $m_n(\hat{Q}) = \sum_k (E_k)^n |\langle\psi_k|\hat{Q}|\psi_0\rangle|^2$. The E_k and ψ_k have been already introduced as eigenvalues and eigenstates of the MNP electron system, respectively (note that the definition provides $E_0 = 0$). The possibility to compute the spectrum of $H_{\rm mnp}$ by the moments m_1 and m_3 is based on the observation that they are identical with particular ground-state expectation values. For example, one obtains (provided $\hat{Q}$ is Hermitian)

$$m_1(\hat{Q}) = \langle\psi_0|\hat{Q}H_{\rm mnp}\hat{Q}|\psi_0\rangle \equiv \frac{1}{2}\langle\psi_0|[\hat{Q}, [H_{\rm mnp}, \hat{Q}]_-]_-|\psi_0\rangle\,. \tag{10.14}$$

This relation has been used in Ref. [18] to compute multipole electron density vibrations of alkali clusters and to demonstrate the recovery of the Mie frequencies in large-diameter clusters. If the operator $\hat{Q}$ is not Hermitian we set $m_1(\hat{Q}) = \langle\psi_0|\hat{Q}^+ H_{\rm mnp}\hat{Q}|\psi_0\rangle$ and derive

$$\begin{aligned} m_1(\hat{Q}) + m_1(\hat{Q}^+) &= \langle\psi_0|[\hat{Q}^+, [H_{\rm mnp}, \hat{Q}]_-]_-|\psi_0\rangle \\ &= \frac{\hbar^2}{m_{\rm el}}\int d^3\mathbf{x}|\nabla q(\mathbf{x})|^2 n_0(\mathbf{x})\,. \end{aligned} \tag{10.15}$$

Accordingly, the first moment could be expressed by the single-electron operator $q(\mathbf{x})$ and the single-electron density of the MNP electronic ground state $n_0(\mathbf{x})$. Higher moments cannot be expressed by the single-electron density in this simple way. Different types of approximations, for example, in the framework of DFT are necessary [18].

A single matrix element of the type $\langle\psi_\kappa|\hat{Q}|\psi_0\rangle$ can be obtained from the first moment m_1 if the so-called *plasmon resonance approximation* is applied [22]. Therefore, we assume that $\hat{Q}$ generates a state $\hat{Q}\psi_0$ which approximates a possible collective excitation of the MNP electron system with state vector ψ_κ. As a consequence we may set

$m_1(\hat{Q}) \approx E_\kappa|\langle\psi_\kappa|\hat{Q}|\psi_0\rangle|^2$. Noting Eq. 10.15 and $m_1(\hat{Q}) = m_1(\hat{Q}^+)$ (be aware of $m_1(\hat{Q}^+) = (m_1(\hat{Q}))^*$ in this connection) the transition matrix element is determined by a single-electron integral including the single-electron quantity $q(\mathbf{y})$ and the MNP ground-state electron density. If E_κ and n_0 are independently known the transition matrix element can be computed (except an unknown phase factor).

In order to specify the plasmon resonance approximation to multipole excitations of a spherical MNP we specify the $q(\mathbf{y})$ by the multipole operator q_{lm}, Eq. 10.13. Then, the plasmon resonance approximation reads $m_1(\hat{Q}_{lm}) \approx E_{lm}|\langle\psi_{lm}|\hat{Q}_{lm}|\psi_0\rangle|^2$ (note $\hat{Q}_{lm} = \sum_j q_{lm}(\mathbf{y}_j)$). By using Eq. 10.15 we arrive at

$$\frac{\hbar^2}{m_{\text{el}}}\int d^3\mathbf{x}|\nabla q_{lm}|^2 n_0(\mathbf{x}) \approx 2E_{lm}|\langle\psi_{lm}|\hat{Q}_{lm}|\psi_0\rangle|^2 . \quad (10.16)$$

For a constant density in the MNP the single-electron integral is easily computed [22], and the left-hand side of Eq. 10.16 takes the form $(\hbar\omega_l^{(M)})^2 a^{2l+1}$ ($\omega_l^{(M)}$ is the Mie frequency, Eq. 10.12 and a denotes the radius of the MNP).

10.3 Molecule Metal Nanoparticle Coupling

Recent experimental work on molecule MNP systems initiated numerous theoretical studies, for example, Refs. [25–35] as well as [36–39] (where, in part, semiconducting nano crystals instead of molecules have been considered). In line with the treatment of isolated MNPs these theories mainly use a macroscopic descriptions introducing dielectric functions for the involved molecules and MNPs and trying to determine the local field induced by the MNPs.

In the present approach which considers the SC–MNP systems as a uniform quantum system the SC–MNP interaction is determined by the respective expression for the Coulomb coupling. We start with the consideration of the Coulomb coupling between a single molecule and the MNP

$$H_{\text{mol-mnp}} = \sum_{j,j'} \frac{q_j q_{j'}}{|\mathbf{x}_j - \mathbf{y}_{j'}|} . \quad (10.17)$$

The molecule is characterized by the charges q_j and the respective position vectors $\mathbf{x}_j$ and the MNP by $q_{j'}$ and $\mathbf{y}_{j'}$, respectively. In general, these quantities refer to electrons and nuclei either of the molecule or of the MNP. Since we have in mind that in the different electronic states of the molecule the charge distribution of electrons and nuclei are well balanced, that is, the molecule should be neutral also locally, mirror charge formation at the MNP is absent. The remaining coupling can be reduced to an expression which only considers the coupling among electrons. Furthcr, all matrix elements of $H_{\text{mol-mnp}}$ have to refer to electronic transitions (note that this observation is also valid for a consideration of the Coulomb coupling among two molecules [12]). Introducing the electronic charge density operator of the molecule $\hat{\rho}_{\text{mol}}(\mathbf{x}) = \sum_j e\delta(\mathbf{x}-\mathbf{x}_j)$ and of the MNP $\hat{\rho}_{\text{mnp}}(\mathbf{x}) = \sum_j e\delta(\mathbf{y} - \mathbf{y}_j)$ the molecule–MNP Coulomb coupling follows as

$$H_{\text{mol-mnp}} = \int d^3\mathbf{x}\, d^3\mathbf{y}\, \frac{1}{|\mathbf{x}-\mathbf{y}|} \hat{\rho}_{\text{mol}}(\mathbf{x})\hat{\rho}_{\text{mnp}}(\mathbf{y}) \,. \tag{10.18}$$

To turn this expression into a form describing excitation energy exchange between the molecule and the MNP the charge density operators are expanded with respect to a complete set of related electronic states. Since we have in mind only weak or moderate (optical) excitation we consider besides the molecular electronic ground state φ_g only the first excited state φ_e. In the case of the MNP also transitions from the ground state ψ_0 to some excited states ψ_k are included. Later the latter are replaced by MNP multipole excitations. The expansion gives

$$\hat{\rho}_{\text{mol}}(\mathbf{x}) = e n_{eg}(\mathbf{x})|\varphi_e\rangle\langle\varphi_g| + \text{H. c.} \tag{10.19}$$

and

$$\hat{\rho}_{\text{mnp}}(\mathbf{y}) = \sum_k e n_{k0}(\mathbf{y})|\psi_k\rangle\langle\psi_0| + \text{H. c.} \,. \tag{10.20}$$

The molecular transition density $n_{eg}(\mathbf{x})$ is $N_{\text{el}} \int dx' \quad \delta(\mathbf{x} - \mathbf{x}')\varphi_e^*(x')\varphi_g(x')$ where N_{el} denotes the number of electrons and x' abbreviates the whole set of electron coordinates. The respective quantity n_{k0} of the MNP is defined in the same way.

If these expansions are inserted into $H_{\text{mol-mnp}}$, Eq. 10.18, we arrive at an energy exchange coupling between the molecule and the MNP, for example, $\sum_k V_{gk,0e}|\varphi_g\rangle|\psi_k\rangle\langle\psi_0|\langle\varphi_e|$. It refers to excitation energy transfer from the molecule to the MNP. The related matrix element is

$$V_{gk,0e} = \int d^3\mathbf{x}\, d^3\mathbf{y}\, \frac{en_{ge}(\mathbf{x})en_{k0}(\mathbf{y})}{|\mathbf{x}-\mathbf{y}|}\,. \tag{10.21}$$

Besides the Hermitian conjugated expression of $V_{gk,0e}$ there also appear terms describing simultaneous excitation and deexcitation of the molecule and the MNP. Those do not describe real processes and can be ignored here.

To stay simple, our discussion will be reduced to a description of the molecule by its transition dipole moment only. Therefore, we write $H_{\text{mol-mnp}}$, Eq. 10.18, as $\int d^3\mathbf{x}\ \ \hat{\rho}_{\text{mol}}(\mathbf{x})\hat{\Phi}_{\text{mnp}}(\mathbf{x})$. Here, $\hat{\Phi}_{\text{mnp}}(\mathbf{x})$ is the operator of the MNP-induced electrostatic potential $\int d^3\mathbf{y}\ \hat{\rho}_{\text{mnp}}(\mathbf{y})/|\mathbf{x}-\mathbf{y}|$. For a sufficiently large distance $\mathbf{X}$ between the center of mass of the molecule and the MNP it can be expanded with respect to $\mathbf{x}' = \mathbf{x}-\mathbf{X}$ and we obtain in first order $[\mathbf{x}'\nabla_{\mathbf{X}}\hat{\Phi}_{\text{mnp}}(\mathbf{X})]$. If we introduce this approximation into $H_{\text{mol-mnp}}$, Eq. 10.18, we get

$$H_{\text{mol-mnp}} = [\hat{\boldsymbol{\mu}}_{\text{mol}}\nabla_{\mathbf{X}}]\hat{\Phi}_{\text{mnp}}(\mathbf{X}) = -[\hat{\boldsymbol{\mu}}_{\text{mol}}\hat{\mathbf{E}}_{\text{mnp}}(\mathbf{X})]\,. \tag{10.22}$$

The molecular dipole operator is given by

$$\hat{\boldsymbol{\mu}}_{\text{mol}} = \int d^3\mathbf{x}'\ \mathbf{x}'\hat{\rho}_{\text{mol}}(\mathbf{x}') = \mathbf{d}_{eg}|\varphi_e\rangle\langle\varphi_g| + \text{H. c.}\,. \tag{10.23}$$

Moreover, we introduced in Eq. 10.22 the operator $\hat{\mathbf{E}}_{\text{mnp}}(\mathbf{X}) = -\nabla_{\text{R}}\hat{\Phi}_{\text{mnp}}(\mathbf{X})$ of the MNP induced electric field (at the position of the molecule):

$$\hat{\mathbf{E}}_{\text{mnp}}(\mathbf{X}) = \sum_k \mathbf{E}_{k0}(\mathbf{X})|\psi_k\rangle\langle\psi_0| + \text{H. c.} \tag{10.24}$$

with the partial field components

$$\mathbf{E}_{k0}(\mathbf{X}) = -\nabla_{\mathbf{X}}\int d^3\mathbf{y}\, \frac{en_{k0}(\mathbf{y})}{|\mathbf{X}-\mathbf{y}|}\,. \tag{10.25}$$

The expectation value of $\hat{\mathbf{E}}_{\text{mnp}}$ put together with the externally applied field gives the MNP-induced local field felt by the molecule. $H_{\text{mol-mnp}}$, Eq. 10.22, is easily generalized to an SC–MNP coupling by introducing a summation with respect to the various molecules forming the SC.

10.4 Description of a Spherical MNP

We specify the formulas presented so far to a spherical MNP and write for the MNP Hamiltonian, Eq. 10.3, $H_{\rm pl} = \sum_{l,m} E_{lm} |\psi_{lm}\rangle\langle\psi_{lm}|$. Note that the multipolar indices lm characterize the collective MNP state and not that of a single electron (see also Fig. 10.2). As already indicated, some similarities of the E_{lm} with the Mie frequencies, Eq. 10.12, have to be expected.

According to Refs. [15–17] the Hamiltonian $H_{\rm pl}$ is understood as a part of a system reservoir description. It is extended in the density matrix description of Section 10.5.1 by the introduction of dephasing rates $\gamma_{\rm pl}$ reflecting the short plasmon lifetime. Here, $\gamma_{\rm pl}$ corresponds to dipole plasmons. The choice for higher multipole excitations is not yet clear. Concerning the MNP excitation energies E_{lm}, that one for the dipole plasmon is known from experiment [17]. We use this measured value to compute higher excitation energies according to the Mie formula as $E_l = \sqrt{3l/(2l+1)}E_{l=1}$. The overestimation of the dipole plasmon resonance by the Mie frequencies, Eq. 10.12, could be related to the effect of spill-out electrons and a coupling to individual electron–hole pair excitations [17].

10.4.1 Coupling of an SC to a Spherical MNP

In the case of a spherical MNP we expect excited electronic states ψ_{lm} of multipolar type. To account for these excitations we start from the energy exchange coupling, Eq. 10.22. If one carries out a multipole expansion of the MNP transition density in the electric field expression, Eq. 10.25, one gets

$$\mathbf{E}_{k0}(\mathbf{X}) = -\nabla_{\mathbf{X}} \sum_{l,m} \sqrt{\frac{4\pi}{2l+1}} \frac{e\mathcal{Q}^*_{lm}(k0) Y_{lm}(\vartheta, \varphi)}{X^{l+1}} \qquad (10.26)$$

with the MNP transition multipole moments

$$\mathcal{Q}_{lm}(k0) = \int d^3\mathbf{x}\; q_{lm} n_{k0}(\mathbf{x}) = \langle\psi_k|\hat{Q}_{lm}|\psi_0\rangle\,. \qquad (10.27)$$

The multipole moment operators q_{lm} have been defined in Eq. 10.13. The matrix element with $\hat{Q}_{lm}$ represents the many-electron variant of the foregoing single coordinate expression. Remembering the plasmon resonance approximation we consider $\mathcal{Q}_{lm}(k0)$ to be nonzero only, if the quantum number k corresponds to a collective state ψ_{lm} with multipolarity according to l and m. So we get $\mathcal{Q}_{lm}(k0) \approx \delta_{k,lm}\langle\psi_{lm}|\hat{Q}_{lm}|\psi_0\rangle$. The right-hand side of this expression, however, is independently known from Eq. 10.16 (except a phase factor). According to this reasoning the SC–MNP coupling reads

$$H^{\text{(inst)}}_{\text{sc-mnp}} = \sum_{n,lm} \left(V_{n,lm}|\varphi_{ne}\rangle|\psi_0\rangle\langle\psi_{lm}|\langle\varphi_{ng}| + \text{H. c.} \right) . \qquad (10.28)$$

Note that the first term describes excitation energy transfer from the MNP to molecule n (in contrast to Eq. 10.21). The involved matrix elements take the form

$$V_{n,lm} = \sqrt{\frac{4\pi}{2l+1}} e\mathcal{Q}^*_{lm}[\mathbf{d}_n \nabla_{\mathbf{X}}] \frac{Y_{lm}(\vartheta,\varphi)}{X^{l+1}} \qquad (10.29)$$

with $\mathcal{Q}_{lm} = \langle\psi_{lm}|\hat{Q}_{lm}|\psi_0\rangle$ (the spherical coordinates X, ϑ, φ belong to molecule n what has not been indicated separately). Following Ref. [22] we estimate $e\mathcal{Q}_{lm}$ in the framework of the plasmon resonance approximation as $\hbar\omega_l^{(\text{M})}\sqrt{a^{2l+1}/2E_l}$ (a is the MNP radius). The expression for $e\mathcal{Q}_{lm}$ is based on a constant ground-state MNP electron density $n_0(\mathbf{x})$ across the MNP. Since this assumption ignores spill-out electrons one may replace E_l by the Mie frequencies $\omega_l^{(\text{M})}$. The coupling potential, Eq. 10.29, averaged with respect to the multipolar index m is shown in Fig. 10.3 versus the multipolar index l and for different distances ΔX of the molecule to the MNP surface.

10.5 Photoinduced Processes in a Supramolecular Complex Coupled to a Metal Nanoparticle

In the subsequent discussion we will apply the theory presented in the earlier sections to a linear SC (polymer strand or aggregate of identical molecules) coupled to a spherical MNP (see Fig. 10.1). So far, such an SC–MNP system was not realized in the experiment. However, a spectroscopic characterization of single polymeric chains showing the formation of Frenkel exciton states, but without

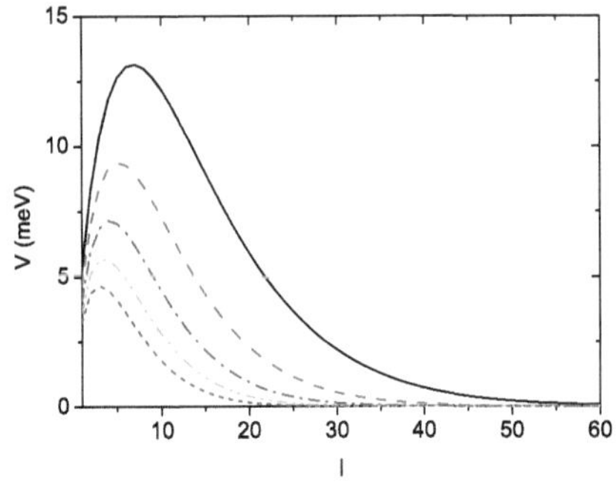

Figure 10.3 Averaged molecule–MNP coupling matrix elements $V(nl) = \sqrt{\sum_m V_{n,lm}^2}$ versus the multipole index l and for the molecule–MNP surface distances $\Delta X = 1.5 \ldots 3.5$ nm (the large numbers of l motivate for a continuous line presentation; molecular dipole moment perpendicular to the plane formed by the SC and the MNP). Solid black line: $\Delta X = 1.5$ nm; dashed red line: $\Delta X = 2$ nm; chain dotted blue line: $\Delta X = 2.5$ nm; chain dotted green line: $\Delta X = 3$ nm; short dashed violet line: $\Delta X = 3.5$ nm. Reprinted from Ref. [7], with permission from John Wiley & Sons, Inc.

the inclusion of an MNP, has been reported recently (see Refs. [40–43]). Our description concentrates on photoinduced excitation energy transfer in the SC affected by the MNP (Section 10.5.2) and on the computation of SC–MNP linear absorption (Section 10.5.3) as well as emission (Section 10.5.4).

We will not directly focus on one of those molecular systems studied in [40–43] but use parameterized models. So, we have a certain freedom in choosing the respective parameters. But in any case we consider a spherical Au MNP with a radius a of 10 nm. Its dipole plasmon energy $E_{l=1}$ is located at 2.595 eV and the transition dipole moment $d_{\rm pl}$ amounts to 2925 D. It enters the general vectorial dipole moment $\mathbf{d}_{lm} = \delta_{l,1}\mathbf{d}_{1m}$ (in Cartesian coordinates) as $\mathbf{d}_{1m=0} = \mathbf{e}_z d_{\rm pl}$ and $\mathbf{d}_{1m=\pm1} = (\mp\mathbf{e}_x + i\mathbf{e}_y)d_{\rm pl}/\sqrt{2}$. The MNP nonradiative decay rate $2\gamma_{\rm pl}$ follows from $\hbar\gamma_{\rm pl} = 28.6$ meV [6]. The molecules of the SC should be identical, they should form a regular chain, and their excitation energy $E_n = E_{\rm mol}$ should lie in the range of the dipole plasmon. The transition dipole moments $d_n = d_{\rm mol}$ are changed in a range less than 10 D. The intermolecular distance $\Delta r_{\rm mol}$ is varied in the nanometer range to cover different strengths of excitonic coupling. H-aggregate and J-aggregate configurations of the linear SC are taken (molecular transition dipole moments are perpendicular and parallel to the chain axis, respectively). The excitonic coupling between molecule n and n' will be calculated in dipole–dipole approximation according to $J_{nn'} = d_{\text{mol}}^2(1-3\cos^2\alpha)/(|n-n'|\Delta r_{\text{mol}})^3$ (α is the angle between the transition dipole moment and the line connecting the two molecules). The optical excitation should be always polarized in the direction of the molecular dipole moment and should be resonant to the SC–MNP excitations ($\hbar\omega_0 \approx E_{l=1}$ or $\hbar\omega_0 \approx E_{\text{mol}}$). To avoid contributions of higher multipoles with respect to the SC–MNP interaction (see Fig. 10.3) we set the closest distance ΔX of a molecule of the SC to the MNP surface to 2.5 nm. Then, the computation of SC–MNP properties is only slightly affected by multipole contributions $l > 1$ (cf. Ref. [6]). Exciton vibrational coupling is only accounted for by a dephasing rate with $\hbar\gamma_{\text{mol}} = 3$ meV.

Our whole treatment should concentrate on a weak optical excitation. Therefore, we chose field strengths for the exciting laser pulse which result in a sufficiently small overall excited-state population of the SC–MNP system. This guarantees that excited-state absorption would be of less importance and allows to ignore higher SC exciton states (simultaneous excitation of two or three molecules). Consequently, it is convenient for a description of the coupled SC–MNP system to introduce product states of these excitations and to carry out a respective expansion of the total Hamiltonian $H(t)$. This standard procedure notices the SC–MNP ground state $|\phi_g\rangle = \prod_n |\varphi_{ng}\rangle|\psi_0\rangle$ and the singly excited states $|\phi_n\rangle = |\varphi_{ne}\rangle \prod_{n'\neq n} |\varphi_{n'g}\rangle|\psi_0\rangle$ as well as $|\phi_{lm}\rangle = \prod_n |\varphi_{ng}\rangle|\psi_{lm}\rangle$ (the φ_{na} as well as the ψ_0 and ψ_{lm} are the earlier introduced single molecule and MNP states, respectively). Higher excited states are neglected. The resulting expansion of $H_{\text{sc}} + H_{\text{mnp}} + H_{\text{sc-mnp}}^{(\text{inst})}$ is written as $H_0 + H_1$. The Hamiltonian of the SC–MNP electronic ground state takes the form $H_0 = E_g|\phi_g\rangle\langle\phi_g|$ with $E_g = \sum_n E_{ng} + E_0$. In the following, E_g is set equal to zero. We get the singly excited-state Hamiltonian as

$$\begin{aligned} H_1 &= \sum_n E_n|\phi_n\rangle\langle\phi_n| + \sum_{n,n'} J_{nn'}|\phi_n\rangle\langle\phi_{n'}| \\ &+ \sum_{l,m} E_l|\phi_{lm}\rangle\langle\phi_{lm}| + \sum_{n,lm} \left(V_{n,lm}|\phi_n\rangle\langle\phi_{lm}| + \text{H. c.}\right). \end{aligned} \quad (10.30)$$

The SC part of this Hamiltonian can be also taken in the exciton representation using exciton states $|\alpha\rangle = \sum_n C_\alpha(n)|\phi_n\rangle$. Then, Fig. 10.2 directly quotes the ingredients of H_1. The coupling to the external field, Eq. 10.7, is determined by the dipole operator of the SC–MNP system

$$\hat{\mu} = \sum_n \mathbf{d}_{\text{mol}}|\phi_n\rangle\langle\phi_g| + \sum_{l,m} \mathbf{d}_{lm}|\phi_{lm}\rangle\langle\phi_g| + \text{H. c.}\,. \quad (10.31)$$

It includes the transition dipole moments $\mathbf{d}_{\text{mol}}$ of the various molecules. The $\mathbf{d}_{lm}$ refer to those of the MNP (l is restricted to 1). The laser pulse excitation is characterized by the electric field pulse written as

$$\mathbf{E}_{\text{ext}}(t) = \mathbf{n}E(t)e^{-i\omega_0 t} + \text{c. c.}\,. \quad (10.32)$$

The expression includes the unit vector $\mathbf{n}$ of field polarization, the envelope $E(t)$, and the carrier frequency ω_0. The envelope takes the form $E(t) = E_0 \exp(-2(t - t_{\text{p}})^2/\tau_{\text{p}}^2)$ with pulse maximum at t_{p} and pulse length τ_{p}. The coupling to transversal photons is specified in Section 10.5.4.

10.5.1 Density Matrix Theory

Photoinduced dynamics in the SC–MNP system will be simulated by using the density matrix theory (cf., for example, [12]). To offer respective equations of motion (quantum master equation) we introduce a common notation of the involved states by ϕ_A with A covering the indices g, n, and lm. Accordingly, the (reduced) density matrix reads

$$\rho_{AB}(t) = \text{tr}\{\hat{W}(t)|\phi_B\rangle\langle\phi_A|\}\,, \quad (10.33)$$

where $\hat{W}(t)$ denotes the overall statistical operator of the system including the environment of individual electronic excitations of the MNP responsible for plasmon decay and the SC molecular vibrational degrees of freedom causing intramolecular vibrational energy redistribution (IVR). To stay simple, we focus on an approximate quantum master equation where the dissipative part does not couple the time evolution of populations and coherences. It takes the form

$$\frac{\partial}{\partial t}\rho_{AB}(t) = -\frac{i}{\hbar}\sum_{C}\left(H_{AC}(t)\rho_{CB}(t) - H_{CB}(t)\rho_{AC}(t)\right)$$
$$- \delta_{A,B}\sum_{C}\left(k_{A\to C}\rho_{AA}(t) - k_{C\to A}\rho_{CC}(t)\right)$$
$$- (1-\delta_{A,B})\gamma_{AB}\rho_{AB}(t)\,. \qquad (10.34)$$

We introduced $H_{AB}(t) = \langle\phi_A|H(t)|\phi_B\rangle$ with off-diagonal matrix elements accounting for the intra-SC coupling (excitonic coupling), the SC–MNP interaction, and the effect of the external laser pulse. The $k_{A\to C}$ are transition rates which only describe the decay of MNP excitations as $k_{N\to 0} = 2\gamma_{\text{pl}}$ (radiative and nonradiative decay of the molecules are of no importance on the time scale of interest here). Any vibrational contribution to the molecular dynamics is condensed in the dephasing rates γ_{mol}. To account for the external laser pulse excitation in an efficient way we apply the RWA (rotating wave approximation).

10.5.2 Photoinduced Excitation Energy Transfer

The subsequent discussion concentrates on laser pulse–induced excitation energy transfer in an SC of 50 molecules coupled to a spherical MNP. The MNP is placed at the center of the chain as shown in Fig. 10.2. Laser pulse excitation results in the formation of a large local field around the MNP. Then, an enhancement of photoabsorption takes place at those molecules of the SC which are placed near the MNP. The optical excitation of the SC becomes localized. If the laser pulse excitation is over, the SC excitation is distributed across the whole SC and is dissipated into the MNP. We will demonstrate this *spatiotemporal* energy localization in the following.

However, the discussion has to notice that the SC excitations are (delocalized) Frenkel exciton states characterized by particular selection rules of the SC optical excitation. We will focus on an H-aggregate configuration in the following (all molecular transition dipole moments are perpendicular to the chain direction) where the highest exciton level possesses the majority of oscillator strength. Interestingly, this selection rule is strongly changed due

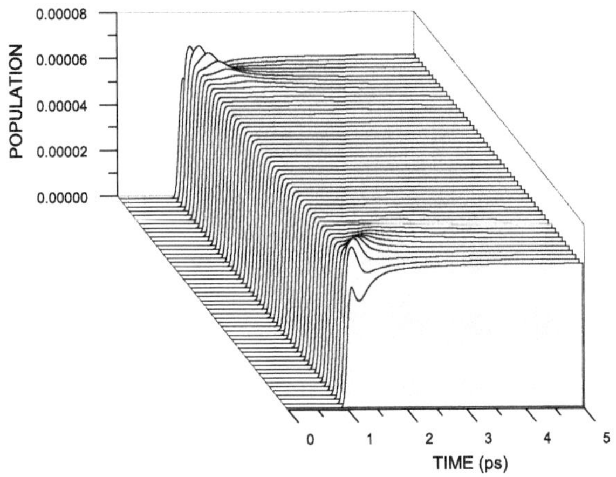

Figure 10.4 Photoinduced molecular populations P_n versus time for an SC with 50 molecules in the absence of the MNP ($d_{\mathrm{mol}} = 4.5$ D, $\Delta r_{\mathrm{mol}} = 1.5$ nm, photoexcitation at $\hbar\omega_0 = E_{\mathrm{mol}}$, laser pulse width $\tau_{\mathrm{p}} = 100$ fs, $E_0 = 5 \times 10^5$ V/m). P_1 to P_{50} are drawn from the background to the foreground. Reprinted from Ref. [7], with permission from John Wiley & Sons, Inc.

to the presence of the MNP by allowing optical absorption by all other exciton levels (see the subsequent section). Now, all exciton levels can be addressed by the exciting laser pulse. The resulting superposition state $\sum_\alpha A_\alpha(t)|\alpha\rangle$ (the exciton states $|\alpha\rangle$ form an excitonic wave packet) probably introduces excitation energy localization. The pulse duration, however, decides whether or not an excitonic wave packet is really formed. Therefore, its spectral broadening roughly given by $\hbar/\tau_{\mathrm{p}}$ has to be compared with the width of the exciton band of about $4|J|$ (four times the nearest-neighbor excitonic coupling). For example, if $\hbar/\tau_{\mathrm{p}} \ll 4|J|$ only a single level might be excited and no wave packet can be formed. In the reverse case all exciton levels are excited simultaneously and contribute to the wave packet. We will discuss this in the following by visualizing photoinduced excitation energy transfer via the probabilities $P_n(t) = \rho_{nn}(t)$ to have the various molecules of the SC in their excited state.

Firstly, Fig. 10.4 displays photoinduced excitation energy transfer in the absence of the MNP. Choosing an optical excitation with a pulse length of 100 fs, a uniform distribution of excitation energy across the whole SC appears. Here, the spectral broadening of the pulse amounts about 6 meV while the exciton band has a width of 15 meV (for the other used parameters see Fig. 10.4). The 100 fs pulse comprises slightly less than half of the band and a certain number of states interfere. Nevertheless, a constant population of the whole SC is obtained except some small oscillations at the chain ends. Note also the small molecular population of about 6×10^{-5} which is due to the smallness of most of the exciton transition dipole moments. But they are also related to the chosen field strength which has the same value as in the following consideration including the presence of the MNP.

If we place an MNP near the SC the temporal behavior of the molecular populations changes drastically, as displayed by Fig. 10.5. Localization of excitation energy appears in that part of the SC positioned near to the MNP (chain center). Decreasing the laser pulse duration results in a slight improvement of excitation energy localization. A closer inspection of the 10 fs case shows that energy is concentrated at one third of the chain (a range of 24 nm, which is somewhat larger than the MNP diameter). These results can be directly confronted with those of Fig. 10.4 which have been obtained in the absence of the MNP. The small molecular excitation probability of about 6×10^{-5} obtained there is changed to values more than 2 orders of magnitude larger if the MNP is present. The molecules at the center with the highest population lose their population in the fastest way. This causes the central dip in the population distribution across the chain. Excitation energy transfer from the central part of the chain to its terminal sites is also visible.

To demonstrate the MNP-induced oscillator strength redistribution among the SC exciton levels one might diagonalize the excited-state Hamiltonian H_1, Eq. 10.30. Since the MNP excitations are characterized by a large lifetime broadening this procedure seems to be questionable. It is more appropriate to directly calculate a quantity measurable in the experiment. Therefore we will discuss in the subsequent section the linear absorption of the SC–MNP system.

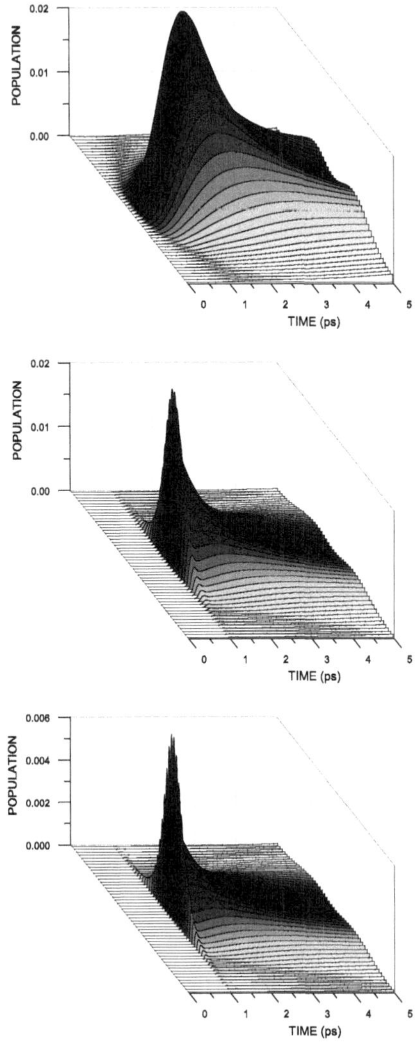

Figure 10.5 Photoinduced molecular populations P_n versus time for the SC–MNP system with 50 molecules (the MNP is placed at the center of the chain as in Fig. 10.1 with its center of mass positioned in the plane defined by the chain; parameters as in Fig. 10.4). The populations P_1 to P_{50} are drawn from the background to the foreground. Variation of the laser pulse width τ_p. Upper panel: $\tau_p = 1$ ps ($E_0 = 2\times10^5$ V/m). Middle panel: $\tau_p = 100$ fs ($E_0 = 5\times10^5$ V/m). Lower panel: $\tau_p = 10$ fs ($E_0 = 10^6$ V/m). Reprinted from Ref. [7], with permission from John Wiley & Sons, Inc.

10.5.3 Linear Absorption

The spectrum of linear absorption is given by $\alpha(\omega) = 4\pi\omega n I(\omega)/\hbar c$. Here, n is the volume density of SC–MNP systems indicating that an ensemble of complexes is detected. The absorption line shape

$$I(\omega) = \mathrm{Re}\int_0^\infty dt\, e^{i\omega t} C_{\text{d-d}}(t) \qquad (10.35)$$

is expressed by the dipole–dipole correlation function, whose resonant part is exclusively of interest in the following. The focus on absorption spectra is justified by the fact that for such small MNPs as discussed here measured extinction spectra are dominated by radiation field absorption and not scattering. Noting the system reservoir description the correlation function takes the form [6]

$$C_{\text{d-d}}(t) = \mathrm{tr}_S\{\hat{\mu}\left[\mathcal{U}(t)\hat{\mu}|\phi_g\rangle\langle\phi_g|\right]\}\,, \qquad (10.36)$$

where $\mathcal{U}$ denotes the time evolution superoperator from which the density matrix equations introduced in Section 10.5.1 result. The trace in Eq. 10.36 accounts for all electronic states included (they form the system S). The initially realized electronic ground state is characterized by $|\phi_g\rangle\langle\phi_g|$. $\hat{\mu}$ is the SC–MNP system dipole operator. It is written here in a way to cover the case of equally oriented absorbing systems. If all SC–MNP systems are identical but also have the same spatial orientation (this would be achievable on a surface and is typical for so-called metamaterials) we have to set $\hat{\mu}\hat{\mu} = [\mathbf{n}\hat{\mu}][\mathbf{n}\hat{\mu}]$ where $\mathbf{n}$ is the polarization unit vector of the external field. In order to compute $C_{\text{d-d}}$ we introduce

$$\hat{\sigma}(t) = \mathcal{U}(t)\Big(\sum_n d_n|\phi_n\rangle\langle\phi_g| + \sum_{l',m'} d_{l'm'}|\phi_{l'm'}\rangle\langle\phi_g|\Big) \qquad (10.37)$$

(note $d_n = [\mathbf{n}\mathbf{d}_n]$ and $d_{l'm'} = [\mathbf{n}\mathbf{d}_{l'm'}]$). The correlation function follows as

$$C_{\text{d-d}}(t) = \sum_n d_n^* \sigma_{ng}(t) + \sum_{l,m} d_{lm}^* \sigma_{lm,g}(t) \qquad (10.38)$$

with matrix elements $\sigma_{ng}(t) = \langle\phi_n|\hat{\sigma}(t)|\phi_g\rangle$ and $\sigma_{lm,g}(t) = \langle\phi_{lm}|\hat{\sigma}(t)|\phi_g\rangle$. Both types of (off-diagonal) matrix elements obey the density matrix equations 10.34 (in the absence of an external field) with initial conditions $\sigma_{ng}(0) = d_n$ and $\sigma_{lm,g}(0) = d_{lm}$. A

Fourier transformation of $C_{\text{d-d}}$ results in the absorption line shape, Eq. 10.35.

To clearly demonstrate the SC oscillator strength redistribution we will focus here on a chain of 10 molecules (resulting in 10 instead of 50 exciton levels as in the earlier section). We also increase the excitonic coupling and place the MNP at the terminal site of the chain-like SC. Respective results are displayed in Fig. 10.6 for the SC

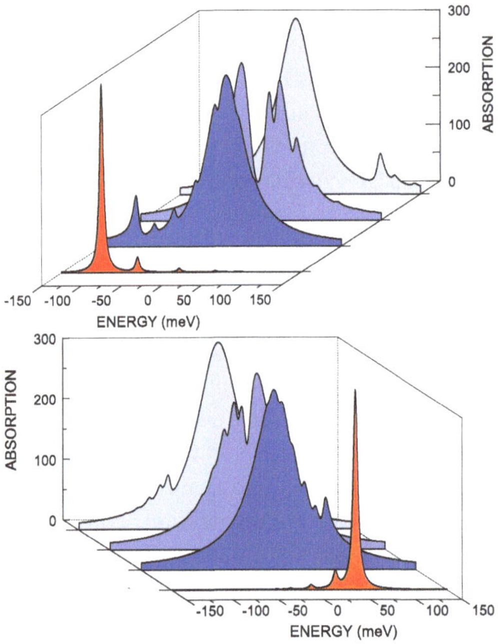

Figure 10.6 SC–MNP absorption line shape $I(\omega)/N_{\text{mol}}d^2_{\text{mol}}$ versus $\hbar\omega - E_{l=1}$ for equally oriented SC–MNP complexes (external field polarization $\mathbf{n} \parallel \mathbf{d}_{\text{mol}}$) and a variation of the molecular excitation energy E_{mol} (SC with 10 molecules, $\Delta r_{\text{mol}} = 1.2$ nm, $d_{\text{mol}} = 8$ D, spherical MNP like in Figs. 10.4 and 10.5 but placed at the terminal site of the molecular chain, $\Delta X = 2.5$ nm). Upper panel: SC in J-aggregate configuration, background-to-foreground curves: $E_{\text{mol}} = E_{l=1} + 200$ meV; $E_{\text{mol}} = E_{l=1} + 100$ meV, $E_{\text{mol}} = E_{l=1}$. Lower panel: SC in H-aggregate configuration, background-to-foreground curves: $E_{\text{mol}} = E_{l=1} - 100$ meV, $E_{\text{mol}} = E_{l=1} - 50$ meV, $E_{\text{mol}} = E_{l=1}$. The foremost red-filled curves show the SC absorption line shape in the absence of the MNP (increased by about 5×10^3, at $E_{\text{mol}} = E_{l=1}$). Reprinted with permission from Ref. [6]. Copyright (2012) American Chemical Society.

in J- and H-aggregate configuration (the molecular excitation energy E_{mol} is moved through the dipole plasmon resonance $E_{l=1}$). For comparison the absorption line shapes of the isolated SC (absence of the MNP) are also drawn. Those show a dominance of the lowest exciton level (J-aggregate configuration) or of the highest exciton level (H-aggregate configuration). The related peak height is more than 3 orders of magnitude smaller demonstrating the MNP-induced enhancement effect when compared with the other curves.

If the MNP is placed near the SC further peaks appear in the absorption line shape, which clearly demonstrates the oscillator strength redistribution. Nearly all 10 exciton levels become visible via so-called Fano resonances which emerge in the broad background of the MNP absorption. The MNP-induced oscillator strength change is strongest if the dominant exciton level (lowest one for the J-aggregate configuration and highest one for the H-aggregate configuration) comes into resonance with the dipole plasmon of the MNP. We also indicate that at the chosen SC–MNP distance ΔX of 2.5 nm the spectra are dominated by the coupling of the molecular transition to the MNP dipole plasmon excitation (for the inclusion of higher multipole moments and other SC–MNP geometries see Ref. [6]).

10.5.4 Emission

One basic observation in the field of plasmonics is the increased photon emission of a molecule if it is placed in the proximity of an MNP [44]. Meanwhile such an MNP-induced emission enhancement (a decreased radiative lifetime) has been demonstrated for a variety of systems [45–53]. As a final application of our theory on coupled SC–MNP systems we present data for the time- and frequency-resolved emission. This includes experimental evidence for the excitation energy localization described in Section 10.5.2.

As already indicated in the introductory part photons are related to the retarded part $H^{(\mathrm{ret})}_{\mathrm{sc\text{-}mnp}}$, Eq. 10.6, of the SC–MNP coupling. Spontaneous photon emission can be considered in second-order perturbation theory and covers the population of free-space photon states. Any effect of the presence of the MNP is accounted for by the nonperturbative consideration of the SC–MNP Coulomb coupling.

This gives a consistent picture but is in contrast to the majority of theoretical studies. Those are based on the solution of Maxwell's equations (determination of the respective Green's function) where the presence of the MNP is considered by an appropriately chosen dielectric function [27, 29, 30, 33, 54, 55].

The emission spectrum is defined as the number of photons emitted per time and frequency interval (idealized emission spectrum)

$$F(\omega; t) = \frac{V\omega^2}{(2\pi c)^3} \sum_{\lambda} \int do\ R_{\lambda}(\omega; t)\,. \tag{10.39}$$

We introduced the rate $R_{\lambda}(\omega; t) = R_{\lambda\mathbf{k}}(t)$ determining the number of photons emitted per time into the state with polarization λ and wave vector $\mathbf{k}$ (V denotes the quantization volume and the solid angle integration is abbreviated by $\int do$). Since emission appears into the photon vacuum, $R_{\lambda\mathbf{k}}(t)$ is obtained as the time derivative of the photon number expectation value $\text{tr}\{\hat{W}(t)a^{+}_{\lambda\mathbf{k}}a_{\lambda\mathbf{k}}\}$. The time-dependent statistical operator $\hat{W}(t)$ covers photon states but is a reduced statistical operator with respect to the other types of reservoirs. Consequently, we consider the operator $\text{tr}_{\text{phot}}\{\hat{W}(t)\}$ where the photon state has been traced out as the reduced statistical operator $\hat{\rho}(t)$ of the SC–MNP system, which also accounts for dissipative effects due to a finite plasmon lifetime and IVR.

To proceed further we have to specify the SC–MNP photon coupling operators entering Eq. 10.6. According to our expansion of the SC–MNP Hamiltonian in terms of product states we arrive at (note here $\phi_A = \phi_n, \phi_{lm}$, that is, the states are restricted to excited SC–MNP ones)

$$\hat{h}^{(\pm)}_{\lambda\mathbf{k}} = \sum_{A} \left(g^{(\pm)}_{\lambda\mathbf{k}}(A, g)|\phi_A\rangle\langle\phi_g| + g^{(\pm)}_{\lambda\mathbf{k}}(g, A)|\phi_g\rangle\langle\phi_A|\right) \tag{10.40}$$

with

$$g^{(\pm)}_{\lambda\mathbf{k}}(A, g) = -\sqrt{\frac{\pi e^2}{2Vm_{\text{el}}^2\hbar\omega_{\mathbf{k}}}}\langle\phi_A|\sum_{j}\left[e^{\pm i\mathbf{k}\mathbf{x}_j}[\mathbf{n}_{\lambda\mathbf{k}}\mathbf{p}_j]\right]_{+}|\phi_g\rangle\,. \tag{10.41}$$

While this expression allows to account for photon emission from MNP multipole excitations ($\mathbf{p}_j$ is the momentum operator of the jth electron) we will stay at dipole excitations in the following

($\exp(\pm i\mathbf{k}\mathbf{x}_j) \approx 1$). Therefore we replace $g^{(+)}_{\lambda\mathbf{k}}(A,g) = g^{(-)*}_{\lambda\mathbf{k}}(g,A)$ by $g_{\lambda\mathbf{k}}(A) = -i\sqrt{2\pi\omega_A^2/V\hbar\omega_{\mathbf{k}}}[\mathbf{n}_{\lambda\mathbf{k}}\mathbf{d}_A]$ ($\hbar\omega_A = E_{\text{mol}}$, E_l is the basic electronic transition energy either in the molecule or the MNP, and $\mathbf{n}_{\lambda\mathbf{k}}$ represents the unit vector of transversal photon polarization).

To determine $R_{\lambda\mathbf{k}}$ in second order with respect to the molecule–MNP photon interaction we note $R_{\lambda\mathbf{k}} = \partial\text{tr}\{\hat{W}(t)a^+_{\lambda\mathbf{k}}a_{\lambda\mathbf{k}}\}/\partial t = i\text{tr}\{\hat{h}^{(+)}_{\lambda\mathbf{k}}a_{\lambda\mathbf{k}} - \hat{h}^{(-)}_{\lambda\mathbf{k}}a^+_{\lambda\mathbf{k}})\hat{W}(t)\}$. Then, $\hat{W}(t)$ is computed in the first order with respect to the molecule–MNP photon interaction. The related expression can be deduced from the integrated equation of motion for $\hat{W}$ (cf. Refs. [56, 57]). Splitting $\hat{W}$ into a photon (vacuum) part $W_{\text{phot}} = |\text{vac}\rangle\langle\text{vac}|$ and a system part $\hat{\rho}$ and focusing on the resonant contribution to the frequency dependence of $R_{\lambda\mathbf{k}}$ we arrive at

$$R_{\lambda\mathbf{k}} = 2\text{Re}\int_{t_0}^{t} d\bar{t}\, e^{-i\omega_{\mathbf{k}}(t-\bar{t})}\sum_{A,B} g_{\lambda\mathbf{k}}(A,g)g^*_{\lambda\mathbf{k}}(g,B)$$
$$\times\langle\phi_g|\mathcal{U}(t,\bar{t})\big(|\phi_g\rangle\langle\phi_B|\hat{\rho}(\bar{t})\big)|\phi_A\rangle\,. \qquad (10.42)$$

The time evolution of the density operator $\hat{\rho}$ is characterized by the time argument $\bar{t}$. Together with $|\phi_g\rangle\langle\phi_B|$ it defines an initial operator to be propagated by $\mathcal{U}(t-\bar{t})$. If the time interval $[t,\bar{t}]$ where $\mathcal{U}$ acts is short compared to characteristic changes of $\hat{\rho}$ we can replace $\hat{\rho}(\bar{t})$ by $\hat{\rho}(t)$. Introducing $\tau = t - \bar{t}$ the resulting upper integration limit $t - t_0$ can be replaced by infinity. Then, the photon wave vector summation is translated to spherical coordinates. We note the identity $\sum_\lambda\int do\,[\mathbf{n}_{\lambda\mathbf{k}}\mathbf{d}_A][\mathbf{n}_{\lambda\mathbf{k}}\mathbf{d}^*_B] = 8\pi/3\times\mathbf{d}_A\mathbf{d}^*_B$ and get the photon emission rate per frequency interval as $F(\omega;t) = 4\omega^3 I(\omega;t)/3\pi c^3\hbar$. The emission line shape takes the form

$$I(\omega;t) = \text{Re}\sum_{A,B}[\mathbf{d}_A\mathbf{d}^*_B]\int_0^\infty d\tau\, e^{-i\omega\tau}\langle\phi_g|\big(\mathcal{U}(\tau)|\phi_g\rangle\langle\phi_B|\big)\hat{\rho}(t)|\phi_A\rangle$$
$$= \text{Re}\sum_{A,B,C}[\mathbf{d}_A\mathbf{d}^*_B]\int_0^\infty d\tau\, e^{-i\omega\tau}\sigma_{gC}(\tau;B)\times\rho_{CA}(t)\,. \qquad (10.43)$$

In the second part we introduced the density matrix $\rho_{CA}(t)$ and the quantity $\sigma_{gC}(\tau;B) = \langle\phi_g|\big(\mathcal{U}(\tau)|\phi_g\rangle\langle\phi_B|\big)|\phi_C\rangle$. It is also determined by the density matrix equations, but the initial value is $\sigma_{gC}(0;B) = \delta_{C,B}$. The τ integral with respect to the (off-diagonal) density matrix type function $\sigma_{gC}(\tau; B)$ constitutes the frequency dependency of the emission spectrum. In contrast, $\rho_{CA}(t)$ reflects the actual excited-state population of the molecule–MNP complex, thus, defines the time dependence of the emission (it is computed by including the external field excitation). Its off-diagonal parts express the SC–MNP coupling, that is, energy exchange between the two constituents of the system. The dipole operator product $[\mathbf{d}_A \mathbf{d}_B^*]$ in Eq. 10.43 includes not only a mixture of molecular and MNP contributions but also terms with different molecular contributions $[\mathbf{d}_n \mathbf{d}_{n'}^*]$. Besides the excitonic coupling in the density matrix equations, those guarantee that emission from exciton levels are accounted for.

Figure 10.7 presents the time- and frequency-resolved emission of an SC–MNP system with a linear SC of 50 molecules following a 100 fs laser pulse excitation. Also shown is the temporal evolution of the molecular excited-state populations P_n. As already discussed in Section 10.5.2 excitation energy localization appears followed by a redistribution at later times due to the excitonic coupling among the molecules. Also shown in Fig. 10.7 are the population P_g of the ground state and the total population $P_{\text{sc}} = \sum_n \rho_{nn}$ of the excited states of the SC and $P_{\text{pl}} = \sum_{l,m} \rho_{lm,lm}$ of the MNP ($l = 1, m = -1, 0, 1$). The short plasmon lifetime let P_{pl} follow rather directly the laser pulse excitation (the asymmetry at later times is due to excitation energy transfer from the SC to the MNP). In contrast, P_{sc} decays much slower (excitation quenching via energy dissipation in the MNP). This behavior is reflected by the time-resolved emission spectra.

For the smallest time differences to the maximum of the exciting laser pulse, $\Delta t = t - t_{\text{p}} = 100$ fs, the broad plasmon emission is modified by a dip indicating the Fano-like interference with the lowest SC exciton level. It is placed at about $2|J| = 15$ meV below the molecular excitation energy and here also below E_{pl} (if nearest-neighbor coupling among molecules is considered only, the exciton spectrum start at $-2|J|$, in the present computation the value lies below $-2|J|$). At larger Δt there remains a single peak (the curves for $\Delta t = 120$ fs and 140 fs just display a small MNP background). However, this peak is not positioned at $\hbar\omega = E_{\text{pl}}$ but at a lower energy. It is the emission due to the remaining chain population but noting the SC–MNP coupling. The

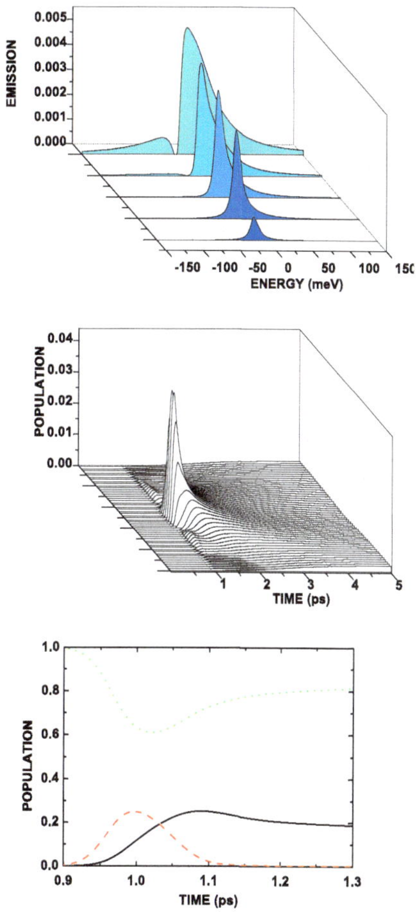

Figure 10.7 Photon emission of an SC (50 molecules, $d_n = 4.5$ D, $\Delta r_{\text{mol}} =$ 1.5 nm) placed in the vicinity of a spherical MNP (a = 10 nm, positioned in front of the SC's central part, cf. Fig. 10.2, $\Delta x = 2.5$ nm) and related photoinduced excitation energy transfer (laser pulse duration $\tau_p = 100$ fs, $E_0 = 5 \times 10^5$ V/m). Upper panel: Time- and frequency-resolved emission versus $\hbar\omega - E_{l=1}$ and for various time differences Δt to the maximum of the exciting laser pulse (localized at $t_p = 1$ ps). From the background to the foreground: $\Delta t = 100$, 120, 140, 160, and 280 fs. Middle panel: Molecular populations $P_1, \ldots P_{50}$ versus time (from the background to the foreground). Lower panel, full black line: Total population of the SC excited states; red dashed line: population of the MNP dipole plasmon; green dotted line: ground-state population. Reprinted with permission from Ref. [8], copyright (2012), SPIE.

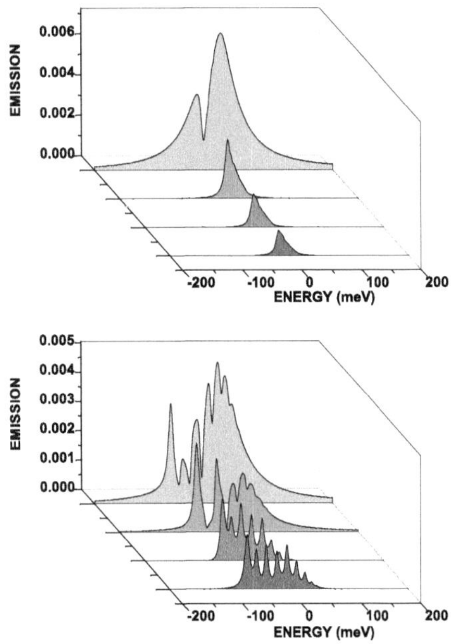

Figure 10.8 Time- and frequency-resolved emission versus $\hbar\omega - E_{l=1}$ for an SC of 10 molecules ($\Delta r_{\text{mol}} = 1.5$ nm) and the MNP placed at the terminal site of the SC ($\Delta X = 2.5$ nm). Spectra for various time differences Δt to the maximum of the exciting laser pulse are given (localized at $t_{\text{p}} = 1$ ps, duration $\tau_p = 100$ fs). Upper panel from the background to the foreground: $\Delta t = 100$, 200, 300, and 400 fs ($E_0 = 10^6$ V/m). Lower panel: Doubling of the molecular dipole moments: From the background to the foreground: $\Delta t = 100$, 120, 200, and 300 fs ($E_0 = 5 \times 10^5$ V/m). Reprinted with permission from Ref. [8], copyright (2012), SPIE.

resulting SC excitation energy dissipation is responsible for the overall decreasing emission intensity. At the same time the emission peak position indicates emission from the lower part of the exciton band. It is necessary to note that this is not an effect of exciton relaxation (the used simple model of dephasing only results in a uniform population of all states). Instead, MNP-induced SC oscillator strength enhancement is stronger for the lower part of the exciton spectrum.

Considering an SC with 50 molecules, single exciton levels could not be resolved. Therefore, it might be appropriate to study a smaller

SC with less exciton levels and a larger energetic distance among them. Figure 10.8 displays respective results. As it becomes obvious a doubling of the molecular transition dipole moment to 9 D is necessary to make the individual exciton levels visible (lower panel). The lowest exciton level is positioned below $E_{\text{mol}}-$ 60 meV and this energy range fixes the low-energy side of the emission band. After plasmon decay in the MNP and the vanishing of the related emission background the exciton levels of the chain are nearly completely present. Again, their overall intensity decreases with increasing Δt. The slight asymmetry with respect to $\hbar\omega = E_{\text{pl}}$ follows from the fact that the exciton band is somewhat asymmetric in relation to the molecular excitation energy if one allows that all molecules are coupled and not only nearest-neighbor ones. The fast plasmon decay results in emission spectra (at later times) which display properties of the SC alone. They are somewhat modified due to the remaining SC–MNP coupling (without noticeable plasmon populations) but in any case enhanced by more than 3 orders of magnitude compared to the isolated SC.

10.6 Concluding Remarks

A uniform quantum description has been presented in order to describe the photoinduced dynamics in an SC interacting with a nearby placed MNP. The approach uses a Frenkel exciton model for the electronic excitations of the SC which are coupled to the multipole collective electronic excitations of the MNP. To simulate the dynamics of the system a density matrix approach has been chosen. It offers a nonperturbative description of the excitation energy transfer in the SC as well as of the excitation energy exchange between the SC and the MNP. Moreover, it accounts for fast MNP plasmon decay and the dephasing of SC excitations due to a coupling to SC vibrational degrees of freedom.

Spatiotemporal localization of electronic excitation energy in a chain-like SC placed in the proximity of a spherical MNP could be demonstrated. Considering optical excitation of an SC–MNP complex one expects the formation of a local field around the MNP much larger than the externally applied one. Consequently, that part of the SC should be preferably excited which feels the enhanced local field. Our computations reproduce this effect, however, with a model which does not cover the computation of a local field. Instead, the Coulombic coupling between the SC and the MNP is considered in a nonperturbative description and excitation energy localization is explained as a combined effect of the SC exciton oscillator strength change and the formation of an excitonic wave packet. The oscillator strength change has been demonstrated more explicitly by presenting spectra of liner absorbance and spontaneous photon emission. While a regular chain-like SC placed far away from the MNP shows spectra with the main contribution of a single exciton level, the whole SC exciton band may become visible if the interaction with a nearby positioned MNP is strong enough. Of course, this is a subtle effect which requires the absence of disorder in the ensemble of SC–MNP complexes. If the exciton parameters and the SC–MNP coupling fluctuate somewhat the presence of various SC exciton levels shall be washed out.

Due to the fast MNP plasmon decay, time-resolved emission spectra display properties of the SC alone if later times relative to the optical excitation are chosen. These spectra include a slight shift of the exciton energies but a large change of the related oscillator strengths. Moreover, an enhancement of the emission by more than 3 orders of magnitude compared to the absence of the MNP is achieved. Overall the spectroscopy of SCs coupled to MNPs offers significantly more insight into the SC level structure than standard techniques without the utilization of MNPs.

Acknowledgments

Financial support by *Deutsche Forschungsgemeinschaft* through Sfb 951 (G. Kyas and Y. Zelinskyy) and by the *China Scholarship Council* (Y. Zhang) is gratefully acknowledged.

References

1. T. W. Odom and G. C. Schatz, *Chem. Rev.*, **11**, special issue, 3667 (2011).

2. N. J. Halas, *Nano Lett.*, **10**, 3816 (2010).
3. G. Kyas and V. May, *J. Chem. Phys.*, **134**, 034701 (2011).
4. Y. Zelinskyy and V. May, *Chem. Phys. Lett.*, **511**, 372 (2011).
5. Y. Zelinskyy and V. May, *Nano Lett.*, **12**, 446 (2012).
6. Y. R. Zelinskyy, Y. Zhang, and V. May, *J. Phys. Chem. A*, **116**, 11330 (2012) Joern-Manz-Festschrift.
7. G. Kyas, Y. Zelinskyy, Y. Zhang, and V. May, *Ann. Phys. (Berlin)*, **525**, 189 (2012), special issue "Ultrafast Phenomena at the Nanoscale".
8. Y. Zhang, Y. Zelinskyy, and V. May, *J. Nanophot.*, **6**, 063533 (2012).
9. Y. Zhang, Y. Zelinskyy, and V. May, *J. Phys. Chem. C*, **116**, 25962 (2012).
10. Y. Zelinskyy, Y. Zhang, and V. May, *J. Chem. Phys.*, **138**, 114704 (2013).
11. Y. Zhang, Y. Zelinskyy, and V. May, *Phys. Rev. B*, **88**, 155426 (2013).
12. V. May and O. Kühn, *Charge and Energy Transfer Dynamics in Molecular Systems* (Wiley-VCH, Weinheim, 2000, 2004, 2011).
13. E. Prodan and P. Nordlaender, *J. Chem. Phys.*, **120**, 5444 (2004).
14. A. Trügler and U. Hohenester, *Phys. Rev. B*, **77**, 115403 (2008).
15. L. G. Gerchikov, C. Guet, and A. N. Ipatov, *Phys. Rev. B*, **66**, 053202 (2002).
16. G. Weick, R. A. Molina, D. Weinmann, and R. A. Jalabert, *Phys. Rev. B*, **72**, 115410 (2005).
17. G. Weick, G.-L. Ingold, R. A. Jalabert, and D. Weinmann, *Phys. Rev. B*, **74**, 165421 (2006).
18. M. Brack, *Phys. Rev. B*, **39**, 3533 (1989).
19. M. Brack, *Rev. Mod. Phys.*, **65**, 677 (1993).
20. G.F. Bertsch and H. Feldmeier, *Phys. Rev. C*, **56**, 839 (1997).
21. K. Hagino, *Phys. Rev. B*, **60**, R2198 (1999).
22. J.-P. Connerade and A. V. Solovyov, *Phys. Rev. A*, **66**, 013207 (2002).
23. K. Lamichhane, M. Brack, and P. Winkler, *Int. J. Quantum Chem.*, **111**, 4363 (2011).
24. J. Jacak, J. Krasnyj, W. Jacak, R. Gonczarek, A. Chepok, and L. Jacak, *Phys. Rev. B*, **82**, 035418 (2010).
25. D. J. Masiello and G. C. Schatz, *Phys. Rev. B*, **78**, 042505 (2008).
26. T. Ambjörnsson, G. Mukhopadhyay, A. P. Apell, and M. Käll, *Phys. Rev. B*, **73**, 085412 (2006).
27. P. Anger, P. Bharadwaj, and L. Novotny, *Phys. Rev. Lett.*, **96**, 113002 (2006).
28. S. Bhowmick, S. Saini, V. B. Shenoy, and B. Bagchi, *J. Chem. Phys.*, **125**, 181102 (2006).

29. J. Vielma and P. T. Leung, *J. Chem. Phys.*, **126**, 194704 (2007).
30. G. Colas des Francs, C. Girar, T. Laroche, G. Leveque, O. J. F. Martin, *J. Chem. Phys.*, **127**, 034701 (2007).
31. A. Myers Kelley, *Nano Lett.*, **7**, 3235 (2007).
32. W. Gu, H. Choiand K. Kim, *J. Phys. Chem. A*, **11**, 8121 (2007).
33. H. Mertens, A. F. Koenderink, and A. Polman, *Phys. Rev. B*, **76**, 115123 (2007).
34. H. Y. Xie, H. Y. Chung, P. T. Leung, and D. P. Tsai, *Phys. Rev. B*, **80**, 155448 (2009).
35. T. J. Davis, D. E. Gomes, and K. C. Vernon, *Phys. Rev. B*, **81**, 045432 (2010).
36. W. Zhang, A. O. Govorov, and G. W. Bryant, *Phys. Rev. Lett.*, **97**, 146804 (2006).
37. A. O. Govorov, J. Lee, and N. A. Kotov, *Phys. Rev. B*, **76**, 125308 (2007).
38. J.-Y. Yan, W. Zhang, S. Duan, X.-G. Zhao, and A. O. Govorov, *Phys. Rev. B*, **77**, 165301 (2008).
39. A. O. Govorov, Z. Fan, P. Hernandez, J. M. Slocik, and R. R. Naik, *Nano Lett.*, **10**, 1347 (2010).
40. E. T. Niles, J. D. Roehling, H. Yamagata, A. J. Wise, F. C. Spano, A. J. Moule, and J. K. Grey, *J. Phys. Chem. Lett.*, **3**, 259 (2012).
41. J. Holcman, A. Al Choueiry, A. Enderlin, S. Hameau, T. Barisien, and L. Legrand, *Nano Lett.*, **11** 4496 (2011).
42. F. Bai, Z. Sun, H. Wu, R. E. Haddad, E. N. Coker, J. Y. Huang, M. A. Rodriguez, and H. Fan, *Nano Lett.*, **11** 5196 (2011).
43. H. Marciniak, X.–Q. Li, F. Würthner, and S. Lochbrunner, *J. Phys. Chem. A*, **115**, 648 (2011).
44. C. D. Geddes, and J. R. Lakowicz, *J. Fluoresc.*, **12**, 121 (2002).
45. A. Bek, R. Jansen, M. Ringler, S. Mayilo, T. A. Klar, and J. Feldmann, *Nano Lett.*, **8**, 485 (2008).
46. M. Ringler, A. Schwemer, M. Wunderlich, A. Nichtl, K. Kürzinger, T. A. Klar, and J. Feldmann, *Phys. Rev. Lett.*, **100**, 203002 (2008).
47. S. Schietinger, M. Barth, T. Aichele, and O. Benson, *Nano Lett.*, **9**, 1694 (2009).
48. Y. Zhang, A. Padhyay, J. E. Sevilleja, R. L. Guerrant, and C. D. Geddes, *J. Phys. Chem. C*, **114**, 7575 (2010).
49. T. Ming, L. Zhao, H. Chen, K. C. Woo, J. Wang, and H.–Q. Lin, *Nano Lett.*, **11**, 2296 (2011).
50. H. Aouani, O. Mahboub, E. Devaux, H. Rigneault, T. W. Ebbesen, and J. Wenger, *Nano Lett.*, **11**, 2400 (2011).

51. A. A. Toropov, T. V. Shubina, K. G. Belyaev, S. V. ivanov, P. S. Kopev, Y. Ogawa, and F. Minami, *Phys. Rev. B*, **84**, 085323 (2011).
52. S. R. Beyer, S. Ullrich, S. Kudera, A. T. Gardiner, R. J. Cogdell, and J. Köhler, *Nano Lett.*, **11**, 4897 (2011).
53. V. J. Sorger, N. Pholchai, E. Cubukcu, R. F. Oulton, P. Kolchin, C. Borschel, M. Gnauck, C. Ronning, and X. Zhang, *Nano Lett.*, **11**, 4907 (2011).
54. V. N. Pustovit and T. V. Shahbazyan, *Phys. Rev. Lett.*, **102**, 077401 (2009).
55. V. N. Pustovit and T. V. Shahbazyan, *Phys. Rev. B*,**82**, 075429 (2010).
56. H. Zhu, B. Röder, and V. May, *Chem. Phys.*, **362**, 19 (2009).
57. L. Wang and V. May, *Phys.Chem.Chem.Phys.*, **13**, 8755 (2011).

Chapter 11

Ultrafast Energy and Charge Transfer in Functional Molecular Nanoscale Aggregates

Hiroyuki Tamura,[a] Keith H. Hughes,[b] Rocco Martinazzo,[c] Jan Wahl,[d] Robert Binder,[d] and Irene Burghardt[d]

[a] *WPI-Advanced Institute for Material Research, Tohoku University, 2-1-1 Katahira, Aoba-ku, Sendai 980-8577, Japan*
[b] *School of Chemistry, Bangor University, Bangor, Gwynedd LL57 2UW, UK*
[c] *Dipartimento di Chimica, Università degli Studi di Milano, v. Golgi 19, 20133 Milano, Italy*
[d] *Institute of Physical and Theoretical Chemistry, Goethe University Frankfurt, Max-von-Laue-Str. 7, 60438 Frankfurt, Germany*
burghardt@chemie.uni-frankfurt.de

11.1 Introduction

Photoinduced energy and charge transfer events in biological and material functional systems are often ultrafast and involve delocalized excitations—that is, molecular excitons—and strong electron–vibration (vibronic) coupling effects which necessitate a nonperturbative treatment. The ultrafast nature of these processes is a result of nonadiabatic interactions by which electronically

Ultrafast Dynamics at the Nanoscale: Biomolecules and Supramolecular Assemblies
Edited by Irene Burghardt and Stefan Haacke

ISBN 978-981-4745-33-8 (Hardcover), 978-981-4745-34-5 (eBook)
www.panstanford.com

excited states undergo nonradiative transfer and decay processes. Examples include biological systems like chromophore–protein complexes, natural and artificial light-harvesting systems, organic materials like semiconducting polymers, and various other molecular assemblies. A number of studies involving these and related systems are presented in this volume.

In line with the quantum dynamical nature of the photoinduced processes, recent experimental observations have provided compelling evidence that quantum coherence plays an important role even in high-dimensional, extended molecular assemblies, and in solvent or protein environments. In particular, ultrafast energy transport, that is, excitation energy transfer (EET) in biological light-harvesting systems and semiconducting polymer materials has been shown to conserve excitonic coherence [1–3], rather than involving a Förster-type hopping [4] between neighboring sites. Contrary to the conventional assumption that environment-induced decoherence sets in within several tens of femtoseconds in condensed-phase systems [5], long-lived coherences have been observed on a timescale of picoseconds [1–3, 6]. Conjectures have been made regarding the possible role of non-Markovian effects and correlated environmental fluctuations in preserving these long-lived coherences [7].

Besides coherent transfer mechanisms, a distinguishing feature of photoexcitations in molecular aggregates is the role of spatial delocalization, which is manifested in the formation of Frenkel-type exciton states [4] as well as delocalized charge transfer (CT) states. These coherent superposition states entail a modification of the energetics—and, hence, of the spectroscopically observable states—as illustrated by the formation of J-type or H-type aggregates exhibiting head-to-tail and face-to-face alignment, respectively [4, 8]. Further, the spatial extension of these states is modulated by the coupling to vibrational (phonon) modes, such that transient vibronic effects resulting from photoexcitation also affect spatial correlations. In particular, trapping effects [9–12] are observed that lead to quasi-stationary polaronic species. Generally, dephasing and relaxation effects take on a characteristic signature that is specific to lattice structures composed of interacting monomer species [11, 13].

From a system–bath theory perspective, the systems of interest fall into a markedly non-Markovian regime [14, 15], since the coupling between the electronic and vibrational degrees of freedom is strong and the environment is neither static nor rapidly fluctuating (Markovian). Instead, photoexcitation of the subsystem (chromophore) induces a nonequilibrium response of the environment that is interleaved with the subsystem evolution, thus generating a dynamical evolution in the high-dimensional system-plus-environment space.

Over the past decade, various theoretical approaches and simulation techniques have been developed to tackle photoinduced dynamics in high-dimensional molecular aggregates. Broadly, two types of approaches can be distinguished. First are approaches which rely on quantum-dynamical calculations in conjunction with model Hamiltonians of vibronic coupling type, or generalized spin-boson models [4, 16, 17]. Here, the phonon modes are either treated explicitly or are (partially) integrated out by a reduced dynamics procedure. The relevant model Hamiltonians are generally parametrized using electronic structure calculations, along with information on spectral densities [14]. The second type of approach relates to an explicit, on-the-fly treatment [18, 19] of the electronic structure *and* dynamics of the high-dimensional, supermolecular system. Here, the dynamics is often approximated by classical trajectory ensembles or Gaussian wave packets [20, 21]. The main strength of the first approach is its accurate description of the underlying quantum dynamics, at the expense of using approximate potential surfaces, while the strength of the second approach lies in the more flexible treatment of the electronic structure properties at all geometries.

Here, we adopt the first approach, which allows insight into the role of coherent features of the transfer processes on ultrafast timescales. To this end, a vibronic coupling model in a generalized electron–hole representation [11, 22–24] is employed, which allows for a unified picture of both exciton and charge transfer. The relevant model Hamiltonian is parametrized using high-level electronic structure calculations or time-dependent density functional theory (TD-DFT), in conjunction with suitable diabatization procedures [16, 25, 26]. Finally, quantum-dynamical calculations are carried out using multiconfigurational wave packet techniques [17, 27] or non-Markovian master equation approaches [28, 29]. In view of reducing the dimensionality of the problem, so-called effective-mode schemes [15, 29] can be employed, which are especially suitable to accurately capture the dynamics on ultrafast time scales.

The remainder of this chapter is organized as follows. Section 11.2 introduces the relevant vibronic Hamiltonian in an electron-hole representation, Section 11.3 gives a brief account of the quantum-dynamical propagation schemes that were used in our recent work, and Section 11.4 and Section 11.5 present two case studies, involving ultrafast exciton migration and CT at a donor-acceptor (D–A) heterojunction. Finally, Section 11.6 summarizes and concludes.

11.2 Electron–Hole Lattice Hamiltonian

To treat EET and CT on the same footing, a generalized excitonic representation in an electron–hole (e–h) quasi-particle picture is employed [11, 30]. This type of basis was proposed by Merrifield [22] in the context of molecular crystals and has since been employed in various later works [23, 24, 31, 32]. In the following, we first describe the formulation of the Hamiltonian in this basis (Section 11.2.1 and Section 11.2.2) and subsequently include vibronic coupling effects (Section 11.2.3). Finally, the vibronic coupling Hamiltonian will be recast in a reduced-dimensional form, using an effective-mode representation, as detailed in Section 11.2.4.

11.2.1 Electron–Hole Basis

In line with the treatment of Refs. [11, 24, 30], a basis of e–h quasi-particle states is introduced, $|\nu_e \mu_h\rangle \equiv |\nu\mu\rangle$, where the electron is located at site $\nu_e = \nu$ with position r_ν^e, while the hole is located at site $\mu_h = \mu$ with position r_μ^h. (In a valence/conduction band picture, electron (hole) states localized at site ν (μ) are created by taking suitable linear combinations of the conventional, extended conduction/valence band states.) All other sites remain in their ground states. We will refer to localized e–h pairs $|\nu\nu\rangle$

as Frenkel configurations while nonlocal e–h states $|\nu\mu\rangle$, $\mu \neq \nu$, correspond to CT configurations. This generalized exciton basis including CT effects is particularly appropriate in cases where the exciton coherence size [23, 33], that is, the e–h separation, is of the order of several monomer units such that the Frenkel picture is not strictly applicable. However, the system is still taken to belong to a small-radius exciton regime, rather than a large-radius Wannier–Mott regime [34].

The e–h states in question can be generated from the ground state $|\,0\,\rangle$ (which is here taken as the vacuum state for e–h formation, that is, an empty conduction band and a filled valence band), by the action of the exciton creation operator $\hat{c}^{\dagger}_{\nu\mu}$ [35–37],

$$|\nu\mu\rangle = \hat{c}^{\dagger}_{\nu\mu}|\,0\,\rangle \tag{11.1}$$

The operator $\hat{c}^{\dagger}_{\nu\mu}$ is connected as follows to electron creation and annihilation operators, $\hat{c}^{\dagger}_{\nu\mu} = \hat{b}^{\dagger}_{\nu}\hat{a}_{\mu}$, that is, an electron is annihilated at site μ (in the valence band) and simultaneously generated at site ν (in the conduction band), while ensuring momentum conservation.

Within the context of many-body perturbation theory, the simplest treatment relating to the e–h state representation is in terms of the Tamm–Dancoff approximation (TDA) [37, 38]. Here, single excitations from the Hartree–Fock closed-shell ground state are considered, such that the excited states $|k\rangle$ can be written as linear combinations of e–h states defined according to Eq. 11.1, that is,

$$\begin{aligned} |k\rangle &= \sum_{\nu\mu} |\nu\mu\rangle\langle\nu\mu|k\rangle \\ &= \sum_{\nu\mu} \rho^{(k)}_{\nu\mu}|\nu\mu\rangle \end{aligned} \tag{11.2}$$

where $\rho^{(k)}_{\nu\mu}$ can be identified as elements of the single-particle transition density matrix [33],

$$\rho^{(k)}_{\nu\mu} = \langle\nu\mu|k\rangle \tag{11.3}$$

This definition carries over to the analysis of high-level wave functions in a basis of localized fragments, as employed, for example, in an excited-state analysis in terms of CT indices [39, 40].

Note that the TDA as such has not proven adequate to describe the strongly bound excitons characteristic of organic semiconductors [41], despite its success in the description of optical excitations in bulk inorganic semiconductors and other solid materials [38]. In the present context, either high-level electronic structure methods or TD-DFT are used to characterize excitonic states of small oligomer fragments, which are subsequently expressed in a fragment-based electron–hole basis by using suitable diabatization procedures [16, 25, 26].

Due to the direct relation between the transition densities Eq. 11.3 and the electron–hole distributions employed in this work, the dynamics in the e–h basis can be interpreted as a time-dependent transition density representation. A related perspective on time-dependent transition density matrices has recently been proposed in Ref. [42].

11.2.2 Wave Functions and Density Matrix Representation

In the following, we will use the e–h representation to expand the wave function of the combined electronic and vibrational (phonon) state,

$$|\psi(\{z_i\}, t)\rangle = \sum_n \sum_{\nu\mu} A_{n,\nu\mu}(t)\, \phi_n(\{z_i\}, t)\, |\nu\mu\rangle \tag{11.4}$$

where the index n labels time-dependent vibrational configurations $\phi_n(\{z_i\})$, a more explicit form of which is given in Section 11.3.

The corresponding time-evolving density operator is given as the projector $\hat{\rho}(t) = |\psi(t)\rangle\langle\psi(t)|$ in the pure-state case, and more generally as

$$\hat{\rho}(t) = \sum_l p_l\, |\psi_l(t)\rangle\langle\psi_l(t)| \tag{11.5}$$

for statistical mixtures, with statistical weights $\{p_l\}$. In practice, mixed states are propagated either as ensembles of pure states or else using an explicit density matrix representation.

For the purpose of analysis, time-evolving density matrix elements are calculated,

$$\langle\rho_{\nu\mu,\nu'\mu'}\rangle(t) = \mathrm{Tr}\{|\nu'\mu'\rangle\langle\nu\mu|\,\hat{\rho}(t)\} \tag{11.6}$$

where the trace (Tr) is taken over the electronic and phonon degrees of freedom of the density operator $\hat{\rho}$. The diagonal elements,

$$P_{\nu\mu}(t) = \delta_{\nu,\nu'}\delta_{\mu,\mu'}\langle\rho_{\nu\mu,\nu'\mu'}\rangle(t) \tag{11.7}$$

correspond to e–h state populations. The off-diagonal elements of Eq. 11.6 with $\nu \neq \nu'$ and/or $\mu \neq \mu'$ represent coherences.

Further, marginal distributions are defined [11],

$$P_{\xi}^{\mathrm{CM}}(t) = \sum_{\substack{\nu,\mu=1 \\ \nu+\mu=2\xi}}^{N_{\mathrm{site}}} P_{\nu\mu}(t) \quad ; \quad P_{\Delta}^{\mathrm{CS}}(t) = \sum_{\substack{\nu,\mu=1 \\ \nu-\mu=\Delta}}^{N_{\mathrm{site}}} P_{\nu\mu}(t) \tag{11.8}$$

where $P_{\xi}^{\mathrm{CM}}(t)$ represents the time-evolving distribution with respect to the center-of-mass (CM) coordinate $\xi = (\nu + \mu)/2$ while $P_{\Delta}^{\mathrm{CS}}(t)$ is the corresponding distribution with respect to the e–h distance coordinate $\Delta = \nu - \mu$. From the distributions P_{ξ}^{CM} and P_{Δ}^{CS}, one can infer the *delocalization length* (spatial extension) and the *coherence size* (e–h separation) of the excitonic state, respectively. The corresponding standard deviations $\sigma_{\xi}^{\mathrm{CM}}$ and $\sigma_{\Delta}^{\mathrm{CS}}$ are measures of the width of these marginal distributions. (Other relevant measures are given in terms of participation ratios, see, for example, Ref. [33]). In the Frenkel case, we have $\sigma_{\Delta}^{\mathrm{CS}} = 0$, that is, an e–h pair is always localized at a given site, and delocalization results entirely from the superposition of localized e–h states $|\nu\nu\rangle$.

11.2.3 Vibronic Hamiltonian in the e–h Basis

In the following, we will use a vibronic coupling Hamiltonian in the e–h quasi-particle basis [11],

$$\hat{H}(\{\hat{z}_i\}) = \sum_{\nu\mu}\sum_{\nu'\mu'} \hat{H}_{\nu\mu,\nu'\mu'}(\{\hat{z}_i\})|\nu\mu\rangle\langle\nu'\mu'| \tag{11.9}$$

whose elements are conveniently written as

$$\hat{H}_{\nu\mu,\nu'\mu'}(\{\hat{z}_i\}) = \delta_{\nu,\nu'}\delta_{\mu,\mu'}\hat{H}^{(0)}_{\nu\mu,\nu'\mu'}(\{\hat{z}_i\}) + \hat{H}^{\mathrm{coup}}_{\nu\mu,\nu'\mu'}(\{\hat{z}_i\}) \tag{11.10}$$

where an electronically diagonal zeroth-order part $(\hat{H}^{(0)}_{\nu\mu,\nu\mu})$ is distinguished from a coupling part $(\hat{H}^{\mathrm{coup}}_{\nu\mu,\nu'\mu'})$ inducing EET and CT, that is, for $\nu = \nu'$ and $\mu = \mu'$,

$$\hat{H}^{(0)}_{\nu\mu,\nu\mu} = \hat{H}^{eh}_{\nu\mu,\nu\mu} + \hat{H}^{\text{field}}_{\nu\mu,\nu\mu} + \hat{H}^{\text{vibr}}_{\nu\mu,\nu\mu} \quad (11.11)$$

and for $\nu \neq \nu'$ and/or $\mu \neq \mu'$,

$$\hat{H}^{\text{coup}}_{\nu\mu,\nu'\mu'} = \hat{H}^{\text{EET}}_{\nu\mu,\nu'\mu'} + \hat{H}^{\text{CT}}_{\nu\mu,\nu'\mu'} \quad (11.12)$$

As already mentioned, the terms of Eq. 11.11 are diagonal in the e–h basis and correspond to (i) the intrinsic electron–hole interaction ($\hat{H}^{eh}$), (ii) the effect of the internal field ($\hat{H}^{\text{field}}$) generated by the electrodes (and possibly an additional external field), and (iii) vibronic contributions ($\hat{H}^{\text{vibr}}$) related to the coupling to the relevant vibrational (phonon) modes. Conversely, the terms of Eq. 11.12, $\hat{H}^{\text{EET}}_{\nu\mu,\nu'\mu'}$ and $\hat{H}^{\text{CT}}_{\nu\mu,\nu'\mu'}$ represent off-diagonal electronic coupling terms describing EET (Frenkel)-type excitonic couplings as well as transfer integrals that couple CT configurations to Frenkel configurations and among each other.

The above form of the Hamiltonian corresponds to a quasi-diabatic vibronic coupling Hamiltonian [16, 43] with a diagonal representation of the nuclear kinetic energy and potential-type electronic couplings. A diabatization procedure [16, 25, 26] or the determination of selected coupling elements [44, 45] is therefore necessary to map electronic structure information onto the Hamiltonian Eqs. 11.9–11.12. As it stands, the present vibronic coupling Hamiltonian is also an extension to exciton transport models like the Holstein Hamiltonian [46–48], which are usually restricted to Frenkel configurations.

We now further specify the terms appearing in Eqs. 11.11–11.12.

In Eq. 11.11, the intrinsic e–h pair interaction determines the on-site potential and is taken as

$$H^{eh}_{\nu\mu,\nu'\mu'} = \delta_{\nu,\nu'}\delta_{\mu,\mu'}\left(e_0 - \frac{1}{4\pi\epsilon_0\epsilon_r r_{\nu\mu}}\right) \quad (11.13)$$

where ϵ_0 and ϵ_r are the vacuum permittivity and relative permittivity, respectively, and $r_{\nu\mu}$ is the e–h distance as a function of the sites ($\nu\mu$), $r_{\nu\mu} = |r^e_\nu - r^h_\mu| \simeq |\nu - \mu| r_{ru} + r_0$, with r_{ru} the repeat unit length, and r_0 the intrinsic e–h distance for a Frenkel exciton. Further, e_0 is the binding energy for a localized e–h pair such that $H_{\nu\nu,\nu\nu} = 0$. (For alternative formulations, see [49]).

Addition of an internal electric field ($\hat{H}^{\text{field}}$), that is, the second term in Eq. 11.11, to the intrinsic electron–hole interaction ($\hat{H}^{\text{eh}}$) of Eq. 11.13 results in an effective Coulomb barrier to e–h separation [25, 50] (see also Section 11.5.3).

Next, the vibronic terms represented by $\hat{H}^{\text{vibr}}_{\nu\mu,\nu'\mu'}$, that is, the third term in Eq. 11.11, are taken to be electronically diagonal and include a dependence on site-local and site-correlated coordinates $z_i \in \{x_{i,\xi}, y_{i,\xi\xi'}\}$ where (ξ, ξ') denote different fragments. The vibronic Hamiltonian takes the following form,

$$\hat{H}^{\text{vibr}}_{\nu\mu,\nu'\mu'}(\{\hat{z}_i\}) = \delta_{\nu,\nu'}\delta_{\mu,\mu'}\sum_i\left(\frac{1}{2}\,\hat{p}_i^2 + \hat{V}_{i,\nu\mu}(\hat{z}_i)\right) \tag{11.14}$$

where $\hat{V}_{i,\nu\mu}(\hat{z}_i)$ corresponds to a state-specific vibrational potential in the e–h representation. (These potentials typically represent excitonic potentials involving combinations of monomer ground-state and excited-state potentials [11, 26]). For a subset of modes—or sometimes for all modes—a linear vibronic coupling (LVC) approximation [16] is often used,

$$\hat{H}^{\text{LVC}}_{\nu\mu,\nu'\mu'}(\{z_i\}) = \delta_{\nu,\nu'}\delta_{\mu,\mu'}\sum_i\left(\frac{1}{2}(\hat{p}_i^2 + \omega_i^2\hat{z}_i^2) + \kappa_{i,\nu\mu}\hat{z}_i\right) \tag{11.15}$$

This form of the vibronic Hamiltonian—here given in mass-weighted coordinates—connects to typical system–bath models and their effective-mode representations, as addressed later (Section 11.2.4).

Turning to Eq. 11.12, the site-to-site couplings are of two types: First, Frenkel-type couplings between localized e–h pairs ($\nu = \mu$, $\nu' = \mu'$), and second, electron/hole transfer integrals between delocalized e–h pairs ($\nu \neq \mu$ and/or $\nu' \neq \mu'$) [11]. For both types of couplings, we will assume that only nearest-neighbor interactions are relevant. Furthermore, we will assume that the transfer integrals are identical for electrons and holes (an assumption that has to be relaxed if carrier transport is studied in detail). The explicit form of these couplings reads as follows [11]:

$$\begin{aligned}\hat{H}^{\text{EET}}_{\nu\mu,\nu'\mu'}(\{\hat{z}_i\}) =&\delta_{\nu,\mu}\delta_{\nu',\mu'}\Big(\delta_{\nu',\nu+1}J_{\text{EET}}(\{\hat{z}_i\}) \\ &+ \delta_{\nu',\nu-1}J_{\text{EET}}(\{\hat{z}_i\})\Big)\end{aligned} \tag{11.16}$$

with the Frenkel couplings $J_{\rm EET}(\{\hat{z}_i\})$, and

$$\hat{H}^{\rm CT}_{\nu\mu,\nu'\mu'}(\{\hat{z}_i\}) = \delta_{\nu,\nu'}\left(\delta_{\mu',\mu+1}t_{\rm CT}(\{\hat{z}_i\}) + \delta_{\mu',\mu-1}t_{\rm CT}(\{\hat{z}_i\})\right) + \delta_{\mu,\mu'}\left(\delta_{\nu',\nu+1}t_{\rm CT}(\{\hat{z}_i\}) + \delta_{\nu',\nu-1}t_{\rm CT}(\{\hat{z}_i\})\right) \quad (11.17)$$

with the transfer integrals $t_{\rm CT}(\{\hat{z}_i\})$.

Specific forms of this Hamiltonian that describe either EET or CT (or both) will be referred to in Section 11.4 and Section 11.5.

11.2.4 Effective-Mode Reduction Techniques

As mentioned above, it is often useful to partition the vibronic coupling part of the Hamiltonian Eq. 11.14 into a portion that involves several dominant coordinates which may be treated anharmonically, and a residual part that is treated according to a simplified system–bath model of LVC form Eq. 11.15. If the frequency distribution of the bath modes is dense, one may sample the bath frequencies ω_i and system–bath couplings κ_i from a spectral density [14],

$$\mathcal{J}(\omega) = \frac{\pi}{2}\sum_{i=1}^{N_B\to\infty}\frac{\kappa_i^2}{\omega_i}\delta(\omega-\omega_i) \quad (11.18)$$

In situations where the vibrational modes couple simultaneously to several electronic states $|\nu\mu\rangle$, one refers to a correlated bath [7].

The spectral density is usually sampled in a discrete fashion, with a frequency domain sampling interval $\Delta\omega$ that guarantees an effectively irreversible dynamics for times up to the Poincaré recurrence period $\tau_P = 2\pi/\Delta\omega$ [14].

Alternatively, an effective-mode decomposition of the spectral density may be used [15, 51–53], which opens the possibility to systematically reduce the dimensionality of the problem. To this end, an orthogonal coordinate transformation is carried out which converts the standard form of the vibronic Hamiltonian Eq. 11.14 to a chain representation. As a simple example, consider a set of site-local modes $\{\hat{x}_i\}$, that couple to a single electron–hole state $|\nu\mu\rangle$,

$$\hat{H} = \hat{H}^{(0)}_{\nu\mu} + \sum_i \frac{1}{2}\left(\hat{p}_i^2 + \omega_i^2\hat{x}_i^2\right)\hat{1} + \sum_i \kappa_i\hat{x}_i|\nu\mu\rangle\langle\nu\mu| \quad (11.19)$$

In this case, the transformed Hamiltonian reads,

$$\hat{H} = \hat{H}^{(0)}_{\nu\mu} + \sum_i \frac{1}{2}\left(\hat{P}_i^2 + \Omega_i^2 \hat{X}_i^2\right)\hat{1} + D_{0,1}\hat{X}_1|\nu\mu\rangle\langle\nu\mu| + \sum_i D_{i,i+1}\hat{X}_i\hat{X}_{i+1} \quad (11.20)$$

where the first collective mode $\hat{X}_1 = (1/D_{0,1})\sum_i \kappa_i \hat{x}_i$ is seen to completely absorb the coupling to the electronic subsystem. Further, the first and second collective modes are bilinearly coupled, as are all remaining (residual) modes $(\hat{X}_i, \hat{X}_{i+1})$.

In the new coordinates, the spectral density takes a Mori-type continued fraction form [15, 51–53]. Truncation of the chain Hamiltonian at the level of the Mth chain member leads to a controlled approximation, generating a series of approximate Mth order spectral densities that take the following form [51],

$$\mathcal{J}^{(M)}_{\text{eff}}(\omega) = \text{Im} \lim_{\epsilon\to 0^+} W^{(M)}_{\text{eff}}(\omega + i\epsilon) \quad (11.21)$$

where $W^{(M)}_{\text{eff}}(z)$ is a continued fraction with closure at the Mth order of the effective-mode hierarchy,

$$W^{(M)}_{\text{eff}}(z) = \cfrac{D^2_{0,1}}{\Omega_1^2 - z^2 - \cfrac{D^2_{1,2}}{\Omega_2^2 - z^2 - \cdots \cfrac{D^2_{M-2,M-1}}{\Omega^2_{M-1} - z^2 - \cfrac{D^2_{M-1,M}}{\Omega^2_M - z^2 - i\gamma z}}}} \quad (11.22)$$

Here, the Mth chain member is taken to undergo Markovian (Ohmic) dissipation with a friction coefficient γ. Decomposition of the continued fraction in terms of its complex poles yields a set of $n = 2M + n_{\text{mat}}$ poles, where n_{mat} corresponds to the Matsubara frequencies [14, 29]. (The associated correlation functions are accordingly expressed in terms of n complex exponentials).

In a complementary fashion to the Mth-order approximate spectral densities $(\mathcal{J}^{(M)}_{\text{eff}})$, residual spectral densities $\mathcal{J}^{(M)}_{\text{res}}$ can be defined that result from successively extracting M effective modes from the original spectral density [15],

$$\mathcal{J}^{(M)}_{\text{res}}(\omega) = \text{Im} \lim_{\epsilon\to 0^+} W_M(\omega + i\epsilon) \quad (11.23)$$

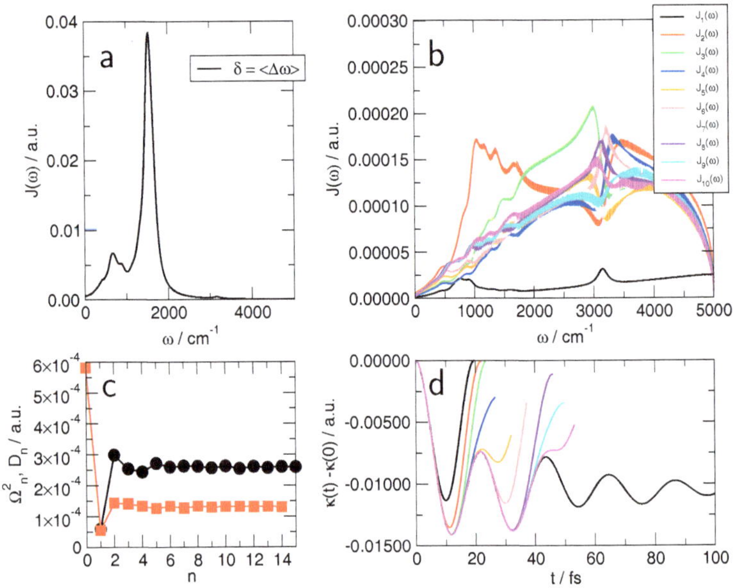

Figure 11.1 Effective-mode decomposition of a spectral density pertaining to the oligothiophene–fullerene donor–acceptor system addressed in Section 11.5. (a) Spectral density obtained from electronic structure data, as detailed in Refs. [54, 55]. (b) Residual spectral densities $\mathcal{J}_{\text{res}}^{(M)}$ approaching a quasi-ohmic form (with cutoff) (see Refs. [15, 53]). (c) Effective-mode parameters, that is, transformed frequencies and couplings. (d) Sequence of approximants to the Langevin memory kernel, including an increasing number of effective modes (see Ref. [53] for details.)

where W_M satisfies the recursion

$$W_M(z) = \Omega_m^2 - z^2 - \frac{D_{M-1,M}^2}{W_{M-1}(z)} \tag{11.24}$$

If the residual spectral densities tend towards a (quasi-)ohmic form at sufficiently high orders, this provides a rationale for the truncation procedure.

Figure 11.1 illustrates the effective-mode decomposition of a spectral density $\mathcal{J}(\omega)$ for a spin-boson-type system representing the charge separation at an oligothiophene–fullerene heterojunction (see Section 11.5.1). This spectral density was obtained from electronic structure calculations of a D–A dimer complex [55], using a shifted harmonic oscillator model in conjunction with a normal-mode expansion. (Other approaches to obtain spectral densities include, in particular, the Fourier transformation of correlation functions obtained from molecular dynamics (MD) simulations [56, 57], besides experimental information, for example, from photon echo peak shift [58] or dynamical Stokes shift [59] experiments).

Figure 11.1 illustrates the convergence of the residual spectral densities $\mathcal{J}_{\rm res}^{(M)}$ to an Ohmic form with cutoff (panel b), along with the distribution of chain parameters $\{\Omega_n, D_n\}$ (panel c). In addition, the expansion of the Langevin memory kernel in terms of the effective modes is illustrated (panel d). This expansion demonstrates that the effective mode series serves to successively unravel the dynamics as a function of time [53].

The truncation of the chain representation corresponds to a rational approximation which yields a sequence of controlled approximants to the short-time bath response, and can be made exact for increasingly longer timescales by increasing the number of modes. The approach is also related to time-dependent density matrix renormalization group (t-DMRG) techniques [60, 61].

11.3 Quantum-Dynamical Calculations Using Multiconfigurational Methods

From a computational viewpoint, the correlated electronic-nuclear dynamics in molecular aggregates poses a formidable challenge. Suitable techniques range from high-dimensional wave function methods and path integral techniques to non-Markovian reduced dynamics approaches. In the applications reported below, two techniques have been employed, that is, the multiconfiguration time-dependent Hartree (MCTDH) method [17, 27, 62, 63] and the hierarchical equations of motion (HEOM) [28, 64] approach. Since the HEOM method is amply commented upon in Chapter 13 by Dykstra and Tanimura, we focus here on the MCTDH method.

In the MCTDH approach [27, 62, 63], (see the Heidelberg MCTDH package [65]), the wave function of the type Eq. 11.4 is expanded in terms of Hartree products, that is, products of time-dependent so-called single-particle functions (spf's) $\varphi_j^{(\kappa)}(z_\kappa, t)$,

$$|\psi(z_1, \ldots, z_f, t)\rangle = \sum_{\mu,\nu}^{N_{\text{site}}} \sum_{j_1}^{n_1} \cdots \sum_{j_f}^{n_f} A_{j_1,\ldots,j_f,\nu\mu}(t) \times \prod_{\kappa=1}^{f} \varphi_{j_\kappa}^{(\kappa)}(z_\kappa, t)|\nu\mu\rangle \tag{11.25}$$

with time-dependent expansion coefficients $A_{j_1,\ldots,j_f,\mu\nu}(t)$ and time-dependent spf configurations, whose evolution is given in terms of nonlinear variational dynamical equations. The spf's are further expanded in a primitive grid basis, typically a discrete-variable representation (DVR) [27]. Equation 11.25 corresponds to the so-called *single-set* form of the wave function [27], where electronic states are treated on equal footing with other degrees of freedom; this form is best adapted to the large number of states that are included in the present systems.

The MCTDH method typically allows one to include up to 50–100 vibrational modes in the dynamical treatment, thus going by far beyond conventional wave function propagation methods. A more recent development, using the so-called multilayer MCTDH (ML-MCTDH) approach [66–68] goes yet further, by introducing a hierarchical scheme that makes the treatment of hundreds or even thousands of modes feasible.

11.4 Exciton Migration and Relaxation at a Torsional Defect Site

Exciton migration is one of the key processes in photoactive nanostructures. As highlighted by recent experiments [2, 3], EET, or exciton migration in functional organic materials can be guided by quantum coherence, despite the presence of a high-dimensional phonon environment. Hence, a conventional Förster theory treatment [4] of EET is not appropriate, and a more detailed quantum-dynamical approach is required. Mixed quantum-classical studies of surface-hopping type [18, 19] provide a step in this

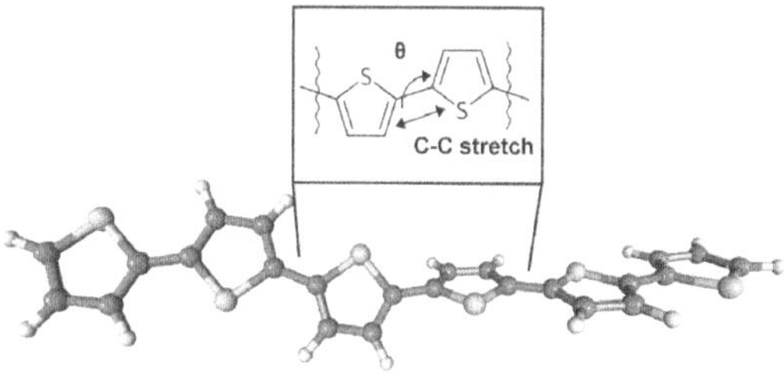

Figure 11.2 Schematic illustration of a minimal hexamer oligothiophene (OT-6) model exhibiting an initial torsional defect at the center of the chain that relaxes in the course of the dynamics. Reprinted from Ref. [30], Copyright (2014), with permission from Elsevier.

direction, but only account in an approximate fashion for quantum coherent effects.

Summarizing our studies of Refs. [11, 30], we address here the quantum dynamics of exciton migration and trapping in a minimal oligomer system—here, an oligothiophene hexamer (OT-6)—involving a torsional defect site (Fig. 11.2). In particular, we aim to understand whether the transfer dynamics falls into a coherent regime, and whether exciton-polaron formation is observable in the minimal system under study.

Since the electronic structure characterization of oligomer species, using a transition density analysis [10, 30, 40], points to the fact that a strict Frenkel picture is not appropriate, the general e–h (Merrifield-type) picture described in Section 11.2 will be employed. An alternative, simplified approach refers to the use of effective Frenkel exciton models [26, 48].

11.4.1 Hamiltonian

The vibronic Hamiltonian $\hat{H}_{\nu\mu,\nu'\mu'} = \sum \hat{H}_{\nu\mu,\nu'\mu'}|\nu\mu\rangle\langle\nu'\mu'|$ of Eqs. 11.9–11.12 now takes the specific form,

$$\hat{H}_{\nu\mu,\nu'\mu'} = \delta_{\nu,\nu'}\delta_{\mu,\mu'}\hat{H}^{(0)}_{\nu\mu,\nu'\mu'}(\{\hat{z}_{\xi,\xi+1},\hat{x}_{B,i}\}) + \hat{H}^{\text{coup}}_{\nu\mu,\nu'\mu'}(\{\hat{z}_i\}) \tag{11.26}$$

with the electronically diagonal part,

$$\hat{H}^{(0)}_{\nu\mu,\nu\mu} = \hat{H}^{eh}_{\nu\mu,\nu\mu} + \hat{H}^{\text{vibr}}_{\nu\mu,\nu\mu}(\{\hat{z}_{\xi,\xi+1}\}) + \hat{H}^{\text{bath}}_{\nu\mu,\nu\mu}(\{\hat{x}_{B,i}\}) \tag{11.27}$$

and electronically off-diagonal terms of EET and CT type (see Eqs. 11.16–11.17),

$$\hat{H}^{\text{coup}}_{\nu\mu,\nu'\mu'} = \hat{H}^{\text{EET}}_{\nu\mu,\nu'\mu'}(\{\hat{z}_{\xi,\xi+1}\}) + \hat{H}^{\text{CT}}_{\nu\mu,\nu'\mu'}(\{\hat{z}_{\xi,\xi+1}\}) \quad (11.28)$$

where $\hat{z}_{\xi,\xi+1} \in \{\hat{x}_{\xi,\xi+1}, \hat{\theta}_{\xi,\xi+1}\}$, correspond to site-correlated CC bond stretch and torsional modes, and $\{\hat{x}_{B,i}\}$ are bath modes coupled to the latter. The EET and CT terms are defined in terms of nearest-neighbor interactions according to Eqs. 11.16–11.17.

As detailed in Refs. [11, 26], CC stretch and torsional potentials have been obtained by a mapping of electronic structure information to a monomer-based exciton model. The Frenkel and CT couplings appearing in Eqs. 11.16–11.17 take the form

$$J_{\text{EET}}(\theta_{\xi,\xi+1}) = a_0^J + \sum_n (a_n^J \cos(n\,\theta_{\xi,\xi+1}) - a_n^J)$$

$$t_{\text{CT}}(\theta_{\xi,\xi+1}) = t\cos\theta_{\xi,\xi+1} \quad (11.29)$$

Further details of the Hamiltonian for the OT oligomer fragment under study have been discussed in Refs. [11, 30].

11.4.2 Ultrafast Exciton Relaxation

As mentioned above, we focus on the relaxation of an initially confined exciton at a torsional defect site. Figure 11.3 and Fig. 11.4 illustrate the exciton dynamics on a minimal e–h lattice for the OT-6 oligomer under study, for an initial condition that is localized to the left of the torsional defect. The initial condition thus corresponds to the relaxed exciton ground state, $|\psi_0\rangle = |\psi_1^L\rangle$, on the left (L) segment of the lattice.

Figure 11.3 shows the time evolving e–h distribution $P_{\nu\mu}(t)$ of Eq. 11.7 for various snapshots during the propagation, while Fig. 11.4 illustrates the corresponding diagonal projection, that is, the marginal distribution $P_\xi^{\text{CM}}(t)$ of Eq. 11.8.

In the course of the propagation, the fastest timescale is defined by the electronic dynamics coupled to a rapid CC stretch evolution. The vibronic wave function is seen to rapidly spread across the entire lattice—denoted $(L + R)$ in the following—even before the torsional coordinate starts to planarize. Since the initial state approximately represents a superposition of the first and second hexamer exciton states, that is, $|\psi_1^{L+R}\rangle$ and $|\psi_2^{L+R}\rangle$, the evolution

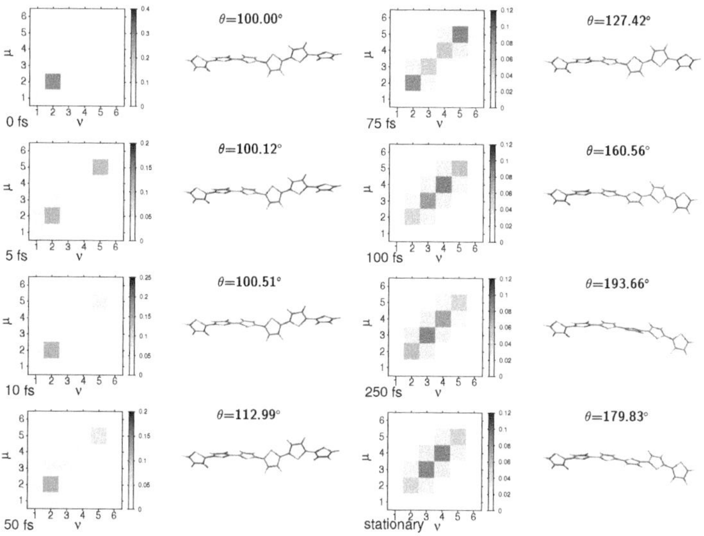

Figure 11.3 Time-evolving e–h distribution $P_{\mu\nu}(t)$ shown at selected times for the minimal OT-6 system under study, along with snapshots of the geometries (expectation values) at the respective times. The fastest timescale is determined by the initial electronic dynamics, preceding a slower timescale of torsion-induced planarization. Adapted from Ref. [30], Copyright (2014), with permission from Elsevier.

manifests the admixture of the second eigenstate, exhibiting a node at the center of the lattice. These observations are analogous to our previous findings for an oligo-phenylene-vinylene hexamer (OPV-6) model [11]. The final stationary state is found to correspond to a local exciton ground state (LEGS) [48], $|\psi^{\text{stationary}}\rangle \sim |\psi_1^{L+R}\rangle$, on the overall lattice.

Figure 11.3 and Fig. 11.4 illustrate that the exciton relaxation process is indeed a two-timescale process, comprising a rapid transient phase up to about 100 fs, followed by a slower relaxation phase, until a quasi-stationary distribution is observed from around 300–500 fs onwards.

The formation of a LEGS state on the overall $(L + R)$ lattice is accompanied by torsional relaxation from an initially displaced state (with $\langle\theta\rangle_0 = 100°$). A harmonic oscillator bath acting on the torsional mode—and therefore acting indirectly on the excitonic

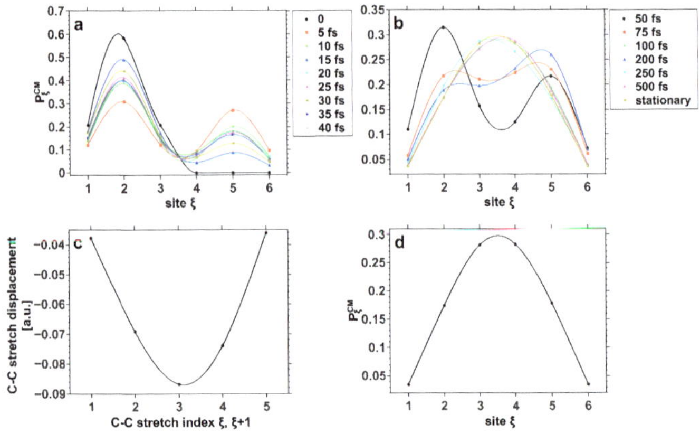

Figure 11.4 (a and b) Reduced center-of-mass distribution $P_{\xi}^{\mathrm{CM}}(t)$ complementary to Fig. 11.3. Note the different scales of the left versus the right panel. (c) CC stretch coordinate displacement $\Delta x_{\xi,\xi'}$ in the quasi-stationary state (averaged over the interval 1400 fs $< t <$ 1500 fs), illustrating the trapping effect. (d) P_{ξ}^{CM} distribution in the quasi-stationary, trapped state. Reprinted from Ref. [30], Copyright (2014), with permission from Elsevier.

subsystem—eventually absorbs the excess energy stored in the torsional mode as well as in the excitonic subsystem. In the final, quasi-stationary state, the system has planarized, even though a nonstationary, oscillatory evolution of the torsional coordinate is still observed. Likewise, the CC stretch coordinates also exhibit a continued oscillatory evolution, but their average displays a characteristic "trapping" effect as illustrated in Fig. 11.4c, that is, the excitonic distribution over the lattice sites (Fig. 11.4d) is mirrored by the lattice distortion which is maximal at the center of the lattice.

11.5 Exciton Dissociation at an Oligothiophene–Fullerene Junction

This section addresses the break-up of excitons at a D–A interface, that is, the crucial step prior to the irreversible separation of charge carriers. In our recent work of Refs. [25, 29, 55, 69] we have addressed two stages of this process, for a representative oligothiophene–fullerene system: First, the initial e–h separation—

creation of the primary CT states—and second, the ensuing formation of charge-separated (CS) states. Recent discussions in the literature have focused, in particular, on the role of delocalized CT states that might lead to an efficient electron–hole separation [69–71].

11.5.1 Hamiltonian

The Hamiltonian describing the D–A CT system is again a special case of the vibronic Hamiltonian $\hat{H} = \sum_{\nu\mu}\sum_{\nu'\mu'}\hat{H}_{\nu\mu,\nu'\mu'}|\nu\mu\rangle\langle\nu'\mu'|$ of Eqs. 11.7–11.8, with

$$\hat{H}_{\nu\mu,\nu'\mu'} = \delta_{\nu\nu'}\delta_{\mu\mu'}\hat{H}^{(0)}_{\nu\mu,\nu'\mu'}(\{\hat{x}_{i,\xi}\},\{\hat{y}_{i,\xi\xi'}\}) + \hat{H}^{\text{coup}}_{\nu\mu,\nu'\mu'}(\{\hat{y}_{i,\xi\xi'}\}) \tag{11.30}$$

where

$$\hat{H}^{(0)}_{\nu\mu,\nu\mu} = \hat{H}^{eh,\text{eff}}_{\nu\mu,\nu\mu} + \hat{H}^{\text{vibr}}_{\nu\mu,\nu\mu}(\{\hat{x}_{i,\xi}\},\{\hat{y}_{i,\xi\xi'}\}) \tag{11.31}$$

along with donor EET and donor-to-acceptor CT terms (see Eqs. 11.16–11.17),

$$\hat{H}^{\text{coup}}_{\nu\mu,\nu'\mu'} = \hat{H}^{\text{EET,D}}_{\nu\mu,\nu'\mu'}(\{\hat{y}_{i,\xi\xi'}\}) + \hat{H}^{\text{CT,DA}}_{\nu\mu,\nu'\mu'}(\{\hat{y}_{i,\xi\xi'}\}) \tag{11.32}$$

where $\hat{H}^{eh,\text{eff}} = \hat{H}^{eh} + \hat{H}^{\text{field}}$ represents an effective e–h potential corresponding to a Coulomb barrier including the effect of the internal electric field due to the electrodes [25]. The vibronic part of the Hamiltonian depends on intrafragment modes $\{\hat{x}_{i,\xi}\}$ as well as interfragment (correlated) modes $\{\hat{y}_{i,\xi\xi'}\}$. Finally, the EET (Frenkel)-type coupling is restricted to the donor species which is initially photoexcited, and the CT coupling refers to the electronic coupling between the donor and acceptor species. The latter in turn depends upon the interfragment modes $\{\hat{y}_{i,\xi\xi'}\}$.

11.5.2 Primary Exciton Break-Up Step

In a first study [54, 55], we considered the initial break-up of the exciton in a minimal D–A dimer complex. The Hamiltonian Eq. 11.32 thus reduces to

$$\hat{H}_{\text{XT,CT}} = \hat{H}_0 + \hat{H}^{\text{vibr}}_{\text{XT}}(\{\hat{x}_i\},y) + \hat{H}^{\text{vibr}}_{\text{CT}}(\{\hat{x}_i\},y) + \hat{H}^{\text{CT,DA}}(y) \tag{11.33}$$

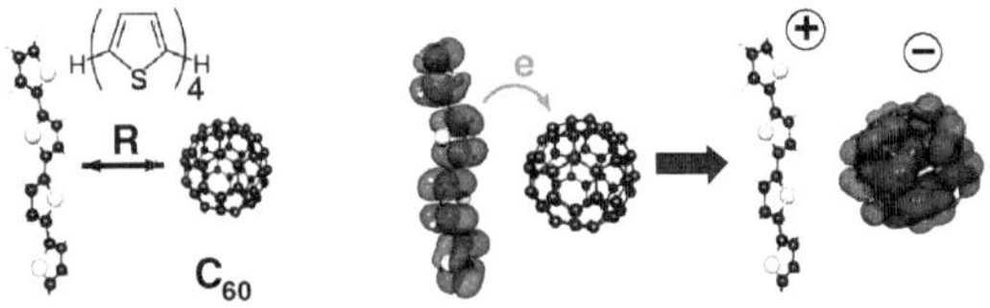

Figure 11.5 Schematic representation of an OT-4-fullerene charge transfer complex. Reprinted with permission from Ref. [55]. Copyright 2012, AIP Publishing LLC.

where y denotes the DA interfragment distance coordinate. Here, the electronic coupling $\hat{H}^{\mathrm{CT,DA}}$ was determined by a diabatization procedure [54], and a linear dependence on the interfragment distance y was found. The vibronic part of the Hamiltonian is of LVC form, where the vibronic couplings are derived from the spectral density shown in Fig. 11.1 of Section 11.2. We employed both standard equidistant sampling and an effective-mode sampling, as discussed in Section 11.2.4, see Ref. [54] and Ref. [55], respectively.

In Ref. [29], we employed the MCTDH and HEOM approaches, respectively. The MCTDH results refer to temperature $T = 0$, unless a Monte Carlo sampling is undertaken to construct a statistical distribution (see Ref. [54]). In the HEOM treatment, the effective-mode series was truncated at the order $M = 5$ (see Eq. 11.21 and Eq. 11.22) while the HEOM hierarchy was taken to the tenth order.

Figure 11.6 shows the decay dynamics of the excitonic (XT) state, which is seen to be ultrafast and oscillatory. From the HEOM calculations at different temperatures, one can infer that the transfer is only weakly temperature dependent. Indeed, this is to be expected for a barrierless, ultrafast process.

The time evolution of the electronic coherence is also monitored in Fig. 11.6 (lower panel). Clearly the oscillatory transients are not of purely vibrational origin—rather, the shortest timescale is determined by electronic coherence. This reflects Rabi-type oscillations which are subsequently damped due to vibronic coupling.

11.5.3 Formation of Charge-Separated States

In the study of Ref. [69], the above DA Hamiltonian was adapted so as to include a series of CS states, $|\,\mathrm{CS}_n\rangle = |\nu, \mu = \mu_n\rangle$ with $\mu_n = \nu +$

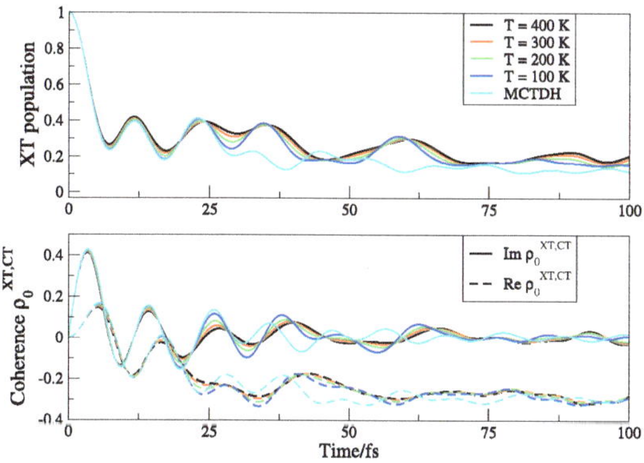

Figure 11.6 Ultrafast decay of the OT-4 exciton and accompanying electronic coherence dynamics. HEOM calculations at different temperatures are compared with reference MCTDH calculations ($T = 0$). Reprinted from Ref. [29], Copyright (2014), with permission from Elsevier.

$(n + 1)$. The Hamiltonian of Eq. 11.33 is thus augmented as follows,

$$\hat{H}_{\mathrm{XT,CS}} = \hat{H}_{\mathrm{XT,CT}} + \hat{H}_{\mathrm{CT,CS}} + \hat{H}_{\mathrm{CS}} \tag{11.34}$$

with $\hat{H}_{\mathrm{XT,CT}}$ of Eq. 11.33, along with the coupling $\hat{H}_{\mathrm{CT,CS}}$ between the primary CT state, with an electron–hole separation $|\nu - \mu| = 1$, and the first CS state $|\,\mathrm{CS}_1\rangle$, with an electron-hole separation $|\nu - \mu| = 2$,

$$\hat{H}_{\mathrm{CT,CS}} = t\,(|\mathrm{CT}\rangle\langle\,\mathrm{CS}_1| + \text{ h.c.}) \tag{11.35}$$

where t is the hole transfer integral. Further, the Hamiltonian in the CS subspace (with $|\nu - \mu| \geq 2$), is given as

$$\hat{H}_{\mathrm{CS}} = \sum_{n} \hat{h}_{\mathrm{CS}_n}(\{x_i\})\,|\mathrm{CS}_n\rangle\langle\,\mathrm{CS}_n| + \sum_{nn'} t\,|\mathrm{CS}_n\rangle\langle\,\mathrm{CS}_{n'}| \tag{11.36}$$

The on-site energies of the CS states, contained in $\hat{h}_{\mathrm{CS}_n}$, increase with the e–h distance according to the Coulomb barrier depicted in Fig. 11.7, as obtained from TDDFT calculations [25]. Hence, it is immediately apparent that following exciton dissociation, the charge carriers may be trapped in the primary CT state. Indeed, according to the conventional interpretation, thermal activation is necessary to overcome the Coulomb barrier, whose height typically exceeds the thermal energy ($kT \approx 0.025$ eV at 300 K) by a factor of ten. Hence, the Onsager–Brown model [72] is conventionally applied

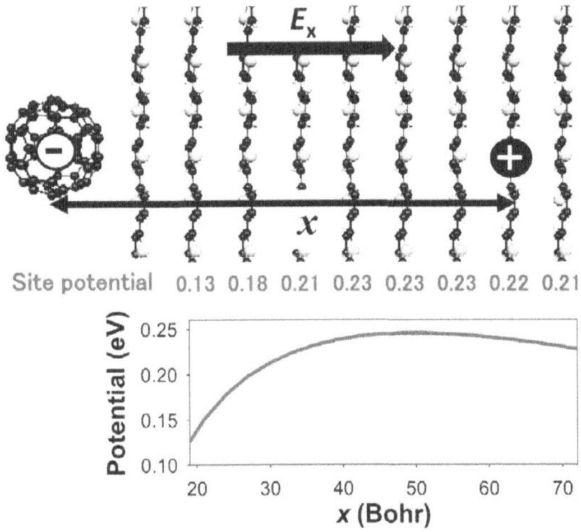

Figure 11.7 Illustration of electron–hole dissociation in a fullerene–oligothiophene donor–acceptor system. The Coulomb barrier that defines the on-site energies of the successive charge-separated states is illustrated in the lower panel. Reprinted with permission from Ref. [25]. Copyright (2013) American Chemical Society.

which predicts the thermal rate of exciton dissociation as a function of the internal electric field. Experimental evidence varies between the observation of long-lived CT (exciplex) states and ultrafast and irreversible charge separation, depending upon the nature of the D–A system [73, 74].

In Ref. [69], we employed the Hamiltonian of Eqs. 11.34–11.36 to investigate the quantum coherent dynamics under the influence of the Coulomb barrier to charge separation. Our main conclusions relate to two factors that facilitate overcoming the Coulomb barrier: first, the delocalization of electron and hole states, and second, vibronic excess energy which produces "hot" CT states.

Figure 11.8 illustrates the effective height of the Coulomb barrier as a function of electron delocalization within an ordered fullerene domain [69]. In addition, hole delocalization results from π-conjugation on the oligothiophene donor side. As a result, the formation of charge separated states is feasible on an ultrafast timescale, as shown in Fig. 11.8. A tunneling-type mechanism is observed, by which a nonnegligible population of charge separated

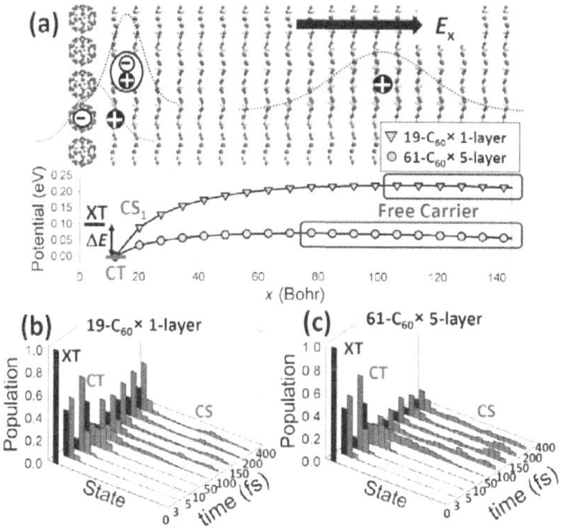

Figure 11.8 Electron–hole separation dynamics in a fullerene–oligothiophene system with increasing degree of delocalization in the fullerene domain, entailing a lowering of the Coulomb barrier (a). From the comparison of the quantum-dynamical simulation results of panels (b and c), it is seen that an increasing amount of charge-separated (CS) states is generated as the barrier height decreases. Reprinted with permission from Ref. [69]. Copyright (2013) American Chemical Society.

states emerges beyond the barrier, even though the primary CT state leads to partial trapping.

Recent experiments lend support to this picture, by demonstrating that fullerene aggregation plays a striking role in the ultrafast generation of charge carriers [75]. Conversely, the experiments reported in Ref. [76] show that the D–A energetics can be unfavorably influenced by aggregation in low-bandgap copolymer blends. A detailed knowlege of the energetics is required to fully characterize charge separation in different types of materials.

11.6 Conclusions and Outlook

By the approach outlined in the present chapter, a systematic attempt is made to combine (i) model Hamiltonians with an ab initio–based parametrization, that are suitable to describe energy and CT in extended molecular aggregates, (ii) state-specific, realistic spectral densities, and (iii) efficient multiconfigurational quantum-dynamical propagation techniques. Overall, this approach bridges the gap between model Hamiltonians and electronic structure analysis, and allows for a quantum-dynamical characterization of coherent short-time transients in large systems. This is of particular importance if the resulting dynamics differs significantly from the results of a kinetic description, as is the case for all examples shown here.

The e–h representation chosen here is particularly useful since Frenkel-type states and CT states are treated on the same footing. As a consequence, this representation naturally accommodates a generalized exciton basis including molecular CT excitons, in accordance with the approach by Merrifield and others [11, 22, 23, 30]. To map electronic structure information on the diabatic e–h Hamiltonian, we used a diabatization procedure by projection onto reference wave functions [54], for example, for the case of D–A CT states, or else a mapping procedure in the context of Frenkel J-aggregate or H-aggregate states [26]. In future work, systematic diabatization approaches should be explored more thoroughly.

Further, the inclusion of vibronic couplings and the nonperturbative treatment of vibronic wave packet dynamics is of particular importance for the systems under study. In some cases, it is useful to construct potential energy surface cuts for the most important vibrational modes [11, 30]. More systematically, spectral densities can be obtained from electronic structure information, using a shifted-harmonic oscillator approximation [54, 55]. Reduced-dimensional models that build upon effective-mode transformations can be constructed that yield simplified spectral densities from successive orders of a truncated chain model with Markovian closure [15, 29]. Importantly, the short-time evolution is accurately reproduced by low-order truncation within this chain representation [53].

Multiconfigurational wave packet propagation, in particular using the MCTDH method [27], is a powerful approach to high-dimensional quantum dynamics in molecular aggregates. Current developments move towards multilayer approaches [67, 68] that allow for propagation of hundreds of degrees of freedom. In many cases, photoinduced dynamics on ultrashort timescales is only weakly temperature dependent, such that wave packet propagation approaches are appropriate. To account for temperature effects, wave function ensembles can be constructed, using Monte Carlo sampling techniques. Alternatively, techniques like the HEOM propagation can be employed, as illustrated in Section 11.4.

The two case studies that have been presented, illustrating energy and charge transfer in semiconducting polymer systems, highlight the occurrence of initial femtosecond-scale transients, reflecting vibronic wave packet motion. This phase of the dynamics can be crucial in mediating ultrafast transfer processes, especially if vibronic excess energy is available. On a somewhat longer timescale (i.e., a few picoseconds), quasi-stationary polaronic, or exciton-polaron states are reached, due to dephasing and relaxation on the molecular lattice structure. Indeed, local exciton ground states (LEGS) [48] are reached on a typical timescale of several hundred femtoseconds.

Importantly, exciton and charge delocalization effects change both the energetics and dynamics. This is illustrated by the dynamics of typical oligothiophene (donor)–fullerene (acceptor) systems (Section 11.5) where the Coulomb barrier to exciton dissociation can be shown to be significantly lowered as a result of charge delocalization [69]. Hence, it is a combination of several effects—charge and exciton delocalization, vibronic excess energy, coherent transient dynamics, and subsequent trapping and polaron formation—that lead to efficient charge separation. The present simulation approach captures all of these effects, and gives a balanced, even though approximate, description of the short-time transients that play a crucial role in the charge separation dynamics.

While the focus of the present chapter has been on functional polymer materials, many conclusions carry over to delocalized excitons in biological systems like DNA and light-harvesting complexes, as illustrated in several chapters of this volume. For example, recent work on DNA highlights the mixing of Frenkel excitons and CT states [77], by immediate analogy with the situations discussed above. Clearly, the design of artificial and biomimetic functional materials will benefit from general approaches to the quantum treatment of extended systems. These can also provide building blocks for the multiscale modeling [78] of functional nanoscale systems over a broad range of spatial and temporal scales.

Acknowledgments

We gratefully acknowledge discussions with Hans Lischka, Felix Plasser, and Miquel Huix-Rotllant. The Advanced Institute for Materials Research (AIMR) is supported by the World Premier International Research Center Initiative (WPI), MEXT, Japan. Support by Deutsche Forschungsgemeinschaft (DFG) in the framework of Project No. BU-1032-2 is gratefully acknowledged.

References

1. H. Lee, Y.-C. Cheng, and G. Fleming, *Science*, **316**, 1462 (2007).
2. E. Collini and G. D. Scholes, *Science*, **323**, 369 (2009).
3. E. Collini and G. D. Scholes, *J. Phys. Chem. A*, **113**, 4223 (2009).
4. V. May and O. Kühn, *Charge and Energy Transfer Dynamics in Molecular Systems*, 3rd ed. (VCH-Wiley, 2011).
5. O. V. Prezhdo and P. J. Rossky, *Phys. Rev. Lett.*, **81**, 5294 (1998).
6. G. Panitchayangkoon et al., *Proc. Natl. Acad. Sci.*, **107**, 12766 (2010).
7. X. Chen and R. J. Silbey, *J. Chem. Phys.*, **132**, 204503 (2010).
8. F. C. Spano and C. Silva, *Annu. Rev. Phys. Chem.*, **65**, 477 (2014).
9. I. H. Nayyar et al., *J. Phys. Chem. Lett.*, **2**, 566 (2011).
10. A. N. Panda, F. Plasser, A. Aquino, I. Burghardt, and H. Lischka, *J. Phys. Chem. A*, **117**, 2181 (2013).
11. R. Binder, J. Wahl, S. Römer, and I. Burghardt, *Faraday Discuss.*, **163**, 205 (2013).
12. W. Barford, I. Boczarow, and T. Wharram, *J. Phys. Chem. A*, **115**, 9111 (2011).
13. V. Pouthier, *J. Chem. Phys.*, **137**, 114702 (2012).
14. U. Weiss, *Quantum Dissipative Systems*, 3rd ed. (World Scientific, Singapore, 2008).
15. R. Martinazzo, B. Vacchini, K. H. Hughes, and I. Burghardt, *J. Chem. Phys.*, **134**, 011101 (2011).
16. H. Köppel, W. Domcke, and L. S. Cederbaum, *Adv. Chem. Phys.*, **57**, 59 (1984).
17. G. A. Worth, H. Meyer, H. Köppel, L. S. Cederbaum, and I. Burghardt, *Int. Revs. Phys. Chem.*, **27**, 569 (2008).

18. T. Nelson, S. Fernandez-Alberti, V. Chernyak, A. E. Roitberg, and S. Tretiak, *J. Phys. Chem. B*, **115**, 5402 (2011).
19. T. Nelson, S. Fernandez-Alberti, A. E. Roitberg, and S. Tretiak, *Phys. Chem. Chem. Phys.*, **15**, 9245 (2013).
20. B. G. Levine and T. J. Martínez, *Annu. Rev. Phys. Chem.*, **58**, 613 (2007).
21. G. A. Worth, M. A. Robb, and I. Burghardt, *Faraday Discuss.*, **127**, 307 (2004).
22. R. E. Merrifield, *J. Chem. Phys.*, **34**, 1835 (1961).
23. S. Mukamel, S. Tretiak, T. Wagersreiter, and V. Chernyak, *Science*, **277**, 781 (1997).
24. S. Karabunarliev and E. R. Bittner, *J. Chem. Phys.*, **118**, 4291 (2003).
25. H. Tamura and I. Burghardt, *J. Phys. Chem. C*, **117**, 15020 (2013).
26. R. Binder, S. Römer, J. Wahl, and I. Burghardt, *J. Chem. Phys.*, **141**, 014101 (2014).
27. M. H. Beck, A. Jäckle, G. A. Worth, and H.-D. Meyer, *Phys. Rep.*, **324**, 1 (2000).
28. Y. Tanimura, *J. Phys. Soc. Jpn.*, **75**, 082001 (2006).
29. K. H. Hughes, B. Cahier, R. Martinazzo, H. Tamura, and I. Burghardt, *Chem. Phys.*, **111**, 442 (2014).
30. J. Wahl, R. Binder, and I. Burghardt, *Comp. Theor. Chem.*, **1040**, 167 (2014).
31. M. Hoffmann, *Chem. Phys.*, **258**, 73 (2000).
32. M. Hoffmann, *Thin Films Nanostruct.*, **31**, 221 (2003).
33. S. Tretiak and S. Mukamel, *Chem. Rev.*, **102**, 3171 (2002).
34. W. Barford, *Electronic and Optical Properties of Conjugated Polymers*, 2nd ed. (Clarendon Press, Oxford, 2013.
35. M. Hawton and D. Nelson, *Phys. Rev. B*, **57**, 4000 (1998).
36. B. Laikhtman, *J. Phys.: Condens. Matter*, **19**, 295214 (2007).
37. A. L. Fetter and J. D. Walecka, *Quantum Theory of Many-Particle Systems* (Dover, New York, 2003).
38. L. X. Benedict, E. L. Shirley, and R. B. Bohn, *Phys. Rev. Lett.*, **80**, 4514 (2003).
39. A. V. Luzanov and O. A. Zhikol, *Int. J. Quant. Chem.*, **110**, 902 (2010).
40. F. Plasser and H. Lischka, *J. Chem. Theory Comput.*, **8**, 2777 (2012).
41. M. Grüning, A. Marini, and X. Gonze, *Nano Lett.*, **9**, 2820 (2009).
42. Y. Li and C. A. Ullrich, *Chem. Phys.*, **391**, 157 (2011).

43. H. Köppel, W. Domcke, and L. S. Cederbaum, 2004, *Conical Intersections*, Vol. 15, W. Domcke, D. R. Yarkony, and H. Köppel, eds. (World Scientific, New Jersey, 2004), p. 323.
44. C.-P. Hsu, *Acc. Chem. Res.*, **42**, 509 (2009).
45. H. Tamura, J.-M. Mallet, M. Oheim, and I. Burghardt, *J. Phys. Chem. C*, **113**, 7548 (2009).
46. T. Holstein, *Ann. Phys. (N. Y.)*, **8**, 325 (1959).
47. W. Barford, E. R. Bittner, and A. Ward, *J. Phys. Chem. A*, **116**, 10319 (2012).
48. O. R. Tozer and W. Barford, *J. Phys. Chem. A*, **116**, 10310 (2012).
49. G. D. Scholes, *ACS Nano*, **2**, 523 (2008).
50. C. Deibel, T. Strobel, and V. Dyakonov, *Adv. Mater.*, **22**, 4097 (2010).
51. K. H. Hughes, C. D. Christ, and I. Burghardt, *J. Chem. Phys.*, **131**, 024109 (2009).
52. K. H. Hughes, C. D. Christ, and I. Burghardt, *J. Chem. Phys.*, **131**, 124108 (2009).
53. R. Martinazzo, K. H. Hughes, and I. Burghardt, *Phys. Rev. E*, **84**, 030102(R) (2011).
54. H. Tamura, I. Burghardt, and M. Tsukada, *J. Phys. Chem. C*, **115**, 9237 (2011).
55. H. Tamura, R. Martinazzo, M. Ruckenbauer, and I. Burghardt, *J. Chem. Phys.*, **137**, 22A540 (2012).
56. S. Mukamel, *Principles of Nonlinear Optical Spectroscopy* (Oxford University Press, New York/Oxford, 1995).
57. A. Nitzan, *Chemical Dynamics in Condensed Phases: Relaxation, Transfer, and Reactions in Condensed Molecular Systems* (Oxford University Press, Oxford, 2006).
58. G. R. Fleming and M. Cho, *Annu. Rev. Phys. Chem.*, **47**, 109 (1996).
59. P. Changenet-Barret, C. T. Choma, E. F. Gooding, W. F. DeGrado, and R. M. Hochstrasser, *J. Phys. Chem. B*, **104**, 9322 (2000).
60. S. R. White and A. E. Feiguin, *Phys. Rev. Lett.*, **93**, 076401 (2004).
61. M. P. Woods, R. Groux, A. W. Chin, S. F. Huelga, and M. B. Plenio, *J. Math. Phys.*, **55**, 032101 (2014).
62. H.-D. Meyer, U. Manthe, and L. S. Cederbaum, *Chem. Phys. Lett.*, **165**, 73 (1990).
63. U. Manthe, H.-D. Meyer, and L. S. Cederbaum, *J. Chem. Phys.*, **97**, 3199 (1992).

64. Y. Tanimura and R. K. Kubo, *J. Phys. Soc. Jpn.*, **58**, 114 (1989).
65. G. A. Worth, M. H. Beck, A. Jäckle, and H. Meyer, 2006, The MCTDH Package, Version 9 (2007); H.-D. Meyer, Version 8.3 (2002). See http://www.pci.uni-heidelberg.de/tc/usr/mctdh/.
66. H. Wang and M. Thoss, *J. Chem. Phys.*, **119**, 1289 (2003).
67. U. Manthe, *J. Chem. Phys.*, **128**, 164116 (2008).
68. O. Vendrell and H. Meyer, *J. Chem. Phys.*, **134**, 044135 (2011).
69. H. Tamura and I. Burghardt, *J. Am. Chem. Soc.*, **135**, 16364 (2013).
70. G. Grancini et al., *Nat. Mater.*, **4**, 1602 (2013).
71. D. Caruso and A. Troisi, *PNAS*, **109**, 13498 (2012).
72. C. L. Braun, *J. Chem. Phys.*, **80**, 4157 (1984).
73. J. Guo, H. Ohkita, H. Benten, and S. Ito, *J. Am. Chem. Soc.*, **132**, 6154 (2010).
74. R. D. Pensack and J. B. Asbury, *J. Am. Chem. Soc.*, **131**, 15986 (2009).
75. S. Gélinas et al., *Science*, **343**, 512 (2014).
76. Z. Guo et al., *J. Am. Chem. Soc.*, **136**, 10024 (2014).
77. D. B. Bucher, B. M. Pilles, T. Carell, and W. Zinth, *PNAS*, **111**, 4369 (2014).
78. S. Donets, A. Pershin, M. Christlmaier, and S. A. Baeurle, *J. Chem. Phys.*, **138**, 094901 (2013).

Chapter 12

Ultrafast Spectroscopy: Quantum Information and Wave Packets

Joel Yuen-Zhou, Jacob J. Krich, Ivan Kassal, and Alán Aspuru-Guzik

Department of Chemistry and Chemical Biology, Harvard University, 12 Oxford Street, Room M113 Cambridge, MA 02138, USA

joelyuen@ucsd.edu

This chapter provides a self-contained introduction to pump–probe (PP') spectroscopy of molecular systems from the perspective of the quantum process matrix χ, a formalism that is used to describe open quantum systems. In doing so, we first present the theory of PP' spectroscopy using an intuitive time-dependent approach based on wave packet methods. The theory is accompanied by explicit analytical calculations using model molecular dimer systems. This paves the way to the derivation of the central result of this Chapter, namely, that a frequency-integrated broadband PP' spectra may be re-expressed as a linear combination of elements associated with the light–matter interaction (e.g., dipole matrix elements) and the process matrix χ describing excited-state system dynamics. We thus offer experimental access to this theoretical formalism. This work serves as a starting point for the reader to understand ultrafast laser experiments in molecular systems and to gain a novel perspective of nonlinear spectroscopy based on quantum information theory.

Ultrafast Dynamics at the Nanoscale: Biomolecules and Supramolecular Assemblies
Edited by Irene Burghardt and Stefan Haacke

ISBN 978-981-4745-33-8 (Hardcover), 978-981-4745-34-5 (eBook)
www.panstanford.com

12.1 Introduction

The study of excited-state dynamics in molecular aggregates, such as photosynthetic complexes, has been of great interest in the last few years due to the advent of novel techniques in ultrafast spectroscopy [1–8]. In these experiments, short pulses of light interact with an ensemble of molecular systems, which subsequently emit light that is captured and analyzed.

A fundamental question is whether the evolution of the molecular excited states can be traced or inferred from these spectroscopic signals [9–12]. This issue is closely related to a protocol called quantum process tomography (QPT), a technique from the quantum information processing community which aims at the complete characterization of dynamics of a quantum device [13, 14]. The basic idea of QPT is that by preparing a small set of input states (e.g., particular excited states) of a system and measuring the resulting output states after free evolution, one can extract a linear input–output map (see Fig. 12.1) called the *process matrix* χ, which allows full determination of the output state given any input state, subject to mild constraints, which we discuss below. The process matrix can be rigorously related to an integrated equation of motion for the open system. In this chapter we demonstrate the connection between χ and a spectroscopic signal stemming from pump–probe (PP') measurements.

The outline of the chapter is as follows. We first provide a short introduction to the χ formalism and the model molecular systems where we will apply it. Then, we introduce a description of PP' spectroscopy that will allow us to make a smooth connection between this spectroscopy and χ. This description is supplemented by explicit calculations on model molecular dimer systems. We conclude by showing how to re-express a broadband frequency-integrated PP' signal as a linear combination of elements of χ. Although we do not delve into details on how to numerically invert χ from this type of data or explicitly perform QPT on molecular aggregates, we briefly mention the general strategies and give the respective references to the reader. This work is designed to give novices to ultrafast spectroscopy some intuition about these techniques and to give experts a new perspective on the topic. This chapter is adapted from a longer work [15], and the interested reader is invited to see more topics and discussion there.

12.2 The Quantum Process Matrix χ

We begin by introducing the formalism of the process matrix χ. We consider a system $\mathcal{S}$ interacting with a bath $\mathcal{B}$, with which energy (but not particles) can be exchanged. This is generically called an open quantum system (OQS). The total Hamiltonian that governs the two as a whole can be written as

$$H_0 = H_{\mathcal{S}} + H_{\mathcal{B}} + H_{\mathcal{SB}}, \tag{12.1}$$

where $H_{\mathcal{S}}$, $H_{\mathcal{B}}$, $H_{\mathcal{SB}}$ depend only on degrees of freedom of $\mathcal{S}$, $\mathcal{B}$, or both, respectively. The state of the $\mathcal{S}$-$\mathcal{B}$ is characterized by a density matrix ρ_{total}. In general, we are interested in the evolution of $\mathcal{S}$ as a function of time T in the form of a reduced system density matrix $\rho(T) = \mathrm{Tr}_{\mathcal{B}}[\rho_{total}(T)]$, where the partial trace is taken over all degrees of freedom in the bath. It is easy to show that if $\rho_{total}(0) = \rho(0) \otimes \rho_{\mathcal{B}}(0)$, i.e., the total initial state has no system–bath correlations, then the state $\rho(T)$ of $\mathcal{S}$ at every time $T > 0$ can be expressed as[a] [19, 20],

$$\rho(T) = \chi(T)\rho(0), \tag{12.2}$$

where the "superoperator" $\chi(T)$ is the quantum process matrix, a linear operator on the space of density matrices. If the Hilbert space of $\mathcal{S}$ is N-dimensional, $\rho(T)$ is an $N \times N$ matrix and $\chi(T)$ an $N \times N \times N \times N$ tensor. Equation 12.2 indicates that $\rho(T)$ is a linear transformation of the initial $\rho(0)$ via the action of $\chi(T)$ on it. Equation 12.2 can be regarded as an integrated equation of motion for every T, and in principle, does not make assumptions about the nature of the bath, and therefore holds for both Markovian and non-Markovian dynamics. Remarkably, given a fixed $\rho_{\mathcal{B}}(0)$, $\chi(T)$ allows us to predict the OQS dynamics of $\mathcal{S}$ for arbitrary initial states of

[a]Note that in general, $\chi(T_1 + T_2) \neq \chi(T_2)\chi(T_1)$. Important exceptions are the case of Markovian baths $\mathcal{B}$ and isolated systems [16–18].

Figure 12.1 A quantum black box is probed by a series of input states $\rho(0)$ which yield output states corresponding to $\rho(T) = \chi(T)\rho(0)$, where $\chi(T)$ is the process matrix that fully characterizes the black box. The goal of QPT is to reconstruct $\chi(T)$ with a finite number of input–output relations. From Ref. [15] (http://iopscience.iop.org/book/978-0-750-31062-8).

the system. Therefore, knowledge of $\chi(T)$ completely characterizes any system (see Fig. 12.1). The reconstruction of the process matrix $\chi(T)$ via a finite number of experiments is the main goal of QPT.

Equation (12.2) can be expressed in terms of a basis for the system:

$$\rho_{ab}(T) = \sum_{cd} \chi_{abcd}(T)\rho_{cd}(0). \tag{12.3}$$

The interpretation of $\chi(T)$ is easy to grasp: If the initial state were prepared in $\rho(0) = |c\rangle\langle d|$ (even if this is not a valid density matrix), $\chi_{abcd}(T)$ would be the value of the entry ab of the density matrix after time T, i.e., $\chi_{abcd}(T) = \langle a|\rho(T)|b\rangle$. Diagonal elements ($\rho_{aa}$) of a density matrix are called *populations,* whereas off-diagonal ones (ρ_{ab} for $a \neq b$) are known as *coherences.*[b] Therefore, $\chi_{abcd}(T)$ denotes a state-to-state amplitude transfer, and indicates a process where a population or a coherence $|c\rangle\langle d|$ at time 0 is transferred to another population or coherence $|a\rangle\langle b|$ at time T.

12.2.1 Properties and Examples

Example 1. Properties of the process matrix $\chi(T)$.

We will now state without proof several properties of the process matrix $\chi(T)$. Derivations of them can be found in [15], among other

[b]Note that $\chi_{abcd}(T)$ and $\rho_{ab}(T)$ depend on the basis.

sources [13, 21]. However, the reader is invited to prove these statements as an exercise. Part 6 is a worked example.

(1) The system and bath together evolve due to H_0. By taking a trace of $\rho_{total}(T)$ with respect to $\mathcal{B}$, one can show that

$$\rho(T) = \sum_i E_i(T)\rho(0)E_i^\dagger(T). \tag{12.4}$$

The terms $E_i(T)$ are called Kraus operators, where i is a multi-index, running over all pairs of $\mathcal{B}$ states. Using the time evolution operator for the entire $\mathcal{S}$ and $\mathcal{B}$ spaces $U(T) = \mathcal{T}(e^{-i\int_0^T H_0(t')dt'})$, where $\mathcal{T}$ is the time-ordering operator (H_0 could in general be time dependent), $E_i(T)$ can be written as $E_i(T) = \langle e_\alpha|U(T)|e_\beta\rangle$ where $i = (\alpha, \beta)$ is a multi-index and $|e_\delta\rangle$ are orthonormal states of the bath.

(2) By introducing a basis for $\mathcal{S}$ in Eq. 12.4, one can show that $\rho_{ab}(T) = \sum_{cd} \chi_{abcd}(T)\rho_{cd}(0)$.

(3) $\chi_{abcd}(T)$ can be reexpressed as

$$\chi_{abcd}(T) = \mathrm{Tr}_B\left\{\langle a|U(T)\left(|c\rangle\langle d| \otimes \rho_B(0)\right)U^\dagger(T)|b\rangle\right\}. \tag{12.5}$$

(4) $\chi(T)$ has the following symmetries:

(a) **Hermiticity:** $\chi_{abcd}(T) = \chi^*_{badc}(T)$. Also, if $\rho(0)$ is Hermitian, then $\rho(T)$ will be too if this symmetry is satisfied.

(b) **Trace preservation:** $\sum_a \chi_{aacd}(T) = \delta_{cd}$. This condition preserves the trace of $\rho(0)$ at all times.

(c) **Positivity:** $\sum_{abcd} z^*_{ac}\chi_{abcd}(T)z_{bd} \geq 0$ for arbitrary complex matrices z. If $\rho(0)$ is positive semidefinite, then this condition guarantees that $\rho(T)$ is too.

(d) Given the properties in a–b, for a system in d-dimensional Hilbert space, there are $d^4 - d^2$ linearly independent real-valued elements that parametrize $\chi(T)$ completely.

(5) Suppose $H_{\mathcal{SB}} = 0$ so $\mathcal{S}$ is a perfectly isolated system. By expressing $\chi(T)$ in the *eigenbasis* of $H_\mathcal{S}$, $\chi_{abcd}(T) = \delta_{ac}\delta_{bd}e^{-i\omega_{ab}T}$, where $\omega_{ab} = E_a - E_b$ and δ_{ij} is the Kronecker delta-function. We see that populations do not evolve and the $e^{-i\omega_{ab}T}$ expresses the evolution of coherences in this system.

(6) A simple model of Markovian OQS dynamics is given by the secular Redfield equations. Consider a two-level system (TLS) $\mathcal{S}$ with eigenstates $|a\rangle$ and $|b\rangle$, which from coupling to $\mathcal{B}$ exhibits the following kinetics,

$$\dot{\rho}_{aa}(T) = -k_{b\leftarrow a}\rho_{aa}(T), \tag{12.6a}$$

$$\dot{\rho}_{bb}(T) = k_{b\leftarrow a}\rho_{aa}(T), \tag{12.6b}$$

$$\dot{\rho}_{ab}(T) = -i\omega_{ab}\rho_{ab}(T) - \kappa_{ab}\rho_{ab}(T), \tag{12.6c}$$

$$\dot{\rho}_{ba}(T) = [\dot{\rho}_{ab}(T)]^*. \tag{12.6d}$$

Derive analytical expressions for $\chi_{abcd}(T)$ for this particular model and interpret their physical significance.

Solution (for part 6)

Integrating Eqs. 12.6a–12.6d, we immediately have the following nonzero elements of $\chi(T)$:

$$\chi_{aaaa}(T) = e^{-k_{b\leftarrow a}T}, \tag{12.7a}$$

$$\chi_{bbaa}(T) = 1 - e^{-k_{b\leftarrow a}T}, \tag{12.7b}$$

$$\chi_{bbbb}(T) = 1, \tag{12.7c}$$

$$\chi_{aabb}(T) = 0, \tag{12.7d}$$

$$\chi_{abab}(T) = e^{-i\omega_{ab}T}e^{-\kappa_{ab}T}, \tag{12.7e}$$

$$\chi_{abab}(T) = [\chi_{baba}(T)]^*. \tag{12.7f}$$

Here, $\chi_{aaaa}(T)$ and $\chi_{bbaa}(T) = 1 - \chi_{aaaa}(T)$ indicate population decay and transfer from state a with a first-order kinetics rate given by $k_{b\leftarrow a}$. Population transfer from $|b\rangle$ to $|a\rangle$ does not occur in our model and therefore if the system is prepared at $|b\rangle$ it will remain there indefinitely. Finally, coherences oscillate at the ω_{ab} frequency but they are damped at a dephasing rate κ_{ab}. No transfers between population to coherence, coherence to population, or coherence to coherence are included in this particular model.

12.2.2 QPT Algorithms

A possible algorithm to perform QPT is the following: (a) we must have the ability to prepare a linearly independent set of states that spans the space of the possible initial density matrices of $\mathcal{S}$. For a

Hilbert space of size d, this requires d^2 states; (b) for each of the prepared states, wait for a free evolution time T and determine the density matrix $\rho(T)$. Any protocol for determining a density matrix for a system is called quantum state tomography (QST) [22–27]. In essence, QPT can be carried out for any system with both selective preparation of initial states and QST. Variants of this methodology exist although all of them operate within the same spirit [13, 14, 28, 29]. QPT has been successfully implemented in a wide variety of experimental scenarios, including nuclear magnetic resonance [30–32], ion traps [33], single photons [34, 35], solid state qubits [36], optical lattices [37], and Josephson junctions [38]. In this chapter, we will sketch how to perform QPT for the excited-state dynamics of a simple model multichromophoric system using methods of ultrafast spectroscopy, although we will not get into the details of the procedure. Further results are available in [9–11, 15].

12.3 The Model System

We consider a model system, the *coupled dimer* [39–42], constructed by coupling two two-level molecules (sites), which we refer to as a and b. Each site has an electronic excited state $|a\rangle$, $|b\rangle$, so the electronic system is spanned by four states: $|g\rangle$, $|a\rangle$, $|b\rangle$, and $|f\rangle$, where $|g\rangle$ is the ground state and $|f\rangle$ is the double-excited state. The Hamiltonian is given by

$$H_0(\boldsymbol{R}) = T_N + H_{el}(\boldsymbol{R}), \tag{12.8}$$

where $\boldsymbol{R} = (x_1, \cdots, x_n)$ denotes the nuclear coordinates of n atoms, $T_N = \sum_i p_i^2/2m_i$ is the nuclear kinetic energy, and $H_{el}(\boldsymbol{R})$ is given by

$$H_{el}(\boldsymbol{R}) = \sum_{i=g,a,b,f} V_i(\boldsymbol{R})|i\rangle\langle i| + J(\boldsymbol{R})(|a\rangle\langle b| + |b\rangle\langle a|). \tag{12.9}$$

$V_i(\boldsymbol{R})$ are the diabatic potential energy surfaces (PES),[c] and $J(\boldsymbol{R})$ is the coupling between the two site excitations $|a\rangle$ and $|b\rangle$ in the SEM. Let us consider two different partitions of $H_0(\boldsymbol{R})$.

[c]See chapter 12 of [43] for a discussion of adiabatic versus diabatic PES. The truncation of $H_0(\boldsymbol{R})$ with a small set of diabatic PES is an approximation to the true molecular Hamiltonian.

(1) **Excitation manifolds.** Defining the Hamiltonian for each surface,

$$H_i(\boldsymbol{R}) = [T_N + V_i(\boldsymbol{R})]|i\rangle\langle i|, \tag{12.10}$$

we may partition $H_0(\boldsymbol{R})$ in Eq. 12.8 as a sum of contributions acting separately on different excitation manifolds,

$$H_0(\boldsymbol{R}) = H_{GSM}(\boldsymbol{R}) + H_{SEM}(\boldsymbol{R}) + H_{DEM}(\boldsymbol{R}), \tag{12.11}$$

$$H_{GSM}(\boldsymbol{R}) = H_g(\boldsymbol{R}), \tag{12.12}$$

$$H_{SEM}(\boldsymbol{R}) = H_a + H_b + J(\boldsymbol{R})(|a\rangle\langle b| + |b\rangle\langle a|), \tag{12.13}$$

$$H_{DEM}(\boldsymbol{R}) = H_f(\boldsymbol{R}), \tag{12.14}$$

where GSM, SEM, and DEM are labels for the ground-state, singly excited, and doubly excited manifolds.
In words, the GSM has only one electronic state $|g\rangle$, the SEM has two states $|a\rangle$ and $|b\rangle$, and the DEM has again only one state $|f\rangle$. Physically, we are considering two chromophores (sites), each with only two electronic states, a ground and an excited state. $|g\rangle$ is the electronic state where both chromophores are in their ground states. $|a\rangle$ and $|b\rangle$ are the electronic states where only one chromophore site (a or b, respectively) is excited. Yet, via Coulomb interaction $J(\boldsymbol{R})$, the excitation in a can be passed onto b (or the other way around) via $J(\boldsymbol{R})$. $|f\rangle$ is the electronic state where both a and b are excited. Notice that we are ignoring coupling terms between the different excitation manifolds, and therefore, we also disregard the possibility of conical intersections between the latter, although not within the SEM itself. Inclusion of substantial nonadiabatic dynamics between excitation manifolds is beyond the scope of this chapter, but is an interesting research topic on its own [44, 45].
In many biologically relevant and artificial systems, the energy difference between subsequent excitation manifolds is on the order of visible-UV frequencies ($\sim$12000–25000 cm^{-1}) and the coupling between chromophores is in a much smaller range of energies ($\sim$50–1000 cm^{-1}).

(2) **Electronic system $\mathcal{S}$ and vibrational bath $\mathcal{B}$.** Alternatively, it is often convenient to rewrite Eq. 12.8 in the language of OQS, Eq. 12.1, where the electrons are the system $\mathcal{S}$ and the nuclei

are the bath $\mathcal{B}$,

$$H_0(\boldsymbol{R}) = H_{\mathcal{S}} + H_{\mathcal{B}}(\boldsymbol{R}) + H_{\mathcal{SB}}(\boldsymbol{R}) \tag{12.15a}$$

$$H_{\mathcal{S}} = H_{el}(\boldsymbol{0}), \tag{12.15b}$$

$$H_{\mathcal{B}}(\boldsymbol{R}) = T_N + \sum_i \frac{m_i \omega_i^2 x_i^2}{2}, \tag{12.15c}$$

$$H_{\mathcal{SB}}(\boldsymbol{R}) = H_{el}(\boldsymbol{R}) - H_{el}(\boldsymbol{0}) - \sum_i \frac{m_i \omega_i^2 x_i^2}{2}. \tag{12.15d}$$

Here, $\mathcal{B}$ has been chosen to be a set of harmonic oscillators (HO) with nuclear coordinates x_i and springs of frequencies ω_i. The choice of a HO bath, instead of say, an anharmonic one, is convenient as analytical results associated with HOs are readily available. The frequencies ω_i can be chosen in a physically motivated way (see Example 2), such as corresponding to a quadratic expansion of $V_g(\boldsymbol{R})$ about its global minimum, or such that $H_{\mathcal{SB}}(\boldsymbol{R})$ is small in some metric, and can be treated as a perturbation on $H_{\mathcal{S}} + H_{\mathcal{B}}$. Note that the right hand side of Eqs. 12.15b–12.15d add up to $H_0(\boldsymbol{R})$. The eigenstates of H_0 are called vibronic states. As one can imagine, for more complicated systems, the system–bath partition is not unique, and it is often nontrivial to draw the border between $\mathcal{S}$ and $\mathcal{B}$ in a way that $H_{\mathcal{SB}}$ is manageable. This is an important problem in the theory of OQS, which we shall not address here. We refer the interested reader to the appropriate sources [17, 18].

To study optical experiments, we use the electric dipole approximation, and we will consider transitions between the different vibronic states of the dimer due to coupling of the dipole operator $\boldsymbol{\mu}$,

$$\boldsymbol{\mu} = \left(\sum_{p=a,b} \boldsymbol{\mu}_{pg} |p\rangle\langle g| + \boldsymbol{\mu}_{fp} |f\rangle\langle p| \right) + \text{c.c.}, \tag{12.16}$$

with a time-dependent electric field. We keep only the off-diagonal portion of μ, which produces transitions between different excitation manifolds, and we ignore the static dipoles $\boldsymbol{\mu}_{ii}$, which are irrelevant to transitions induced by weak intensity visible-UV light. Importantly, throughout this chapter, we will assume the Condon approximation, namely, that the dipole elements μ_{ij}

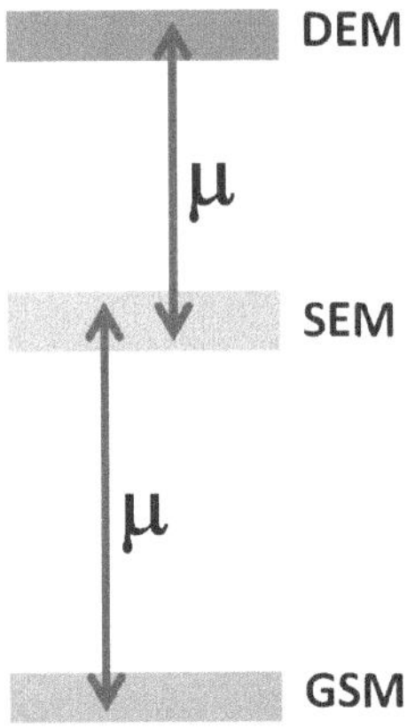

Figure 12.2 Energy-level diagram of a model molecular system. It consists of a ground-state manifold (GSM), a singly excited manifold (SEM), and a doubly excited manifold (DEM). Dipole-allowed transitions are indicated by arrows. From Ref. [15] (http://iopscience.iop.org/book/978-0-750-31062-8).

between different electronic states do not depend on $\boldsymbol{R}$. In terms of OQS, this means $\boldsymbol{\mu}$, which only acts on $\mathcal{S}$, does not depend on the state of $\mathcal{B}$. Also, we will assume that the values of $\boldsymbol{\mu}_{ij}$ are known.

Example 2. Partition of a molecular Hamiltonian $H_0(\boldsymbol{R})$ as an Open Quantum System.

Let us motivate the partition from Eqs. 12.15b–12.15d a bit further. We consider N_A atoms in monomer a and N_B atoms in monomer b. A quadratic expansion of the PES $V_i(\boldsymbol{R})$ about their minima yields[d]

$$V_g(\boldsymbol{R}) = \omega_g + \sum_{i=1}^{N_a+N_b} \frac{m_i\omega_i^2 {x_i}^2}{2}, \tag{12.17}$$

$$V_a(\boldsymbol{R}) = \omega_a + \sum_{i=1}^{N_a} \frac{m_i\omega_i^2 (x_i - \Delta_i)^2}{2} + \sum_{i=N_a+1}^{N_a+N_b} \frac{m_i\omega_i^2 {x_i}^2}{2}, \tag{12.18}$$

[d]Here ω_i, where i are integers, are oscillator frequencies, but ω_g, ω_a, and ω_b are base electronic transition frequencies of the different PES.

$$V_b(\boldsymbol{R}) = \omega_b + \sum_{i=1}^{N_a} \frac{m_i \omega_i^2 {x_i}^2}{2} + \sum_{i=N_a+1}^{N_a+N_b} \frac{m_i \omega_i^2 (x_i - \Delta_i)^2}{2}, \tag{12.19}$$

$$V_f(\boldsymbol{R}) = \omega_a + \omega_b + \sum_{i=1}^{N_a+N_b} \frac{m_i \omega_i^2 (x_i - \Delta_i)^2}{2}, \tag{12.20}$$

$$J(\boldsymbol{R}) = J. \tag{12.21}$$

These PES represent a simple model for a coupled dimer. The nuclear coordinates from $i = 1$ to $i = N_a$ are localized in molecule a, whereas the ones from $i = N_a + 1$ to $i = N_a + N_b$ are in molecule b. The PES are quadratic with respect to all the nuclei, and there is a single frequency ω_i associated with each nuclear coordinate (which holds for every electronic state). In the ground state, the potential energy minimum is located at $\boldsymbol{R} = \boldsymbol{0}$. The displacements Δ_i indicate changes in the equilibrium geometry of the sites (in the absence of coupling J) upon excitation, so that $V_a(\boldsymbol{R})$ only exhibits these changes for the modes of molecule a, and similarly for $V_b(\boldsymbol{R})$. Hence, the minima of $V_a(\boldsymbol{R})$ and $V_b(\boldsymbol{R})$ are at $\boldsymbol{R} = (\Delta_1, \cdots, \Delta_{N_a}, 0, \cdots, 0)$ and $\boldsymbol{R} = (0, \cdots 0, \Delta_{N_a+1}, \cdots, \Delta_{N_a+N_b})$, respectively. We define the energy minimum of each of these potentials to be at ω_a and ω_b, respectively. V_f describes the potential energy when both molecules are excited, so that the potential energy minimum is at $\omega_a + \omega_b$ and $\boldsymbol{R} = (\Delta_1, \cdots, \Delta_{N_a+N_b})$. This approximation ignores the interaction between the two excitations (sometimes called binding energy), and so considers that $|f\rangle$ is simply a product state of two excitations, one in each molecule. In this approximation,

$$\boldsymbol{\mu}_{ag} = \boldsymbol{\mu}_{fb}, \tag{12.22}$$

$$\boldsymbol{\mu}_{bg} = \boldsymbol{\mu}_{fa}, \tag{12.23}$$

that is, the transition from $|b\rangle$ or $|a\rangle$ to $|f\rangle$ has the same amplitude as the excitation of $|a\rangle$ or $|b\rangle$ from $|g\rangle$, respectively. Since $H_0(\boldsymbol{R})$ is time-reversal invariant, we may choose the dipole matrix elements as well as the coupling J to be real-valued. For simplicity, we also assume that J is independent of $\boldsymbol{R}$.

Substitute Eqs. 12.17–12.21 into Eqs. 12.15b–12.15d and interpret the resulting expressions.

Solution

The Hamiltonian for the electronic system $\mathcal{S}$ is given by,

$$\begin{aligned} H_{\mathcal{S}} &= \omega_g|g\rangle\langle g| + (\omega_a + \Lambda_a)|a\rangle\langle a| \\ &+ (\omega_b + \Lambda_b)|b\rangle\langle b| + J\left(|a\rangle\langle b| + |b\rangle\langle a|\right) \\ &+ (\omega_f + \Lambda_f)|f\rangle\langle f| \end{aligned} \tag{12.24}$$

where $\Lambda_a = \sum_{i=1}^{N_a} m_1\omega_i^2{\Delta_i}^2/2$ and $\Lambda_b = \sum_{i=N_a+1}^{N_b} m_1\omega_i^2{\Delta_i}^2/2$ are the *reorganization energies* in the absence of coupling J, $\Lambda_f = \Lambda_a + \Lambda_b$, and $\omega_f = \omega_a + \omega_b$. Physically, the reorganization energy of each site corresponds to the energy the nuclei lose after vertical excitation (i.e., excitation keeping fixed nuclear coordinates) in order to settle in the equilibrium geometry of the excited state within the nanosecond timescale. Hence, $H_{\mathcal{S}}$ is determined not only by the energy offsets E_i of the PES but also by the reorganization energies Λ_i.

$H_{\mathcal{B}}$ is exactly given by Eq. 12.15c,

$$H_{\mathcal{B}} = \sum_{i=1}^{N_a+N_b} \frac{p_i^2}{2m_i} + \frac{m_i\omega_i^2 x_i^2}{2}. \tag{12.25}$$

That is, the bath is composed of a discrete set of harmonic oscillators of frequencies ω_i. Finally, $H_{\mathcal{SB}}$ is given by,

$$H_{\mathcal{SB}}(\boldsymbol{R}) = -\sum_{i=1}^{N_a} m_i\omega_i^2\Delta_i x_i|a\rangle\langle a| - \sum_{i=N_a+1}^{N_a+N_b} m_i\omega_i^2\Delta_i x_i|b\rangle\langle b| \tag{12.26}$$

which indicates that the coupling between the system and the bath is linear in the bath coordinates and diagonal in the electronic site basis. If $\{\Delta_i\}$ are small, the dynamics of the electronic system $\mathcal{S}$ is largely described by $H_{\mathcal{S}}$, and $H_{\mathcal{SB}}$ can be regarded as a perturbation. This model describes a multilevel electronic system coupled to a bath of harmonic oscillators and is a good starting point to develop theories of relaxation of OQSs. When a large number of bath coordinates is included, $\mathcal{B}$ can be approximated as a continuum, and the physics of the relaxation of $\mathcal{S}$ depends on the distribution of the couplings (i.e., displacements Δ_i) as a function of frequency ω_i (spectral density) [17, 18].

In the next section, we will show that *PP′* spectroscopy reports directly on $\chi(T)$ for the states in the SEM.

12.4 Frequency-Integrated Pump–Probe Spectroscopy

Our goal in this section is to introduce the basics of nonlinear spectroscopy by focusing on the *PP′* experiment. An understanding of this experiment provides a valuable introduction to more complicated spectroscopic techniques. For a detailed exposition of the topic and related spectroscopies we refer the reader to [46, 47] and our own work on the topic [15].

Let us consider a *PP′* experiment like the one illustrated in Fig. 12.3. In it, a weak intensity ultrafast resonant *pump* pulse P traveling with wavevector $\boldsymbol{k}_P$ excites the molecules from their ground to their bright excited states via the dipole operator. Nonequilibrium dynamics ensues, such as the activation of certain vibrational modes of the bath and the transfer of amplitude to other bright and dark states. How do we detect these processes? After a waiting time T, a second short weak intensity resonant *probe* pulse P' traveling in the $\boldsymbol{k}_{P'}$ direction transmits through the sample and its photon number gain/loss is measured in a photodetector. We are interested in a differential signal $S_{PP'}$, namely, the number of photons lost (regardless of color) by the probe P' after the pump P has acted ($S_{P'}$(with P)), minus the ones lost by P' had P not been present ($S_{P'}$(without P)). This technique is known by a variety

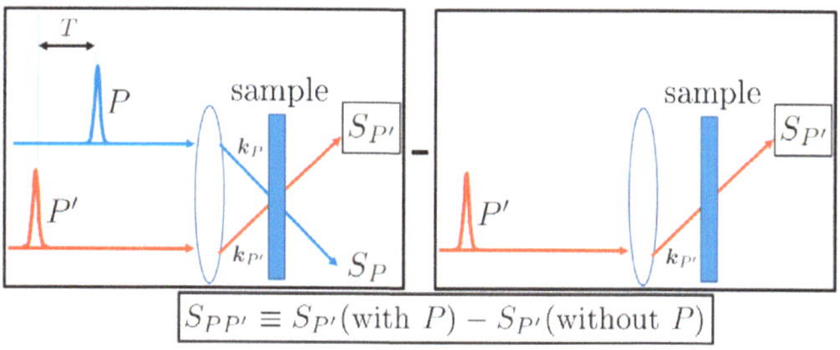

Figure 12.3 Pump–probe (*PP′*) spectroscopy. A pump P excites the sample. After a waiting time T, the absorption of a probe P' is measured. To obtain a differential measurement, the background absorption of P' in the absence of P is subtracted. From Ref. [15] (http://iopscience.iop.org/book/978-0-750-31062-8). © IOP Publishing. Reproduced with permission. All rights reserved.

of names, including transient or differential absorption, or dynamic absorption.

In order to describe this experiment, we need to add a light–matter perturbation term to our molecular Hamiltonian of Eq. 12.8, so that the total Hamiltonian H reads:

$$H(\mathbf{r}, t) = H_0 + H'(\mathbf{r}, t). \tag{12.27}$$

Here,

$$H'(\mathbf{r}, t) = -\boldsymbol{\mu} \cdot \boldsymbol{\epsilon}(\mathbf{r}, t) \tag{12.28}$$

is the light–matter interaction in the dipole approximation, which couples the molecular dipole with the electric field of the two light pulses. We approximate both P and P' as Gaussians,

$$\boldsymbol{\epsilon}(\mathbf{r}, t) = \sum_{n=P,P'} [\epsilon_n(t - t_n)e^{i\mathbf{k}_n\cdot\mathbf{r}+i\phi_n}\mathbf{e}_n + \text{c.c.}], \tag{12.29}$$

$$\epsilon_n(t) = \frac{\lambda e^{-t^2/2\sigma_n^2}e^{-i\omega_n t}}{\sqrt{2\pi\sigma_n^2}}. \tag{12.30}$$

Let us first express the PP' signal in a physically intuitive way. We define projectors onto the unexcited and doubly excited excitation manifolds,

$$\mathbb{P}_{\text{GSM}} = \sum_{m\in GSM} |m\rangle\langle m|,$$

$$\mathbb{P}_{\text{DEM}} = \sum_{m\in DEM} |m\rangle\langle m|.$$

We use the rotating wave approximation (RWA) to describe the perturbative wave function, including interactions with the optical fields to first, second, and third order, where ϵ_n and ϵ_n^* induce absorption and emission processes, respectively [15],

$$|\Psi_0(t)\rangle = e^{-iH_0t}|\Psi_0(0)\rangle, \tag{12.31}$$

$$|\Psi_n(t)\rangle = i\int_{-\infty}^{t} dt' e^{-iH_0(t-t')}[\boldsymbol{\mu}\cdot\mathbf{e}_n(\epsilon_n(t'-t_n)+\text{c.c.})]|\Psi_0(t')\rangle \quad \text{for } n = P, P', \tag{12.32}$$

$$|\Psi_{Pn}(t)\rangle = i\int_{-\infty}^{t} dt' e^{-iH_0(t-t')}[\boldsymbol{\mu}\cdot\mathbf{e}_n(\epsilon_n(t'-t_n)+\text{c.c.})]|\Psi_P(t')\rangle \quad \text{for } n = P, P', \tag{12.33}$$

$$|\Phi_{PPP'}(t)\rangle = i\int_{-\infty}^{t} dt' e^{-iH_0(t-t')}[\boldsymbol{\mu}\cdot\mathbf{e}_{P'}(\epsilon_{P'}(t'-t_{P'})+\text{c.c.})] \times \mathbb{P}_{\text{GSM}}|\Psi_{PP}(t')\rangle. \tag{12.34}$$

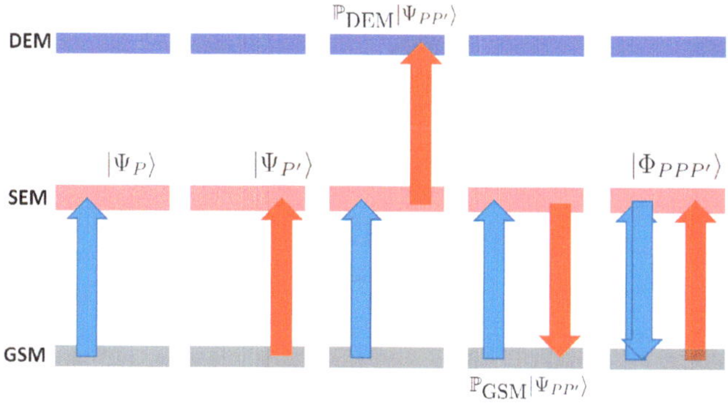

Figure 12.4 Wave packets involved in the calculation of the frequency-integrated PP' signal. The blue and the red arrows denote actions of the pump P and the probe P', respectively. From Ref. [15] (http://iopscience.iop.org/book/978-0-750-31062-8).

Here, $|\Psi_0(t)\rangle$ (Eq. 12.31) is the zeroth-order wave function evolution in the absence of light perturbation. Typically, it corresponds to a stationary GSM wave function (evolving with a trivial phase). In the RWA, $|\Psi_n(t)\rangle$ (Eq. 12.32) is a first-order wave packet in the SEM created by the action of ϵ_n alone; the c.c. ϵ_n^* contribution is negligible, but we keep it in the formal structure of the wave function definitions. Focusing on $|\Psi_P(t)\rangle$, this SEM wave function can be promoted to the DEM by ϵ_n or back to the GSM by ϵ_n^*; these are the two contributions of the second order $|\Psi_{Pn}(t)\rangle$ (Eq. 12.33). Finally, let us project out the GSM contribution to $|\Psi_{PP}(t)\rangle$ and excite it to the SEM via $\epsilon_{P'}$; this gives the third order $|\Phi_{PPP'}(t)\rangle$. See Fig. 12.4 for diagrams for each of these wave functions. The RWA is highly accurate when the electronic excitations are of much higher frequency than the vibrational modes and the optical pulses are resonant with the electronic frequencies. We define asymptotic states, which do not oscillate as $t \to \infty$,

$$|\Psi_{P'}\rangle \equiv \lim_{t\to\infty} e^{iH_0t}|\Psi_{P'}(t)\rangle, \tag{12.35}$$

$$|\Psi_{Pn}\rangle \equiv \lim_{t\to\infty} e^{iH_0t}|\Psi_{Pn}(t)\rangle \quad \text{for } n = P,\ P', \tag{12.36}$$

$$|\Phi_{PPP'}\rangle \equiv \lim_{t\to\infty} e^{iH_0t}|\Phi_{PPP'}(t)\rangle. \tag{12.37}$$

The PP' signal can be written as

$$S_{PP'} \equiv S_{P'}(\text{with } P) - S_{P'}(\text{without } P)$$
$$= S_{ESA} + S_{SE} + S_{GSB}, \tag{12.38}$$

where each of the terms corresponds to excited state absorption (ESA), stimulated emission (SE), and ground-state bleach (GSB) processes, given by the wave function overlaps [15],

$$S_{ESA} \equiv \langle \Psi_{PP'} | \mathbb{P}_{DEM} | \Psi_{PP'} \rangle, \tag{12.39a}$$
$$S_{SE} \equiv -\langle \Psi_{PP'} | \mathbb{P}_{GSM} | \Psi_{PP'} \rangle, \tag{12.39b}$$
$$S_{GSB} = -2\Re\{\langle \Phi_{PPP'} | \Psi_{P'} \rangle\}. \tag{12.39c}$$

The interpretation of these equations is as follows. P promotes amplitude from $|g\rangle$ to $|\Psi_P(s)\rangle$, a wave packet consisting of amplitude in the SEM. P' acts on this state, creating $|\Psi_{PP'}(s)\rangle$, a superposition of wave packets in the GSM and DEM. Naturally, the number of photons *absorbed* by the material through ESA correspond to the population transferred to the DEM, that is, the square amplitude of $\mathbb{P}_{\mathrm{DEM}}|\Psi_{PP'}(s)\rangle$. Analogously, the squared amplitude of $\mathbb{P}_{\mathrm{GSM}}|\Psi_{PP'}(t)\rangle$ corresponds to the photons *gained* by P' via SE. Hence, $S_{ESA}(T)$ and $S_{SE}(T)$ are always positive and negative contributions to $S_{PP'}(T)$. $S_{GSB}(T)$ seems a bit trickier to interpret, but it can be similarly analyzed with respect to the other processes. Consider the wave packet left in the ground state after P has acted but before P' has. It reads:

$$\mathbb{P}_{\mathrm{GSM}}|\Psi(t)\rangle = \mathbb{P}_{\mathrm{GSM}}(|\Psi_0(t)\rangle + |\Psi_{PP}(t)\rangle), \tag{12.40}$$

where we have used the fact that the only wave packets that have amplitude in the GSM are $|\Psi_0(t)\rangle$[e] and $|\Psi_{PP}(t)\rangle$. This wave packet can absorb energy from P' to transition to the SEM, creating the wave function,

$$i\int_{-\infty}^{\infty} dt'[\epsilon^*_{P'}(t'-t_{P'})\boldsymbol{\mu}\cdot \boldsymbol{e}^*_{P'}]\mathbb{P}_{\mathrm{GSM}}|\Psi(t')\rangle = |\Psi_{P'}(t)\rangle + |\Phi_{PPP'}(t)\rangle. \tag{12.41}$$

The number of photons absorbed through this process is equal to the population created in the SEM once P' has been fully absorbed,

$$\text{Number of photons of } P' \text{absorbed to go from GSM to SEM}$$
$$= \lim_{t\to\infty}[(\langle\Psi_{P'}(t)| + \langle\Phi_{PPP'}(t)|)(|\Psi_{P'}(t)\rangle + |\Phi_{PPP'}(t)\rangle)]$$
$$= \langle\Psi_{P'}|\Psi_{P'}\rangle + 2\Re\{\langle\Phi_{PPP'}|\Psi_{P'}\rangle\} + O(\lambda^6)$$
$$\approx S_{P'}(\text{without } P) + S_{GSB} \tag{12.42}$$

---

[e] Note that $|\Psi_0(t)\rangle = \mathbb{P}_{\mathrm{GSM}}|\Psi_0(t)\rangle$.

where we have identified $\langle \Psi_{P'} | \Psi_{P'} \rangle$ as the linear absorption of P' (in the absence of P) and used Eq. 12.39c for S_{GSB}. Also, we have ignored $\langle \Phi_{PPP'} | \Phi_{PPP'} \rangle$ as an $O(\lambda^6)$ contribution. Hence, the total absorption of P' is given by

$$S_{P'}(\text{with } P) \approx S_{SE} + S_{ESA} + S_{P'}(\text{without } P) + S_{GSB}. \qquad (12.43)$$

However, by definition, $S_{PP'}$ denotes only the differential absorption. Subtracting the linear absorption $S_{P'}$ (without P) from Eq. 12.43, finally yields Eq. 12.38. Even though Eq. 12.42, being the norm squared of a wave packet, is always positive, S_{GSB} does not need to be positive due to the possibility of interferences in the overlap $\langle \Phi_{PPP'} | \Psi_{P'} \rangle$.

The contributions due to Eqs. 12.39a–12.39c clearly depend on the waiting time T, which may be sampled by repeating the experiment multiple times. Since $|\Psi_P(t)\rangle$ describes a wave function in the SEM, $|\Psi_{PP'}\rangle$ (and consequently $S_{SE}(T)$ and $S_{ESA}(T)$) directly report on dynamics in the SEM as a function of T. Qualitatively, P launches amplitude from the GSM to the SEM, and coherent dynamics in the SEM follows throughout the waiting time T. Via P', we probe these amplitudes by transferring them into the DEM (S_{ESA}) or back into the GSM (S_{SE}). The number of photons of P' that are absorbed is equal to the population created in the DEM minus that which ended in the GSM. These populations depend on the interferences between the different transferred amplitudes, and hence, directly report on the dynamics in the SEM, and hence, on the $\chi(T)$ elements of the SEM, as we shall see in the last section of this chapter.

Finally, note that $S_{GSB}(T)$ reports on GSM dynamics rather than on the SEM. However, under certain conditions, it is possible to make its contribution a T-independent background, which can be eliminated by appropriate subtraction from the total $S_{PP'}(T)$ signal. We shall address this issue more carefully in the last section.

Example 3. PP' signal for a vibrationless dimer.

Evaluate Eqs. 12.39a–12.39c for the simple case where the electronic system $\mathcal{S}$ of the dimer does not interact with vibrations $\mathcal{B}$, that is, take $H_0 = H_S$ for H_S in Eq. 12.24. Discuss the results.

Solution

Let $|g\rangle$, $|\alpha\rangle$, $|\beta\rangle$, $|f\rangle$ be eigenstates of H_S in Eq. 12.24 ($|\alpha\rangle$ and $|\beta\rangle$ are linear combinations of $|a\rangle$ and $|b\rangle$) . Invoking the RWA, and defining the "upward" $|i\rangle \leftarrow |j\rangle$ transition amplitude $\Omega^n_{ij} = \bar{\epsilon}_n(\omega_{ij})\boldsymbol{\mu}_{ij} \cdot \boldsymbol{e}_n$, its "downward" counterpart $|j\rangle \leftarrow |i\rangle$ $\Omega^{\bar{n}}_{ji} = (\Omega^n_{ij})^*$ (assuming $\omega_i > \omega_j$), and the nth pulse Fourier transform $\bar{\epsilon}_n(\omega) = \int_{-\infty}^{\infty} dt e^{i\omega t}\epsilon_n(t) = \lambda e^{-(\omega-\omega_n)^2\sigma_n^2/2}$, we get,

$$|\Psi_{P'}\rangle = i \sum_{q=\alpha,\beta} \Bigg(\underbrace{e^{i\omega_q t_{P'}}}_{\text{SEM evolution}} \Bigg) \Bigg(\underbrace{|q\rangle}_{\text{final SEM state}} \Bigg) \times \Bigg(\underbrace{\Omega^{P'}_{qg}}_{\text{SEM}\leftarrow\text{GSM transition}} \Bigg) \Bigg(\underbrace{e^{-i\omega_g t_{P'}}}_{\text{GSM evolution}} \Bigg), \tag{12.44}$$

$$\begin{aligned} |\Psi_{Pn}\rangle = & -e^{i\omega_g t_n}|g\rangle \sum_{q=\alpha,\beta} \Omega^{\bar{n}}_{gq} e^{-i\omega_q(t_n - t_P)} \Omega^P_{qg} e^{-i\omega_g t_P} \\ & -e^{i\omega_f t_n}|f\rangle \sum_{q=\alpha,\beta} \Omega^n_{fq} e^{-i\omega_q(t_n - t_P)} \Omega^P_{qg} e^{-i\omega_g t_P}, \\ & \text{for } n = P,\ P', \end{aligned} \tag{12.45}$$

$$\begin{aligned} |\Phi_{PPP'}\rangle = & -i \sum_{i=\alpha,\beta} e^{i\omega_i t_{P'}} |i\rangle \Omega^{P'}_{ig} e^{-i\omega_g(t_{P'} - t_P)} \\ & \times \left(\sum_{q=\alpha,\beta} \frac{\Omega^{\bar{P}}_{gq}\Omega^P_{qg}}{2} \right) e^{-i\omega_g t_P}. \end{aligned} \tag{12.46}$$

The intuition behind Eqs. 12.44–12.46 is the following: The free evolution of the system (which is trivial in this example) is interrupted by transitions Ω_{ji} that depend on the amplitude of the electric field $\bar{\epsilon}_n(\omega_{ij})$ at the given transition energy and the corresponding dipole element $\boldsymbol{\mu}_{ij}$ projected onto the pulse polarization $\boldsymbol{e}_n$. Note the definitions of the asymptotic wave functions (Eqs. 12.35–12.37) when carrying out these evaluations.

In the vibrationless dimer model, $\mathbb{P}_{\text{GSM}} = |g\rangle\langle g|$ and $\mathbb{P}_{\text{DEM}} = |f\rangle\langle f|$, we can readily substitute Eqs. 12.44–12.46 into Eqs. 12.39a–

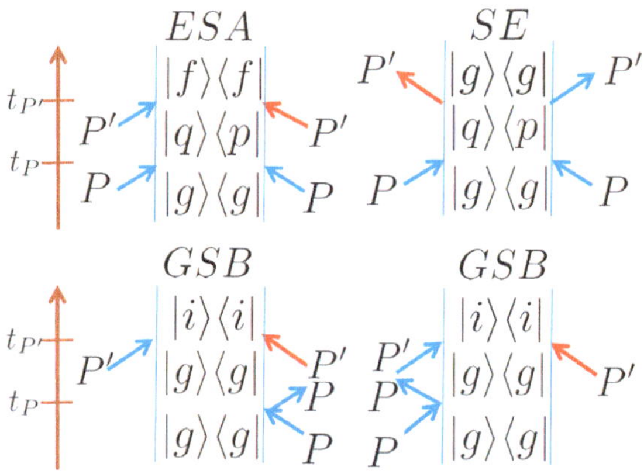

Figure 12.5 DS-FDs for vibrationless dimer PP' spectra. Note that coherences or populations do not decay in this model, as coupling to the bath is not present in this model. From Ref. [15] (http://iopscience.iop.org/book/978-0-750-31062-8).

12.39c to obtain,

$$S_{ESA}(T) = \sum_{p,q=\alpha,\beta} \Omega^{P'}_{fq}\Omega^{P}_{qg}\Omega^{\bar{P}}_{gp}\Omega^{\bar{P}'}_{pf}e^{-i\omega_{qp}T}, \tag{12.47}$$

$$S_{SE}(T) = -\sum_{p,q=\alpha,\beta} \Omega^{\bar{P}'}_{gq}\Omega^{P}_{qg}\Omega^{\bar{P}}_{gp}\Omega^{P'}_{pg}e^{-i\omega_{qp}T}, \tag{12.48}$$

$$S_{GSB}(T) = -\sum_{p,q=\alpha,\beta} \Omega^{P'}_{qg}\Omega^{\bar{P}}_{gp}\Omega^{P}_{pg}\Omega^{\bar{P}'}_{gq}. \tag{12.49}$$

As expected, notice that $S_{ESA}(T)$ and $S_{SE}(T)$ directly inform on the dynamics of the SEM (in this simple case, the $e^{-i\omega_{qp}T}$), reporting on populations and coherences, which evolve unitarily due to the simplicity of the dimer model in the absence of vibrations.

The reader that is well versed in double-sided Feynman diagrams (DS-FDs) [15, 46, 47] might find it useful to look at Fig. 12.5 for the diagrams corresponding to Eqs. 12.47–12.49. As the reader becomes more familiar with the material, she will realize that one can bypass large amounts of algebra by noticing the patterns of the resulting expressions, which are encoded in the DS-FDs.

12.5 Relationship between Frequency-Integrated PP' Spectra and the Process Matrix $\chi(T)$

As shown in Example 3, for the case of a vibrationless dimer, S_{ESA} and S_{GSB} report on its SEM dynamics as a function of waiting time T, while S_{GSB} yields a T-independent background. We shall see that this result can be readily generalized to arbitrary dimers in the presence of vibrations if the optical pulses are impulsive, i.e., ideally broadband with $\sigma_n \to 0$, $\epsilon_n = \lambda\delta(t - t_n)$ or $\bar{\epsilon}_n(\omega) = \lambda$ for $n = P, P'$ (Eq. 12.30) [12, 15].

Let the initial state of the dimer before photoexcitation be $|\Psi_0(0)\rangle = |g\rangle|\nu_n^{(g)}\rangle$, where $|\nu_n^{(g)}\rangle$ is a vibrational eigenstate of H_{GSM} (Eq. 12.12). Proceeding as in Example 3, the asymptotic wave packets from Eqs. 12.35–12.37, generalize for the dimer with vibrations to give

$$|\Psi_{P'}\rangle = i \sum_{i,q=\alpha,\beta} e^{iH_0 t_{P'}}|i\rangle\langle i|e^{-iH_0T}|q\rangle|\nu_n^{(g)}\rangle\Omega_{qg}^{P}e^{-i\omega_{gn}t_P}, \quad (12.50)$$

$$\begin{aligned} |\Psi_{PP'}\rangle = &- e^{iH_0 t_{P'}}|g\rangle \sum_{i,q=\alpha,\beta} \Omega_{gi}^{\bar{P}'}\langle i|e^{-iH_0T}|q\rangle|\nu_n^{(g)}\rangle\Omega_{qg}^{P}e^{-i\omega_{gn}t_P} \\ &- e^{iH_0 t_{P'}}|f\rangle \sum_{i,q=\alpha,\beta} \Omega_{fi}^{P'}\langle i|e^{-iH_0T}|q\rangle|\nu_n^{(g)}\rangle\Omega_{qg}^{P}e^{-i\omega_{gn}t_P}, \end{aligned} \quad (12.51)$$

$$|\Phi_{PPP'}\rangle = -i \sum_{i,q=\alpha,\beta} e^{iH_0 t_{P'}}|i\rangle\Omega_{ig}^{P'}e^{-i\omega_{gn}T}|\nu_n^{(g)}\rangle\frac{\Omega_{gq}^{\bar{P}}\Omega_{qg}^{P}}{2}e^{-i\omega_{gn}t_P}. \quad (12.52)$$

and $|\alpha\rangle$ and $|\beta\rangle$ are again (electronic) eigenstates of H_S alone. Note that as opposed to the analogous equations for the vibrationless dimer (Eqs. 12.44–12.45), in Eqs. 12.50–12.51 we allow for the possibility of nonzero amplitude transfers from say, α to β, $\langle\beta|e^{-iH_0T}|\alpha\rangle \neq 0$, since in general, there is no strictly adiabatic electronic basis $\{\alpha, \beta\}$ in the presence of a vibrational bath $\mathcal{B}$ coupling to the electronic system $\mathcal{S}$. Further note that e^{-iH_0T} in Eqs. 12.50, 12.51 propagates the vibrational state $|\nu_n^{(g)}\rangle$.

On the other hand, Eq. 12.52 is quite similar to Eq. 12.46 for $|\Phi_{PPP'}\rangle$. The rationale for this coincidence is as follows. In the impulsive pulse limit, P takes $|g\rangle$ to $|q\rangle \in \{|\alpha\rangle, |\beta\rangle\}$ and "immediately" brings it back to $|g\rangle$, without letting $|q\rangle$ transfer to another electronic state. During the waiting time T, the electronic state is in $|g\rangle$, which does not transfer to other electronic states, even in the presence of vibrations. Next, P' comes in and excites $|g\rangle$ into $|i\rangle \in \{-\alpha\rangle, -\beta\rangle\}$. The distinction between Eqs. 12.52) and (12.46 is, however, that once $|i\rangle$ is prepared, $e^{iH_0 t_{P'}}|i\rangle \neq e^{i\omega_i t_{P'}}|i\rangle$ for a dimer with vibrations.

Let us evaluate $S_{ESA}(T)$. This time, we include finite temperature by performing a classical average of the signal over the different initial states $|g\rangle|\nu_n^{(g)}\rangle$, drawn from a thermal ensemble with Boltzmann probability $p_n = \exp(-\omega_{g_n}/k_B\mathcal{T})/\mathrm{Tr}\left[\exp(-H_g/k_B\mathcal{T})\right]$ where k_B is Boltzmann constant and $\mathcal{T}$ is the temperature of the ensemble. Equation (12.39a) then reads,

$$\begin{aligned} S_{ESA}(T) &= \sum_n p_n \mathrm{Tr}_{\mathcal{B}}[\mathbb{P}_{DEM}|\Psi_{PP'}\rangle\langle\Psi_{PP'}|\mathbb{P}_{DEM}] \\ &= \sum_n p_n \mathrm{Tr}_{\mathcal{B}}\left[\left(\sum_{iq}\Omega_{fi}^{P'}\langle i|e^{-iH_0T}|q\rangle|\nu_n^{(g)}\rangle\Omega_{qg}^{P}\right)\right. \\ &\quad \times\left.\left(\sum_{pj}\Omega_{gp}^{\bar{P}}\langle\nu_n^{(g)}|\langle p|e^{iH_0T}|j\rangle\Omega_{jf}^{\bar{P}'}\right)\right] \\ &= \sum_{ijpq}\Omega_{fi}^{P'}\Omega_{qg}^{P}\Omega_{gp}^{\bar{P}}\Omega_{jf}^{\bar{P}'}\chi_{ijqp}(T), \end{aligned} \tag{12.53}$$

where we have used the definition of $\chi(T)$ for the SEM from Eq. 12.5 and identified the initial density matrix of $\mathcal{B}$ as $\rho_{\mathcal{B}}(0) = p_n|\nu_n^{(g)}\rangle\langle\nu_n^{(g)}|$. Similarly, the SE and GSB contributions to $S_{PP'}(T)$ are given by,

$$S_{SE}(T) = -\sum_{ijpq}\Omega_{gi}^{\bar{P}'}\Omega_{qg}^{P}\Omega_{gp}^{\bar{P}}\Omega_{jg}^{P'}\chi_{ijqp}(T), \tag{12.54}$$

$$S_{GSB}(T) = -\sum_{ip}\Omega_{ig}^{P'}\Omega_{gp}^{\bar{P}}\Omega_{pg}^{P}\Omega_{gi}^{\bar{P}'}, \tag{12.55}$$

where $i, j, p, q \in \{\alpha, \beta\}$. Equations 12.53–12.55 constitute the core of this chapter and show the linear relation between a broadband PP' signal and the process matrix $\chi(T)$ for the electronic system in the SEM [11, 12]. These equations generalize Eqs. 12.47–12.49 to situations where the bath $\mathcal{B}$ induces transfers between populations and coherences of the system $\mathcal{S}$. The signal PP' is a reporter on $\chi(T)$ since P and P' prepare and measure states in the SEM. Less trivial is the fact that the signals only report on the reduced state of the electronic system $\mathcal{S}$ and not of the entire molecule including vibrations. This result can be traced back to a combination of the use of broad bandwidth as well as the Condon approximation. The DS-FDs associated with each of these contributions are depicted in Fig. 12.6, which is a generalization of Fig. 12.5.

Equation (12.50) describes the situation where the nuclear wave packet $|\nu_n^{(g)}\rangle$ is promoted from $|g\rangle$ to $|q\rangle$ by P and after time T, via free evolution generated by H_{SEM}, ends up in electronic state $|i\rangle$, which may be different from $|q\rangle$. The information of this state gets transferred via P' into a superposition of wave packets in $|f\rangle$ and $|g\rangle$ (Eq. 12.51), which give $S_{ESA}(T)$ and $S_{SE}(T)$, respectively. These contributions depend on products of four dipole matrix elements associated with the different transitions between the excitation manifolds. Recall that QPT is well-defined if the initial state of the bath $\mathcal{B}$ is identical for all initial electronic system $\mathcal{S}$ states. In this experiment, at $T = 0^+$ (immediately following the action of P), the nuclei can be considered to be in state $\rho_B(0)$ for every initial state $|q\rangle\langle p|$ of the electrons. This is due to the Condon approximation and the extremely short (impulsive) perturbation. In this limit, the reduced electronic dynamics is given by Eq. 12.2.

Notice that $S_{GSB}(T)$, being in general a probe for GSM dynamics, is in this case T independent. The way to understand it is as follows. Consider one member of the initial thermal ensemble, $|\Psi_0(0)\rangle = |g\rangle|\nu_n^{(g)}\rangle$. In the broadband limit and Condon approximation, after P acts, $\mathbb{P}_{GSM}|\Psi_{PP}\rangle \propto |\nu_n^{(g)}\rangle|g\rangle = |\Psi_0\rangle$ (Eq. 12.40); that is, P takes $|\nu_g^{(n)}\rangle$ to the SEM and instantaneously brings it down before it allows the shape of the wave packet to distort via evolution under H_{SEM}. Hence, $\mathbb{P}_{GSM}|\Psi_{PP}\rangle$ is proportional to $\Omega_{gq}^{\bar{P}}\Omega_{qg}^{P}/2$ (the same transition is probed in the two steps). Subsequently, since $|\Psi_0\rangle$ is an eigenstate of H_0, it remains stationary. $|\Psi_{P'}\rangle$ and $|\Phi_{PPP'}\rangle$ are the result of acting P' on $|\Psi_0(t)\rangle$ and $\mathbb{P}_{GSM}|\Psi_{PP}(t)\rangle \propto |\Psi_0(t)\rangle$, respectively, and therefore, are scaled copies of each other. Hence $\langle\Psi_{P'}|\Phi_{PPP'}\rangle$ is T independent, and so is $S_{GSB}(T)$. This result is true while spontaneous emission is negligible, generally up to the nanosecond timescale. Importantly, once the states in the SEM decay back to the GSM, $S_{GSB}(T)$ vanishes since it corresponds to a differential signal (Eq. 12.38). This argument remains valid upon averaging over the initial thermal ensemble.

The result that the PP' signal is a linear combination of SEM elements of $\chi(T)$ is the central result of the chapter. It provides us with a link between the OQS formalism of $\chi(T)$ with an experimental observable given by $S_{PP'}(T)$. A more ambitious result is to invert $\chi(T)$ itself from $S_{PP'}(T)$. This can be achieved by generating a large enough system of equations to extract all the elements of $\chi(T)$. For the two-level system in the SEM, extraction requires 12 linearly independent signals (see Exercise 1, part 4d). From Eq. 12.28, one can see that, in the dipole approximation, the only way to generate linearly independent signals $S_{PP'}(T)$ is by varying the polarizations or the frequencies of the pulses. However, it can be shown that a simple PP' experiment with two pulse shapes alone cannot generate a sufficient number of independent signals, and one needs to work with PP' variants that involve at least four different pulse shapes.

Although Eqs. 12.53–12.55 have been formally derived in the ideal broadband limit, it is possible to argue that they also hold in a quasi-broadband regime where the pulses have a finite time duration (allowing selectivity of particular *electronic* transitions), but the vibrational bath $\mathcal{B}$ does not change much during the time windows in which the pulses act. This observation provides a way to carry out QPT using ultrafast spectroscopy, where one varies the frequencies of the various pulses to selectively prepare and detect states in the SEM. Even though, in principle, using polarization control alone can generate enough linearly independent signals $S_{PP'}$, this method breaks down in typical experimental situations where the collected signal results form an isotropically averaged ensemble of molecules [15]. For more details on the QPT procedure using ultrafast spectroscopy, we refer the reader to our work [9–11], and specifically, to [15], from which this chapter has been adapted.

As an example of the result just derived, we now demonstrate the utility of PP' spectroscopy to detect electronic coherences using computational simulations which take realistic values for pulse durations, which we choose to be in the broadband regime for

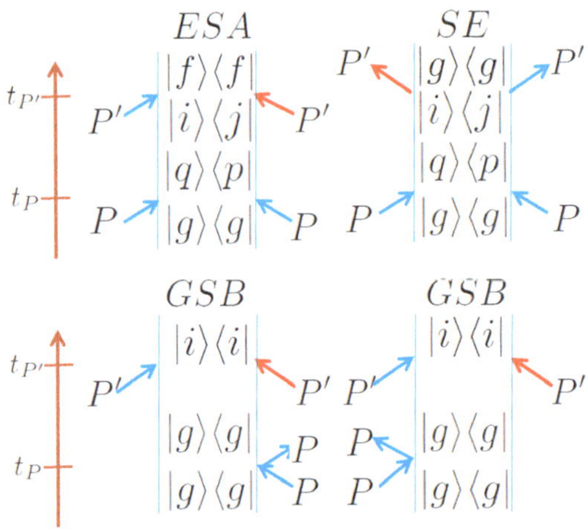

Figure 12.6 DS-FDs for a general broadband PP' signal. The ESA and SE pathways report on $\chi_{ijqp}(T)$; namely if $|q\rangle\langle p|$ is the initial state in the waiting time, what is the probability amplitude that the final state in the waiting time at time T will be $|i\rangle\langle j|$? Contrast these pathways with the ones in Fig. 12.5, where $|q\rangle\langle p|$ stays as $|q\rangle\langle p|$ because no bath is present such that it can induce a transfer to $|i\rangle\langle j|$. The GSB pathways provide a time independent background as long as both pulses are sufficiently broadband. From Ref. [15] (http://iopscience.iop.org/book/978-0-750-31062-8).

the prototypical organic dyes we are interested in (in the regime of full-width at half- maximum [FWHM] of the intensity $|\epsilon_n(t - t_n)|^2$, FWHM=$2\sqrt{\ln 2}\sigma_n$ $\approx$ 7–15 fs). Details of the simulations can be found in [12]. Figure 12.7 shows, from left to right, calculated absorption spectra for a monomer with a single excited electronic state $|\alpha\rangle$, a "coherent" dimer with strong $J(\boldsymbol{R})$ coupling between chromophores, and an "incoherent" dimer with weak $J(\boldsymbol{R})$. The monomer has only a single-electron excited state and therefore no coherence in the SEM. These systems are modeled by Eqs. 12.17–12.20 with $N_A = N_B = 1$. In addition to homogeneous broadening of the electronic levels due to vibrations, the spectra have been inhomogeneously broadened by choosing a range of ω_a and ω_b values (site energy disorder) to simulate a realistic ensemble. The power spectra of the pulse shapes used are plotted above the absorption spectra in Fig. 12.7, showing that they cover all the

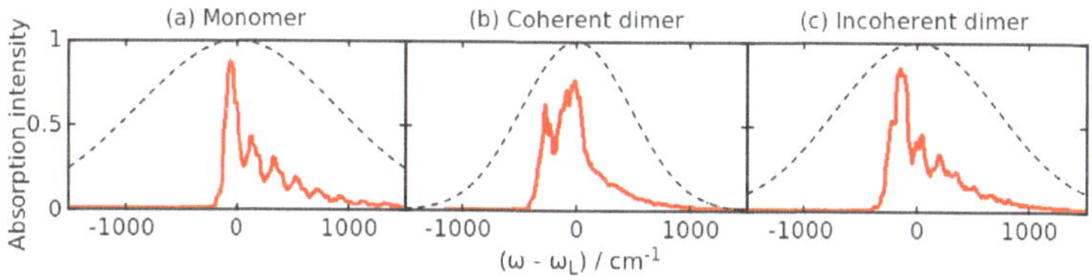

Figure 12.7 Inhomogeneously broadened absorption spectra of chromophores (solid curve) centered at $\omega = \omega_L$, with pulse spectral profiles $|\bar{\epsilon}_P(\omega)|^2 = |\bar{\epsilon}_n(\omega)|^2$ on top (dashed curve). Since there is only one electronic excited state in the monomer, the peaks in its absorption spectrum must correspond to vibrations. This conclusion does not necessarily follow for the dimers because of the presence of multiple excited states in the SEM. From Ref. [15] (http://iopscience.iop.org/book/978-0-750-31062-8). © IOP Publishing. Reproduced with permission. All rights reserved.

transitions almost uniformly (this is the physical criterion of a broadband pulse for our purposes).

Figure 12.8, top panel, shows inhomogeneously broadened $S_{PP'}(T)$ resulting from these simulations. The bottom panel shows a few time traces of elements of $\chi(T)$. In order to plot the different elements of $\chi(T)$, we have chosen the electronic eigenbasis of H_S, given by Eq. 12.24, but this choice is arbitrary. Note that the dynamics of the elements of $\chi(T)$ translates into the dynamics of $S_{PP'}(T)$ in the broadband limit ($\sigma_n \to 0$), as Eqs. 12.53–12.55 show. The oscillations of the PP' signal of the coherent dimer are much larger in magnitude than those of the monomer and the incoherent dimer. In the monomer, oscillations vanish in the ideal broadband limit, as there is only one element of $\chi(T)$ in the SEM for that situation, and population does not spontaneously decay back to the GSM or go up to the DEM in our model. This holds true *even* if *vibrational coherent dynamics* persist in the SEM. The simulations include finite duration pulses, so remnants of oscillations survive, as can be seen in Fig. 12.7. Ways to calibrate the magnitude of these oscillations have been proposed in [12]. Either way, the message is that, in the Condon approximation, oscillations in broadband PP' spectra directly reflect *electronic* dynamics (instead of the full vibronic dynamics), as $S_{PP'}(T)$ is a direct reflection of $\chi(T)$. Hence, oscillations in these measurements are strong signatures

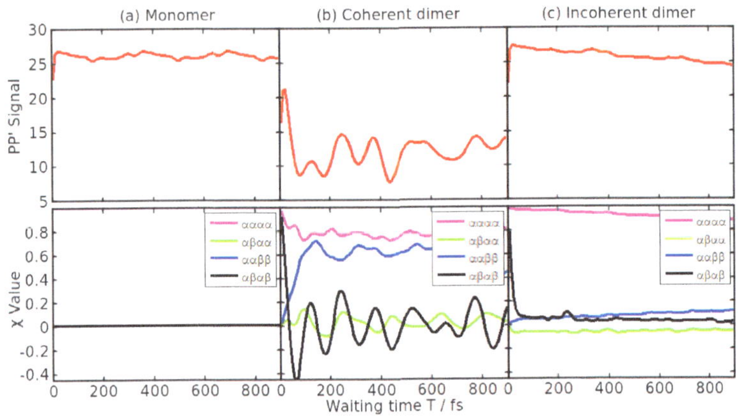

Figure 12.8 (Top) Broadband PP' spectra as a function of waiting time T as a witness for electronic coherences. The small oscillations in (a) and (c) are due to finite pulse durations. (Bottom) The witness is a linear combination of elements of the process matrix $\chi(T)$. Traces of a few representative elements of $\chi(T)$ are shown for illustration. From Ref. [15] (http://iopscience.iop.org/book/978-0-750-31062-8).

for dynamical electronic coherences, in some basis, and distinguish between such electronic coherences and coherences that are purely vibrational in nature.

12.6 Conclusions

In this chapter, we have provided a pedagogical introduction to nonlinear spectroscopy in light of the quantum information theory concept of the process matrix $\chi(T)$. We have first built our intuition on PP' spectroscopy by expressing the spectroscopic observable in terms of wave packet overlaps. We have noticed that the spectroscopic measurements give explicit information on the quantum dynamics induced in the systems of interest. In particular, we have shown that the broadband $S_{PP'}(T)$ contains the process matrix $\chi(T)$ elements from the SEM of a molecular dimer, and illustrated this concept by developing a tool that allows us to witness electronic coherence in a molecular ensemble. More sophisticated multidimensional spectroscopies can be entirely understood using this framework. The interested reader can consult [15].

Understanding the PP' signal in terms of $\chi(T)$ offers numerous conceptual advantages if one is interested in studying the OQS dynamics of the excited states of a molecular system. The process matrix χ is important to unravel the mechanisms of light harvesting and excitation energy transfer at the nanoscale. Even from the broadband $S_{PP'}(T)$ alone, we can rigorously report on timescales of decoherence, nature of the vibrational bath coupling to the electronic degrees of freedom, and so on, by making associations of $S_{PP'}(T)$ with $\chi(T)$. Once these issues are addressed, interesting questions of control and manipulation of excited states [48, 49] as well as quantum computing with molecular systems [50, 51] can be asked. We believe the reader who has mastered the presented formalism will be able to learn other techniques of nonlinear spectroscopy rather quickly, by appealing to the intuition on wave packets and the concept of $\chi(T)$, both of which we have reviewed in this chapter.

References

1. G. S. Engel et al., *Nature*, **446**, 782 (2007).
2. E. Collini and G. D. Scholes, *J. Phys. Chem. A*, **113**, 4223 (2009).
3. J. M. Womick and A. M. Moran, *J. Phys. Chem. B*, **113**, 15747 (2009).
4. G. Panitchayangkoon et al., *Proc. Natl. Acad. Sci. U S A*, **107**, 12766 (2010).
5. E. Collini et al., *Nature*, **463**, 644 (2010).
6. E. Harel, A. F. Fidler, and G. S. Engel, *Proc. Natl. Acad. Sci. U S A*, **107**, 16444 (2010).
7. K. W. Stone et al., *Science*, **324**, 1169 (2009).
8. M. Khalil, N. Demirdöven, and A. Tokmakoff, *J. Chem. Phys.*, **121**, 362 (2004).
9. P. Rebentrost, S. Shim, J. Yuen-Zhou, and A. Aspuru-Guzik, *Procedia Chem.*, **3**, 332 (2011).
10. J. Yuen-Zhou and A. Aspuru-Guzik, *J. Chem. Phys.*, **134**, 134505 (2011).
11. J. Yuen-Zhou, J. J. Krich, M. Mohseni, and A. Aspuru-Guzik, *Proc. Nat. Acad. Sci. U S A*, **108**, 17615 (2011).

12. J. Yuen-Zhou, J. J. Krich, and A. Aspuru-Guzik, *J. Chem. Phys.*, **136**, 234501 (2012).
13. M. A. Nielsen and I. L. Chuang, *Quantum Computation and Quantum Information* (Cambridge University Press, 2000).
14. M. Mohseni and D. A. Lidar, *Phys. Rev. Lett.*, **97**, 170501 (2006).
15. J. Yuen-Zhou, J. J. Krich, I. Kassal, A. Johnson, and A. Aspuru-Guzik, *Ultrafast Spectroscopy: Quantum Information and Wavepackets* (IOP Press, 2014).
16. H.-P. Breuer and F. Petruccione, *The Theory of Open Quantum Systems* (Oxford University Press, 2002).
17. A. Nitzan, *Chemical Dynamics in Condensed Phases* (Oxford University Press, Oxford, 2006).
18. V. May and O. Kuhn, *Charge and Energy Transfer Dynamics in Molecular Systems* (Wiley-VCH, 2004).
19. M. Choi, *Linear Algebra Appl.*, **10**, 285 (1975).
20. E. C. G. Sudarshan, P. M. Mathews, and J. Rau, *Phys. Rev.*, **121**, 920 (1961).
21. B. Schumacher and M. Westmoreland, *Quantum Processes, Systems, and Information* (Cambridge University Press, Cambridge, 2010).
22. U. Leonhardt, *Phys. Rev. Lett.*, **74**, 4101 (1995).
23. T. J. Dunn, I. A. Walmsley, and S. Mukamel, *Phys. Rev. Lett.*, **74**, 884 (1995).
24. C. Leichtle, W. P. Schleich, I. S. Averbukh, and M. Shapiro, *Phys. Rev. Lett.*, **80**, 1418 (1998).
25. T. S. Humble and J. A. Cina, *Phys. Rev. Lett.*, **93**, 060402 (2004).
26. J. A. Cina, *J. Chem. Phys.*, **113**, 9488 (2000).
27. Z. H. Loh et al., *Phys. Rev. Lett.*, **98**, 143601 (2007).
28. I. L. Chuang and M. A. Nielsen, *J. Mod. Opt.*, **44**, 2455 (1997).
29. M. Mohseni, A. T. Rezakhani, and D. A. Lidar, *Phys. Rev. A*, **77**, 032322 (2008).
30. Y. S. Weinstein et al., *J. Chem. Phys.*, **121**, 6117 (2004).
31. A. M. Childs, I. L. Chuang, and D. W. Leung, *Phys. Rev. A*, **64**, 012314 (2001).
32. H. Kampermann and W. S. Veeman, *J. Chem. Phys.*, **122**, 214108 (2005).
33. M. Riebe et al., *Phys. Rev. Lett.*, **97**, 220407 (2006).
34. J. B. Altepeter et al., *Phys. Rev. Lett.*, **90**, 193601 (2003).
35. M. W. Mitchell, C. W. Ellenor, S. Schneider, and A. M. Steinberg, *Phys. Rev. Lett.*, **91**, 120402 (2003).
36. M. Howard et al., *New J. Phys.*, **8** (2006).

37. S. H. Myrskog, J. K. Fox, M. W. Mitchell, and A. M. Steinberg, *Phys. Rev. A*, **72**, 013615 (2005).
38. R. C. Bialczak et al., *Nat. Phys.*, **6**, 409 (2010).
39. T. Forster, *Delocalized Excitation and Energy Transfer*, Vol. 3 (Academic Press, New York, 1965), pp. 93–137.
40. J. A. Cina, D. S. Kilin, and T. S. Humble, *J. Chem. Phys.*, **118**, 46 (2003).
41. J. D. Biggs and J. A. Cina, *J. Chem. Phys.*, **131**, 224101 (2009).
42. J. D. Biggs and J. A. Cina, *J. Chem. Phys.*, **131**, 224302 (2009).
43. D. J. Tannor, *Introduction to Quantum Mechanics: A Time Dependent Approach* (University Science Books, 2007).
44. J. C. Tully, *J. Chem. Phys.*, **93**, 1061 (1990).
45. D. R. Yarkony, *Rev. Mod. Phys.*, **68**, 985 (1996).
46. S. Mukamel, *Principles of Nonlinear Optical Spectroscopy* (Oxford University Press, 1995).
47. M. Cho, *Two Dimensional Optical Spectroscopy* (CRC Press, 2009).
48. P. W. Brumer and M. Shapiro, *Principles of the Quantum Control of Molecular Processes* (Wiley-Interscience, 2003).
49. S. Rice and M. Shao, *Optical Control of Molecular Dynamics* (Wiley-Interscience, 2000).
50. V. V. Lozovoy and M. Dantus, *Chem. Phys. Lett.*, **351**, 213 (2002).
51. R. R. Zaari and A. Brown, *J. Chem. Phys.*, **137**, 104306 (2012).

Chapter 13

Simulating the Nonlinear Optical Response of Multichromophore Complexes

Arend G. Dijkstra[a] **and Yoshitaka Tanimura**[b]

[a] *Max-Planck Institut für Struktur und Dynamik der Materie Luruper Chaussee 149, CFEL, 22761 Hamburg, Germany*

[b] *Department of Chemistry, Graduate School of Science, Kyoto University, Kyoto 606-8502, Japan*

a.g.dijkstra@leeds.ac.uk, tanimura@kuchem.kyoto-u.ac.jp

In this chapter, we explain the theory used to treat electronic excitations in multichromophoric systems like biological light-harvesting complexes and DNA as well as vibrations in peptides. Although these systems appear very different, they can be treated on a common theoretical footing. Two main issues will appear throughout this chapter. First, the systems under study are composed of multiple chromophores interacting with light. These building blocks are, however, not independent, but coupling between them leads to collective behavior observable in the optical response. The second important ingredient of the theory is interaction of the chromophores with a complex environment, which tends to destroy the cooperativity.

Ultrafast Dynamics at the Nanoscale: Biomolecules and Supramolecular Assemblies
Edited by Irene Burghardt and Stefan Haacke

ISBN 978-981-4745-33-8 (Hardcover), 978-981-4745-34-5 (eBook)
www.panstanford.com

Secondly, in all examples that we will discuss we will not only be interested in the dynamics of excitations, but also in the experimentally observable nonlinear response [1] where quantum coherence plays a major role. The aim of this chapter is, thus, to introduce models that describe interactions between local excitations and their environment to various degrees of sophistication and to study the nonlinear response predicted by these models. We will first give a phenomenological introduction to 2D spectra and discuss the approximation of a fast bath. We will then go beyond this approximation, and first focus on vibrations in peptides, where the environment can be modeled using classical physics. Then, we will move on to models where the environment is a quantum system and apply that theory to electronic excitations.

13.1 Two-Dimensional Spectra in the Limit of a Fast Environment

It is often assumed in theories of open quantum systems, which are systems interacting with their environment, that the environment is evolving on a timescale that is much faster than any relevant dynamics in the system. This is convenient because it allows for the derivation of time-local equations of motion for the systems degrees of freedom. The Haken Strobl model, as well as Lindblad and Redfield master equations are in this category. We will see, however, that the assumption of a fast environment, in addition of not predicting the correct dynamics [2], is extremely limiting in the calculation of the nonlinear response.

Linear absorption spectroscopy, which measures the absorption of light by a sample as a function of frequency is a common way to probe the properties of a material. The response is linear in the sense that the polarization P created in the sample is linear in the applied electric field E, $P = \chi^{(1)}E$, where $\chi^{(1)}$ is the (linear) susceptibility. The light interacts with chromophores and the response contains information about their electronic transition states. However, the measurement only detects the coherence of electronic excitations and their dynamics cannot be accessed directly. This limitation is overcome in nonlinear optical experiments. These experiments probe nonlinearities in the light matter interaction. The polarization is no longer linear in the applied field, but can be written as an expansion $P = \chi^{(1)}E + \chi^{(3)}E^3 + \ldots$ [3]. Here, the second order term was discarded because we assume that the system under study is symmetric under coordinate inversion (as is the case, for example, for an assembly of molecules randomly oriented in solution) in which case the second order vanishes. The quantity $\chi^{(3)}$ is the third-order nonlinear susceptibility. An example of a technique that measures the third-order nonlinearity is pump–probe spectroscopy. An excitation is created in the sample, and its state is probed after an experimentally controllable waiting time. By repetition of the probing as a function of waiting time the dynamics in the system is observed.

A more advanced technique, which has been developed mostly over the past 20 years [1], is 2D optical spectroscopy. By using this technique all information contained in the third-order nonlinear response is accessible experimentally. The name two-dimensional refers to the 2D plots in which results are typically presented. The third-order signal is measured after exposing the system to three short laser pulses. In a simplified picture, which neglects certain coherent contributions to the spectrum, the state of the system after interaction with the first pulse is plotted on the horizontal axis of the 2D spectrum, while the state after the third pulse is shown on the vertical axis. The variables that appears on the axis are the Fourier transforms of the time periods following pulses one (t_1) and three (t_3). The time interval between the second and the third pulse, called the waiting time or t_2, is left as a controllable parameter. The 2D spectrum, thus, shows correlations between the state of the system before and after the waiting time, as a function of frequency.

With this brief introduction to 2D spectroscopy, we are in a position to discuss the effect of the fast bath approximation. Let's consider a single two-level chromophore system, coupled to a bath. The time evolution of the chromophore density matrix in the absence of light is given by the free propagator $G(t)$ as $\rho(t) = G(t - t_0)\rho(t_0)$. Because the bath is fast and contains no memory, the propagator depends only on the time difference $t - t_0$. The propagators are operators on the bath degrees of freedom, and, denoting the state of the bath by indices α and β we should write in a more complete notation $\rho_A(t) = \mathrm{Tr_B} G^{\beta\alpha}(t - t_0)\rho_A(t_0)\rho_B^{\alpha}(t_0)$, where $\mathrm{Tr_B}$ denotes the partial trace over the bath degrees of freedom and ρ_B is the density matrix for the bath. In this expression, we have assumed that the system and the bath are not correlated at time t_0, which is valid in the fast bath approximation. Even when the fast bath approximation breaks down, a factorized initial condition at time t_0 can be used in optical experiments because of the Frank Condon principle, which states that optical excitation is so fast that there is no time for the environment to readjust. However, correlations between the system and the environment are essential at the moment of interaction with the later pulses.

The 2D spectrum can be calculated as the sum of several contributions, which all have a similar form. They contain interactions with the light through the dipole operator μ, which operates on the system degrees of freedom and the propagators G which are operators on the system as well as on the environment. One of the contributions can be written as (complete expressions for the spectra will be given in Section 13.4)

$$S^{(3)} = \mathrm{Tr}\mu G^{\alpha\gamma}(t_3)\left(\left\{G^{\gamma\beta}(t_2)\left[\mu G^{\beta\alpha}(t_1)\left(\rho_A(t_0)\rho_B^{\alpha}(t_0)\mu\right)\right]\right\}\mu\right), \tag{13.1}$$

where the trace is over system as well as bath degrees of freedom. The quantity μ is the transition dipole of the system. Now, if the dynamics in the environment is fast and the state of the system does not influence the dynamics in the bath, the trace over the bath degrees of freedom can be taken for each propagator separately. If we denote $\mathrm{Tr_B} G^{\beta\alpha}(t) = g(t)$, the expression for the signal becomes

$$S^{(3)} = \mathrm{Tr_S}\mu g(t_3)\left(\{g(t_2)\left[\mu g(t_1)\left(\rho_A(t_0)\mu\right)\right]\}\mu\right). \tag{13.2}$$

The g's are now no longer operators with respect to the bath, although they are still operators on the system. As a result, the signal no longer contains memory in the bath. The evolution during the times t_1, t_2 and t_3 are now independent of each other. This has important consequences for the 2D signal. As mentioned before, the 2D plot correlates the state of the system during t_3 with the state during t_1. Because there is no memory in the current description, these two states are uncorrelated. The 2D signal will thus be symmetric and almost round. If we assume a pure dephasing model

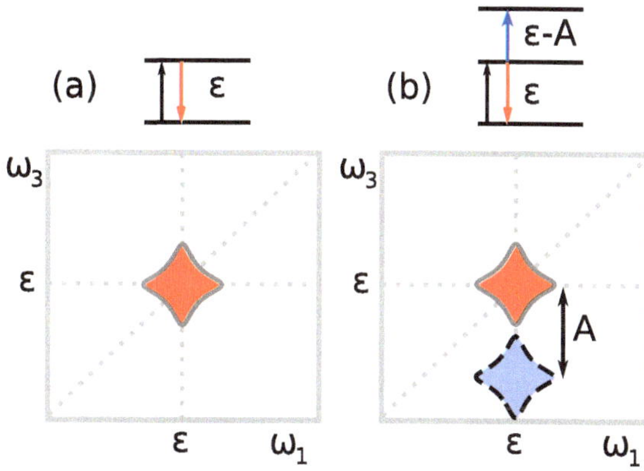

Figure 13.1 Two-dimensional line shape in the fast bath approximation for exponentially decaying coherence for (a) a single two-level system and (b) a single three-level system. The signals are obtained after a double Fourier transformation of the third-order response function over t_1 and t_3. Red and blue parts represent the negative (bleaching and stimulated emission) and positive (induced absorption) parts of the response function.

for the environment, where the coherence in the system decays exponentially, the line shape will be a 2D Lorentzian. The 2D line shapes for a single chromophore in the fast bath approximation are shown in Fig. 13.1. Similar line shapes will result from theories that do not properly include correlations between system and bath states at the moment of interaction with external pulses, even if they do treat the timescale of the bath and the strength of the system–bath interaction correctly. The difference between the correct approach and the approximation of no system–bath correlations at these moments, which is valid only in the fast bath approximation, is shown in Fig. 13.2.

Because 2D spectroscopy is used to study very fast processes, the dynamics in the environment often occur on the same timescale as in the system, and the fast bath approximation breaks down. This can be seen in the 2D spectrum as an elongation of the line shape along the diagonal, which shows the presence of correlations between t_1 and t_3. As the waiting time is increased, these correlations disappear and the line shape becomes round. The 2D spectrum is, thus, also a good measure to test the validity of theories that correctly deal

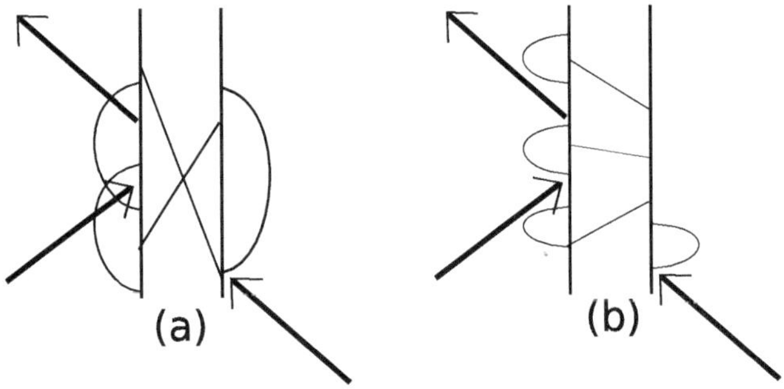

Figure 13.2 Double-sided Feynman diagrams showing a cartoon of the interaction of the system with external pulses (arrows) and with the bath (thin lines). Time runs from bottom to top. The correct calculation includes memory that extends from before until after the interaction with the pulses, as shown in (a). (b) Calculation using master equations, where all correlations between the system and the bath that occur during free propagation can be included, but correlations at the moment of interaction with the pulses are absent. The before-mentioned memory effect is not taken into account in this approximation.

with the timescale of the bath. Such theories must be able to include correlations between system and bath states at the moments of interaction with the external pulses. In the following sections, we will discuss such methods.

13.2 Classical and Adiabatic Environment: Vibrations in Peptides

Natural as well as synthetic peptides and proteins are composed of a chain of amino acids bound together by peptide bonds. The chain, called the backbone, consists of a repeating series of atoms with varying side groups that determine the chemical and biological activity of the molecule. Each peptide bond contributes an amide group to the backbone, which contains a carbon atom bound to a nitrogen neighbor and to an oxygen atom. We will be interested in the stretching vibrations of the bonds between these atoms. These stretching vibrations can be excited by infrared light and have been the topic of much research because they are sensitive to the structure of the peptide as well as to the dynamics of the surrounding solvent, which is most often (heavy) water. In particular, the so-called amide I vibration, which is mainly the stretch of the C=OO bond has been intensely studied [4]. To model the properties of this vibrational mode, we need to include each C=O bond in the peptide. In a peptide that contains N amide groups, there will be N such bonds, which form the local oscillators in the theory. Crucially, as can be ascertained from the line shapes observed in experimental spectra, these oscillators are not independent. Electrostatic as well as mechanical coupling exists between them which leads to collective properties.

Usually the analysis of such coupled oscillators starts from an adapted Frenkel exciton Hamiltonian. It is written in terms of the operators $b_n^\dagger$ and b_n, which create respectively annihilate an excitation on the nth oscillator in the chain. The excitations are Bosonic and together form what is called the system. An excitation has an energy ϵ_n, which is the frequency of a single excitation. Because the potential is (be it only slightly) anharmonic, the state where two excitations are present on a single oscillator has a slightly smaller energy than expected from twice the anharmonic frequency. It can be written as $2\epsilon - A$, where A is the anharmonicity. The coherent interactions between oscillator n and m are called J_{nm}. The Hamiltonian is then written

$$H = \sum_{n=1}^{N} \epsilon_n(X_n) b_n^\dagger b_n + \sum_{nm} J_{nm} b_n^\dagger b_m - \sum_{n} \frac{A_n}{2} b_n^\dagger b_n^\dagger b_n b_n. \qquad (13.3)$$

Equation 13.3 includes the couplings between different local oscillators. The second important ingredient in the theory is the interaction with the environment. Electrodynamic interactions between the atoms in the amide group and the surrounding solvent, as well as the peptide, change the frequency of the amide oscillator. In the Hamiltonian we have indicated this effect by writing $\epsilon_n(X_n)$, where X_n are the coordinates of the environment. Because the solvent is fluctuating on timescales as short as femtoseconds, the influence of the environment is highly dynamic. If a model can be made of the electrostatic influence of the environment on the local mode excitation energies ϵ_n can be predicted as a function of the configuration of the environment. We now make the crucial approximation that the dynamics of the environment is independent from the exciton dynamics and that it can be described using classical mechanics. In that case, the coordinates X_n are well-defined functions of time. Implicitly, this means that the excitation energies ϵ_n are also functions of time. The complete system, including the interaction with the environment, can thus be described as a time-dependent Hamiltonian for the degrees of freedom of the amide vibrations only. The effect of the dynamic solvent and peptide environment is included in the time dependence of the parameters $\epsilon_n(t)$ as adiabatic modulation. Of course, this type of modelling presumes that the dynamical time evolution of the environment is known. This information can be obtained from molecular dynamics simulations. Furthermore, in this model energy flow from the system to the environment, which relates to thermal dissipation, is neglected.

The properties of the time-dependent frequencies can be analyzed by studying time correlation functions. In the commonly valid approximation of Gaussian fluctuations, only the two-point correlation functions are important, while higher-order correlation functions must be studied if the dynamics are non-Gaussian. The two-point correlation functions are defined as

$$C_{nm}(t) = \langle \epsilon_n(t) \epsilon_m(0) \rangle, \tag{13.4}$$

where the brackets denote an average over a long trajectory. The functions $C_{nn}(t)$ are autocorrelation functions, while functions with $n \neq m$ represent cross-correlations. By studying the correlation functions, information can be obtained on the size of the fluctuations induced by the environment, as well as on their typical timescale. Furthermore, the precise shape of the correlation functions encodes the details of the system–environment interaction. Note that the correlation functions are complex functions of time due to the presence of dissipation. However, the functions we have used here have only real parts corresponding to the fluctuations, because we assumed adiabatic modulation from the bath. In a more general model, the fluctuation (real part) and dissipation (imaginary part) have to satisfy either classical or quantum fluctuation-dissipation theorems depending on how the bath is treated. We will encounter a quantum environment later.

Once the time-dependent Hamiltonian is established, the dynamics of the system degrees of freedom can be obtained from the time-dependent Schrödinger equation,

$$\frac{\hbar}{i}\frac{\mathrm{d}}{\mathrm{d}t}\psi_n(t) = \sum_m H_{nm}(t)\psi_m(t). \tag{13.5}$$

A time-dependent wave function $\psi_n(t)$, where n runs from 1 to N contains all the necessary information of the system. Once it is obtained by (numerically) solving the Schrödinger equation, the dynamics of excitations as well as all linear and nonlinear response functions can be calculated. To make the connection with experiments, where an ensemble of many copies of the system is present, it is necessary to average the results over many realizations of the dynamics of the environment. The wave function can not be averaged, therefore the quantity to calculate is the reduced density matrix

$$\rho^A_{nm}(t) = \psi_n(t)\psi^*_m(t). \tag{13.6}$$

This method than allows the calculation of excitation dynamics and (nonlinear) spectra without the need for Markovian or Gaussian approximations [5]. An example of 2D spectra is shown in Fig. 13.3. The peaks are clearly somewhat extended along the diagonal, showing that memory effects in the environment are included in the calculation. The presence of intensity away from the diagonal, the so-called cross peaks, show that the wave functions of the two modes overlap in space. Analysis of the cross peak intensity as a function of waiting time gives information about vibrational relaxation.

At this point it is worthwhile to think about the approximation we made in setting up this theory. Although the system degrees of freedom—the vibrations observed directly in experiment—are treated fully quantum mechanically, the dynamics of all other degrees of freedom were approximated using classical mechanics. The bath dynamics is also independent from the system dynamics, and, thus, the system is adiabatically modulated by the bath. One obvious way to improve upon this approximation is to include more degrees of freedom in the system instead of in the environment. This, however, makes the numerical calculations more expensive. One observation of the limitations of the classical approximation

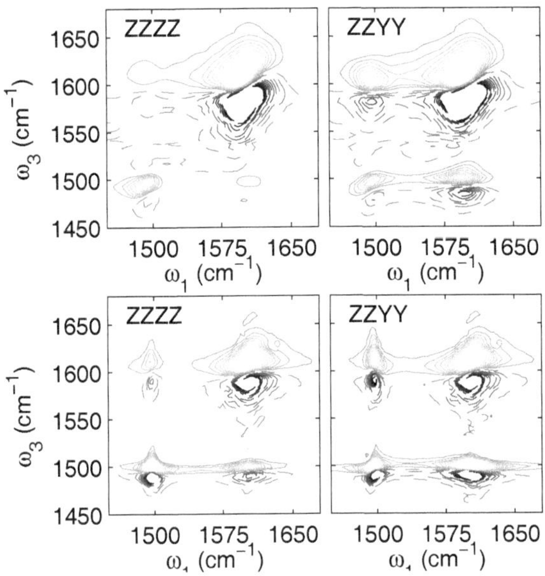

Figure 13.3 Two-dimensional spectra of the amide I and II modes in NMA-d7 in solution for two different experimentally controllable polarization directions ($ZZZZ$ and $ZZYY$). Experimental spectra are shown at the top [6], while theoretical spectra are plotted at the bottom row. The spectra show clear cross-peaks (away from the diagonal), which indicated that the wave functions of the two amide modes overlap in space. Reprinted with permission from Ref. [7]. Copyright (2011) American Chemical Society.

is the equilibrium state, which is not correct because there is no energy flow from the system to the bath. After propagation from an initial state for a sufficient amount of time each eigenstate of the system should be populated in line with its energy, i.e. high energy states should have less population than lower energy eigenstates. However, in the model of a classical environment it turns out that each eigenstate in equilibrium has the same population. Even if the system of interest is not a two-level system but a harmonic oscillator, the system heats up if only adiabatic fluctuations are present. The physical explanation for this observation is that fluctuations drive the system to ever higher temperatures, but no dissipation is present to counteract this. If some sort of energy backflow from the system to the environment is added, one may obtain a semiclassical equilibrium state. We will later see that dissipation is present in a fully quantum mechanical theory as an imaginary part in the correlation functions.

Thus, the model of a dissipation free environment corresponds to an effective high-temperature approximation. Only the combination $\beta\hbar$ occurs in the full theory of a harmonic bath, thus, taking the classical limit $\hbar \to 0$ corresponds to taking the high-temperature limit $\beta \to 0$. To understand whether this approximation is valid, one compares the total bandwidth of the system excitations with the thermal energy. In the case of a single excitation of the amide I vibrations, the bandwith typically is a few 10's of wavenumbers, while the thermal energy at room temperature is 200 cm^{-1}. When both the amide I and the amide II (mainly C-N stretch) vibrations are included in the system Hamiltonian, the bandwidth increases to more than 100 cm^{-1} and the approximation is only just valid. If, however, we want to apply a similar model to electronic excitations in molecular aggregates the approximation breaks down, because typical bandwidths in those systems are of the order of 100's to over 1000 cm^{-1}.

13.3 Quantum Environment: Electronic Excitations in Molecular Aggregates

The Hamiltonian introduced in the previous section can be applied to electronic excitations in molecular aggrates with a small modification. The common characteristic of these systems is that they consist of multiple closely spaced chromophores. It is usual to model each chromophore as a two-level system, with a ground state and an electronic excited state. If we denote the creation and annihilation operators for a single two-level system as $c^\dagger$ and c, the Hamiltonian is given by

$$H = H_S + H_{SB} = \sum_{n=1}^{N} \epsilon_n(X_n) c_n^\dagger c_n + \sum_{nm} J_{nm} c_n^\dagger c_m, \qquad (13.7)$$

where again J_{nm} denotes the coherent coupling and X_n the coordinate of the environment. We now introduce a model for the environment that can be treated fully quantum mechanically and describe it as a set of infinitely many harmonic oscillators with masses m_α, coordinates x_α, momenta p_α and frequencies ω_α. The collective bath coordinate that influences the system is

$$X_n = \sum_\alpha g_{n\alpha} x_\alpha, \tag{13.8}$$

with $g_{n\alpha}$ the coupling strength of chromophore n to bath oscillator α. We choose the dependence of the chromophore excitation energy to be linear in these collective bath coordinates (any proportionality constant can be absorbed in the g's),

$$\epsilon_n(X_n) = \epsilon_n - X_n. \tag{13.9}$$

The Hamiltonian for the bath is given by

$$H_\mathrm{B} = \sum_\alpha \left(\frac{p_\alpha^2}{2m_\alpha} + \frac{1}{2} m_\alpha \omega_\alpha^2 x_\alpha^2 \right). \tag{13.10}$$

Because of the model of linear coupling to a harmonic bath, several standard results can be used. First of all, all necessary information about the bath is contained in the spectral densities

$$\mathcal{J}_{nm}(\omega) = \sum_\alpha \frac{g_{n\alpha} g_{m\alpha}}{2m_\alpha \omega_\alpha} \delta(\omega - \omega_\alpha), \tag{13.11}$$

which therefore, rather than the coupling constants g, become the parameters of the theory. The spectral density gives the density of bath states at frequency ω, weighted by their coupling to the system. Secondly, the correlation functions (Tr_B denotes the trace over the bath degrees of freedom and $\beta = 1/k_\mathrm{B}T$ is the inverse temperature)

$$L_{nm}(t) = \mathrm{Tr}_\mathrm{B} e^{-\beta H_\mathrm{B}} X_n(t) X_m(0) / \mathrm{Tr}_\mathrm{B} e^{-\beta H_\mathrm{B}} \tag{13.12}$$

can be calculated in terms of the spectral densities as

$$\begin{aligned} L_{nm}(t) &= \frac{1}{\pi} \int_0^\infty \mathrm{d}\omega J_{nm}(\omega) (\coth \frac{\beta \hbar \omega}{2} \cos \omega t - i \sin \omega t) \\ &= \mathcal{F}^{-1} \left[J_{nm}(\omega) \coth \frac{\beta \hbar \omega}{2} \right] + \mathcal{F}^{-1} \left[J_{nm}(\omega) \right]. \end{aligned} \tag{13.13}$$

In the second line, $\mathcal{F}^{-1}$ denotes the inverse Fourier transform. The real part of these correlation functions are the quantum equivalent of the classical correlation functions introduced in the previous section. Here, the correlation functions are complex quantities. Since dissipation originates from the total Hamiltonian, they have an imaginary part which corresponds to this dissipation. As is clear by Fourier transforming the second line, the Fourier transforms of the real and imaginary parts of the correlation function are related by a factor coth $\beta\hbar\omega/2$. This is a reflection of the fluctuation dissipation theorem, which shows that fluctuations and dissipation are balanced to achieve equilibrium at a finite temperature.

If the interaction between the system and the bath commutes with the system Hamiltonian, the dynamics of this model can be solved exactly using the cumulant expansion or path integral methods. However because each chromophore in general feels a different environment and coherent coupling between chromophores exists, this assumption is normally not valid. We therefore need a (numerical) method to describe the quantum dynamics generated by this model.

The difficulty in deriving an equation of motion for the model presented here is the presence of memory in the quantum environment. If the system interacts with the environment at a certain moment in time, information about this interaction is stored in the bath and can influence the system at a later point in time. An equation of motion for the system degrees of freedom can therefore be expected to contain cumbersome convolution integrals. Simplification of course occurs when the timescale of the environment, parameterized by the correlation functions, is much faster than any dynamics in the system. A time-local equation of motion can then be derived. Although this is the approximation behind many master equations, this approach is invalid for the description of electronic excitation dynamics in molecular aggregates where environment timescales similar to the ones in the system play an important role.

An approach to treat the memory in the environment exactly, while maintaining time-local equations of motion, is given by the hierarchy of equations of motion [8, 9]. The reason that the equations can be time local, while still describing a memory effect, is that extra degrees of freedom are introduced through auxiliary density matrices. These take care of the memory stored in the environment.

The hierarchy of equations of motion can be derived if the correlation functions are a sum of exponential functions. In principle, any spectral density can be expanded such that the correlation functions are in this form. However, for an efficient numerical implementation it is desirable to have few terms in the expansion. The simplest implementation of the hierarchy of equations of motion is a single exponential correlation function. This form can be obtained by choosing the Drude model spectral density and working at high temperature. Although the general form of the correlation functions in Eq. 13.13 include the possibility of correlated fluctuations between chromophores, it is often assumed that the site energy of each chromophore is coupled to its own separate environment. If we furthermore assume that the environments of each chromophore are equal, only a single correlation function $L(t)$ and a single spectral density $\mathcal{J}(\omega)$ are required. The Drude model for the spectral density is

$$\mathcal{J}(\omega) = 2\lambda\gamma\frac{\omega}{\gamma^2+\omega^2}, \tag{13.14}$$

where λ is the reorganization energy and γ is the inverse timescale of the environment. The correlation function in the high-temperature approximation is calculated using Eq. 13.13 to be

$$L(t) = ce^{-\gamma|t|}, \tag{13.15}$$

with the prefactor $c = \lambda(-i\gamma + 2/\beta)$.

The high-temperature limit turns out to be valid for electronic excitations in molecular aggregates at room temperature [10]. Here, high temperature means that the temperature is high compared to the time scale of the environment, $\beta\hbar\gamma < 1$. For systems where this approximation breaks down, the hierarchy can be extended with low-temperature correction terms [11].

We give the derivation of the hierarchy of equations of motion [8, 9, 12, 13] for a single bath at high temperature. Extensions to multiple baths or low temperatures, in which case the correlation functions are multiexponential, are possible. The density matrix at time t can be written in path integral form as

$$\rho(q, q'; t) = \int \mathcal{D}[q]\mathcal{D}[q'] \int \mathrm{d}q_i \mathrm{d}q_i' \\ e^{iS[q;t]/\hbar} F(q, q'; t) e^{-iS[q';t]/\hbar} \rho(q_i, q_i'; 0), \tag{13.16}$$

where S is the action corresponding to the system Hamiltonian and F is the Feynman–Vernon influence functional. This can be written in terms of the correlation function and the system–bath interaction as

$$F(q,q';t) = \exp - \int_0^t \mathrm{d}t' \int_0^{t'} \mathrm{d}t'' \\ \hat{V}^{\times}(t')\left(L_2(t'-t'')\hat{V}^{\times}(t'') + iL_1(t'-t'')\hat{V}^{\circ}(t'')\right), \tag{13.17}$$

where $L_2(t)$ and $L_1(t)$ denote the real and imaginary parts of the correlation function and $\hat{V}^{\times}(t) = V[q(t)] - V[q'(t)]$ and $\hat{V}^{\circ}(t) = V[q(t)] + V[q'(t)]$. The time derivative of the reduced density matrix is given by

$$\frac{\mathrm{d}}{\mathrm{d}t}\rho(q,q';t) = -i\hat{H}_S^{\times}\rho(q,q';t) + \int \mathcal{D}[q]\mathcal{D}[q'] \int \mathrm{d}q_i \mathrm{d}q_i' \\ e^{iS[q;t]/\hbar}\frac{\mathrm{d}}{\mathrm{d}t}F(q,q';t)e^{-iS[q;t]/\hbar}\rho(q_i,q_i';0). \tag{13.18}$$

The time derivative of the influence functional can be evaluated as

$$\frac{\mathrm{d}}{\mathrm{d}t}F(q,q';t) = -\hat{V}^{\times}(t)\int_0^t \mathrm{d}t''\left(L_2(t-t'')\hat{V}^{\times}(t'') \right. \\ \left. + iL_1(t-t'')\hat{V}^{\circ}(t'')\right)F(q,q';t) \\ = -i\hat{V}^{\times}(t)F^{(1)}(q,q';t), \tag{13.19}$$

where we defined the auxiliary influence functional $F^{(1)}$. Introducing also the auxiliary density matrix $\rho^{(1)}(t)$, the equation of motion can be rewritten as

$$\frac{\mathrm{d}}{\mathrm{d}t}\rho(t) = -i\hat{H}_S^{\times}\rho(t) - i\hat{V}^{\times}(t)\rho^{(1)}(t). \tag{13.20}$$

This is a recurring trick in the derivation of the hierarchy. The time derivative of an (auxiliary) density matrix is written and the terms which are easily evaluated are retained while complicated terms are dealt with by introducing additional auxiliary functions. The resulting hierarchy of equations of motion is fully time local, that is, only the time t appears. Memory effects, through which the density matrix at time t depends on earlier times, are dealt with through the auxiliary functions.

To proceed, we write the equation of motion for the auxiliary density matrix $\rho^{(1)}$,

$$\frac{\mathrm{d}}{\mathrm{d}t}\rho^{(1)}(q,q';t) = -i\hat{H}_S^{\times}\rho^{(1)}(q,q';t) + \int \mathcal{D}[q]\mathcal{D}[q'] \int \mathrm{d}q\mathrm{d}q' \\ e^{iS[q;t]/\hbar}\frac{\mathrm{d}}{\mathrm{d}t}F^{(1)}(q,q';t)e^{-iS[q;t]/\hbar}\rho(q_i,q_i';0). \tag{13.21}$$

The time derivative of the auxiliary influence functional can be evaluated as

$$
\begin{aligned}
\frac{\mathrm{d}}{\mathrm{d}t}F^{(1)}(q,q';t) &= -i(L_2(0)\hat{V}^{\times}(t) + iL_1(0)\hat{V}^{\circ}(t))F(q,q';t) \\
&\quad - \hat{V}^{\times}(t)\left(\int_0^t \mathrm{d}t'' L_2(t-t'')\hat{V}^{\times}(t'')\right. \\
&\quad \left. + iL_1(t-t'')\hat{V}^{\circ}(t'')\right) F^{(1)}(q,q';t) \\
&\quad - i\int_0^t \mathrm{d}t'' \frac{\mathrm{d}}{\mathrm{d}t} L_2(t-t'')\hat{V}^{\times}(t'') \\
&\quad + i\frac{\mathrm{d}}{\mathrm{d}t} L_1(t-t'')\hat{V}^{\circ}(t''))F(q,q';t) \\
&= -i(c_2\hat{V}^{\times}(t) + ic_1\hat{V}^{\circ}(t))F(q,q';t) \\
&\quad - i\hat{V}^{\times}(t)F^{(2)}(q,q';t) - \gamma F^{(1)}(q,q';t). \quad (13.22)
\end{aligned}
$$

In the last equality, we have used $L(t) = ce^{-\gamma t}$ and $c = c_2 + ic_1$. Introducing the auxiliary density matrix $\rho^{(2)}$, the equation of motion for $\rho^{(1)}$ becomes

$$
\frac{\mathrm{d}}{\mathrm{d}t}\rho^{(1)}(t) = (-i\hat{H}_S^{\times} - \gamma)\rho^{(1)}(t) - i\hat{V}^{\times}\rho^{(2)}(t) - i(c_2\hat{V}^{\times} - ic_1\hat{V}^{\circ})\rho(t). \quad (13.23)
$$

In the same manner, an equation of motion can be derived for $\rho^{(2)}$. The result of this procedure is a hierarchy of coupled equations of motion for the physical reduced density matrix and the auxiliary density matrices.

Extending the previous derivation to multiple baths, the hierarchy of equations of motion is given by

$$
\begin{aligned}
\dot{\rho}^{\{p\}}(t) &= -\left(i\hat{H}_S^{\times} + \sum_n p_n\gamma\right)\rho^{\{p\}}(t) \\
&\quad - i\sum_n p_n\left(cV_n\rho^{p_n^-} - c^*\rho^{p_n^-}V_n\right) - i\sum_n \hat{V}_n^{\times}\rho^{p_n^+}. \quad (13.24)
\end{aligned}
$$

In this equation, the auxiliary matrices are labeled by a set of indices denoted $\{p\}$. Each of the N indices p_n runs from zero to a maximum value set by the predetermined depth of the hierarchy. In fact, the hierarchy is terminated by choosing a depth T and including only those auxiliary matrices with a sum of indices $\sum_n p_n \leq T$. The physical reduced density matrix is the matrix with all indices equal to zero. $\rho^{p_n^+}$ and $\rho^{p_n^-}$ denote an increase and decrease of the relevant index. Furthermore, we defined $V_n = c_n^\dagger c_n$ as the system part of the system–bath interaction and $H_S = \sum_{n=1}^{N} \epsilon_n c_n^\dagger c_n + \sum_{nm} J_{nm} c_n^\dagger c_m$ is the system Hamiltonian. Finally, the constant c is the prefactor of the correlation function defined below Eq. 13.15.

13.4 Nonlinear Spectra: Correlations and Line Shape

In this section, we will introduce the reader to the calculation of 2D spectra and explain how they encode information about the dynamics as it was presented in the previous sections.

The third-order contribution to the polarization P [14] can be written in terms of the incident electric field E as

$$P^{(3)}(t) = \int \mathrm{d}\tau_3 \mathrm{d}\tau_2 \mathrm{d}\tau_1 S^{(3)}(\tau_1, \tau_2, \tau_3) \\ E(t-\tau_3)E(t-\tau_3-\tau_2)E(t-\tau_3-\tau_2-\tau_1), \quad (13.25)$$

where the three time integrals run from zero to infinity and $S^{(3)}$ is the third-order response function. It is given by

$$S^{(3)}(\tau_1, \tau_2, \tau_3) = -i^3 \mathrm{Tr} \mu G(\tau_3) \hat{\mu}^\times G(\tau_2) \hat{\mu}^\times G(\tau_1) \hat{\mu}^\times \rho(t_0). \quad (13.26)$$

Here, $\hat{\mu}^\times \rho = \mu\rho - \rho\mu$. The propagators G represent the dynamics of the system in the absence of light,

$$\rho(t) = G(t; t_0)\rho(t_0), \quad (13.27)$$

and $G(\tau_3)$, $G(\tau_2)$ and $G(\tau_1)$ are shorthand notation for $G(t; t-\tau_3)$, $G(t-\tau_3; t-\tau_3-\tau_2)$ and $G(t-\tau_3-\tau_2; t-\tau_3-\tau_2-\tau_1)$. Note that the shorthand notation is a bit misleading in the sense that the propagators depend on both time arguments, and not only on the difference between them. In particular, the propagators G are still operators on the bath degrees of freedom and correlations between the chromophore system and the bath formed by their environment exist at the moments of interaction with the external electric fields.

In experiment, the third-order polarization is normally measured after exposing the system to three ultrashort laser pulses, which furnish the three fields E. The pulses are separated by experimentally controllable times t_1, t_2 and t_3. The response is obtained by mixing the signal radiated from the sample with a fourth pulse, whence the technique is called four-wave mixing. To obtain the signal the rotating wave approximation has to be performed. The final signal can be measured in different phase matched directions. In each direction, a subset of the eight terms in Eq. 13.26 is obtained. The two most common measurable response functions are the so-called rephasing and nonrephasing signals. The response functions are both given as the sum of three terms, which can be interpreted as a ground-state bleach (GSB) contribution, a stimulated emission (SE) contribution, and an excited-state absorption (ESA) contribution. The response function for the rephasing or photon echo geometry is the sum of

$$\begin{aligned} S_{\mathrm{GB}}(t_1, t_2, t_3) &= i^3 \langle \vec{\mu}_{m'} \vec{\mu}_m G_{m'0;m0}(t; \tau_3) \sigma \rangle_O, \\ S_{\mathrm{SE}}(t_1, t_2, t_3) &= i^3 \langle \vec{\mu}_{m'''} \vec{\mu}_{n''} G_{m'''0;m''0}(t; \tau_3) \rho_{m''n''} \rangle_O, \\ S_{\mathrm{IA}}(t_1, t_2, t_3) &= -i^3 \langle \vec{\mu}_{w'n'''} \vec{\mu}_{wm''} G_{w'n''';wn''}(t; \tau_3) \rho_{m''n''} \rangle_O. \end{aligned} \tag{13.28}$$

In these expressions, the indices n, m and their primed variants run over the chromophores in the system and are understood to be summed. The matrix element $\mu_{m'}$ denotes a transition between the ground state and a state where a single excitation is present in the system, while $\mu_{w'n'''}$ is the matrix element between a one-exciton state n''' and a state with two excitations w'. The notation $\langle \cdots \rangle_O$ denotes averaging over orientations of the system with respect to the laboratory frame. We have introduced the auxiliary functions

$$\sigma = \vec{\mu}_{n'} \vec{\mu}_n G_{0n';0n}(\tau_2; \tau_1) \tag{13.29}$$

and

$$\rho_{m''n''} = \vec{\mu}_{m'} \vec{\mu}_n G_{m''n'';m'n'}(\tau_3; \tau_2) G_{0n';0n}(\tau_2; \tau_1). \tag{13.30}$$

We reiterate the important point that the propagators G are still operators on the bath degrees of freedom. If the system and bath are decoupled at the moments of interaction with the external pulses, a completely different 2D spectrum is obtained. Finally, the 2D spectrum can be calculated for a fixed waiting time t_2 by performing a Fourier transform of the coherence times t_1 and t_3 and plotting the result as a function of the conjugate variables ω_1 and ω_3.

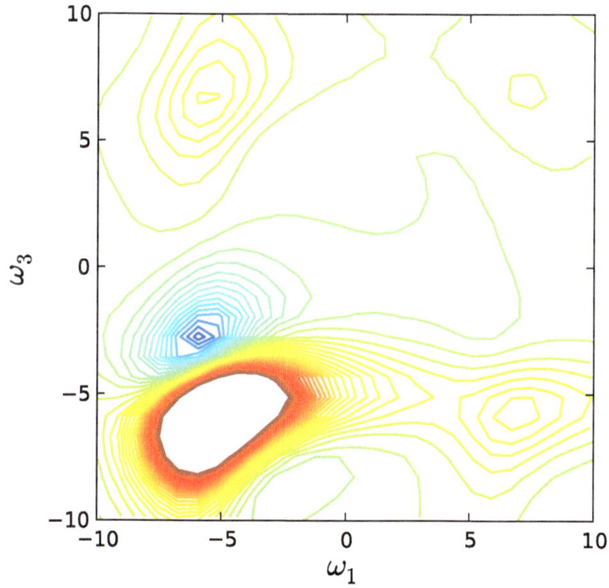

Figure 13.4 Two-dimensional spectrum of a dimer of two-level systems calculated with the hierarchy of equations of motion method.

The system's dynamics appears in the expressions for the nonlinear response through the propagators G. These can be obtained using the numerical integration of the Schrödinger equation or hierachy of equations of motion methods discussed before by propagating the wave function or density matrix, respectively. At times t_1, $t_1 + t_2$ and $t_1 + t_2 + t_3$ the dipole operator is applied to the system, while correlations between the system and the bath are maintained.

An example of a 2D spectrum calculated with the hierarchy of equations of motion method [15, 16] is shown in Fig. 13.4. The system contains two chromophores, each independently coupled to a bath of harmonic oscillators. The site energies of the chromophores are $\epsilon_1 = -3$ and $\epsilon_2 = 3$, while the coupling is $J_{12} = -5$. Parameters for the system–bath coupling are $\lambda = 0.5$, $\gamma = 2.0$ and the inverse temperature is $\beta = 0.13$.

Using the same third-order response theory, several other experimental observable quantities can be calculated besides the

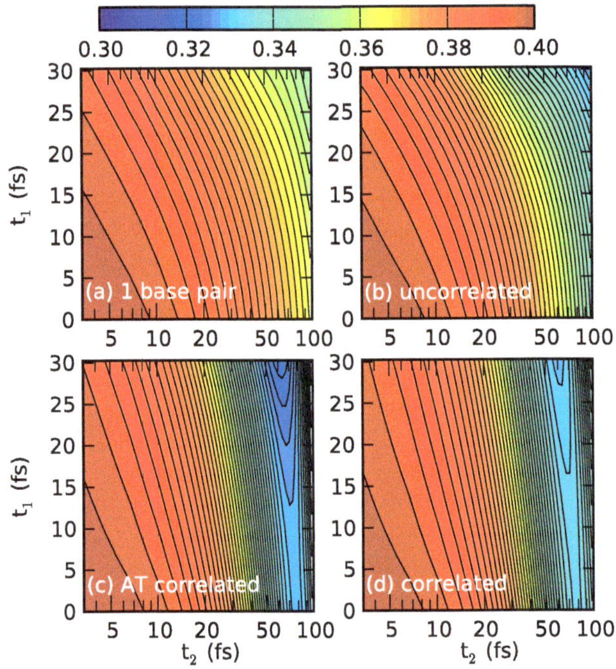

Figure 13.5 Two-time anisotropy decay in a model DNA helix for different models of the environment. The colors indicate the anisotropy as a function of a coherence time t_1 and a population time t_2. Correlations between the two time variables are clearly present, indicating the relevance of memory effects. Adapted from Ref. [17].

2D spectrum. In Fig. 13.5 we show the anisotropy as a function of the coherence time t_1 and the population time t_2 with $t_3 = 0$. The anisotropy shows the rotation of the effectively excited dipole in the sample as a function of the two times. The system is a 6 base pair helix of poly-A poly-T DNA. Because of the helical structure, the direction of the transition dipole is different for each base. Therefore, when excitation is transferred from one base to another, its dipole rotates, which is observed as a decay in the anisotropy. In this way, the energy transfer process can be followed experimentally. The figure shows clear correlations between the relaxation along t_1 and t_2, revealing the presence of memory in the environment.

13.5 Conclusion

In conclusion, we have presented the theory used to model vibrational excitations in peptides as well as electronic excitations in multichromophore systems and their infrared, visible, and UV optical spectra. Numerical integration of the Schrödinger equation and the hierarchy of equations of motion were introduced as methods to model the dynamics of the system in between the laser interactions. An important element of the theory required to properly model the nonlinear optical response are memory effects that extend from before to after an interaction with light. Theories that do not include this memory effect cannot predict the line shapes observed in experiment. Using the methods presented in this chapter the nonlinear response can be calculated and a direct comparison with experiment is possible. Development of these methods is a topic of ongoing research. The reader is referred to Refs. [18, 19] and references therein for the latest developments.

References

1. S. Mukamel, Y. Tanimura, and P. Hamm, *Acc. Chem. Res.*, **42**, 1207 (2009).
2. A. Ishizaki and G. R. Fleming, *J. Chem. Phys.*, **130**, 234110 (2009).
3. R. Boyd, *Nonlinear Optics* (Academic Press, 2008).
4. Z. Ganim et al., *Acc. Chem. Res.*, **41**, 432 (2008).
5. T. l. C. Jansen and J. Knoester, *Acc. Chem. Res.*, **42**, 1405 (2009).
6. L. P. DeFlores, Z. Ganim, S. F. Ackley, H. S. Chung, and A. Tokmakoff, *J. Phys. Chem. B*, **110**, 18973 (2006).
7. A. G. Dijkstra, T. l. C. Jansen, and J. Knoester, *J. Phys. Chem. B*, **115**, 5392 (2011).
8. Y. Tanimura and R. Kubo, *J. Phys. Soc. Jpn.*, **58**, 101 (1989).
9. Y. Tanimura, *J. Phys. Soc. Jpn.*, **75**, 082001 (2006).
10. A. Ishizaki and G. R. Fleming, *Proc. Natl. Acad. Sci. U S A*, **106**, 17255 (2009).
11. A. Ishizaki and Y. Tanimura, *J. Phys. Soc. Jpn.*, **74**, 3131 (2005).
12. R. Xu, P. Cui, X. Li, and Y. Yan, *J. Chem. Phys.*, **122**, 041103 (2005).
13. J. Strümpfer and K. Schulten, *J. Chem. Phys.*, **131**, 225101 (2009).

14. S. Mukamel, *Principles of Nonlinear Optical Spectroscopy* (Oxford University Press, 1995).
15. L. Chen, R. Zheng, Q. Shi, and Y. Yan, *J. Chem. Phys.*, **132**, 024505 (2010).
16. B. Hein, C. Kreisbeck, T. Kramer, and M. Rodríguez, *New J. Phys.*, **14**, 023018 (2012).
17. A. G. Dijkstra and Y. Tanimura, *New J. Phys.*, **12**, 055005 (2010).
18. Y. Tanimura, *J. Chem. Phys.*, **141**, 044114 (2014).
19. A. G. Dijkstra and Y. Tanimura, *J. Chem. Phys.*, **142**, 212423 (2015).

Index

An environmentally friendly book printed and bound in England by www.printondemand-worldwide.com

PEFC Certified

This product is from sustainably managed forests and controlled sources

www.pefc.org

PEFC/16-33-415

This book is made of chain-of-custody materials; FSC materials for the cover and PEFC materials for the text pages.